Intelligent Data Mining and Analysis in Power and Energy Systems

IEEE Press
445 Hoes Lane
Piscataway, NJ 08854

Intelligent Data Mining and Analysis in Power and Energy Systems

Models and Applications for Smarter Efficient Power Systems

Edited by

Zita Vale
GECAD Research Group on Intelligent Engineering and Computing for Advanced Innovation and Development (GECAD) Polytechnic Institute of Porto (ISEP/IPP) Porto, Portugal

Tiago Pinto
GECAD Research Group on Intelligent Engineering and Computing for Advanced Innovation and Development (GECAD) Polytechnic Institute of Porto (ISEP/IPP) Porto, Portugal and University of Trás-os-Montes e Alto Douro Vila Real, Portugal

Michael Negnevitsky
School of Engineering, University of Tasmania Hobart, Tasmania, Australia

Ganesh Kumar Venayagamoorthy
Holcombe Department of Electrical and Computer Engineering, Real-Time Power and Intelligent Systems Laboratory Clemson University Clemson, SC, USA and School of Engineering University of KwaZulu-Natal Durban, South Africa

IEEE Press Series on Power and Energy Systems
Ganesh Kumar Venayagamoorthy, Series Editor

IEEE PRESS

WILEY

For general information on our other products and services or for technical support, please contact our Customer Care Department within the United States at (800) 762-2974, outside the United States at (317) 572-3993 or fax (317) 572-4002.

Wiley also publishes its books in a variety of electronic formats. Some content that appears in print may not be available in electronic formats. For more information about Wiley products, visit our web site at www.wiley.com.

Library of Congress Cataloging-in-Publication Data

Names: Vale, Zita, editor. | Pinto, Tiago, PhD, editor. | Negnevitsky, Michael, editor. | Venayagamoorthy, Ganesh Kumar, editor.
Title: Intelligent data mining and analysis in power and energy systems : models and applications for smarter efficient power systems / edited by Zita Vale, Tiago Pinto, Michael Negnevitsky, Ganesh Kumar Venayagamoorthy.
Description: Hoboken, New Jersey : Wiley-IEEE Press, [2023] | Series: IEEE press series on power and energy systems
Identifiers: LCCN 2022043311 (print) | LCCN 2022043312 (ebook) | ISBN 9781119834021 (cloth) | ISBN 9781119834038 (adobe pdf) | ISBN 9781119834045 (epub)
Subjects: LCSH: Electric power systems. | Data mining.
Classification: LCC TK1001 .I577 2023 (print) | LCC TK1001 (ebook) | DDC 621.31–dc23/eng/20220909
LC record available at https://lccn.loc.gov/2022043311
LC ebook record available at https://lccn.loc.gov/2022043312

Cover Design: Wiley
Cover Image: © metamorworks/Shutterstock

Set in 9.5/12.5pt STIXTwoText by Straive, Chennai, India

"To our dear Parents and Children"

Contents

About the Editors

Zita Vale, IEEE Senior Member, graduated in electrical engineering in 1986, received the PhD degree in electrical and computer engineering in 1993, and the Agregação title (Habilitation) in 2003 from the University of Porto, Portugal. She is a full professor in the School of Engineering, Polytechnic of Porto. She leads the research activities on Intelligent Power and Energy Systems at GECAD – Research Group on Intelligent Engineering and Computing for Advanced Innovation and Development. She has been involved in more than 60 R&D projects and published more than 200 papers in international scientific journals. Her scientific research activities mainly focus on Power and Energy Systems Operation, Electricity Markets, Demand Response, Renewables, Electric Vehicles, and Distributed Generation and Storage. She has been developing artificial-intelligence-based models, methods, and applications for power and energy, using agents and multiagent systems, knowledge-based systems, semantics, machine learning, data mining, and evolutionary computation.

She actively participates in several technical working groups and committees. She is the chair of the IEEE PES Intelligent Data Analysis and Mining (IDMA) Working Group and of the Open Data Sets (ODS) Task Force. She is the chair of the board of directors of ISAP – Intelligent Systems Application to Power Systems. She is involved in editing activities for different journals and books and is a regular reviewer and evaluator for papers and for project proposals and monitoring from different funding agencies around the world.

Tiago Pinto is an assistant professor at the Universidade de Trás os Montes e Alto Douro (UTAD), Portugal, and researcher at INESC-TEC. He has concluded the BSc and MSc, both at the School of Engineering of the Polytechnic of Porto, where he has also developed his research work for more than 10 years, namely at GECAD – Research Group on Intelligent Engineering and Computing for Advanced Innovation and Development. He got the PhD from UTAD in 2016, after which he engaged in a postdoc at the University of Salamanca, in collaboration with the University of Oxford and ENGIE. His main research interests focus on artificial intelligence and its application in power and energy systems, particularly machine learning, multi-agent systems, decision support systems, and metaheuristic optimization. Through the involvement in more than 30 research projects in these fields, he has authored over 200 publications in international journals and conferences, and has co-edited several books and special issues in journals related to power and energy systems and artificial intelligence.

Michael Negnevitsky received the BE (Hons) in Electrical Engineering degree and PhD degree from Byelorussian University of Technology, Minsk, Belarus, in 1978 and 1983, respectively. From 1978 to 1980, he worked at the Electrical Maintenance, Construction and Commissioning Company, and from 1984 to 1991, he was Senior Research Fellow at the Department of Electrical Engineering, Byelorussian University of Technology, Minsk. After his arrival to Australia, he worked at Monash University, Melbourne, and then the University of Tasmania. Currently he is Professor and Chair in Power Engineering and Computational Intelligence, and Director of the Centre for Renewable Energy and Power Systems.

His research interests include power system analysis and control, micro-grids with distributed and renewable energy resources, smart grids, power quality and applications of artificial intelligence in power systems. He has published more than 450 papers in high-quality journals and refereed conference proceedings, authored 14 chapters in several books, and received 4 patents for inventions. His book *Artificial Intelligence* (Addison Wesley 2002, 2005, 2011) has been translated into Mandarin, Cantonese, Korean, Greek, and Vietnamese and adopted in many universities around the world.

He is Fellow of Engineers Australia, Fellow of the Japan Society for the Promotion of Science, Member of CIGRE AP C4 (System Technical Performance) and AP C6 (Distribution Systems and Dispersed Generation), Australian Technical Committee. Dr. Negnevitsky has served as Secretary and Deputy Chair of IEEE PES Energy Development and Power Generation Committee, Chair of IEEE PES International Practices Subcommittee, and Chair of the IEEE PES Working Group on High Renewable Energy Penetration in Remote and Isolated Power Systems. In 2018, he received the Joint Australasian Committee for Power Engineering and CIGRE Australia Award for "outstanding career-long contributions to research and teaching in electric power engineering as well as contribution to industry and CIGRE activities."

Ganesh Kumar Venayagamoorthy is the Duke Energy Distinguished Professor of Power Engineering and Professor of Electrical and Computer Engineering at Clemson University since January 2012. Prior to that, he was a Professor of Electrical and Computer Engineering at the Missouri University of Science and Technology (Missouri S&T), Rolla, USA, where he was from 2002 to 2011. Dr. Venayagamoorthy was a Senior Lecturer in Department of Electronic Engineering, Durban University of Technology, Durban, South Africa, where he was from 1996 to 2002. Dr. Venayagamoorthy is the Founder and Director of the Real-Time Power and Intelligent Systems Laboratory at Missouri S&T and Clemson University.

Dr. Venayagamoorthy received his PhD and MScEng degrees in Electrical Engineering from the University of Natal, Durban, South Africa, in April 2002 and April 1999, respectively. He received his BEng degree with a First-Class Honors in Electrical and Electronics Engineering from Abubakar Tafawa Balewa University, Bauchi, Nigeria, in March 1994. He holds an MBA degree in Entrepreneurship and Innovation from Clemson University, SC (August 2016).

Dr. Venayagamoorthy's interests are in research, development and innovation of power systems, smart grid, and artificial intelligence technologies. Dr. Venayagamoorthy is a Fellow of the IEEE, IET (UK), the South African Institute of Electrical Engineers (SAIEE) and Asia-Pacific Artificial Intelligence Association (AAIA), and a Senior Member of the International Neural Network Society.

List of Contributors

A. Ahmed
Smart Grid Demonstration and Research Investigation Lab
Washington State University
Pullman, WA
USA

Hirohisa Aki
Faculty of Engineering, Information and Systems
University of Tsukuba
Tsukuba, Ibaraki
Japan

Philipp Andelfinger
Institute for Visual and Analytic Computing
University of Rostock
Rostock
Germany

Victor Andrean
Department of Electrical Engineering
National Taiwan University of Science and Technology
Taipei
Taiwan

Ramón Aranda
Tepic Technology Transfer Unit, Center for Scientific Research and Higher Education of Ensenada
Tepic, Nayarit, Mexico

and

National Council of Science and Technology
Mexico City, Mexico City
Mexico

Nelson F. Avila
Independent Electricity System Operator
Toronto, ON
Canada

Wenlei Bai
Oracle Energy and Water
Austin, TX
USA

Rúben Barreto
Research Group on Intelligent Engineering and Computing for Advanced Innovation and Development (GECAD) Institute of Engineering Polytechnic of Porto
Porto
Portugal

Miguel A. Carmona
Tepic Technology Transfer Unit, Center for Scientific Research and Higher Education of Ensenada
Tepic
Nayarit
Mexico

and

National Council of Science and Technology
Mexico City
Mexico City
Mexico

Rajeev K. Chauhan
Department of Electrical Engineering, Faculty of Engineering
Dayalbagh Educational Institute
Agra, Uttar Pradesh
India

Chia-Chi Chu
Department of Electrical Engineering
National Tsing Hua University
HsinChu
Taiwan, R.O.C.

Juan M. Corchado
BISITE research group
University of Salamanca
Salamanca
Spain

Pavel Etingov
Pacific Northwest National Laboratory
Richland, WA
USA

Pedro Faria
GECAD – Research Group on Intelligent Engineering and Computing for Advanced Innovation and Development
Porto
Portugal

Gerardo Figueroa
Sentiance NV
Antwerpen
Belgium

Zahra Forouzandeh
GECAD – Research Group on Intelligent Engineering and Computing for Advanced Innovation and Development, Polytechnic of Porto, School of Engineering (ISEP)
Porto
Portugal

Reza Ghorbani
Renewable Energy Design Laboratory (REDLab), Department of Mechanical Engineering
University of Hawaii at Manoa
Honolulu, HI
USA

James Hamilton
School of Engineering, University of Tasmania
Hobart
Tasmania
Australia

Jens B. Holm-Nielsen
Department of Energy Technology
Center for Bioenergy and Green Engineering
Aalborg University
Esbjerg
Denmark

Zhangshuan Hou
Pacific Northwest National Laboratory
Richland, WA
USA

Qiuhua Huang
Pacific Northwest National Laboratory
Richland, WA
USA

Xinda Ke
Pacific Northwest National Laboratory
Richland, WA
USA

Irfan Khan
Supreme & Co. Pvt. Ltd.
Kolkata, West Bengal
India

Rahmat Khezri
College of Science and Engineering
Flinders University
Adelaide, SA
Australia

Foreword

Recent machine learning and data analytics methods have proliferated into most areas of science, engineering, and commerce. There are excellent reasons for their increasing popularity and applications. Many real-world problems are too complex to come up with closed-form analytical solutions. However, such challenges did not make practitioners idle; instead, they have created working models, prototypes and even built systems with a careful understanding of critical components of the systems as a first step. The data generated from such systems are then analyzed by machine learning and data analytics methods to have a more comprehensive understanding of the systems.

This book titled *Intelligent Data Mining and Analysis in Power and Energy Systems* makes a huge leap in this direction in providing a better understanding of power and energy systems. Compiled by Zita Vale, Tiago Pinto, Michael Negnevitsky, and Ganesh Kumar Venayagamoorthy, the book begins with an introduction to machine learning and data analytics methods and then lays out state-of-the-art methods in addressing various topics in power and energy system design, clustering, classification, forecasting, and analysis with latest machine learning and data analytics methods.

The book is self-contained and written for both novice and experts on the topic. The topics are discussed in simple manner with adequate references and details, so that readers can understand the current state-of-the-art and also find relevant past studies in a single volume.

If you are working in power and energy systems either as a researcher or a practitioner, this is a must-have book to stay ahead in the game. Authors are experts in their own fields. The book will save your efforts in searching for materials on the topic, provide you with the latest methodologies, and direct you to other similar past studies.

Kudos to the editors for this compilation and authors for their contributions.

Kalyanmoy Deb
University Distinguished Professor
Withrow Senior Distinguished Research Scholar
Koenig Endowed Chair Professor
IEEE CIS Evolutionary Computation Pioneer
IEEE Fellow

Department of Electrical and Computer Engineering
Michigan State University, East Lansing, MI, USA

Introduction

In an era of ever-increasing data, there is also an increasing need for the development of suitable intelligent data mining and analysis solutions that enable taking the value out of these data. Fostered by this increasing need and also boosted by the recent worldwide boom in artificial intelligence interest and development, we have been witnessing a significant development of a wide array of new advanced data mining models and methods. These models and methodologies have been instrumental in dealing with real problems, especially in highly complex domains such as power and energy systems [1].

The challenges in power and energy systems have changed completely during the past years, especially because of the increase in the distributed renewable energy sources and the consequently required transformations in power systems' operation, management, and planning, and also in electricity markets [2]. New players are emerging, such as prosumers, electric vehicles, new types of aggregators, energy communities, new local market operators, energy managers of different kinds, among many others [3, 4]. Consequently, new business models are also being proposed, experimented, and implemented as the way to involve such new players in the sector in an active way while creating a new value for these players and for the system, e.g. through the enhancement of the use of local generation and the fostering of consumption and generation flexibility trading [5].

Such significant changes in a traditionally conservative sector require an unprecedented adaptation and foresight capacity from the entire energy value chain, including from policy makers, regulators, operators, planners, and even from the smaller players. This is where the role of new and intelligent data mining and analysis models and methodologies become crucial, contributing to overcoming multiple problems with distinct characteristics. Some relevant examples are power system planning, state estimation, energy resource profiling, aggregation and forecasting, market negotiation, and energy management at multiple levels, including building, microgrid, smart grid, energy community, and distribution grid levels.

This book provides a comprehensive review of intelligent data mining and analysis applications in power and energy systems. This book is organized in six complementary parts, each focusing on a specific topic within the data mining and analysis domain, namely, data mining and analysis fundamentals, clustering, classification, forecasting, data analysis, and other machine learning applications. Each of the six parts is briefly described as follows:

Part I: Data mining and analysis fundamentals provide an overview on data mining and analysis foundations as a means to introduce the reader to the main concepts of the domain and facilitate the deeper understanding of the works described in the rest of the book. Besides the main concepts behind data mining and analysis, the first chapter is dedicated to highlighting the importance of data pre-processing and feature engineering as a means to enable a suitable

Intelligent Data Mining and Analysis in Power and Energy Systems: Models and Applications for Smarter Efficient Power Systems, First Edition. Edited by Zita Vale, Tiago Pinto, Michael Negnevitsky, and Ganesh Kumar Venayagamoorthy.

application of the state-of-the-art models dedicated to the diverse traditional problems related to data mining. This introductory overview provides the means for a deeper understanding of data mining and analysis, bridging the reader into the power and energy systems application domain through the presentation of two systematic reviews, namely, on data mining and analysis applications in power and energy systems and on the contributions of deep learning in power system problems. These two chapters address different problems within power and energy systems, which benefit from the advances of data mining models related to clustering, classification, forecasting, and other common approaches.

Part II: Clustering presents a description of works that apply clustering models and methods to address power and energy system problems. These include standard clustering approaches, as well as the combination of clustering with other models, e.g. classification-based, to solve different types of problems. Specifically, the power and energy system problems addressed by this part include consumer-directed problems related to consumer clustering and demand profiling. Aggregation problems considering not only the aggregation of consumers but also of their consumption flexibility are explored as a means to solve demand response challenges in wider scales. Synchrophasor data analytics taking advantage on clustering models and their combination with other approaches are also addressed, aiming at anomaly detection, localization, and classification.

Part III: Classification includes the description of works that apply classification models such as artificial neural networks, support vector machines, K-nearest neighbors, among others, to solve problems using labeled data. The application cases of these works are related to non-technical loss detection in electric distribution systems, electrical vehicle integration in the power system under multiple worldwide perspectives and considering different types of technologies, and electricity market participation and decision support, namely, in the scope of bilateral contract negotiations using historic and overserved data from multiple negotiators' negotiation process.

Part IV: Forecasting is devoted to the description of works related to the forecasting of energy resources with distinct characteristics. These works are mainly focused on the forecasting of highly variable renewable energy sources; besides, the application of traditional regression-based algorithms describes the advantages of specific approaches such as multivariate stochastic models, spatiotemporal models, and decomposition-based models. The forecasting models are applied to solar irradiance and temperature estimation and to wind and solar power forecasting under distinct scenarios regarding historical data, power system characteristics, and overall renewable energy penetration.

Part V: Data analysis presents the application of data analysis models of distinct natures to address different types of power system-related problems. These problems concern issues such as the vibration of transmission line conductors through the analysis of harmonic dynamic response and the design of power distribution network in hilly areas with the purpose of enabling off-grid electrification. The application of intelligent demand response models as part of microgrid planning is another of the addressed problems. This part is finalized with a chapter focusing on socioeconomic analysis of renewable energy interventions toward affordable and sustainable household technologies.

Part VI: Other machine learning applications describe applications that use distinct types of machine learning approaches such as reinforcement learning, federated learning, and probabilistic modeling, addressing a varied set of challenges of natures. Such challenges include the state estimation of power electronic converters, using both white box and black box approaches. The problem of intelligent building energy management and control is addressed using reinforcement learning. Federated deep learning is applied to generate global supermodels for power system data analysis, and risk assessment of power system outages is performed through probabilistic modeling, considering weather and climate extremes.

Overall, this book comprises the description of a wide set of intelligent data mining and analysis models, methodologies, and applications, addressing problems of distinct natures within the field of power and energy systems, while highlighting the advantages of the already achieved breakthroughs in the domain and pointing out the main gaps that have not yet been solved, as pointers for future paths of continuous research and development.

Zita Vale
Polytechnic of Porto, Portugal
Tiago Pinto
Polytechnic of Porto, Portugal
University of Trás-os-Montes e Alto Douro
Portugal
Michael Negnevitsky
University of Tasmania, Australia
Ganesh Kumar Venayagamoorthy
Clemson University, USA

References

1 Ibrahim, M.S., Dong, W., and Yang, Q. (2020). Machine learning driven smart electric power systems: current trends and new perspectives. *Applied Energy* 272: 115237.
2 Pinto, T., Vale, Z., Widergren, S., and editors. (2021). *Local Electricity Markets*, 1e. Academic Press 384 pp. https://www.elsevier.com/books/local-electricity-markets/pinto/978-0-12-820074-2.
3 Koirala, B.P., Koliou, E., Friege, J. et al. (2016). Energetic communities for community energy: a review of key issues and trends shaping integrated community energy systems. *Renewable and Sustainable Energy Reviews* 56: 722–744.
4 de São, J.D., Faria, P., and Vale, Z. (2021). Smart energy community: a systematic review with metanalysis. *Energy Strategy Reviews* 36: 100678.
5 Hall, S. and Roelich, K. (2016). Business model innovation in electricity supply markets: the role of complex value in the United Kingdom. *Energy Policy* 92: 286–298.

Part I

Data Mining and Analysis Fundamentals

1

Foundations

Ansel Y. Rodríguez-González[1], Angel Díaz-Pacheco[2], Ramón Aranda[3], and Miguel Á. Álvarez-Carmona[4]

[1] *Unidad de Transferencia Tecnológica Tepic, Centro de Investigación Científica y de Educación Superior de Ensenada, Nayarit, México*
[2] *Departamento de Ingeniería en Electrónica, Campus Irapuato-Salamanca, Universidad de Guanajuato, Guanajuato, México*
[3] *Unidad Mérida, Centro de Investigación en Matemáticas, Yucatán, México*
[4] *Unidad Monterrey, Centro de Investigación en Matemáticas, Nuevo León, México*

Acronyms

ANN	artificial neural networks
IoT	Internet of Things
KDD	knowledge discovery in databases
KNN	k-nearest neighbors
PCA	principal component analysis
SVM	support vector machine
SVR	support vector regression
tSNE	T-distributed stochastic neighbor embedding
TWD	three-way decision

1.1 Data Mining: Why and What?

Storing data and analyzing it to make better decisions are a process that humanity has performed at least since the creation of mathematics for commerce, particularly double-entry book-keeping, initially known in the Renaissance as book-keeping "*alla veneziana*" [1]. This accounting system consisted of recording transactions using a general memorandum and a second, more detailed, and organized record. Additionally, transactions were recorded twice, on the one hand, the ledgers, on the other debtors. Since each income must be balanced with a counterpart, the system allows to find errors, understand where the costs and profits come from, and thus guide decision-making.

With the development of larger and more complex business relationships stemming from the first and second industrial revolutions, the need for more powerful analysis tools arose. Thus, econometrics emerged, a branch of economics that uses mathematical and statistical models to analyze, interpret and make predictions about economic systems, predict variables, and find relationships between them and trends [2]. In general, econometrics is based on the construction of formal models with which it is possible to verify hypotheses, measure statistical variables, and carry out simulation tests.

Intelligent Data Mining and Analysis in Power and Energy Systems: Models and Applications for Smarter Efficient Power Systems, First Edition. Edited by Zita Vale, Tiago Pinto, Michael Negnevitsky, and Ganesh Kumar Venayagamoorthy.

The computing revolution made it inexpensive to carry out multiple (many) hypothesis tests, and, as a consequence, the search for the model that best fits the data was encouraged. To describe this process, terms such as data mining, data dredging, data snooping, and data fishing emerged [3, 4]. Additionally, the term data miner was coined to name the researchers that given a set of data, fit alternative equations since there are alternative subsets of possible explanatory variables and chose the best equation. Also, the term data miner was used to differentiate them from classical statistics researchers [5].

However, it is possible to find a model that fits a data set well, even if it is false (i.e. a model obtained from completely random data). This situation complicates the interpretation of the test results of hypotheses (significance levels) [6]. Due to this problem, the 1980s of the last century were a dark decade for the term data mining as it had a negative connotation in econometrics.

Even so, a new dawn for the term data mining turned up in the early 1990s [7–9]. Computer scientists begin to use the term to describe algorithms that search for previously unexpected structures and patterns in data. Usually, the data sets involved are large. Moreover, data sets are getting bigger and bigger and continue to grow with the growth of the internet, storage and processing capabilities, the widespread use of personal computers and mobile devices, and more recently, the Internet of Things (IoT). In this context, data mining is characterized by focusing on algorithmic and computational problems in data analysis, such as memory requirements and algorithm response time (computational cost), and not so much on more statistical problems, such as inference and estimation.

1.2 Data Mining into KDD

A very related research field with data mining is knowledge discovery in databases (KDD). On the one hand, data mining consists in "*applying data analysis and discovery algorithms that, under acceptable computational efficiency limitations, produce a particular enumeration of patterns (or models) over the data*". A pattern is an expression that describes a subset of the data using some language. On the other hand, KDD is defined as "*the non-trivial process of identifying valid, novel, potentially useful and ultimately understandable patterns in data*". In other words, KDD consists of "*mapping low-level data into other forms that might be more compact, more abstract, or more useful*" [10].

Thus, the KDD is a more general interactive and iterative process that includes data mining at its core. In addition, KDD includes other steps like data selection, data preprocessing, data transformation, and pattern interpretation/evaluation. Figure 1.1 presents the outline of the KDD process from de data viewpoint.

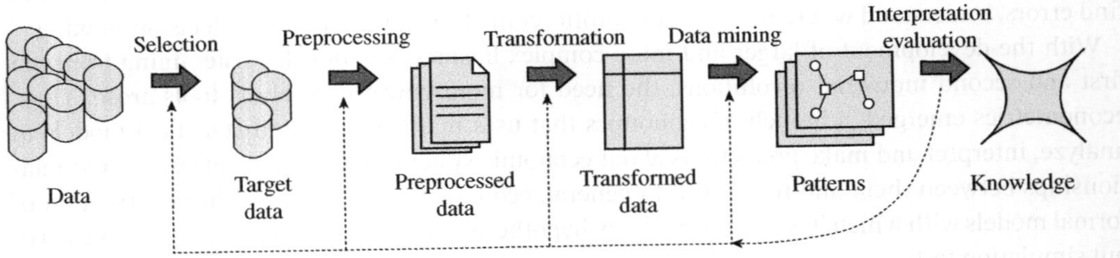

Figure 1.1 The outline of the KDD process. Source: Adapted from Fayyad et al. [10].

However, from a more general viewpoint than the data viewpoint, the KDD involves the following nine steps:

1. **Understanding the application domain:** Understanding the relevant prior knowledge is developed. The goal of the KDD process is defined from the customer's viewpoint.
2. **Creating a target data set:** A data set to perform the discovery is selected, focusing on a subset of variables or data samples.
3. **Data cleaning and preprocessing:** The data set is preprocessed to identify and clean inconsistent and noisy data. Also, strategies for handling noisy data and missing data are developed.
4. **Data reduction and projection:** Dataset is transformed in order to facilitate data mining. The effective number of variables under consideration or invariant representations for the data, depending on the goal of the KDD, process is found by means of dimensionality reduction or transformation methods.
5. **Choosing the suitable data mining task:** The data mining task (e.g. summarization, classification, regression, and clustering) is chosen according to the defined goals for the KDD process.
6. **Choosing the suitable data mining algorithm:** The data mining algorithm is chosen. Also, the appropriate parameters for the said algorithm are decided.
7. **Employing data mining algorithm:** The patterns of interest (e.g. associations, clusters, and classification or regression rules) are discovered by employing the chosen algorithm.
8. **Interpreting mined patterns:** The patterns mined are interpreted. In addition, they can be visualized, and irrelevant or redundant patterns can be removed. Patterns can also be translated in terms understandable to users. The process can continue to the next step or return to any previous step.
9. **Using discovered knowledge:** The knowledge discovered is used in at least one of the following three ways: it is documented and reported to interested parties, it is incorporated into another system that uses it to make decisions, or it is compared with previously believed knowledge to check and resolve potential conflicts.

1.3 The Data Mining Process

Although the data mining core consists of choosing the task and algorithm and extracting interesting patterns, other relevant previous and subsequent steps conform to the data mining process. Thus, the data mining process comprises three blocks of steps: data preprocessing, data mining core, and post-data mining.

Data preprocessing aims to obtain a yield quality data from the raw data. This block includes data cleaning, data integration, data reduction, and data transformation.

On the other hand, postdata mining focused on pattern evaluation and representing knowledge from the patterns mined. Pattern evaluation is focused on identifying interesting patterns that represent the knowledge using interestingness measures evaluated over the mined patterns. While knowledge representation is focused on summarizing and visualizing the data mining result in reports, tables, set of rules, and graphs.

The following subsections (from 1.3.1 to 1.3.4) address the steps into the data pre-processing, while Section 1.4 addresses data mining tasks and techniques involved in the data mining core block.

1.3.1 Data Cleaning

Data cleaning aims to detect and correct corrupt or inaccurate data set records and deals with missing values and noisy data.

Missing values can be handled in the following ways:

- **Ignore de tuple:** It is commonly used when the data mining task is the classification, and the value of the class label does not exist in the tuple or when the tuple has several missing values.
- **Fill the missing value manually:** It is time-consuming and requires human effort. Consequently, it is not used on large data sets.
- **Use a constant global value:** All missing values are replaced by the same constant value. Commonly the value refers to the concept "Unknown." However, it can be a problem because the data mining algorithms used later must know the meaning of the term and incorporate a mechanism to deal with it.
- **Use the mean attribute value:** Each missing value is replaced by the average value of its attribute. Another choice, when the data mining task is the classification, is replacing each missing value of a given tuple by the average value of its attribute only over the subset of tuples with the same class label of the given tuple.
- **Use estimated values:** Each missing value is replaced by an estimated value. The estimated value can be the most probable value of its attribute or the result of regression techniques.

Noise data is understood as an error in the data or the values that deviate from normal. It can be handled in the following ways:

- **Use noise filters:** The noise tuples are identified and removed from the data set. Removing noisy tuples is advantageous [11], but their attributes contain valuable information that can help build a model (e.g. a classifier) [12]. On the other hand, distinguishing between noisy examples and true exceptions is difficult.
- **Use data polishing method:** Considering that a subset of the attributes is a good predictor of the value of the remaining attributes, then inappropriate values in a particular tuple (values far from the predicted) are identified and replaced by non-noisy values (the predicted value) [13]. The main weakness of these methods is the time complexity involved.
- **Use robust algorithms:** The dealing of the noise data is transferred to the data mining core block, and a data mining algorithm is used, little influenced by noise data. For example, if the data mining task is the classification, the Credal-C4.5 algorithm [14] is a choice.

1.3.2 Data Integration

Data integration consists of combining data from multiple heterogeneous data sources and providing a unified view of the data [15]. Having a unified view of the data helps data mining algorithms discover useful information. However, the unified view must be free of inconsistencies, discrepancies, redundancies, and disparities. Otherwise, the knowledge extracted could be wrong, negatively impacting decision-making.

The main issues to deals with the data integration are the following:

- **Entity identification problem:** Two or more tuples from the data sources can refer to the same real-world entity but using different id attributes (e.g. customer_id, customer_number, or passport_number). Then the question is: How to match real-world entities from data? The metadata of each attribute can be used to avoid errors in schema integration.
- **Redundancy and correlation analysis:** Redundant data means data that can be derived or calculated from one another. For example, if a data source provides the attribute age and another

provides the attribute date of birth, the attribute age will be redundant in the integrated data set. Redundancy can be discovered by analyzing the correlation between attributes.

- **Tuple duplication:** Duplicate tuples can occur when some tables in the data sources are denormalized.
- **Data conflict detection and resolution:** Conflicts between values that do not match, but represent the same, can be detected. It can be because the same information may be defined differently in the data sources. For example, inconsistency between two attributes could be detected because, although they represent the same magnitude, they use different metric systems (e.g. km and miles, degrees Fahrenheit to degrees Celsius, dollars, and euros).

The following techniques are used to address the problems mentioned above:

- **Manual integration:** No automation is used for data integration, and all the integration is done manually by the data analyst. This technique is time-consuming and requires much human effort. It is used to integrate small data sources.
- **Middleware integration:** Data from data sources are collected, normalized, and merged by middleware software. This technique is commonly used to integrate data from legacy systems into a unified dataset for a modern system.
- **Application-based integration:** A specific software application is designed and developed to integrate the data sources into a unified dataset. Although this technique saves more time and effort than manual integration, the development time and the technical knowledge required should be considered.
- **Uniform access integration:** A unified view is designed and created. This view represents the integrated data, but the location of the data does not change. It is only a view; data stay in the original data source.
- **Common data storage:** A system that keeps a copy of the data from the data source to store and manage it independently is used. Although this technique can integrate very different data sources like relational databases and flat files, the main problem is how to handle the vast volumes of data.

1.3.3 Data Reduction

When the instances of the data have a large number of variables (features or characteristics) are said to have high dimensionality. Due to the "curse of dimensionality," there are many difficulties to work with high-dimensionality data, such as visualization, assumptions validation, and computational processing time, among others [16].

To deal with the high-dimensionality issues, there are different dimensionality reduction techniques. The idea behind all techniques is to generated a new data set with the same number of instances but with a reducing dimension (lower number of variables). The new low-dimensionality data must preserve some aspects or properties of the original data, for example, two distant instances in the original data must remain distant in the new data, and two nearby instances from the original space must remain nearby in the new data [17]. Some popular techniques in the literature are:

- **Principal Component Analysis (PCA):** By far the most popular method used for dimensionality reduction. It was developed by Hotelling [18] and Pearson [19]. This method is a form of unsupervised learning and is based on the covariance analysis between all the variables from the data. The new data set is generated by a linear transformation that projects the data on the coordinates with the greatest variances.

- **T-distributed Stochastic Neighbor Embedding (tSNE):** This technique was developed by van der Maaten and Hinton [20]. This method is a nonlinear dimensionality reduction that transforms the original instances of the data into a two or three-dimensional graph, where are coded the probabilities of observing other instances in a certain distance.
- **Autoencoders:** The autoencoders are based on artificial neural networks (ANNs) [21, 22]. These methods use the idea of encoding the data by training a neural network. The network inputs are given by each data variable (each input neuron is associated with each variable). The next layers of the network have less neurons than the previous. The output of the last layer returns the encoding data, and the dimension is given by the number of neurons in it.

1.3.4 Data Transformation

Data transformation aims to convert raw data into a suitable format for data mining algorithms to work efficiently. There are several data transformation techniques, which are presented below. Some of them have already been mentioned in previous data mining steps.

- **Data smoothing:** It is used to remove noise from the dataset. methods such as binning, regression, and clustering are used. Binning splits the sorted data into bins and smoothens the data values considering their neighborhood values. Regression identifies relations between attributes, and it uses them to predict an attribute from the other attribute. Clustering obtains groups of similar values, allowing finding outliers, values that do not belong to any group.
- **Attribute construction:** New attributes that can help the data mining are created (computed) from the existing attributes. For example, a new area attribute can be created from the height and width attributes.
- **Data aggregation:** Summary or aggregation operations are applied to the data. For example, the number of transactions per hour can be aggregated to get a daily summary.
- **Data normalization:** Data values are scaled to a smaller range such as [−1, 1] or [0.0, 1.0].
- **Data discretization:** Continuous data are converted into a set of data intervals. When discretizing, the cardinality of the resulting set (i.e. the number of intervals) is much smaller than the cardinality of the original set. Then data mining algorithms work faster. However, care must be taken because discretizing can change the nature of the data. Continuous data are converted into a set of data intervals.
- **Data generalization:** Low-level data attributes are converted to high-level data attributes using concept hierarchy. For example, the attribute city can be generalized as a country.

1.4 Data Mining Task and Techniques

Data mining comprehends a set of techniques and methods that can be applied to a wide range of purposes and goals. To understand this complex paradigm, we need to establish a few categories. Broadly speaking, data mining is used to achieve two main goals, discover interesting patterns, and test hypotheses. For ease of understanding, Figure 1.2 provides a diagram of the data mining tasks in a hierarchical order.

A **verification** task consists of classical statistical methods which try to test a predefined hypothesis. Methods under this category include analysis of variance, T-Test, and goodness of fit, among others. The **discovery** category expands through more levels than verification. Within this task, it does not need a hypothesis to begin with, instead, analytical techniques are tasked with finding the

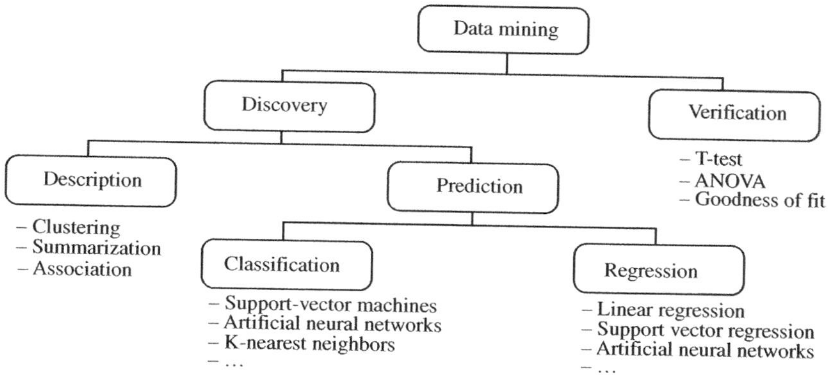

Figure 1.2 Hierarchy of data mining tasks.

right ones from the data. Discovery methods automatically find interesting patterns in the data, so the data provide us with facts that, on occasions, are not obvious.

In this category, there are two main families, description and prediction. **Descriptive** data mining attempts to understand relations among data instances, uncovering hidden patterns that provide the analyst with useful information for the task at hand. For some analytical tasks, when information is scarce or there is no information at all, these methods are preferred as the first approach. Two important tasks in this category are frequent pattern mining and association rules. **Frequent patterns** [23] are substructures appearing in a data set with a certain frequency, such as milk and bread in a grocery transaction. Finding those patterns is very useful for many data mining applications [24]. Regarding the **association rules** [25], it aims to capture hidden patterns by capturing the co-occurrence of events in the data and then trying to establish a correlation between them. These rules are given as "*If X, then Y is likely to occur 30% of the time*" [26].

As for the **predictive** branch, the methods in this category analyze the data to construct a model that imitates the unknown model assumed was used to generate that information. With this model, we can automatically predict new instances from one or more characteristics related to samples. The family of prediction is divided into classification and regression tasks. Both are forms of forecasting, but the target variable to be predicted is different in form and, therefore, its purposes also diverge.

A **regression** is a method for investigating functional relationships among variables. This relationship is expressed as a model linking the response variable and one or more explanatory variables. Formally let y be the response variable and a set of explanatory variables $x_1, x_2, \ldots, x_p$ (with p = number of explanatory variables). The real relationship between y and $x_1, x_2, \ldots, x_p$, can be approximated by the regression model $y = f(x_1, x_2, \ldots, x_p) + \varepsilon$ where ε is a random error representing the discrepancy in approximation [27].

On the other hand, **classification** is a subtype of regression because the response variable is discrete rather than continuous. In a classification task, the problem is defined by a distribution D over $x \times y$ where $y = \{0, 1, \ldots, L\}$ (with L = number of labels). The goal is to find a model $h : x \rightarrow y$ minimizing the error rate on D [28].

From this point, the reader can appreciate the wide range of tasks that data mining can address; however, we still need to give some background about some of the most representative techniques under its umbrella.

1.4.1 Techniques

Having defined the main data mining branches, we can provide some details about the main techniques in each of them. Much has been written on classical statistical techniques, so the approaches under the "*Verification*" branch will not be explained in this book.

1.4.1.1 Techniques in the "Description" Branch

Clustering is one of the most representative methods in this category. This technique aims to group objects in a manner that objects within a group are more similar among them than those in the other groups (see Figure 1.3).

Formally, given a dataset $D = \{t_1, t_2, \dots, t_n\}$ of tuples and an integer number of clusters to create k partitions, we want to define a mapping $f : D \rightarrow (1, 2, \dots, k)$ where each t_i is assigned to one cluster K_j, $1 \leq j \leq k$. A cluster K_j contains just those tuples mapped to it; that means that $K_j = \{t_i \mid f(t_i) = K_j,$ $1 \leq j \leq n\}$ and $t_i \in D$ [29].

Since clustering can reduce the dimensionality of data points, we can use this technique to produce compact representations of such data. On the other hand, using this approach on textual data as a collection of documents, we can extract the most representative terms from each group to **summarize** this collection [30].

1.4.1.2 Regression Techniques

Predicting how much a real estate property could cost or future stock prices are the kind of problems regression algorithms aim to solve.

Linear regression is a powerful technique used to uncover relationships between the response variable (they can be more than one) and the considered set of variables (see Figure 1.4). The basic form of this technique is used to model the relationship between the response variable y and the descriptor variable x that can represent concepts like caloric intake (x) and weight (y) or time for study (x) and grades (y).

This relationship is written $y = \beta_0 + \beta_1 x + \varepsilon$. The response variable is y, β_0 is the y-intercept, β_1 is the slope, x is the descriptor, and ε is the random error [31]. An extension of this approach considers more than one independent variable, and it is known as multiple linear regression.

Two well-known techniques for regression tasks are support vector regression (SVR) and ANN. These approaches are also present in the classification branch and will be discussed in Section 1.4.1.3.

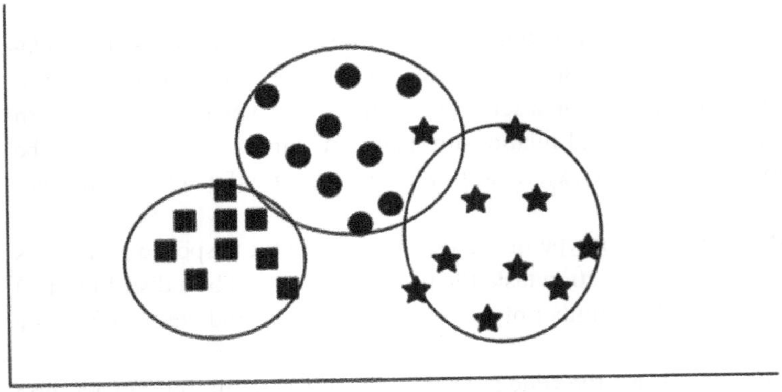

Figure 1.3 Graphical representation of a clustering task.

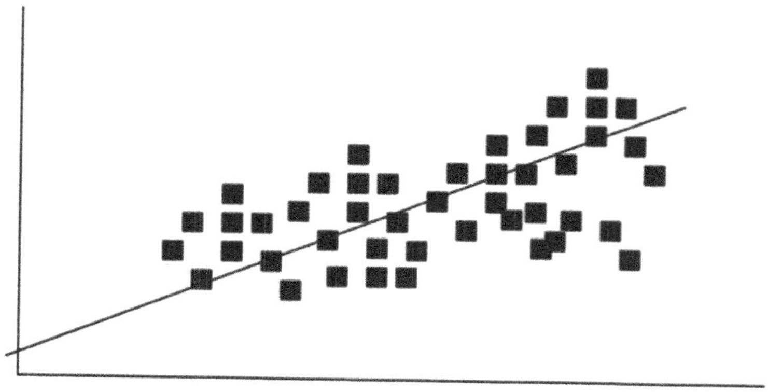

Figure 1.4 Graphical representation of a regression task.

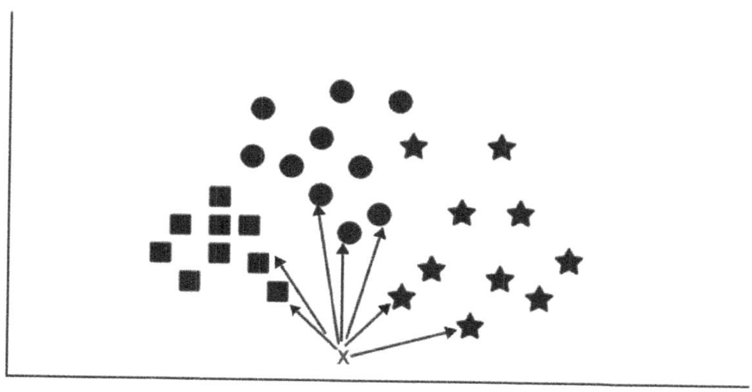

Figure 1.5 Graphical representation of the KNN algorithm for classification.

1.4.1.3 Classification Techniques

One of the simplest and most powerful techniques for classification tasks is the one known as "k" nearest neighbors (KNN). It operates over the intuitive principle that samples from the same class should be closer in a dimensional space and further away otherwise. Given the above, for an unknown sample, it simply calculates its distance to the rest of the data points and its class is determined as that of their KNN (see Figure 1.5). Formally, consider a dataset of n points with their classes $\{(x_1, y_1), (x_2, y_2), \ldots, (x_n, y_n)\}$, where (x_i, y_i) represent the feature vector and their corresponding target class (y_i). Then for a new data point, considering $k = 1$, the most likely class should be computed by $knn(x, 1) = y_p$ where $p = \{1, 2, \ldots, m\}$ and $m =$ number of classes. P is determined as $p = \text{argmin}_i \|x - x_i\|^2$ [32].

Support Vector Machine (SVM) is another widely studied algorithm to address classification tasks, proposed by Vapnik [33]. The basic notion behind this approach is to try to maximize the distance between two classes, which is usually defined by their nearest samples. Given a problem of two classes $z = \{x_i, y_i\}^m{}_{i=1}$, where $y_i \in \{-1, 1\}$. Then z consists of two set of indexes, one for each class, $I = \{i \mid y_i = 1\}$ and $II = \{i \mid y_i = -1\}$. Considering a hyperplane $\boldsymbol{H}$ given by $w^T x + b = 0$, we say that I and II are separable by $\boldsymbol{H}$ if for $i = 1, \ldots m$, $w^T x_i + b > 0$, $\forall_i \in I$ or $w^T x_i + b < 0$, $\forall_i \in II$. Then the distance of each class to hyperplane $\boldsymbol{H}$ is defined as,

$$t_I(w, b) = \min_{i \in I}\{y_i(w^T x_i + b)\}$$

$$t_{II}(w, b) = \min_{i \in II}\{y_i(w^T x_i + b)\}$$

The corresponding classification hyperplane is obtained by $\max_{\|w\|=1,\,b}\{t_I(w, b) + t_{II}(w, b)\}$ [34] (see Figure 1.6).

An interesting category within classification methods is for those that attempt to imitate nature, which is the case of ANN. ANN are inspired by the functioning of the human brain, whose interconnected neurons are at the origin of different processes such as learning. For this reason, this brain abstraction is well suited for classification and regression tasks, among others. There are several types of ANNs according to their complexity. However, one of the most basic forms is the one known as perceptron. It was created by Rosenblatt [35], and it represents the abstraction of a single neuron (see Figure 1.7).

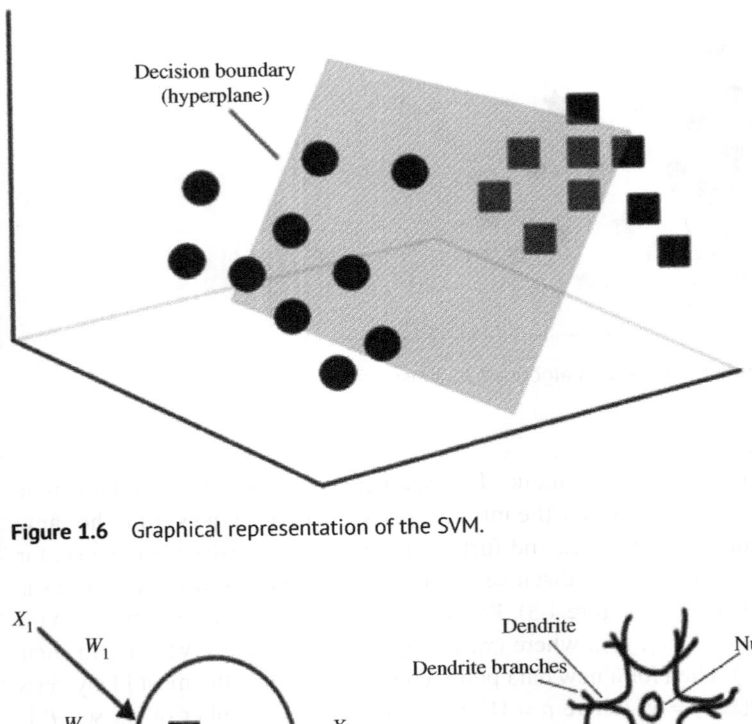

Figure 1.6 Graphical representation of the SVM.

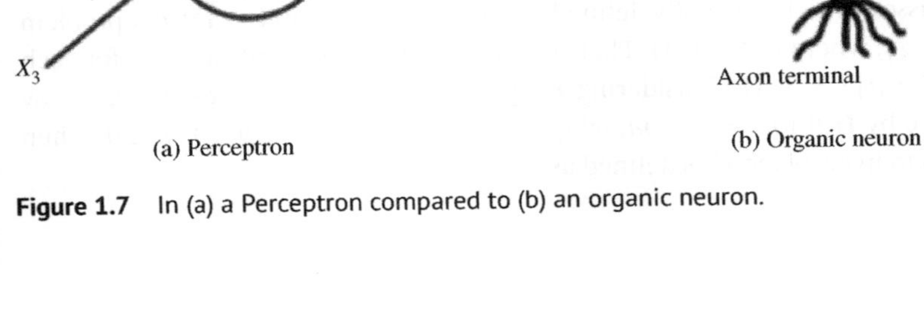

Figure 1.7 In (a) a Perceptron compared to (b) an organic neuron.

The perceptron is made up of two layers of input and output nodes. The first layer $x = (x_1, x_2, \ldots, x_n)$ is for input values, while in the output nodes $(y_1, y_2, \ldots, y_m)$ the result of calculations is returned. The function used to calculate the results is called the activation function, and the sign function is one of the most used. The sign function is given by $\hat{y} = sign\left(\sum_{i=1}^{n} w_i x_i\right)$ where $w = (w_1, w_2, \ldots, w_n)$ is the vector of the weights assigned to the connections between the input to output nodes [36]. The weight vector stores the strength of the connection between the nodes. These values are automatically determined during the learning process on labeled data.

1.4.2 Applications

So far, we have discussed some important aspects of data mining, but to offer a better understanding to the reader, we would like to discuss a central aspect that highlights the importance of this discipline and its applications. Although the multiple purposes of these tools are limited only by our imagination, data mining has been used in a wide range of applications. Some of the most representatives are described below.

Marketing: The possibility to automate these techniques to analyze data-seeking patterns enables to harvest information in massive collections of unstructured data. In the marketing sector, data mining has been used successfully for different objectives such as customer segmentation and the range of approaches known as business intelligence, among others. **Customer segmentations** aim to separate customers into different and uniform groups based on their characteristics. This analysis helps organizations to have a better understanding of their customers to design better strategies based on the common features of their different segments. **Business intelligence** helps organizations sort and structure the information extracted from raw data into tables, graphs, and various interactive dashboards. These tools enable managers to overcome the complexity of the different business aspects and make strategic decisions with real-time, updated information [37].

Fraud detection: In recent years, the detection of financial fraud has received a great deal of attention. With the availability of vast amounts of data from electronic transactions, customer preferences, and patterns, the use of data mining techniques had proved to be adequate for the detection, and therefore prevention of fraud. The most successful fraud detection systems are based on extensive customer historical data where patterns of both fraudulent and nonfraudulent customers are characterized, allowing the data mining algorithms to learn how to differentiate between them [38].

Applications to medicine: Data mining has been used for different applications in the medical field. Using the extensive historical data from patients, noninvasive analyses are conducted to predict the existence of different conditions such as cancer [39, 40], diabetes [41–43], and different disorders in new-borns [44–46]. Furthermore, concerning the patient's historical records, text mining techniques are applied to them in order to automatize the search process for specific data such as prescribed procedures, conditions, and drugs [47]. On the other hand, approaches known as text summarization have been employed to produce compact representations of medical records to facilitate its time-consuming analysis by doctors [48].

Tourism data analysis: In recent years, data mining has become essential in tourism [49]. Mainly the data used are the movements of travelers and the probabilities of movement of the flow of visitors in specific destinations [50]. Also, the management of different types of information has given important advantages to the tourism sector; for example, the use of text has been used to a large extent to determine the collective polarity of visitors to a site in order to detect if tourists had a good experience or bad and based on this make decisions for improvement [51]. On the other hand, image processing has become an essential tool to detect important events that could affect the destination's image [52]. Finally, various tourist recommenders obtain advantages from data mining in order to generate recommendations aimed at a better target [53].

1.5 Data Mining Issues and Considerations

From everything discussed in this chapter, it can agree that data mining is a remarkable tool for many fields. However, many aspects still need to be improved or maintained in a permanent upgrade cycle. These challenges vary from one study to another, but what it considers that the main issues to be addressed are presented below.

1.5.1 Scalability of Algorithms

Over the past decade, the popularization of social networks, the development of affordable storage technologies, the IoT, and many other factors have led to unprecedented data growth. It was estimated that in 2015, 8 zettabytes were created mainly from unstructured data such as Facebook posts, emails, and Tweets, among others [54]. This abundance of data implies benefits for organizations that can obtain business intelligence through its analysis, but conventional data mining algorithms are designed for data that can be loaded completely in main memory. In response to this problem, Dean and Ghemawat [55] introduced the MapReduce programming model. MapReduce is based on the paradigm of divide and conquer to parallel data analysis operations on processing nodes. Although MapReduce is currently the industry standard for the management and analysis of massive datasets, this problem is not solved and at some point, in the future, data overflow will exceed the capabilities of our latest solution, so this challenge is part of the ones that need to be periodically revisited.

1.5.2 High Dimensionality

As mentioned above, the curse of dimensionality is a recurring problem in data mining. In such types of scenarios, many irrelevant dimensions can be present, hindering the task to find patterns. A common step in overcoming this setback is the use of dimensional reduction techniques, such as those mentioned in Section 1.3.3. They transfer the original data from a higher dimensional space by mathematical transformation and preserve as much of the original essential features in this low-dimensional space [56]. Although there are many approaches to overcoming this limitation, there is no universal solution and typically requires testing many techniques in the dataset that is analyzed without guaranteed results.

1.5.3 Improving Interpretability

Data mining techniques have traditionally been compared to black boxes because we feed them with, data and as an output, we get the result we want but without knowledge of the process that occurs inside. However, for many domains, this obscure functioning is not ideal because of their lack of intuitiveness and explanation of model predictions. This lack of explanation represents a major shortcoming in critical decision-making processes because models must have two characteristics: understandability and interpretability. Understandability is related to the question of how an explanation is comprehended by the observer [57]. Interpretability, on the other hand, is related to the intuition behind the results generated by the model. Many efforts have been made to alleviate this defect. According to Linardatos et al. [58], interpretable approaches can be categorized based on different features. If their application is restricted to a specific domain, they are called model-specific. The agnostic methods are those approaches that can be applied to every domain. The scale of interpretation is also taken into account, if the model explains only for specific instances, it is a local one, and if it explains the entire model, it is said to be global. Another category is for the data types in which the model is applied, and the last one is for the purposes of interpretability, as to explain existing models (explain complex black-box models) [58].

1.5.4 Handling Uncertainty

Uncertainty is a common problem in all stages of the data mining process. There are two categories of uncertainty: aleatoric and epistemic. The first one refers to the noise in the data captured by the model, while the epistemic represents the ambiguity that the model presents when dealing with operational inputs [59]. Different techniques have been proposed to deal with the uncertainty. One of the most widespread is the Fuzzy Logic. This theory was introduced by Zadeh in 1975 and uses fuzzy sets to try to reproduce human behavior under certain conditions [60]. Over the past decade, many studies have used type-1 fuzzy sets to address the uncertainty, but significant improvements have been achieved through type-2 fuzzy sets in many recent last-generation works. Another important effort to control the uncertainty is given by the three-way decision (TWD) model. TWD is used to identify and consider uncertain instances (in the input data) and manage uncertainty in the output of the data mining models [61]. Despite the many studies to address uncertainty, this challenge is far from over and will require much more effort to accomplish an optimal solution to this problem.

1.5.5 Privacy and Security Concerns

One of the main concerns of analyzing of data sets consisting of personal information on online transactions, social interactions, and preferences, among others, is the possibility of recovering fragments of sensitive personal information. Although the data can be encrypted, data mining algorithms require decrypted data to work, but organizations are not willingly sharing sensitive information (like medical records) decrypted and ready for analyzing/snooping. To overcome this shortcoming, many studies have been based on an approach known as homomorphic encryption. This technique is a form of encryption that allows certain types of analysis on encrypted data and to generate a ciphered result which, once decrypted, gives the same results that performing the analysis on the simple text [62], this system has been successfully applied in the medical field in combination with the naive Bayes classifier [63]. Also, to classify encrypted communications (SSH and Skype) [64] and classify encrypted images [65]. This approach consumes a lot of computing power, so more work needs to be done to overcome this problem.

1.6 Summary

With the ease of modern computers, countless models and predictions have been generated from various data sets. This has given rise to what we now know as data mining. This term became popular in the 1990s. From this point, the main efforts to describe search algorithms for discovering unexpected structures and patterns in data were carried out. One of the keys to this impact is that data sets are getting bigger and bigger and continue to grow with the growth of the internet, storage and processing capabilities, the widespread use of personal computers and mobile devices, and more recently, the IoT. In this context, data mining is characterized by focusing on algorithmic and computational problems in data analysis, such as memory requirements, and not so much on more statistical problems, such as inference and estimation.

A very related research field to data mining is knowledge discovery in databases. It consists of "low-level mapping data into other forms that might be more compact, more abstract, or more useful." In general, the KDD process involves the following nine steps: (i) understanding of the application domain, (ii) creating a target data set, (iii) data cleaning and preprocessing, (iv) data reduction and projection, (v) choosing the suitable data mining task, (vi) choosing the suitable data mining algorithm, (vii) employing data mining algorithm, (viii) interpreting mined patterns, and (ix) using discovered knowledge.

On the other hand, data mining comprehends a set of techniques and methods that can be applied to a wide range of purposes and goals. Among the most important, it is possible to find two important categories that allow applying data mining to various problems. The first category is descriptive power, which attempts to understand relations among data instances, uncovering hidden patterns that provide the analyst with helpful information for the task at hand. Here, it is possible to find, for example, the clustering algorithms. The second category involves predictive power, which is the ability to analyze the data to construct a model that imitates the unknown model assumed to generate that information. The family of predictions is divided into classification, with algorithms like KNN, SVM, or ANN, and regression tasks as linear regression.

Finally, it is crucial to remark on critical issues. Several areas are constantly improving in different areas such as research or implementation, and whenever data mining processes are going to be carried out, the implementation should be considered. Among these considerations, the main ones are scalability of algorithms, high dimensionality, improving interpretability, handling uncertainty, and privacy and security concerns. Each of these issues could significantly affect the success of the data mining process, so much emphasis is always placed on paying attention to each of these details.

References

1 Pacioli, L. (1494). Summa de Arithmetica geometria proportioni: et proportionalita... N.p.: Paganino de paganini.

2 Goldberger, A.S. (1964). Econometric theory. *Journal of the American Statistical Association* 60 (312): 818–887.

3 Lovell, M.C. (1983). Data mining. *Review of Economics and Statistics* 65 (1): 12.

4 Selvin, H. and Stuart, A. (1966). Data dredging procedures in survey analysis. *American Statistician* 20 (3): 20–23.

5 Denton, F.T. (1985). Data mining as an industry. *The Review of Economics and Statistics* 67 (1): 124–127.

6 Caudill, S.B. (1988). The necessity of mining data. *Atlantic Economic Journal* 16 (3): 11.

7 Agrawal, R., Imielinski, T., and Swami, A. (1993a). Database mining: a performance perspective. *IEEE Transactions on Knowledge and Data Engineering* 5 (6): 914–925.

8 Frawley, W.J., Piatetsky-Shapiro, G., and Matheus, C.J. (1992). Knowledge discovery in databases: an overview. *AI Magazine* 13 (3): 57–70.

9 Michalski, R.S., Kerschberg, L., and Kaufman, K.A. (1992). Mining for knowledge in databases: the INLEN architecture, initial implementation and first results. *Journal of Intelligent Information Systems* 1: 85–113.

10 Fayyad, U., Piatetsky-Shapiro, G., and Smyth, P. (1996). From data mining to knowledge discovery in databases. *AI Magazine* 17 (3): 37–54.

11 Gamberger, D., Boskovic, R., Lavrac, N., and Groselj, C. (1999). Experiments with noise filtering in a medical domain. *Proceedings of the 16th International Conference on Machine Learning*. Morgan Kaufmann Publishers, pp. 143–151.

12 Zhu, X. and Wu, X. (2004). Class noise vs. attribute noise: a quantitative study. *Artificial Intelligence Review* 22: 177–210.

13 Teng, C.M. (1999). Correcting noisy data. *ICML*, pp. 239–248.

14 Mantas, C.J. and Abellán, J. (2014). Credal-C4.5: decision tree based on imprecise probabilities to classify noisy data. *Expert Systems with Applications* 41 (10): 4625–4637.

15 Lenzerini, M. (2002). Data integration: a theoretical perspective. *Proceedings of the 21st ACM SIGMOD-SIGACT-SIGART Symposium on Principles of Database Systems*, pp. 233–246.

16 Altman, N. and Krzywinski, M. (2018). The curse(s) of dimensionality. *Nature Methods* 15: 399–400. https://doi.org/10.1038/s41592-018-0019-x.

17 Lee, J.A. and Verleysen, M. (2014). Two key properties of dimensionality reduction methods. *2014 IEEE Symposium on Computational Intelligence and Data Mining (CIDM)*, pp. 163–170. https://doi.org/10.1109/CIDM.2014.7008663.

18 Hotelling, H. (1933). Analysis of a complex of statistical variables into principal components. *Journal of Educational Psychology* 24: 417–441.

19 Pearson, K. (1901). On lines and planes of closest fit to systems of points in space. *Philosophical Magazine* 2 (11): 559–572.

20 van der Maaten, L.J. and Hinton, G.E. (2008). Visualizing high-dimensional data using t-SNE. *Journal of Machine Learning Research* 9: 2579–2605.

21 Kramer, M. (1991). Nonlinear principal component analysis using autoassociative neural networks. *AIChE Journal* 37: 233–243. https://doi.org/10.1002/aic.690370209.

22 Yang, H. and Schell, K.R. (2022). QCAE: a quadruple branch CNN autoencoder for real-time electricity price forecasting. *International Journal of Electrical Power & Energy Systems* 141: 108092.

23 Agrawal, R., Imieliński, T., and Swami, A. (1993b). Mining association rules between sets of items in large databases. *1993 ACM SIGMOD International Conference on Management of Data*, pp. 207–216.

24 Han, J., Cheng, H., Xin, D., and Yan, X. (2007). Frequent pattern mining: current status and future directions. *Data Mining and Knowledge Discovery* 15 (1): 55–86.

25 Rodríguez-González, A.Y., Martínez-Trinidad, J.F., Carrasco-Ochoa, J.A., and Ruiz-Shulcloper, J. (2013). Mining frequent patterns and association rules using similarities. *Expert Systems with Applications* 40 (17): 6823–6836.

26 Tamayo, P., Berger, C., Campos, M. et al. (2005). Oracle data mining. In: *Data Mining and Knowledge Discovery Handbook*, (ed. Maimon, O. and Rokach, L.), 1315–1329. Boston, MA: Springer. https://doi.org/10.1007/0-387-25465-X_63.

27 Chatterjee, S. and Hadi, A.S. (2006). *Regression Analysis by Example*, Wiley Series in Probability and Statistics. Wiley. ISBN: 9780470055458. https://books.google.com.mx/books?id=uiu5XsAA9kYC.

28 Liu, Z. and Xia, C.H. (2008). *Performance Modeling and Engineering*. New York, NY: Springer. ISBN: 9780387793610. https://books.google.com.mx/books?id=Zda0LxbnCgoC.

29 Jose, J. (2017). *Customer Payment Trend Analysis Based on Clustering for Predicting the Financial Risk of Business Organizations*. Anchor Academic Publishing. ISBN: 9783960671046. https://books.google.com.mx/books?id=WT7vDQAAQBAJ.

30 García-Hernández, R.A., Montiel, R., Ledeneva, Y., Rendón, E. et al. (2008). Text summarization by sentence extraction using unsupervised learning. *Mexican International Conference on Artificial Intelligence*. Berlin, Heidelberg: Springer, pp. 133–143.

31 Yan, X. and Su, X. (2009). *Linear Regression Analysis: Theory and Computing*. World Scientific.

32 Rajaguru, H. and Prabhakar, S.K. (2017). KNN classifier and K-means clustering for robust classification of epilepsy from EEG signals. A detailed analysis.diplom.de.

33 Vapnik, V. (1995). *The Nature of Statistical Learning*. New York: Springer-Verlag.

34 Huang, X., Shi, L., and Suykens, J.A. (2013). Support vector machine classifier with pinball loss. *IEEE Transactions on Pattern Analysis and Machine Intelligence* 36 (5): 984–997.

35 Rosenblatt, F. (1958). The perceptron: a probabilistic model for information storage and organization in the brain. *Psychological Review* 65 (6): 386.

36 Hassanien, A.E., Abraham, A., and Herrera, F. (ed.) (2009). *Foundations of Computational Intelligence Volume 2: Approximate Reasoning*, vol. 202. Springer.

37 Shmueli, G., Bruce, P.C., Yahav, I. et al. (2017). *Data Mining for Business Analytics: Concepts, Techniques, and Applications in R*. Wiley.

38 Nisbet, R., Elder, J., and Miner, G. (2009). *Handbook of Statistical Analysis and Data Mining Applications*. Academic Press.

39 Cochran, A.J. (1997). Prediction of outcome for patients with cutaneous melanoma. *Pigment Cell Research* 10 (3): 162–167.

40 Exarchos, K.P., Goletsis, Y., and Fotiadis, D.I. (2011). Multiparametric decision support system for the prediction of oral cancer reoccurrence. *IEEE Transactions on Information Technology in Biomedicine* 16 (6): 1127–1134.

41 Hasan, M.K., Alam, M.A., Das, D. et al. (2020). Diabetes prediction using ensembling of different machine learning classifiers. *IEEE Access* 8: 76516–76531.

42 Sarwar, M.A., Kamal, N., Hamid, W., and Shah, M.A. (2018). Prediction of diabetes using machine learning algorithms in healthcare. *2018 24th International Conference on Automation and Computing (ICAC)*. IEEE, pp. 1–6.

43 Sisodia, D. and Sisodia, D.S. (2018). Prediction of diabetes using classification algorithms. *Procedia Computer Science* 132: 1578–1585.

44 Díaz-Pacheco, A., Reyes-García, C.A., and Chicatto-Gasperín, V. (2021). Granule-based fuzzy rules to assist in the infant-crying pattern recognition problem. *Sādhanā* 46 (4): 1–15.

45 Savareh, B.A., Hosseinkhani, R., Jafari, M. et al. (2020). Infant crying classification by using genetic algorithm and artificial neural network. *Acta Medica Iranica* 58 (10): 531–539.

46 Sutanto, E., Fahmi, F., Shalannanda, W., and Aridarma, A. (2020). Cry recognition for infant incubator monitoring system based on internet of things using machine learning. *The Intellingent Networks and Systems Society* 14 (1): 444–452.

47 Thompson, P., Batista-Navarro, R.T., Kontonatsios, G. et al. (2016). Text mining the history of medicine. *PLoS One* 11 (1): e0144717. https://doi.org/10.1371/journal.pone.0144717.

48 Moravvej, S.V., Mirzaei, A., and Safayani, M. (2021). Biomedical text summarization using Conditional Generative Adversarial Network (CGAN). arXiv preprint arXiv:2110.11870.

49 Li, J., Xu, L., Tang, L. et al. (2018). Big data in tourism research: a literature review. *Tourism Management* 68: 301–323.

50 Han, Q., Novais, M.A., and Zejnilovic, L. (2021). Toward travel pattern aware tourism region planning: a big data approach. *International Journal of Contemporary Hospitality Management* 33 (6): 2157–2175.

51 Guerrero-Rodriguez, R., Álvarez-Carmona, M.Á., Aranda, R., and López-Monroy, A.P. (2021). Studying Online Travel Reviews related to tourist attractions using NLP methods: the case of Guanajuato, Mexico. *Current Issues in Tourism* 1–16.

52 Arefieva, V., Egger, R., and Yu, J. (2021). A machine learning approach to cluster destination image on Instagram. *Tourism Management* 85: 104318.

53 Arce-Cardenas, S., Fajardo-Delgado, D., Álvarez-Carmona, Á.M., and Ramírez-Silva, J. (2021). A tourist recommendation system: a study case in Mexico. *Mexican International Conference on Artificial Intelligence*. Cham: Springer, pp. 184–195.

54 Rajaraman, V. (2016). Big data analytics. *Resonance* 21 (8): 695–716. https://doi.org/10.1007/s12045-016-0376-7.

55 Dean, J. and Ghemawat, S. (2010). MapReduce: a flexible data processing tool. *Communications of the ACM* 53 (1): 72–77.

56 Huang, X., Wu, L., and Ye, Y. (2019). A review on dimensionality reduction techniques. *International Journal of Pattern Recognition and Artificial Intelligence* 33 (10): 1950017.

57 Stiglic, G., Kocbek, P., Fijacko, N. et al. (2020). Interpretability of machine learning-based prediction models in healthcare. *Wiley Interdisciplinary Reviews: Data Mining and Knowledge Discovery* 10 (5): e1379.

58 Linardatos, P., Papastefanopoulos, V., and Kotsiantis, S. (2021). Explainable AI: a review of machine learning interpretability methods. *Entropy* 23 (1): 18.

59 Hüllermeier, E. and Waegeman, W. (2021). Aleatoric and epistemic uncertainty in machine learning: an introduction to concepts and methods. *Machine Learning* 110 (3): 457–506.

60 Zadeh, L. (1975). The concept of a linguistic variable and its application to approximate reasoning—I. *Information Sciences* 8 (3): 199–249. https://doi.org/10.1016/0020-0255(75)90036-5.

61 Campagner, A., Cabitza, F., and Ciucci, D. (2020). Three-way decision for handling uncertainty in machine learning: a narrative review. *International Joint Conference on Rough Sets*. Cham: Springer, pp. 137–152.

62 Yi, X., Paulet, R., and Bertino, E. (2014). Homomorphic encryption. In: *Homomorphic Encryption and Applications*, (ed. Ahn, G.-J., Thuraisingham, B., Kantarcioglu, M., and Krishna, R.), 27–46. Cham: Springer.

63 Park, H., Kim, P., Kim, H. et al. (2018). Efficient machine learning over encrypted data with non-interactive communication. *Computer Standards & Interfaces* 58: 87–108.

64 Alshammari, R. and Zincir-Heywood, A.N. (2009). Machine learning based encrypted traffic classification: Identifying SSH and Skype. *2009 IEEE Symposium on Computational Intelligence for Security and Defense Applications*, IEEE, pp. 1–8.

65 Hesamifard, E., Takabi, H., and Ghasemi, M. (2019). Deep neural networks classification over encrypted data. *Proceedings of the 9th ACM Conference on Data and Application Security and Privacy*, pp. 97–108.

2

Data Mining and Analysis in Power and Energy Systems: An Introduction to Algorithms and Applications*

Fernando Lezama

GECAD – Research Group on Intelligent Engineering and Computing for Advanced Innovation and Development, LASI – Intelligent Systems Associate Laboratory, Polytechnic of Porto, Porto, Portugal

Acronyms

AMI	advanced metering infrastructure
ANN	artificial neural networks
ARIMA	autoregressive integrated moving average
ARM	association rule mining
FDD	fault detection and diagnosis
HVAC	heating, ventilation, and air conditioning
KDD	knowledge discovery in databases
KNN	K-nearest neighbors
LASSO	the least absolute shrinkage and selection operator
PCA	principal component analysis
PES	power and energy systems
STLF	short-term load forecasting
SVM	support vector machines
SVR	support vector regression

2.1 Introduction

Power and energy systems (PES) are vertiginously evolving nowadays, taking advantage of the ever-increasing adoption of smart grid technologies. With the appearance and installation of new technologies searching to put the "smart" in the electrical grid, utilities, operators, and even end-users have access to a massive amount of information collected from diverse parts of the grid. This information, coming for instance, from sensors and advanced metering infrastructure (AMI), provides a unique opportunity to understand end-users consumption habits better, identify faults, predict consumption and generation patterns, and other tasks that have a positive impact on the stability, operation, and management of the grid [1].

* This is a sample for chapter footnote.

Intelligent Data Mining and Analysis in Power and Energy Systems: Models and Applications for Smarter Efficient Power Systems, First Edition. Edited by Zita Vale, Tiago Pinto, Michael Negnevitsky, and Ganesh Kumar Venayagamoorthy.

However, it is the same reason, the access to an overwhelming amount of available data that makes it challenging to extract valuable information in a way that humans can understand, interpret, and analyze to make use of it efficiently. Hopefully, data mining, a part of the knowledge discovery in databases (KDD) process, is an extensive line of research devoted to finding relationships between multiple variables and even discovering the intrinsic structure and interconnection of available data [2]. Therefore, this chapter aims at providing a comprehensive view of data mining applications in PES. Through a systematic analysis, we will go from the definition and categorization of data mining algorithms (for instance belonging to the category of supervised and unsupervised learning), up to diverse applications on power systems, including profiling, forecasting, and fault detection and diagnosis. While this chapter does not provide an exhaustive list of data mining applications in PES (in part due to the overwhelming amount of literature available on the subject), the reader will have a glimpse into diverse areas of interest. Needless to say, we hardly advise the reader to go beyond the applications presented here to fully understand the importance of such an area of research. Hopefully, the information provided in this chapter will be used as a good starting point to design new data mining algorithms, or to discover new areas of application of those, either in power systems or in related areas of science.

2.2 Data Mining Technologies

Data mining can be defined as a process or technique part of the KDD process. In summary, the KKD process includes the following steps: Step 1 – data selection; Step 2 – data preprocessing; Step 3 – data transformation; **Step 4 – data mining**; Step 5 – data interpretation.

This section focuses on Step 4, data mining, and its application to power systems. In fact, data mining can be a very extensive block of the KDD, having different reaches and focus, and it is interrelated to other research areas such as statistical analysis, machine learning, database engineering, and visualization, among others [3–5].

In a general and broad sense, data mining can be classified into two categories, i.e. supervised and unsupervised data mining technologies. Supervised data mining is typically linked to learning relationships between multiple variables (for example, prediction or classification tasks). On the other hand, unsupervised techniques are linked to discovering intrinsic structures, relations, and interconnection of the data (used for example, to discover patterns or groups sharing similarities) [5]. Each of these two categories comprises different algorithms and methods, which in turn are connected to KKD tasks in multiple domains (e.g. power systems in our case). We briefly comment on the most common algorithms used in each category in this section. The reader can be referred to Chapter 1 for more insight on the methods.

2.2.1 Supervised Methods

Supervised learning is usually related to learning the values of one or more output variables for a given set of input or predictor variables [6]. We could say, thus, that this type of data mining is referred to "learning activities with a teacher." The algorithms belonging to this category are designed commonly to find relations in the data, allowing to perform tasks of classification or prediction. In power systems, for instance, supervised learning can be used for the prediction of energy consumption [7, 8], quality analysis and disturbance classification [9], price forecasting [10], among several other applications. The methods available in the literature are vast, and many others (including hybridization of existing ones) appear daily. We provide here a brief list of some of the most common methods, encouraging the reader to pursue information regarding recent advances in the area, an area that is rapidly evolving toward more sophisticated methods.

2.2.1.1 Regression-Based Methods

This book chapter considers regression-based methods to those belonging to a subset of supervised machine learning, including generalized linear models, logic regression, and regularized regression [11]. The techniques belonging to this category try to predict any continuous valued attribute, analyzing the relationship between an independent variable (i.e. the predictor) and a dependent one (i.e. the target). The algorithms in this category spam from the simple linear (represented for instance by $\mathbf{Y} = a + b\mathbf{X} + e$, with $\mathbf{X}$ and $\mathbf{Y}$ as the predictor and target variables respectively) or polynomial regression models, up to sophisticated artificial neural networks (ANN) [12, 13]. Other popular regression methods include the least absolute shrinkage and selection operator (LASSO) regression [14], the autoregressive integrated moving average (ARIMA) model [15], support vector regression (SVR) [16], deep neural networks [17], and other types of regressions such as multiple linear and nonlinear regression, nonlinear autoregressive exogenous model, or ridge regression [18–20]. Details on each of these methods can be found in specialized chapters in this book and the provided references, while applications of these methods in power systems are summarized in Section 2.3 of this chapter.

2.2.1.2 Classification-Based Methods

Putting classification in a nutshell, we refer to algorithms used to predict a target class by analyzing a training dataset. As the reader might already guess, regression models can also fit this definition since they are used to predict the value of a variable. However, we would like to make a distinction between classification and regression by differentiating algorithms predicting a quantity (regression) and algorithms predicting a label (classification) [21]. Popular examples of classification algorithms include logistic regression (notice that despite being a regression, the purpose of this type of algorithm is to predict labels) [22], support vector machines (SVMs) [23], decision trees [24], K-nearest neighbors (KNN) [25], and ANN. Furthermore, with recent advances in machine learning and data mining, hybrid methods combining cleverly existing ones, or improved versions of existing algorithms, are proposed rapidly, so following this area of research has a lot of potential in power system applications.

2.2.2 Unsupervised Methods

Unlike supervised methods, unsupervised methods try to find the underlying structure or distribution of input data without knowing in advance the corresponding output variables [26]. We might say then that these methods correspond to "learning activities without a teacher" since having a set of observations, the goal is to directly infer some properties of the probability density function without the help of training or a supervisor. One difficulty of these types of methods is the lack of a direct measure of the success of the inferences provided by the algorithms. Having somehow a subjective interpretation of the results (even based on given metrics that do not apply in a general sense to all the problems), many methods are proposed in this field with tons of applications. In PES, for instance, unsupervised methods are used for load profiling [27], identification of energy consumption patterns [28], or fault diagnosis location [29], to name a few. We can classify unsupervised methods into two groups, clustering base approaches, and association-based approaches. In Sections 2.2.2.1 and 2.2.2.2, we define the difference between these two groups and list some of the most popular algorithms belonging to each class.

2.2.2.1 Association Rule Mining

Association rule mining (ARM) refers to algorithms aiming at discovering rules (if-then statements) that describe large portions of data [30], for example, people that likes **X** tend to also like **Y**. Another general example would be algorithms whose goal is to find common values of the variables in the data that appear with a large degree of frequency. The first algorithms proposed for association rules were the AIS algorithm in 1993 [31] and the set-oriented mining algorithm [32] just one year later. These two strategies still have relevance nowadays as a basis for further developments in the area. Another popular line of research for ARM was the one following the Apriori and AprioriTid algorithms, which were designed to avoid the effort wastage of counting candidate item sets that are known to be infrequent [33]. Taking these approaches as a base, several algorithms have been proposed to overcome limitations, such as scanning the whole data set many times or handling data sets with a large number of item sets. To avoid providing a huge list of algorithms, we can summarize the techniques by grouping them in, for example, frequent pattern tree algorithms, rapid association rule mining algorithms, and frequent pattern mining algorithms, to name a few [34, 35]. The ARM-based methods can also be classified into weighted ARM, quantitative ARM, and temporal ARM, in general a sense.

2.2.2.2 Clustering-Based Methods

Clustering methods (sometimes also referred to as data segmentation methods) have different goals related to grouping or segmenting the objects or items into groups (also called "clusters") in which the objects belonging to a given cluster are more closely related given a certain metric than objects assigned to a different group [26]. One of the issues that clustering analysis presents is that groups are formed based on a notion of similarity, defined by a given criterion that directly impacts the way groups are formed. For instance, one of the most popular algorithms for clustering, the k-means algorithm [36, 37], groups the data in function of the Euclidean distance to the cluster center. Another drawback of this algorithm is related to the number of clusters that are formed, being a parameter selected by the user. This drawback was later addressed with automatic clustering algorithms that find a fair number of clusters without a specific definition of the user. Examples of these algorithms are the mean-shift clustering algorithms [38], the X-means algorithm [39], and the G-means algorithm [40]. Other popular clustering algorithms include hierarchical clustering (considered in fact a complete line of research with several algorithms available) [41], the Fuzzy C-Means algorithm [42], support vector clustering [43], decision trees clustering [44], and evolutionary clustering [45].

Clustering is one of the most popular research areas in data mining, so the reader should take the list as a small representative sample of the methods in the field. Please see Chapter 1 for more details on clustering algorithms or follow the avenues market by the revised methods.

2.3 Data Mining Applications in Power Systems

After a brief introduction to data mining and some of the most popular algorithms used in the field, we are ready to present some applications or "tasks" of these techniques in power systems. Since the KDD process has been widely applied in such systems, and due to the complexity of the current multidisciplinary environment (considering not only a wide variety of actors with different objectives but also an extensive range of technologies and engineering fields interconnected), we will limit to provide insights to some of the most common areas of applications, again preaching to the curious reader to go beyond the applications shown here, in order to understand the importance of such an area of research fully. Hopefully, the information provided here will be a good starting point to realize the reach of data mining in power systems.

2.3.1 Profiling

With the smart grid era, the electrical grid has suffered several technological changes, including the addition of a two-way flow of power and information. This has been achieved by installing sensors and an AMI that provides to different stakeholders a vast amount of information from the demand side. If used properly and efficiently, the information available allows a better understanding of the electrical consumption of end-users and can help improve the operation of the energy grid and enhance the grid stability [1].

Profiling in power systems is one of the most successful data mining applications due to the diverse tasks and benefits that it can provide. It achieved significant attention due to the large amount of available data in this new paradigm of smart grids and the impossibility of analyzing these data from the perspective of every single user. Therefore, and for the sake of having a common definition, load profiling was defined in 2001 by the International Energy Agency as [46]: "*The study of the consumption habits of consumers to estimate the amount of power they use at various times of the day and for which they are billed. Load profiling is an alternative to precise metering.*" Of course, this definition is somehow limited these days since other considerations have emerged. Examples of these new considerations include the growing number of users equipped with AMI, consumers that are able to generate their own energy (i.e. prosumers), the emergence of local electricity markets, management of different time intervals and end users of different sizes, among others [47]. For the sake of coming into a common ground, we will refer to profiling as an electricity consumption pattern (either of consumption, generation, or net load) of end-users (independent of their size, e.g. residential, industrial, etc.) over a given period of time. In a more detailed and systematic view, load profiling can be divided into different stages, starting from load data preparation, applying clustering techniques to form groups according to their characteristic load pattern (see for instance Figure 2.1), evaluating such groups to come with some customer segmentation, to finally use the resulting profiles in a wide variety of scopes and applications [48, 49].

The different scopes of profiling applications have been evolving along with the available data, technological infrastructure, and new participants and stakeholders. For instance, profiling can play a key role in the settlement process by identifying the difference between the energy supplied and the actual energy used by end customers or determining such differences in the balancing process (i.e. for a balance responsible entity) [47]. These activities become relevant in networks in which not all end users are equipped with adequate AMI; therefore, classification of the data with alternative solutions is needed. Another example of profiling applications is tariff definition, an activity in which the profiles are used to formulate tariff structures applicable to the defined macro-clusters according to common patterns [50, 51]. Other scopes of applications include forecasting [52], demand response [1], or aggregated load modeling [53].

Going more in detail about applications that have emerged over the years, in Ref. [27], the authors proposed a new methodology (notice that some years have passed since this work was published) to create classes defined in the billing information space. The methodology used billing data from low voltage distribution clients (usually available in databases of distribution companies) and sample load diagrams only available at the time after a dedicated measurement campaign (a situation that might change with the adoption of AMI). In this methodology, the problem was modeled as an optimization problem with the objective of improving given criteria as a function of the obtained clusters. A similar methodology for load pattern shape clustering was later proposed in

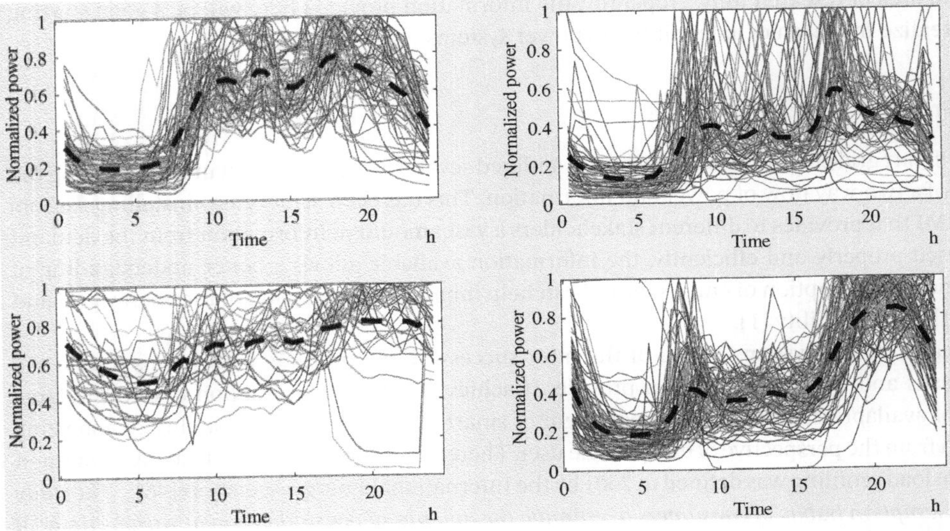

Figure 2.1 An example of load pattern shape clustering. Using k-means algorithm, the daily consumption of 250 end-users is grouped into four clusters (i.e. $k = 4$). The dotted line in each figure shows the resulting characteristic profile (i.e. the centroid of the cluster). Source: Adapted from Lezama et al. [48].

Refs. [48, 51]. In those works, new clustering algorithms were proposed to optimize a given "metric" or criteria rather than using existing clustering algorithms. These works were motivated by previous studies in which the characterization of the electrical patterns was done using various unsupervised and supervised learning techniques, namely k-means, fuzzy k-means, self-organizing maps, hierarchical clustering, etc. [54, 55]. The idea was that, despite the effectiveness of such available algorithms, the electrical load patterns have some unique characteristics (for instance, number of features, overlapping of patterns, etc.) that make their analysis different from the more general pattern recognition algorithms. Inspired by this idea, and following the trends in algorithmic design and technology, new proposals of novel algorithms, such as weighted Fuzzy C-Means [56], enhanced symbolic aggregate approximation [57], online (k-means) adaptive clustering [58], concurrent k-means and spectral clustering [59], just to name a few, have been proposed for customer characterization. It is also important to highlight the presence of the k-means algorithm in many of the studies that have clustering in their frameworks. Also, notice that while the main application of profiling was "load profiling" for customer characterization [28], this trend has changed over the years. With consideration of more sophisticated AMI, and the emergence of new actors in the energy chain such as prosumers, in current days profiling is not limited to load characterization [47]. Other applications of profiling can be, for instance, the determination of heating patterns in buildings [60], retail real pricing determination [61], and user classification considering electric vehicles [62].

Finally, we have summarized some of the mentioned applications in Table 2.1. The reader, however, is advised into the discovery of new techniques emerging in the context of power systems, taking into consideration that such systems are still experiencing vertiginous changes, evolving toward a complex socioeconomic environment in which synergy from different energy sources and stakeholders are not as clear and evident as they were in the past. Overall, load profiling will remain a crucial application of data mining in this new context, supporting the pursuit of an integrated solution that benefits energy efficiency and sustainability.

Table 2.1 Selected load profiling tasks and algorithms in power systems.

Year	Task	Algorithm	References
2012	Load profiling based on billing data	Clustering optimization	[27]
2012	Heating load patterns in buildings	k-means clustering	[60]
2016	Electrical load pattern shape clustering	Evolutionary clustering	[51]
2016	Clustering of smart grid customers	Weighted Fuzzy C-Means	[57]
2017	Energy patterns of office buildings	Fuzzy C-Means clustering	[28]
2018	Retail real-time pricing determination	Ensemble clustering	[61]
2018	Characterization of energy patterns	Enhanced sax	[57]
2019	Profiling of electricity and district heating	Online (k-means) adaptive clustering	[58]
2020	Household electricity consumption patterns	Concurrent k-means and spectral clustering	[59]
2021	User classification considering electric vehicles	k-means	[62]

2.3.2 Forecasting

Similar to "load" profiling, the application of forecasting in PES has become broader, including not only the forecasting of renewable generation (e.g. photovoltaic or wind-based generation) but also the forecast of load demand, market prices, or even electric vehicles trips based on human behavior [63–66]. The new challenges emerging in this domain are linked again to a variety of causes, passing from the need in the adoption of variable and uncertain renewable generation due to environmental issues (a need that is supported by governments worldwide) [52], the adoption of new technologies that allow access to a major amount of data from consumers [67], the deregulation of the power energy industry giving birth to new market structures and roles [65], just to name a few.

Moreover, forecasting techniques are usually inserted in more complex methodologies, such as a block of an energy management system. Therefore, the capability to deal with uncertainty directly impacts the risk and benefits that companies can assume. Another example of this interconnection of tasks can be a retailer's participation in the electricity market. In such a case, a comprehensive and accurate forecast of load demand, power generation, and electricity prices will provide a competitive advantage for the retailer, resulting in better prices for the end-user [68].

Despite the so-far efforts in the area, it is understandable that the present and future power grid requires more accurate forecasting models and tools to achieve optimal coordination between the different distributed resources. Thus, the research in such an area is vast, also supported by the appearance of novel algorithms belonging to the realm of artificial intelligence (see Figure 2.2). One of the most common forecasting applications is, in fact, load forecasting. In Ref. [70], a deep forest regression algorithm is used for short-term load forecasting (STLF) to mitigate the effect of hyper-parameters affecting deep learning algorithms. The authors recognize a large number of studies using ANN for load forecasting but also identify some limitations of ANN-based approaches related, for instance, to the long time required for model training, the configuration of many hyper-parameters, and the high probability of getting stuck in local optimum resulting

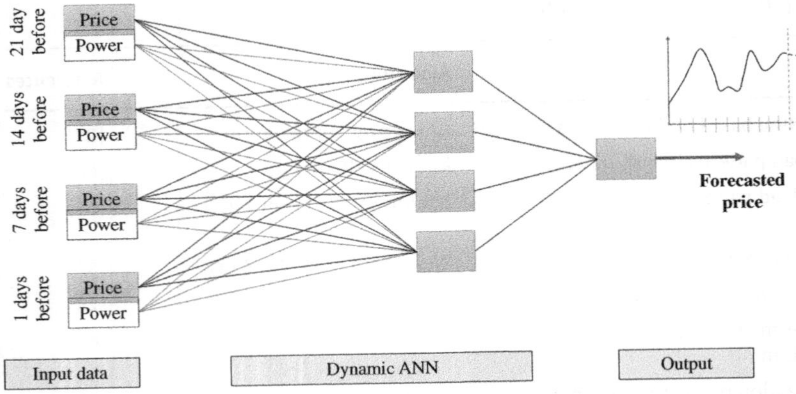

Figure 2.2 An example of an ANN that receives as inputs market prices and energy negotiated in the market (from one day, one week, two weeks, and three weeks before to the desired forecasted day). The ANN has four nodes in the intermediate layer and one output – the forecasted market price. Source: Adapted from Pinto et al. [69].

in high forecasting errors. While new techniques based on deep neural networks solve some of the aforementioned issues, the training speed is still low for short-term forecasts, and the need for parameter configuration is not avoided. More recently, in Ref. [71], a hybrid method combining machine learning techniques and evolutionary computation, named intelligent machine learning with an evolutionary algorithm-based short-term load forecasting (IMLEA-STLF) model, was proposed. This comprehensive approach makes use of other popular techniques in machine learning, such as wavelet transform for the decomposition of time series, oppositional artificial fish swarm optimization algorithm for feature selection, water wave optimization with Elman neural networks model for the prediction process, and finally inverse wavelet transform to obtain the resulting forecast. Results of such a complex method were satisfactory, beating other well-known algorithms commonly used for forecast, namely ARIMA, SVR, ARIMA-NN, and KNN. In the same line, [72] proposes a multi-temporal–spatial-scale method for STLF, using convolutional networks. The approach is compared with 22 forecasting algorithms (the reader can appreciate the vast number of approaches existing out there), demonstrating its capabilities to provide accurate results for the task.

Another popular task in which forecast is applied is the prediction of renewable generation. Renewables represent a sustainable and low-cost form of electricity, but their variability dependent on weather conditions (impossible to fully predict) pose some challenges in their adoption. It is interesting to find some initiatives such as the "global energy forecasting competition" [73], attracting hundreds of participants willing to test their algorithms aiming at achieving a more accurate forecast. From the different sources of renewable energy, wind and solar power are the most popular, and several works can be found in the literature in this regard. For instance, in Ref. [74], an ANN is used to forecast day-ahead wind power and locational marginal prices information. The consideration of wind power and locational marginal prices information, together with the management of energy storage systems, proved to be financially profitable for a wind farm. In Ref. [75], the authors assessed the effectiveness of an SVR model for solar power forecasting of a PV plant. A similar study was conducted in [76], comparing several forecasting algorithms such as KNN, ANN, SVR, etc. to forecast the power generation of a whole year of PV production of 32 plants with a wide variety of sizes and technologies. Other applications of data mining for forecast include the use of deep ensemble learning for probabilistic load forecast [77], the use of ANN, SVM, and radial basis

Table 2.2 Selected forecasting tasks and algorithms in power systems.

Year	Task	Algorithm	References
2015	Wind power forecast	ANN	[54]
2016	Prediction of net demand	Levenberg–Marquardt learning algorithm	[63]
2017	Solar power forecast	SVR	[75]
2018	Photovoltaic generation	KNN; ANN; SVR; and Quantile random forest	[76]
2019	Probabilistic load forecast	Deep ensemble learning	[77]
2019	Price and load forecast	ANN, SVM, and radial basis function neural network	[78]
2020	STLF	Deep forest regression	[70]
2021	STLF	Machine learning and evolutionary algorithms	[71]
2021	STLF	Convolutional network	[72]
2022	Renewable energy forecast	Normalizing flows	[79]

function neural network to provide a multi-block engine for load and price prediction [78], or the use of a novel probabilistic deep learning technique, normalizing flows, to compute accurate PV, wind, and load forecasting [79].

Finally, we summarize the analyzed applications in Table 2.2. Once more, the reader is encouraged to take this as an initial point in search of novel applications in the area, remembering that despite the recent advances, the perfect forecast in any of the domains (load, price, renewable generation, etc.) is still an open issue. While 100% of forecast accuracy is impossible to guarantee, many of the management systems, and why not to say the entire operation and interconnections of the energy grid in the smart grid era, depend on a large degree on the effectiveness of forecasting systems, so that designing better tools always gives an added value to the system as a whole.

2.3.3 Fault Detection and Diagnosis

As one of the most complex systems created by humans, the electrical grid suffers from many undesired variations in voltage and current due to a diverse number of factors such as impulses, notches, glitches, harmonic distortions, and of course, system faults. Moreover, with the emergence of new technologies, supported by the smart grid era, and the penetration of distributed generation (with a large load of uncontrollable renewables into the mix), concerns regarding power quality have grown recently [80].

System faults can be caused by different reasons and factors in the current era. For instance, up to 85% of system faults can be caused by issues occurring in transmission lines [29]. Other types of faults are also related to the penetration of new devices with a high voltage variability into the network. For instance, in 2016, the global installed capacity of PV generation worldwide was reported to be around 310 GW [81]. Therefore, with more penetration of these devices, fault detection and diagnosis (FDD) techniques of PV arrays became a relevant issue to tackle [29]. The same reasoning can be applied to almost all electrical grid components, making FDD a key application of data mining and power engineering in general.

Artificial intelligence and data mining (our topic of interest in this section) have also contributed with tools and techniques for fault detection and diagnosis. For instance, in buildings, a sector that accounts for approximately 39% of energy consumption in the United States and 40% in the European Union, automatic FDD is crucial to reduce service cost monitoring, equipment downtime, or energy penalties due to malfunction of a wide variety of devices, namely economizers, chillers, air handling units, heating, ventilation, and air conditioning (HVAC) systems, among others [82]. As it can be seen, FDD is a key application of data mining and artificial intelligence, as these two fields can provide effective tools for automated analysis to detect, locate, and diagnose faults of diverse types [83].

Now, analyzing in more detail some tasks and algorithms applied to this problem, we can cite, for instance, the work from Ref. [84], in which decision trees and SVM were used to detect and locate in real-time electricity theft in different levels of the power chain, from transmission to distribution. The application was tested showing that the number of false positives can be reduced to a large extent. Another task in which data mining can be used in this context is the detection of high impedance faults in distribution networks. The works of Refs. [85, 86] are a good example of these applications, using SVM, decision trees, and minimum description length-based approaches for high impedance faults detection. In a similar line of research, [87] proposes the use of Aprion, an algorithm for mining frequent item sets of Boolean association rules, with information collected in a fault search table for fault location. Switching a bit in the tasks, in Ref. [88], a stacked sparse autoencoder is proposed for line trip fault diagnosis. The authors claim that the stacked sparse autoencoder has a deep learning structure, using principal component analysis and SVM, that is more effective due to its network structure and layer-wise training mechanism. In a more specific work, Qiqi et al. [89] focuses on the detection of branch lines in 10 kV distribution networks, using principal component analysis (PCA) and the frequent pattern-growth algorithm to determine whether or not the line disconnection has occurred. It can be appreciated that the association of rules mining plays a key role in the fault location and diagnosis tasks, with application that goes from the location of line disconnections, but can be extended to, for instance, the work in Ref. [90], in which association of rules is used for the fault detection of building HVAC systems, or [91], in which rule mining is used to diagnose the outage root cause.

Finally, we summarize some selected publications working on FDD in Table 2.3. The table includes tasks and algorithms used in the context, but the reader is encouraged to take this just as a representative sample of the application. Also, while most of the analyzed works focused on distribution networks (mainly in the detection of faults in lines), there are many other tasks to be explored depending on the context of interest, such as the automatic fault detection of chillers in buildings [92] or the fault detection of single phase-earth feeders [93], just to name a few others. With this summary of the application, we hope to awaken the reader's curiosity, but at the same time, to make the reader realize about the importance and reach that FDD plays in the power system field.

2.3.4 Other Applications

As the reader can guess by now, the applications summarized and grouped by the authors in profiling, forecasting, and fault detection and diagnosis, are just a representative part of the reaches of data mining in power energy systems. This section will help the reader go beyond the categories mentioned above and appreciate other important applications in which studies and new approaches appear day by day.

For instance, security assessment of system stability is one of the more explored and studied topics in power systems, mainly because it is a useful task for many other applications such as day-ahead scheduling, real-time operation, and long-term planning [94]. While traditional methods for stability assessment rely on time-domain simulation requiring huge computational resources,

Table 2.3 Selected FDD tasks and algorithms in power systems.

Year	Task	Algorithm	References
2016	Detection and location electricity theft	Decision trees and SVM	[84]
2018	High impedance faults detection in distribution networks	SVM and decision trees	[86]
2018	Power system line trip fault diagnosis	PCA and SVM	[88]
2019	High impedance faults detection in distribution networks	Minimum description length-based discretization	[23]
2019	FDD of line ungrounded disconnection	PCA and frequent pattern-growth algorithm	[89]
2019	FDD of HVAC systems	Association rule comparison-based approach	[90]
2020	Smart FDD of outage root cause	Association rule-based method	[91]
2020	FDD in distribution networks	Aprior mining algorithm	[87]
2020	Automatic FDD for chillers	Generative adversarial network	[92]
2021	Single-phase earth fault feeder detection	Fuzzy C-Means, fuzzy measure fusion criterion	[93]

data mining opens the possibility of exploring data-driven approaches, using the measurements in the network and artificial intelligence to overcome this situation. Since early year, till now, the use of pattern recognition and data mining has opened a world of possibilities for security assessment [95–98]. Other examples of these alternative applications are the use of SVM for power system stability [99], or transient stability assessment [100]. Clustering techniques have also been used for operational performance in buildings [101], or for economic dispatch [102], and we can even cite applications of machine learning for the design of autonomous brokers participating in electricity markets [103].

To be coherent with the structure of this chapter, we summarize selected task applications in power systems in Table 2.4. It is perhaps pertinent to recall that, as the reader has already noticed, the line dividing the tasks and application is a bit gray since all the devices and systems coexist in the current electricity system. Therefore, it is interesting to do an exercise trying to position the diverse approaches and methodologies exposed here in the big panorama of power energy systems. In fact, power system visualization [106, 107], another field of application in the field, uses a wide variety of data mining algorithms to simplify models and give a better presentation and interpretation of the huge amount of available data nowadays.

2.4 Discussion and Final Remarks

We have analyzed in this chapter several applications of data mining in power systems and highlighted the crucial role that this data science plays in the future of the electrical grid. Accordantly, we put together some of the most representative algorithms in the area, realizing that some applications require even combinations of approaches that respond to a bigger framework. For instance,

Table 2.4 Selected applications in diverse topics on power systems.

Year	Task	Algorithm	References
1989	Security assessment	Adaptive pattern recognition, ANN	[96]
2009	Security assessment	Decision trees	[95]
2014	Building operational performance	Clustering analysis	[101]
2018	Power system stability	Modified SVM	[99]
2018	Protection of distribution networks	Decision trees and adaptive NN	[104]
2018	Power quality problems	Decision trees algorithms in WEKA	[105]
2018	Transient stability assessment	Convolutional SVM	[100]
2019	Autonomous broker for energy markets	Reinforcement learning	[103]
2021	Security assessment	Random bits forest	[97]
2021	Stability and security management	Machine learning and classification models	[98]
2022	Economic load dispatch	Hierarchical clustering	[102]

imagine an energy management system in which a block is in charge of forecasting renewable generation, another block in charge of user profiling, another block in charge of market participation, another block in charge of network reliability and assessment, and so on.

Since the information provided in this chapter is extensive, and we recognize that it could be easy for the reader to get lost in the concepts provided, we have put graphically, in Figure 2.3, the taxonomy resulting from this study. In here, we can first appreciate in the highest level of the KDD process, with its five steps, being data mining a key step of the process. After that, in the second level from top to button, we divided data mining into two categories: supervised and unsupervised data mining algorithms. Supervised (learning) is referred to algorithms that require "a teacher" meaning that these algorithms determine outputs variables for a given set of input or predictor variables.

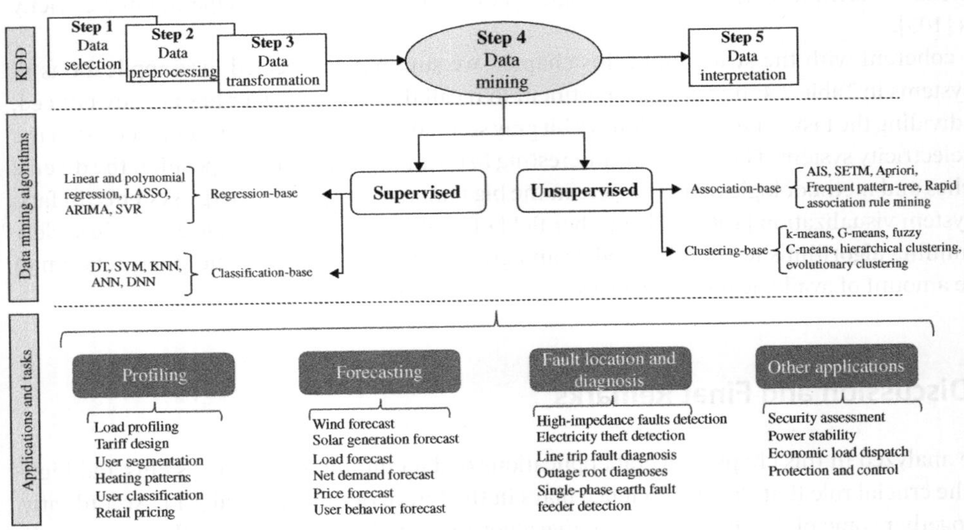

Figure 2.3 Taxonomy of data mining applications in PES.

The two main groups of algorithms belonging to supervised data mining are regression-based methods (e.g. linear regression, LASSO, SVR) and classification-based methods (e.g. decision trees, SVM, KNN). On the other hand, unsupervised (learning) is referred to algorithms that do not require "a teacher" meaning that these algorithms find (or try to find) the underlying structure of input data without knowing the corresponding output variables. Unsupervised data mining algorithms can be classified into association rule-based approaches (e.g. Apriori algorithm, and several types of ARM), and clustering-based approaches (e.g. k-means, G-means, hierarchical clustering). After putting in order some of the (several) available algorithms for data mining, we proceeded to a third level to the application of these methods in power systems. Again, due to the extensive available literature, we came up with some categories that we believe are still relevant in the design and management of power systems, namely profiling, forecasting, and fault detection and diagnosis. For each of these categories, we provided a summary of selected applications, emphasizing the algorithms used and the purpose of the task. We also included in Section 2.3.4, other applications out of these categories (e.g. security assessment, network reliability, etc.) for the reader to have a glimpse into the broader picture of the role of data mining in power systems.

To conclude, we would like to remind the reader about the important role that PES play in the world. Despite the technological developments in the area, improper operation, control, and equipment faults represent a serious energy waste that needs to be handled to pursue a sustainable energy road. The recent advances in knowledge discovery, and the current trends of automation systems, putting the "smart" in the electrical grid, have led to a massive amount of available data. Thus, the use and extraction of knowledge from available data become fundamental. As was previously mentioned, data mining plays a crucial role in the KDD process and is a very promising tool to extract and synthesize information coming from the complex socio-economical system that power systems represent. We certainly hope that this chapter has provided the reader the means and information to pursue this line of research and devise novel applications resulting from the existing ones, an avenue we believe is far from reaching its maximum potential.

References

1 Yi Wang, Qixin Chen, Chongqing Kang, Mingming Zhang, Ke Wang, and Yun Zhao. 2015. "Load profiling and its application to demand response: a review." *Tsinghua Science and Technology* 20 (2): 117–29. https://doi.org/10.1109/TST.2015.7085625.

2 Witten, I.H., Frank, E., and Hall, M.A. (2011). *Data Mining: Practical Machine Learning Tools and Techniques*. Elsevier https://doi.org/10.1016/C2009-0-19715-5.

3 Ibrahim, Muhammad Sohail, Wei Dong, and Qiang Yang. 2020. "Machine learning driven smart electric power systems: current trends and new perspectives." *Applied Energy* 272 115237. https://doi.org/10.1016/j.apenergy.2020.115237.

4 Overbye, T.J. and Weber, J.D. (2000). Visualization of power system data. In: *Proceedings of the 33rd Annual Hawaii International Conference on System Sciences*, vol. 1, 7. IEEE https://doi.org/10.1109/HICSS.2000.926744.

5 Zhao, Yang, Chaobo Zhang, Yiwen Zhang, Zihao Wang, and Junyang Li. 2020. "A review of data mining technologies in building energy systems: load prediction, pattern identification, fault detection and diagnosis." *Energy and Built Environment* 1 (2): 149–64. https://doi.org/10.1016/j.enbenv.2019.11.003.

6 Saravanan, R., and Pothula Sujatha. 2018. "A state of art techniques on machine learning algorithms: a perspective of supervised learning approaches in data classification." In *2018 Second*

International Conference on Intelligent Computing and Control Systems (ICICCS), 945–49. IEEE. https://doi.org/10.1109/ICCONS.2018.8663155.

7 Amasyali, Kadir, and Nora M. El-Gohary. 2018. "A review of data-driven building energy consumption prediction studies." *Renewable and Sustainable Energy Reviews* 81 1192–1205. https://doi.org/10.1016/j.rser.2017.04.095.

8 Zhao, Hai-Xiang, and Frédéric Magoulès. 2012. "A review on the prediction of building energy consumption." *Renewable and Sustainable Energy Reviews* 16 (6): 3586–92. https://doi.org/10.1016/j.rser.2012.02.049.

9 Granados-Lieberman, D., R.J. Romero-Troncoso, R.A. Osornio-Rios, A. Garcia-Perez, and E. Cabal-Yepez. 2011. "Techniques and methodologies for power quality analysis and disturbances classification in power systems: a review." *IET Generation Transmission and Distribution* 5 (4): 519. https://doi.org/10.1049/iet-gtd.2010.0466.

10 García-Martos, C. and Conejo, A.J. (2013). Price forecasting techniques in power systems. In: *Wiley Encyclopedia of Electrical and Electronics Engineering*. Hoboken, NJ: Wiley https://doi.org/10.1002/047134608X.W8188.

11 Dasgupta, Abhijit, Yan V. Sun, Inke R. König, Joan E. Bailey-Wilson, and James D. Malley. 2011. "Brief review of regression-based and machine learning methods in genetic epidemiology: the genetic analysis workshop 17 experience." *Genetic Epidemiology* 35 (S1): S5–11. https://doi.org/10.1002/gepi.20642.

12 Ghiassi, M., H. Saidane, and D.K. Zimbra. 2005. "A dynamic artificial neural network model for forecasting time series events." *International Journal of Forecasting* 21 (2): 341–62. https://doi.org/10.1016/j.ijforecast.2004.10.008.

13 Murata, N., S. Yoshizawa, and S. Amari. 1994. "Network information criterion-determining the number of hidden units for an artificial neural network model." *IEEE Transactions on Neural Networks* 5 (6): 865–72. https://doi.org/10.1109/72.329683.

14 Tibshirani, Robert. 1996. "Regression shrinkage and selection via the Lasso." *Journal of the Royal Statistical Society: Series B: Methodological* 58 (1): 267–88. https://doi.org/10.1111/j.2517-6161.1996.tb02080.x.

15 Box, G. E. P., and David A. Pierce. 1970. "Distribution of residual autocorrelations in autoregressive-integrated moving average time series models." *Journal of the American Statistical Association* 65 (332): 1509–26. https://doi.org/10.1080/01621459.1970.10481180.

16 Smola, Alex J., and Bernhard Schölkopf. 2004. "A tutorial on support vector regression." *Statistics and Computing* 14 (3): 199–222. https://doi.org/10.1023/B:STCO.0000035301.49549.88.

17 Miikkulainen, R., Liang, J., Meyerson, E. et al. (2019). Evolving deep neural networks. In: *Artificial Intelligence in the Age of Neural Networks and Brain Computing* (ed. Kozma, R., Alippi, C., Choe, Y., and Morabito, F.C.), 293–312. Academic Press, Elsevier. https://doi.org/10.1016/B978-0-12-815480-9.00015-3.

18 Adamowski, Jan, Fung Chan, H. Prasher, S. O. Ozga-Zielinski, B. and Sliusarieva, A. 2012. "Comparison of multiple linear and nonlinear regression, autoregressive integrated moving average, artificial neural network, and wavelet artificial neural network methods for urban water demand forecasting in Montreal, Canada." *Water Resources Research* 48 (1). https://doi.org/10.1029/2010WR009945.

19 Boussaada, Zina, Octavian Curea, Ahmed Remaci, Haritza Camblong, and Najiba Mrabet Bellaaj. 2018. "A nonlinear autoregressive exogenous (NARX) neural network model for the prediction of the daily direct solar radiation." *Energies* 11 (3): 620. https://doi.org/10.3390/en11030620.

20 Marquardt, Donald W., and Ronald D. Snee. 1975. "Ridge regression in practice." *The American Statistician* 29 (1): 3–20. https://doi.org/10.1080/00031305.1975.10479105.

21 Matloff, N. (2017). *Statistical Regression and Classification: From Linear Models to Machine Learning*. CRC Press.

22 Hosmer, D.W. Jr., Lemeshow, S., and Sturdivant, R.X. (2013). *Applied Logistic Regression*. Wiley.

23 Noble, William S. 2006. "What is a support vector machine?" *Nature Biotechnology* 24 (12): 1565–67. https://doi.org/10.1038/nbt1206-1565.

24 Quinlan, J.R. 1990. "Decision trees and decision-making." *IEEE Transactions on Systems, Man, and Cybernetics* 20 (2): 339–46. https://doi.org/10.1109/21.52545.

25 Cunningham, Pádraig, and Sarah Jane Delany. 2022. "K-nearest neighbour classifiers – a tutorial." *ACM Computing Surveys* 54 (6): 1–25. https://doi.org/10.1145/3459665.

26 Hastie, Trevor, Robert Tibshirani, and Jerome Friedman. 2009. "Unsupervised learning." In *The Elements of Statistical Learning*, 1–101. New York, NY: Springer. https://doi.org/10.1007/b94608_14.

27 Fidalgo, José Nuno, Manuel António Matos, and Luís Ribeiro. 2012. "A new clustering algorithm for load profiling based on billing data." *Electric Power Systems Research* 82 (1): 27–33. https://doi.org/10.1016/j.epsr.2011.08.016.

28 Qin, J., Zhang, J. 2017. "Sampling for building energy consumption with fuzzy theory." *Energy and Buildings* 156 78–84. https://doi.org/10.1016/j.enbuild.2017.09.047.

29 Mellit, A., G.M. Tina, and S.A. Kalogirou. 2018. "Fault detection and diagnosis methods for photovoltaic systems: a review." *Renewable and Sustainable Energy Reviews* 91 1–17. https://doi.org/10.1016/j.rser.2018.03.062.

30 Agrawal, Rakesh, Tomasz Imieliński, and Arun Swami. 1993. "Mining association rules between sets of items in large databases." In *Proceedings of the 1993 ACM SIGMOD International Conference on Management of Data – SIGMOD '93*, 207–16. New York, NY: ACM Press. https://doi.org/10.1145/170035.170072.

31 Agrawal, R., T. Imieliński, and A. Swami. 1993. "Database mining: a performance perspective." *IEEE Transactions on Knowledge and Data Engineering* 5 (6): 914–25. https://doi.org/10.1109/69.250074.

32 Houtsma, M. and Swami, A. (1995). Set-oriented mining for association rules in relational databases. In: *Proceedings of the Eleventh International Conference on Data Engineering*, 25–33. IEEE https://doi.org/10.1109/ICDE.1995.380413.

33 Agrawal, R. and Ramakrishnan, S. (1994). Fast algorithms for mining association rules. In: *Proc. 20th Int. Conf. Very Large Data Bases, VLDB*, 487–499. Morgan Kaufmann.

34 Rodríguez-González, A.Y., Lezama, F., Iglesias-Alvarez C.A., et al. 2018. "Closed frequent similar pattern mining: reducing the number of frequent similar patterns without information loss." *Expert Systems with Applications* 96 271–83. https://doi.org/10.1016/j.eswa.2017.12.018.

35 Zhao, Q. and Bhowmick, S.S. (2003). Association rule mining: a survey. Technical report 135, Nanyang Technological University, Singapore.

36 Likas, Aristidis, Nikos Vlassis, and Jakob J. Verbeek. 2003. "The global K-means clustering algorithm." *Pattern Recognition* 36 (2): 451–61. https://doi.org/10.1016/S0031-3203(02)00060-2.

37 Peña, J.M, J.A Lozano, and P Larrañaga. 1999. "An empirical comparison of four initialization methods for the K-means algorithm." *Pattern Recognition Letters* 20 (10): 1027–40. https://doi.org/10.1016/S0167-8655(99)00069-0.

38 Wu, Kuo-Lung, and Miin-Shen Yang. 2007. "Mean shift-based clustering." *Pattern Recognition* 40 (11): 3035–52. https://doi.org/10.1016/j.patcog.2007.02.006.

39 Pelleg, D. and Moore, A.W. (2000). X-means: extending K-means with efficient estimation of the number of clusters. In *Proceedings of the Seventeenth International Conference on Machine Learning (ICML '00)*, 727–734. San Francisco, CA: Morgan Kaufmann Publishers Inc.

40 Hamerly, G. and Elkan, C. (2003). Learning the k in K-means. In: *Proceedings of the 16th International Conference on Neural Information Processing Systems*, 281–288. MIT Press.

41 Murtagh, Fionn, and Pedro Contreras. 2011. "Algorithms for hierarchical clustering: an overview." *Wiley Interdisciplinary Reviews: Data Mining and Knowledge Discovery* 2 (1): 86–97. https://doi.org/10.1002/widm.53.

42 Dunn, J. C. 1973. "A fuzzy relative of the ISODATA process and its use in detecting compact well-separated clusters." *Journal of Cybernetics* 3 (3): 32–57. https://doi.org/10.1080/01969727308546046.

43 Ben-Hur, A., Horn, D., Siegelmann, H.T., and Vapnik, V. (2001). Support vector clustering. *Journal of Machine Learning Research* 2: 125–137.

44 Liu, B., Xia, Y., and Philip, S.Y. (2000). Clustering through decision tree construction. In: *Proceedings of the Ninth International Conference on Information and Knowledge Management*, 20–29. Association for Computing Machinery.

45 Handl, Julia, and Joshua Knowles. 2007. "An evolutionary approach to multiobjective clustering." *IEEE Transactions on Evolutionary Computation* 11 (1): 56–76. https://doi.org/10.1109/TEVC.2006.877146.

46 IEA – International Energy Agency (2001). *Competition in Electricity Markets*. Organisation for Economic Co-operation and Development (OECD).

47 Chicco, Gianfranco, and Andrea Mazza. 2021. "Load profiling revisited: prosumer profiling for local energy markets." In *Local Electricity Markets*, 215–42. Elsevier. https://doi.org/10.1016/B978-0-12-820074-2.00004-6.

48 Lezama, Fernando, Ansel Y. Rodriguez-Gonzalez, and de Cote, E.M. 2016. "Load pattern clustering using differential evolution with pareto tournament." In *2016 IEEE Congress on Evolutionary Computation (CEC)*, 241–48. IEEE. https://doi.org/10.1109/CEC.2016.7743801.

49 Chicco, G., R. Napoli, F. Piglione, P. Postolache, M. Scutariu, and C. Toader. 2004. "Load pattern-based classification of electricity customers." *IEEE Transactions on Power Systems* 19 (2): 1232–39. https://doi.org/10.1109/TPWRS.2004.826810.

50 Chicco, G., R. Napoli, P. Postolache, M. Scutariu, and C. Toader. 2003. "Customer characterization options for improving the tariff offer." *IEEE Transactions on Power Systems* 18 (1): 381–87. https://doi.org/10.1109/TPWRS.2002.807085.

51 Lezama, F., A.Y. Rodríguez, E.M. de Cote, and L.E. Sucar. 2016. Electrical load pattern shape clustering using ant colony optimization. *Lecture Notes in Computer Science (Including Subseries Lecture Notes in Artificial Intelligence and Lecture Notes in Bioinformatics)* Vol. 9597. https://doi.org/10.1007/978-3-319-31204-0_32.

52 Ahmed, Adil, and Muhammad Khalid. 2019. "A review on the selected applications of forecasting models in renewable power systems." *Renewable and Sustainable Energy Reviews* 100 9–21. https://doi.org/10.1016/j.rser.2018.09.046.

53 Lu, Ning. 2012. "An evaluation of the HVAC load potential for providing load balancing service." *IEEE Transactions on Smart Grid* 3 (3): 1263–70. https://doi.org/10.1109/TSG.2012.2183649.

54 Figueiredo, V., Rodrigues, F., Vale, Z., and Gouveia, J.B. (2005). An electric energy consumer characterization framework based on data mining techniques. *IEEE Transactions on Power Systems* 20 (2): 596–602.

55 Chicco, Gianfranco. 2012. "Overview and performance assessment of the clustering methods for electrical load pattern grouping." *Energy* 42 (1): 68–80. https://doi.org/10.1016/j.energy.2011.12.031.

56 Lu, Yun, Tiankui Zhang, and Zhimin Zeng. 2016. "Adaptive weighted fuzzy clustering algorithm for load profiling of smart grid customers." In *2016 IEEE/CIC International Conference on Communications in China (ICCC)*, 1–6. IEEE. https://doi.org/10.1109/ICCChina.2016.7636874.

57 Capozzoli, Alfonso, Marco Savino Piscitelli, Silvio Brandi, Daniele Grassi, and Gianfranco Chicco. 2018. "Automated load pattern learning and anomaly detection for enhancing energy management in smart buildings." *Energy* 157 336–52. https://doi.org/10.1016/j.energy.2018.05.127.

58 Ray, G. Le, and P. Pinson. 2019. "Online adaptive clustering algorithm for load profiling." *Sustainable Energy, Grids and Networks* 17 100181. https://doi.org/10.1016/j.segan.2018.100181.

59 Sun, Li, Kaile Zhou, and Shanlin Yang. 2020. "An ensemble clustering based framework for household load profiling and driven factors identification." *Sustainable Cities and Society* 53 101958. https://doi.org/10.1016/j.scs.2019.101958.

60 Nikolaou, Triantafyllia G., Dionysia S. Kolokotsa, George S. Stavrakakis, and Ioannis D. Skias. 2012. "On the application of clustering techniques for office Buildings' energy and thermal comfort classification." *IEEE Transactions on Smart Grid* 3 (4): 2196–2210. https://doi.org/10.1109/TSG.2012.2215059.

61 Joseph, Shibily, and Jasmin Erakkath Abdu. 2018. "Real-time retail price determination in smart grid from real-time load profiles." *International Transactions on Electrical Energy Systems* 28 (3): e2509. https://doi.org/10.1002/etep.2509.

62 Ahmed, Saeed, Zafar Ali Khan, Noor Gul, Junsu Kim, and Su Min Kim. 2021. "Machine learning-based clustering of load profiling to study the impact of electric vehicles on smart meter applications." *2021 Twelfth International Conference on Ubiquitous and Future Networks (ICUFN)*, 444–47. IEEE. https://doi.org/10.1109/ICUFN49451.2021.9528396.

63 Abedinia, O., and Amjady, N. 2016. "Net demand prediction for power systems by a new neural network-based forecasting engine." *Complexity* 21 (S2): 296–308. https://doi.org/10.1002/cplx.21807.

64 Morsalin, Sayidul, Khizir Mahmud, and Graham Town. 2016. "Electric vehicle charge scheduling using an artificial neural network." In *2016 IEEE Innovative Smart Grid Technologies – Asia (ISGT-Asia)*, 276–80. IEEE. https://doi.org/10.1109/ISGT-Asia.2016.7796398.

65 Negnevitsky, M., Mandal, P., and Srivastava, A.K. (2009). An overview of forecasting problems and techniques in power systems. In *2009 IEEE Power & Energy Society General Meeting*, 1–4. IEEE. https://doi.org/10.1109/PES.2009.5275480.

66 Rashid, M.H. (2018). AMI smart meter big data analytics for time series of electricity consumption. In: *In 2018 17th IEEE International Conference on Trust, Security and Privacy in Computing and Communications/12th IEEE International Conference on Big Data Science and Engineering (TrustCom/BigDataSE)*, 1771–1776. IEEE https://doi.org/10.1109/TrustCom/BigDataSE.2018.00267.

67 Quilumba, Franklin L., Wei-Jen Lee, Heng Huang, David Y. Wang, and Robert L. Szabados. 2015. "Using smart meter data to improve the accuracy of intraday load forecasting considering customer behavior similarities." *IEEE Transactions on Smart Grid* 6 (2): 911–18. https://doi.org/10.1109/TSG.2014.2364233.

68 Catalão, J.P. (ed.) (2017). *Electric Power Systems: Advanced Forecasting Techniques and Optimal Generation Scheduling*. CRC Press.

69 Pinto, T., Sousa, T.M., and Vale, Z. (2012). Dynamic artificial neural network for electricity market prices forecast. In: *2012 IEEE 16th International Conference on Intelligent Engineering Systems (INES)*, 311–316. IEEE.

70 Yin, Linfei, Zhixiang Sun, Fang Gao, and Hui Liu. 2020. "Deep forest regression for short-term load forecasting of power systems." *IEEE Access* 8: 49090–99. https://doi.org/10.1109/ACCESS.2020.2979686.

71 Mehedi, Ibrahim M., Hussain Bassi, Muhyaddin J. Rawa, Mohammed Ajour, Abdullah Abusorrah, Mahendiran T. Vellingiri, Zainal Salam, and Md. Pauzi Bin Abdullah. 2021. "Intelligent machine learning with evolutionary algorithm based short term load forecasting in power systems." *IEEE Access* 9: 100113–24. https://doi.org/10.1109/ACCESS.2021.3096918.

72 Yin, Linfei, and Jiaxing Xie. 2021. "Multi-temporal-spatial-scale temporal convolution network for short-term load forecasting of power systems." *Applied Energy* 283 116328. https://doi.org/10.1016/j.apenergy.2020.116328.

73 Hong, Tao, Pierre Pinson, and Shu Fan. 2014. "Global energy forecasting competition 2012." *International Journal of Forecasting* 30 (2): 357–63. https://doi.org/10.1016/j.ijforecast.2013.07.001.

74 Liu, Meng, Franklin L. Quilumba, and Wei-Jen Lee. 2015. "Dispatch scheduling for a wind farm with hybrid energy storage based on wind and LMP forecasting." *IEEE Transactions on Industry Applications* 51 (3): 1970–77. https://doi.org/10.1109/TIA.2014.2372043.

75 Antonanzas, J., D. Pozo-Vázquez, L.A. Fernandez-Jimenez, and F.J. Martinez-de-Pison. 2017. "The value of day-ahead forecasting for photovoltaics in the Spanish electricity market." *Solar Energy* 158 140–46. https://doi.org/10.1016/j.solener.2017.09.043.

76 Gigoni, Lorenzo, Alessandro Betti, Emanuele Crisostomi, Alessandro Franco, Mauro Tucci, Fabrizio Bizzarri, and Debora Mucci. 2018. "Day-ahead hourly forecasting of power generation from photovoltaic plants." *IEEE Transactions on Sustainable Energy* 9 (2): 831–42. https://doi.org/10.1109/TSTE.2017.2762435.

77 Yang, Yandong, Weijun Hong, and Shufang Li. 2019. "Deep ensemble learning based probabilistic load forecasting in smart grids." *Energy* 189 116324. https://doi.org/10.1016/j.energy.2019.116324.

78 Gao, Wei, Ayda Darvishan, Mohammad Toghani, Mohsen Mohammadi, Oveis Abedinia, and Noradin Ghadimi. 2019. "Different states of multi-block based forecast engine for price and load prediction." *International Journal of Electrical Power & Energy Systems* 104 423–35. https://doi.org/10.1016/j.ijepes.2018.07.014.

79 Dumas, Jonathan, Antoine Wehenkel, Damien Lanaspeze, Bertrand Cornélusse, and Antonio Sutera. 2022. "A deep generative model for probabilistic energy forecasting in power systems: normalizing flows." *Applied Energy* 305 117871. https://doi.org/10.1016/j.apenergy.2021.117871.

80 Aleem, Saad Abdul, Nauman Shahid, and Ijaz Haider Naqvi. 2015. "Methodologies in power systems fault detection and diagnosis." *Energy Systems* 6 (1): 85–108. https://doi.org/10.1007/s12667-014-0129-1.

81 Appiah, Albert Yaw, Xinghua Zhang, Ben Beklisi Kwame Ayawli, and Frimpong Kyeremeh. 2019. "Review and performance evaluation of photovoltaic array fault detection and diagnosis techniques." *International Journal of Photoenergy* 2019 1–19. https://doi.org/10.1155/2019/6953530.

82 Zhao, Yang, Tingting Li, Xuejun Zhang, and Chaobo Zhang. 2019. "Artificial intelligence-based fault detection and diagnosis methods for building energy systems: advantages, challenges and the future." *Renewable and Sustainable Energy Reviews* 109 85–101. https://doi.org/10.1016/j.rser.2019.04.021.

83 Furse, Cynthia M., Moussa Kafal, Reza Razzaghi, and Yong-June Shin. 2021. "Fault diagnosis for electrical systems and power networks: a review." *IEEE Sensors Journal* 21 (2): 888–906. https://doi.org/10.1109/JSEN.2020.2987321.

84 Jindal, Anish, Amit Dua, Kuljeet Kaur, Mukesh Singh, Neeraj Kumar, and S. Mishra. 2016. "Decision tree and SVM-based data analytics for theft detection in smart grid." *IEEE Transactions on Industrial Informatics* 12 (3): 1005–16. https://doi.org/10.1109/TII.2016.2543145.

85 Cui, Qiushi, Khalil El-Arroudi, and Yang Weng. 2019. "A feature selection method for high impedance fault detection." *IEEE Transactions on Power Delivery* 34 (3): 1203–15. https://doi.org/10.1109/TPWRD.2019.2901634.

86 Mohammadnian, Youness, Turaj Amraee, and Alireza Soroudi. 2019. "Fault detection in distribution networks in presence of distributed generations using a data mining-driven wavelet transform." *IET Smart Grid* 2 (2): 163–71. https://doi.org/10.1049/iet-stg.2018.0158.

87 Xue, Bing, Qing Chen, and Wudi Huang. 2020. "A fault diagnosis method of active distribution network based on fault search table and data mining technology." *2020 IEEE Sustainable Power and Energy Conference (ISPEC)*, 2512–18. IEEE. https://doi.org/10.1109/iSPEC50848.2020.9350932.

88 Wang, Yixing, Meiqin Liu, Zhejing Bao, and Senlin Zhang. 2019. "Stacked sparse autoencoder with PCA and SVM for data-based line trip fault diagnosis in power systems." *Neural Computing and Applications* 31 (10): 6719–31. https://doi.org/10.1007/s00521-018-3490-5.

89 Qiqi, Z., Siyi, L., and Yan, Z. 2019. "Power distribution network disconnection fault diagnosis using data mining method." In *2019 IEEE Innovative Smart Grid Technologies – Asia (ISGT Asia)*, 407–11. IEEE. https://doi.org/10.1109/ISGT-Asia.2019.8881320.

90 Zhang, Chaobo, Xue Xue, Yang Zhao, Xuejun Zhang, and Tingting Li. 2019. "An improved association rule mining-based method for revealing operational problems of building heating, ventilation and air conditioning (HVAC) systems." *Applied Energy* 253 113492. https://doi.org/10.1016/j.apenergy.2019.113492.

91 Dehbozorgi, Mohammad Reza, and Mohammad Rastegar. 2020. "Association rule mining application to diagnose smart distribution power system outage root cause." In *2020 10th Smart Grid Conference (SGC)*, 1–6. IEEE. https://doi.org/10.1109/SGC52076.2020.9335746.

92 Yan, Ke, Adrian Chong, and Yuchang Mo. 2020. "Generative adversarial network for fault detection diagnosis of chillers." *Building and Environment* 172 106698. https://doi.org/10.1016/j.buildenv.2020.106698.

93 Yu, Kun, Hao Zou, Xiangjun Zeng, Yueyu Li, Hao Li, Chao Zhuo, and Zhan Wang. 2021. "Faulty feeder detection of single phase-earth fault based on fuzzy measure fusion criterion for distribution networks." *International Journal of Electrical Power & Energy Systems* 125 106459. https://doi.org/10.1016/j.ijepes.2020.106459.

94 You, Shutang, Yinfeng Zhao, Mirka Mandich, Yi Cui, Hongyu Li, Huangqing Xiao, Summer Fabus, et al. 2020. "A review on artificial intelligence for grid stability assessment." In *2020 IEEE International Conference on Communications, Control, and Computing Technologies for Smart Grids (SmartGridComm)*, 1–6. IEEE. https://doi.org/10.1109/SmartGridComm47815.2020.9302990.

95 Diao, Ruisheng, Vijay Vittal, and Naim Logic. 2010. "Design of a real-time security assessment tool for situational awareness enhancement in modern power systems." *IEEE Transactions on Power Systems* 25 (2): 957–65. https://doi.org/10.1109/TPWRS.2009.2035507.

96 Sobajic, D.J., and Y.-H. Pao. 1989. "Artificial neural-net based dynamic security assessment for electric power systems." *IEEE Transactions on Power Systems* 4 (1): 220–28. https://doi.org/10.1109/59.32481.

97 Liu, Songkai, Lihuang Liu, Nan Yang, Dan Mao, Lei Zhang, Jiangzhou Cheng, Tianliang Xue, et al. 2021. "A data-driven approach for online dynamic security assessment with spatial-temporal dynamic visualization using random bits Forest." *International Journal of Electrical Power & Energy Systems* 124 106316. https://doi.org/10.1016/j.ijepes.2020.106316.

98 Zhang, C., Yuan, Z., and Yan, P. (2021). Analysis of smart grid stability and security management based on data mining. *IOP Conference Series: Earth and Environmental Science* 651: 022049.

99 Hou, Kaiyuan, Guanghui Shao, Haiming Wang, Le Zheng, Qiang Zhang, Shuang Wu, and Wei Hu. 2018. "Research on practical power system stability analysis algorithm based on modified SVM." *Protection and Control of Modern Power Systems* 3 (1): 11. https://doi.org/10.1186/s41601-018-0086-0.

100 Bashiri Mosavi, Alireza, Ali Amiri, and Seyed Hadi Hosseini. 2018. "A learning framework for size and type independent transient stability prediction of power system using twin convolutional support vector machine." *IEEE Access* 6: 69937–47. https://doi.org/10.1109/ACCESS.2018.2880273.

101 Xiao, Fu, and Cheng Fan. 2014. "Data mining in building automation system for improving building operational performance." *Energy and Buildings* 75: 109–18. https://doi.org/10.1016/j.enbuild.2014.02.005.

102 Dai, Bangwu, Fuli Wang, and Yuqing Chang. 2022. "Multi-objective economic load dispatch method based on data mining technology for large coal-fired power plants." *Control Engineering Practice* 121 105018. https://doi.org/10.1016/j.conengprac.2021.105018.

103 Rodríguez González, A.Y., M. Palacios Alonso, F. Lezama, L. Rodríguez, E. Muñoz de Cote, E.F. Morales, L. Enrique Sucar, and D.D. Crockett. 2019. "A competitive and profitable multi-agent autonomous broker for energy markets." *Sustainable Cities and Society* 49. https://doi.org/10.1016/j.scs.2019.101590.

104 Tang, Wen-Jun, and Hong-Tzer Yang. 2018. "Data mining and neural networks based self-adaptive protection strategies for distribution systems with DGs and FCLs." *Energies* 11 (2): 426. https://doi.org/10.3390/en11020426.

105 Asha Kiranmai, S., and A. Jaya Laxmi. 2018. "Data mining for classification of power quality problems using WEKA and the effect of attributes on classification accuracy." *Protection and Control of Modern Power Systems* 3 (1): 29. https://doi.org/10.1186/s41601-018-0103-3.

106 Talaat, M., Abdulaziz S. Alsayyari, Adel Alblawi, and A.Y. Hatata. 2020. "Hybrid-cloud-based data processing for power system monitoring in smart grids." *Sustainable Cities and Society* 55 102049. https://doi.org/10.1016/j.scs.2020.102049.

107 Zhu, Jun, Eric Zhuang, Chavdar Ivanov, and Ziwen Yao. 2011. "A data-driven approach to interactive visualization of power systems." *IEEE Transactions on Power Systems* 26 (4): 2539–46. https://doi.org/10.1109/TPWRS.2011.2119499.

3

Deep Learning in Intelligent Power and Energy Systems

Bruno Mota, Tiago Pinto, Zita Vale, and Carlos Ramos

GECAD Research Group on Intelligent Engineering and Computing for Advanced Innovation and Development (GECAD), Polytechnic Institute of Porto (ISEP/IPP), Porto, Portugal

Acronyms

AE	autoencoder
AI	artificial intelligence
ANN	artificial neural network
CNN	convolution neural network
DBN	deep belief network
DDPG	deep deterministic policy gradient
DL	deep learning
DNN	deep neural network
DQN	deep Q network
DR	demand response
DRL	deep reinforcement learning
EMS	energy management system
FD	fault detection
GRU	gated recurrent unit
HVAC	heating, ventilation, and air conditioning
IoT	Internet-of-Things
LSTM	long short-term memory
MDP	Markov decision process
ML	machine learning
PES	power and energy systems
PV	photovoltaic
RBM	restricted Boltzmann machine
RNN	recurrent neural network
SVM	support vector machine

Intelligent Data Mining and Analysis in Power and Energy Systems: Models and Applications for Smarter Efficient Power Systems, First Edition. Edited by Zita Vale, Tiago Pinto, Michael Negnevitsky, and Ganesh Kumar Venayagamoorthy.

3.1 Introduction

The ever-increasing development of applications in the Internet space has led to an explosive growth of data scale in the digital world. This is occurring due to the fast development of technologies such as cloud computing, big data, and the Internet-of-Things (IoT) [1]. In comparison, global data in 2020 was expected to be 22 times that of 2011, reaching about 35.2 zettabytes, according to a report by the International Data Corporation in 2012 [2]. This boom of digital data brings huge opportunities and transformative potential for a wide variety of fields, such as in the healthcare industry, energy sector, enterprises, manufacturing, and educational services [3, 4]. As a result of the information overload across fields, artificial intelligence (AI) techniques such as deep learning (DL) have made breakthroughs in natural language processing, speech recognition, image processing, forecasting, and many other fields [5].

Technological companies such as Google, Apple, Twitter, and Facebook collect and analyze huge amounts of data, being one of the biggest investors in projects related to DL. For instance, Google employs DL algorithms on colossal amounts of data acquired from the Internet, for their services, such as Google Translator, Google Street View, the image search engine, and Android speech recognition [6]. Furthermore, Apple's virtual assistant, Siri, provides a wide range of services such as personalized news, reminders, and weather forecasts, among others, by combining DL and the increasing amount of available data [7].

DL approaches are also fundamental for dealing with rising relevant problems with high impact in societies, such as those related to power and energy systems (PES) [8]. Environmental problems and rising energy prices have contributed to a surge of research regarding energy consumption. Accordingly, renewable energy resources are fundamental for grid stability as well as their environmental benefits against climate change and global warming [9]. Renewable energy is expected to account for 40% by 2040 and accounts for 25% of today's electricity production [10]. As a result, there is an abundance of DL works that aim at forecasting photovoltaic (PV) energy, electricity prices, and wind power.

When applied to energy management systems (EMSs), DL can bring many benefits both in energy use and cost reductions, in residential spaces and workplaces. Several important works related to EMSs can be found in the literature, addressing topics such as heating, ventilation, and air conditioning (HVAC) systems, building energy consumption predictors, power quality disturbances detection in smart grids, fault detection (FD) and classification on generators, automatic feature engineering, energy management operational control, demand response (DR), transactions in an electricity market, among others [11].

In 1943 it was first published the McCulloch–Pitts model [12] by Warren McCulloch and Walter Pitts, the first artificial neuron, achieved by analyzing and summarizing the characteristics of neurons. Further research that led to neural networks was conducted in Ref. [13], which focused on a cell assembly theory in order to explain the adaptation of cerebral neurons during the learning process. Only in 1958 was the first perceptron invented by Frank Rosenblatt in the proposed work [14], which consisted of a kind of supervised learning with binary classifiers. In 1969, Marvin Minsky and Seymour Papert wrote an article citing that a perceptron could not learn the XOR function, which lead to a stagnation of neural network research for almost a decade. However, in 1982 it was proposed the Hopfield network [15], which led to a small resurgence in the field. The revival of artificial neural networks (ANNs) was done by David Rumelhart, Geoffrey Hinton, and Ronald Williams with the proposed paper [16], which introduced back-propagating in the learning process of neural networks, through a gradient descent method. Also noteworthy is the work done by David Ackley on Boltzmann machines by using simulated annealing with the proposed paper in

Ref. [17]. ANNs hit another nadir due to the appearance of various shallow machine learning (ML) approaches that had advantages both in theory and application. The concept of DL was only introduced after Geoffrey Hinton put forward the concept in 2006 in the journal *Science* [18], starting an ever-increasing trend of deep ANN research and application in the real world.

The first relevant works in DL consist of deep kernel machines for meta-heuristic searching [19], autoencoders (AEs) for acoustic processing [20], a recurrent neural network (RNN) using long short-term memory (LSTM) for natural language processing [21], a restricted Boltzmann machine (RBM), and a deep belief network (DBN) applied to robotics and linguistics [22], a DBN for image matching [23], a deep convolutional network for object recognition [24], a convolutional DBN for cluster formation and learning [25], and a stacked AEs for denoising [26].

In the field of PES, some of the first neural network approaches [27] propose an ANN to correct the results of a neuro-fuzzy forecast model to improve wind power forecasting, [28] employs an ANN for long-term electricity demand forecasting, [29] uses a dynamic RNN to optimize long-term energy performance forecasting of integrated generation systems, Mellit [30] proposes an RNN for daily PV generation forecasting, and Refs. [31, 32] applies an ANN model to predict electricity prices.

Considering the recent trend and relevance of DL models, this chapter provides an overview of the most relevant and recent research on DL in PES applications, addressing works related to regression, classification, and decision-making problems. The reviewed works consider models that provide contributions to the evolution of EMSs, energy resources forecasting, DR, electricity market, and power quality, among other relevant topics in the PES field.

For the present chapter literature review, several inclusion and exclusion criteria were considered, as shown in Tables 3.1 and 3.2, respectively. As a result, there are a total of 67 publications included in the review, from 2016 to 2021, as represented in Figure 3.1. Of all the publications included, 26 depict regression problems, 24 represent classification problems, and 17 describe decision-making problems, accounting for, 39%, 36%, and 25% of the overall publications included, respectively. Also noteworthy is that there is a larger number of publications included in the review that represent more recent studies (i.e. from 2019 to 2021), as shown in Figure 3.2. It demonstrates that in 2016, 2017, 2018, 2019, 2020, and 2021, there are a total of 2, 6, 10, 18, 16, and 15 publications included in the review, respectively. Therefore, even though the present review includes publications from 2016, most of the publications are from the most recent state-of-the-art literature.

This chapter structure is divided into four main sections. The first section introduces and contextualizes the objective of the chapter and the associated history regarding DL, and the specific application in PES. Section 3.2 presents the state-of-art literature review regarding DL in PES by subdividing it into regression, classification, and decision-making problems. Section 3.3 discusses the accomplishments, limitations, and challenges of DL applications in PES. Finally, Section 3.4 summarizes the chapter by describing the most promising and expected future advancements in the field.

Table 3.1 Inclusion criteria.

	Inclusion criteria
IC1	The source focuses on the application in intelligent power and/or energy systems
IC2	The source belongs to the field of deep learning in computer science
IC3	The source is peer-reviewed, except in cases that the source shows a relevant contribution to the area
IC4	The source describes a significant contribution to the fields of study

Table 3.2 Exclusion criteria.

	Exclusion criteria
EC1	The source is a duplicate
EC2	The source is published before 2016
EC3	The source is not written in English
EC4	The source is either a book chapter, dissertation, review, or thesis
EC5	The source does not present studies related to deep learning in intelligent power and/or energy systems

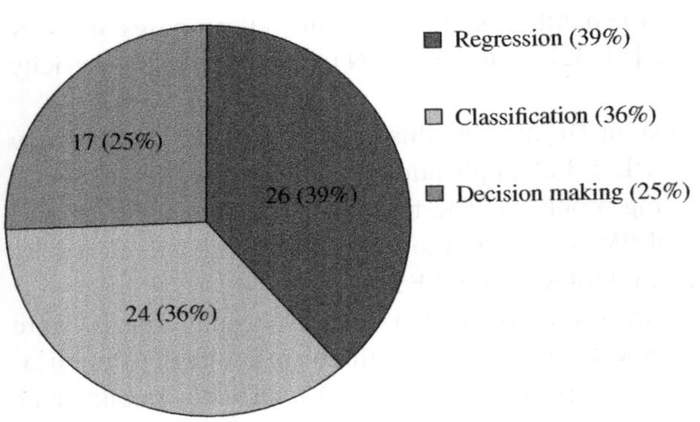

Figure 3.1 Number and percentage of publications per category (i.e. regression, classification, and decision-making) included in the review.

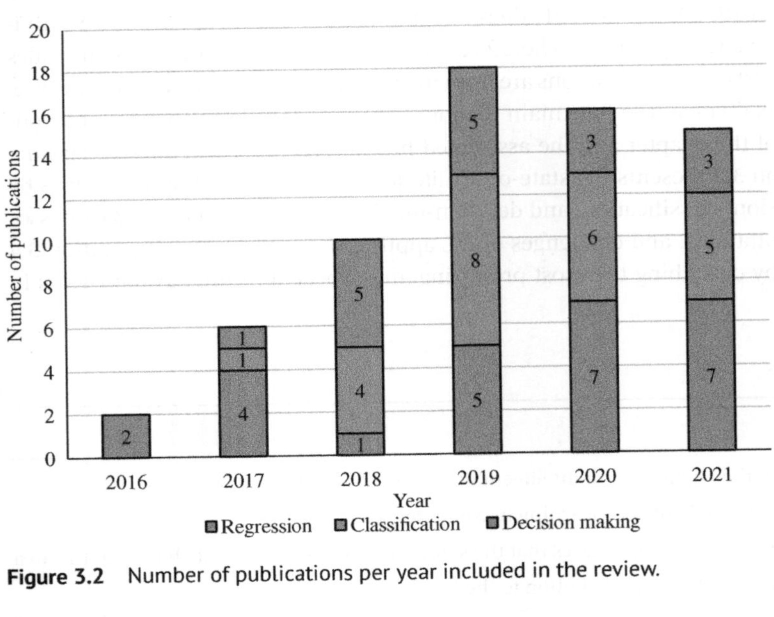

Figure 3.2 Number of publications per year included in the review.

3.2 Deep Learning

DL is a complex ML approach for pattern and structure recognition in very large datasets. These computational models are composed of multiple processing layers that allow visual object recognition, speech recognition, data forecast, and solutions in many other domains, to be greatly improved when compared to other ML approaches [5].

When considering PES, DL is widely used both on regression and classification problems, having a wide range of applications on smart grids, microgrids, smart buildings, energy market, and much more [11]. The regression problems addressed in this domain are mainly related to PV forecast [33], electricity prices forecast [34], predicting power flows [35], energy consumption prediction [36], abnormal energy consumption detection [37], and wind power forecast [38]. Regarding classification problems in PES, these are less common, but they still play a big role in detecting power quality disturbances [39], FD [40], and feature engineering in EMSs [41]. Furthermore, there are also decision-making problems that rely heavily on deep reinforcement learning (DRL), such as energy management [42], DR [43], and electricity market models [44], among others.

Sections 3.2.1–3.2.3 explore the existing solutions for these types of problems in PES, i.e. regression, classification, as well as decision-making problems. For each subchapter, the most relevant works in the literature are reviewed, with a focus on their contributions to PES.

3.2.1 Regression Problems

Sections 3.2.1.1–3.2.1.4 describe the most common regression problems in PES that are addressed using DL, by reviewing the most relevant and recent related works.

3.2.1.1 Photovoltaic Energy Forecast

The increasing penetration of renewable energy sources, and the associated uncertainty brought by the dependency on natural conditions such the wind and solar intensity, required the development of new advanced models that are able to provide good forecasts for this type of generation. Regarding PV generation, in specific, in Ref. [33] it is proposed a novel DL architecture, presented as PV-Net, which is capable of short-term forecasting of day-ahead PV energy. The PV-Net architecture focuses on the redesign of the gates on the gated recurrent unit (GRU) using convolutional layers, originating a so-called convolutional GRU. Furthermore, each convolutional GRU cell is stacked with each other in bidirectional blocks. This innovative approach allows for efficient extraction of temporal and positional characteristics in the PV power sequences, while also avoiding information loss across layers and improving gradient flow during training.

Also regarding PV forecast, Luo et al. [45] proposes an innovative DL approach that incorporates domain knowledge into LSTM to forecast the hourly day-ahead PV power generation. The proposed solution aims at overcoming the shortcoming of ML algorithms that are only dependent on large quantities of data. This solution proposes a physics-constrained long short-term memory, named PC-LSTM, composed of three different modules that represent the imposed constraints/knowledge: the data filtering module, the clipping module, and the loss penalty module. The first module focuses on the world/general knowledge of PV, by filtering unreasonable forecasts, for example, high PV generation during the night. The second module aims to represent natural science knowledge, by restricting the output to a reasonable range. For PV generation in specific, according to the physical law, the output needs, e.g. to be greater or equal to zero. Finally, the last module penalizes the LSTM whenever there are constraints that are being violated.

The work presented in Ref. [46] focuses on identifying the best algorithm for predicting solar energy. It compares three different approaches: an RNN, an LSTM, and a GRU. Moreover, the proposed case study uses real meteorological data, from 2016 to 2018. Results found that the RNN and LSTM have really good performance for this type of problem, showing very high levels of accuracy and low errors, with RNN coming slightly on top in terms of performance. On the other hand, the proposed GRU showed worse performance, lagging behind both RNN and LSTM.

Other noteworthy works on this topic are [47] which focuses on a hybrid DL approach, based on wavelet packet decomposition and LSTM, for short-term PV power forecasting, and Ref. [48] which combines LSTM, stacked AE, DBN, and multilayer perceptron techniques to forecast solar power in a smart grid.

PV forecasting approaches using DL have the ability to deal with the uncertainty related to this type of generation. Some of the considered advances are promising regarding the application to the forecasting of other types of variable generation. However, there are specific characteristics, e.g. concerning wind power forecast, which require the development of specific dedicated models [49].

3.2.1.2 Wind Power Forecast

The work of Wang et al. [38] proposes a novel DL-based ensemble approach for probabilistic wind power forecast. The proposed approach uses wavelet transform, CNN, and ensemble techniques to achieve a system capable of advanced point forecast. The wavelet transform method is used to decompose the raw wind power data into different frequencies; the nonlinear features are then used on the CNN for training. Furthermore, in order to deal with uncertainties in wind power data, i.e. data noise and model misspecification, these are identified separately in a later stage.

In Ref. [50], it is proposed a bidirectional LSTM model for wind power forecast of an Urban Regional Energy Internet. By using a bidirectional LSTM model, the proposed solution is capable of having better accuracy than a normal LSTM, since the latter only considers previous information, while the bidirectional LSTM can comprehensively consider historical and future information.

3.2.1.3 Building Energy Consumption Prediction

Not only renewable generation is associated with significant uncertainty, but also energy consumption [51]. It is therefore necessary that the prediction of energy consumption is enhanced by new models, benefiting from the advances in DL. For example, the work in Ref. [36] aims to create a model capable of predicting the energy consumption of a building while taking into account the randomness and noisy disturbances in data in this type of problem. The proposed approach combines extreme learning machine and stacked AEs, to create a novel approach called extreme stacked AE. The extreme learning machine is used to predict the building's energy consumption, while the stacked AEs are utilized to extract the building energy consumption features. Moreover, the usage of multiple AEs as building blocks, in the stacked AE, allows for the construction of a deep neural network (DNN). From the results of the proposed model, the paper claims that the proposed extreme stacked AE, when compared to other recent techniques, performs best when dealing with uncertainty in building energy consumption data.

Another building energy consumption predictor is proposed in Ref. [52], which focuses on comparing different models for predicting consumers' energy consumption in a grid. The validated methods are an ANN, support vector machine (SVM), RNN, and two newly developed stochastic models, specifically a conditional RBM and a factored conditional RBM. Results from the study showed that the factored conditional RBM outperforms every other method tested.

An innovative DL model for predicting energy consumption is proposed in Ref. [53] which incorporates IoT networks. The proposed solution called Energy-Net is composed of multiple stacked spatial-temporal modules, each one consisting of a temporal transformer and spatial transformer submodules. The temporal transformer submodule's purpose is to learn and extract long-range and short-range sequential energy characteristics along with the temporal domain. The spatial transformer submodule focuses on extracting hidden spatial information by integrating convolutional layers. Moreover, the later submodule includes an improved self-attention mechanism that has been demonstrated to be effective in learning the interactions among different elements.

Also on this topic, Berriel et al. [54] proposes a monthly energy consumption predictor that utilizes an LSTM with data normalization, using a standardization technique, and by utilizing metadata. When compared to other baseline methods that are employed by power companies, and models such as fully connected neural network, and convolution neural network (CNN), the proposed model by the paper outperforms all of them, confirming that LSTMs are more suitable for temporal task analysis.

Building energy consumption forecasting can also benefit from disaggregating consumption data to enable predictions of consumption per device, especially considering devices with large associated consumption and variation, such as HVAC and heat pumps. A DL approach using a feed-forward neural network with back propagation to predict energy consumption of HVAC systems is proposed in Ref. [55]. The study proposes a new variable selection process that aims at improving the model's accuracy when compared to other more traditional ML and statistical approaches. It was found that operational variables in the heating and cooling system, for example, the temperature at the top of the compressors, high pressure, low pressure, etc., were more important than temperature variables, such as outdoor temperature and indoor temperature.

An abnormal energy consumption detection system for ground source heat pumps is proposed in Ref. [37]. The proposed solution works by utilizing a mode decomposition-based LSTM to predict energy consumption, and then through the difference between the predicted and actual values, the anomaly is detected, by using the statistical model Grubbs' test. The system is capable of detecting three categories of anomalies: parabola anomaly, abrupt anomaly, and time-related anomaly.

3.2.1.4 Electricity Price Forecast

On the market side, Marszałek and Burczynski [34] proposes a novel approach to forecasting hourly day-ahead electricity prices, through the combination of multiple techniques, such as RNN, LSTM, attention mechanism, and clustering. Contrary to other proposed approaches in recent years, the paper claims that it focuses on the construction of a fair and unbiased model, which aims to increase accuracy and decrease categorical bias between the different clusters. To achieve such claims, the proposed model calculates the attention weights of the LSTM hidden states by considering a context vector given for each sample individually as the cluster center to which the sample belongs.

An innovative methodology for day-ahead electricity price forecast is presented in Ref. [56]. The proposed method utilizes Bayesian DL techniques for probabilistic energy price forecast and, in order to guarantee scalability in complicated network architectures, a special training method that focuses on processing mini-batches in parallel. Furthermore, the proposed model supports heteroscedasticity, eliminating the conventional homoscedastic assumption and the associated preprocessing cost. To validate the solution, case studies were conducted on two separate day-ahead marketplaces with varied behavior. Results show that the proposed methodology is capable of achieving strong performance in out-of-sample settings while also offering forecast uncertainty signals.

Another day-ahead electricity price forecasting system is proposed in Ref. [57] based on a novel hybrid DL framework. It considers the residuals between the actual electricity price and forecasting electricity price to improve training performance, uncertainties in electricity price, and it proposes a new feature preprocessing approach that allows for better identification of correlated features and outliners. The proposed methodology is composed of four modules: feature preprocessing module, DL-based point prediction module, error compensation module, and probabilistic prediction module. The first module is used to discover outliers and select the associated features of the electricity price series, through the usage of the isolation forest algorithm and the least absolute shrinkage and selection operator method. Then, in the point prediction module, complex nonlinear features are extracted, achieved by combining RNN, LSTM, DBN, and CNN. In the error compensation module, the residual error between the predicted and actual electricity prices is further reduced if possible. Finally, in the probabilistic prediction module, a quantile regression approach is used to quantify uncertainty at varying degrees of confidence. From the results, it is clear that the proposed methodology has better performance in point and probabilistic forecasting, when compared to methods such as light gradient boosting machine, back-propagation neural networks, support vector regression, and k-nearest neighbor.

3.2.1.5 Other Regression Works

Sections 3.2.1.1–3.2.1.4 discuss the most relevant works related to the most widely addressed regression problems in PES. Nevertheless, there are other, less explored regression problems, which are also noteworthy. The work in Ref. [58] proposes a distributed deep network structure for power system security knowledge discovery based on multitask learning to monitor and control power grids. The solution proposed aims to keep power systems secure, stable, and economical. A DBN is implemented to nonlinearly extract deep and abstract features layer-by-layer for total transfer capability regression tasks. Furthermore, it is proposed a new distributed training algorithm for the deep structure to accelerate the training process.

An optimal load dispatch with DL in a community microgrid is proposed in Ref. [59], by taking into account solar power and load demand forecast. In this study, a novel deep RNN with a LSTM model is implemented in a community microgrid to forecast aggregated power load and PV energy over a short-term period. According to the present study, the proposed deep RNN with a LSTM model has fewer limitations in modeling complex nonlinear problems and time dependencies. Therefore, having the potential for making more accurate short-term forecasting to support economical load dispatch systems, when compared to more traditional forecasting methods. Moreover, an optimal load dispatch model for grid-connected community microgrid considers residential power load, PV energy, electric vehicles, and energy storage systems. For load dispatch optimization, a particle swarm optimization algorithm is proposed.

For predicting optimal power flows on air conditioners, Fioretto et al. [35] proposes an innovative DL approach to the problem, by combining a feed-forward DNN and the Lagrangian dual method. The DNN is used to predict the generator setpoints for the optimal power flow of the air conditioner, while the Lagrangian dual method exploits the physical and engineering constraints by using violation degrees.

Other noteworthy works in the field, relating to regression problems in DL, are presented in Table 3.3.

3.2.2 Classification Problems

Sections 3.2.2.1–3.2.2.3 discuss the most prevalent classification problems in PES using DL, by reviewing the most recent and relevant research.

Table 3.3 Summary of other existing work for regression problems concerning DL approaches.

Article	Objective^a)	Method^b)	Application^c)
Soleimanzade and Sadrzadeh [60]	ES, PVF	LSTM, CNN	Smart grid
Pramono et al. [61]	LDF, DR	LSTM, CNN	Smart grid
Wu and Wang [62]	AN, CN	FNN	Microgrids
Huang et al. [63]	EO	CDNN	MIMO-NOMA
Bugbee et al. [64]	UI, IV, S	DNN	ESM

a) ES: energy storage, PVF: photovoltaic forecast, LDF: load demand forecast, AN: action network, CN: critic network, EO: energy optimization, UI: user interface, IV: immersive visualization, S: statistics.
b) CDNN: communication deep neural network, FNN: feed-forward neural network.
c) MIMO-NOMA: multiple-input-multiple-output non-orthogonal multiple access, ESM: energy system models.

3.2.2.1 Power Quality Disturbances Detection/Classification

The paper in Ref. [39] implements a novel stacked AE for feature reduction and power quality disturbances detection and classification in the industrial sector. The proposed solution is able to classify multiple combinations of disturbances at the same time, up to 17 were validated in the paper. It works by feature extracting using a domain fusion approach, in order to characterize the electrical power grid. Then, using a stacked AE, an adaptive pattern characterization is carried out. Finally, a neural network structure is used to identify disturbances.

A sequence-to-sequence DL model is proposed in Ref. [65] that aims at detecting the type and time location of power quality disturbances. It achieves these results by using a bidirectional GRU. The input sequence is first normalized and batched. Then, deep features are retrieved from the input sequence using a bidirectional GRU RNN with several layers stacked together in both the forward and backward directions. Afterward, based on the retrieved features, a fully connected neural network layer and Softmax are used to determine the associated probability indicating the category to which each element in the input sequence is categorized. The Argmax or Top K operation is used to identify the category of each element in the input sequence by picking the highest likelihood. Finally, the category is recognized, and the starting and ending times of disturbances are also located just as the category is changed. Contrary to existing sequence-to-sequence models that employ an encoder-decoder network, the solution proposed in the paper is capable of capturing intrinsic temporal information of power quality disturbances, mainly the starting and ending times of disturbances.

Also regarding power disturbances detection, Mohan et al. [66] proposes an innovative CNN-LSTM to characterize and classify power quality disturbances. The proposed hybrid model contains two CNN layers followed by max pooling in order to extract low-level spatial dependencies among the data. Then, an LSTM layer extracts the long-term temporal dependencies and generates a more abstract feature map. Finally, a fully connected layer with a softmax function is used to calculate the probability of each class of disturbance. Furthermore, the proposed solution was compared to other known models, such as CNN, RNN, identity-RNN, LSTM, and GRU, demonstrating that it outperforms all of them. Also, it was found to be accurate for real-time characterization and classification of power quality disturbances.

In Ref. [67] it is proposed a new classification approach for power quality complex disturbances by combining an independent component analysis and a sparse AE. Also, the solution allows the detection and classification of multiple disturbances at the same time, with high accuracy. The sparse AE is used to extract the value of the features of a single disturbance. In addition, the proposed sparse AE is designed for multiple-label power quality disturbances classification, mainly for double and quadruple disturbances.

A new approach using DBN is proposed in Ref. [68] to classify power quality disturbances. The DBN proposed in the paper consists of several stacked RBMs for unsupervised learning that focuses on classifying seven types of disturbance signals: interruption, sag, swell, harmonic, oscillatory, sag-harmonic, and swell-harmonic. Results show that the proposed approach achieves much better classification rates than more conventional algorithms, such as wavelet transform and SVM.

Other relevant works that address problems related to power quality are Sindi et al. [69] which proposes the combination of 1D and 2D CNN features for power quality analysis, Ref. [70] aims to detect frequency disturbances through two CNNs and classifier fusion, and in Ref. [71] it is proposed an automatic system for complex power quality disturbance identification based on a multi-fusion CNN.

Power quality is hugely influenced by faults in the system. Fault detection and classification are, thereby, a significant problem that required dedicated solutions. For instance, condition monitoring, through FD and classification, in smart grids can be a great way to improve the overall power quality in the grid and reduce power quality disturbances [72].

3.2.2.2 Fault Detection/Classification

A DL-based fault detection and classification system for real-time diagnosis of air-handling units are proposed in Ref. [40]. It focuses on improving the operational efficiency of air-handling units to reduce the energy consumption of HVAC systems in buildings. In addition, the solution was validated using the EnergyPlus simulation platform in order to reduce the complexity of data processing and effectively ensure data integrity, thus improving the reliability of the proposed diagnostic model. With the usage of a DNN with back propagation, the proposed solution is capable of detecting up to five different fault types, operating in real-time, and providing better diagnostic ability than other more conventional approaches.

The work in Ref. [73] proposes a fault diagnosis system for wind turbine gearboxes, by using a stacked AE and SVM for frequency analysis. The stacked AE-based multiclass SVM classifier is used to learn higher level representation of input features in order to classify the fault. The proposed solution in the paper not only outperforms traditional multiclass SVM methods but is also nonintrusive to wind turbines. Also noteworthy, the solution has no or very low hardware cost since it does not require additional sensors that other traditional methodologies might need.

An intelligent fault diagnosis system using DL for solid oxide fuel cell systems is proposed in Ref. [74], which aims to successfully detect multiple faults simultaneously. To achieve these results, a novel simultaneous fault diagnosis model based on a stacked sparse AE is implemented. The proposed model automatically captures essential features from the original input data, thus reducing time on heavily hand-crafted features. Also, the proposed methodology does not require extensive knowledge of the physical model, thus having great generalization capabilities. When compared to other classical data-based diagnosis methods, the proposed solution has better performance on solid oxide fuel cell systems. In fact, the process of capturing essential features from the original input data as well as creating new features through the combination of original features is a relevant topic that enables the improvement of the learning process. Feature engineering works are providing significant contributions in this domain [75].

Other works worth noting regarding this topic are [76] which focuses on FD, classification, and location prediction in power systems by using an LSTM model, Afrasiabi et al. [77] proposes an FD system that integrates a CNN and a light-GRU applied to power transformers, and finally, a power cables FD and classification, by using a DBN, is proposed in Ref. [78].

3.2.2.3 Feature Engineering

In Ref. [41] it is proposed an innovative automatic feature engineering method using DL models to derive high-quality features for building energy prediction. It aims at reducing data dimensionality, decreasing prediction model complexity, and tackling the problem of corrupted and noisy data; thus improving building energy prediction performance. Three types of models are proposed to achieve such results, a fully connected AE, a convolutional AE, and a generative adversarial network. In addition, the models are compared to conventional feature engineering methods, for example, principal components and summarizing statistics. From the results, DL methods outperform all the conventional feature engineering methods, with generative adversarial networks coming on top of the DL models. In short, the generative adversarial network model, proposed in the paper, for both the generator and discriminator uses a convolutional AE as the basic operation unit. Also, since the discriminator only aims to distinguish whether a sample is real or synthetic, the output layer has only one neuron and the *sigmoid* function for classification.

The model proposed by Wang and Chen [79] focuses on a novel full closed-loop approach to detect and classify power quality disturbances through automatic feature extraction. To achieve these results, a CNN is designed to capture multi-scale features and reduce overfitting. Furthermore, the proposed closed-loop feedback, contrary to conventional methods which are divided into three phases: signal analysis, feature selection, and classification, allows for a unified whole that can be solved automatically by a universal DNN. When compared with other algorithms such as wavelet transform, SVM, particle swarm optimization, etc., under no noise, low-level noise, and high-level noise conditions, the accuracy of the proposed method outperforms all other algorithms; thus having a better noise immunity compared to conventional methods.

Another relevant work on this topic is presented in Ref. [80] which focuses on a multi-energy load prediction model. The main contribution of this work is the usage of a hybrid model, based on CNN and GRU, for effective feature extraction and dynamic modeling of time series.

3.2.2.4 Other Classification Works

In the paper [81] an innovative approach is introduced with aims at integrating DL and IoT for energy consumption reduction, by controlling the air conditioners. A people detection system, using a camera, is proposed in order to count the number of persons in a specific area, through a known CNN-based object detection network called You Look Only Once, abbreviated as YOLO, version 3. Accordingly, based on the number of people in the area, the air conditioners are optimized cost- and energy wise. However, the proposed work is designed to recognize human faces, instead of other more privacy-oriented approaches, which is a big red flag since it is to be implemented in homes and workplaces.

A novel architecture for feature learning and classification of voltage dips is proposed in Ref. [82]. An LSTM model is employed to extract voltage dip features from root-mean-square voltage sequences. Then, through a fully connected neural network layer with a *softmax* activation function, voltage dips are classified according to seven different types of dips, by using the feature provided by the LSTM outputs. Therefore the proposed solution is able to learn three-phase voltage dip patterns from time-dependent root-mean-square sequences. Contrary to conventional ML methods, the solution proposed in the paper allows learning dip features without requiring human knowledge or expert rules, transition-event segmentation, and selecting thresholds, while having better performance.

Table 3.4 Summary of other existing work for classification problems concerning DL approaches.

Article	Objective[a)]	Method[b)]	Application[c)]
Li et al. [83]	ETD	CNN-RF	Smart grid
Lu et al. [84]	FD, ETD	SSAE	Smart grid
Feng et al. [85]	BOD	CNN, BiLSTM	SB

a) ETD: electricity theft detection, BOD: building occupancy detection.
b) CNN-RF: convolutional neural network random forest, SSAE: semi-supervised autoencoder, BiLSTM: bidirectional long short-term memory.
c) SB: smart buildings.

Other noteworthy works in the field, relating to classification problems in DL, are described in Table 3.4.

3.2.3 Decision-Making Problems

Although regression and classification problems are the most widely addressed types of problems in DL within PES, DL is also used to address decision-making problems in this domain. Sections 3.2.3.1–3.2.3.3 present the most relevant and recent state-of-art literature review of decision-making problems in PES using DL.

3.2.3.1 Energy Management

An EMS using reinforcement learning based on Deep Q Network (DQN) is proposed in Ref. [42] for cost minimization on hybrid systems, which provide electricity, heating, cooling, and water for homes. The proposed method does not require online trial and error or historical data of system performance to determine the operational strategy. Furthermore, the developed reinforcement learning DQN takes into account hybrid systems with PV collectors, energy storage collectors, wind turbines, combined heating and power units, water heaters, hot water storage tanks, and batteries. When compared to two rule-based methods, Load Following and Full State of Charge, the proposed novel model outperforms both of them.

The paper in Ref. [86] proposes a real-time EMS to reduce energy costs of energy purchased for a storage PV system in a microgrid. To achieve this, a novel approach to the problem is presented, in which the focus of the cost optimization is on the grid energy purchases rather than on a direct storage control. A DRL is employed with a Q-learning algorithm to learn the optimal decision policy. Furthermore, many enhancements to improving learning speed and stability are implemented, that rely on experience replay, target network, and increasing discount factor.

Reference [87] focuses on a real-time dynamic EMS that considers the uncertainties of renewable energy availability and load demand, in a wind-solar-diesel-battery-reverse osmosis hybrid energy system. The proposed solution utilizes DRL with proximal policy optimization to obtain the optimal control policy that aims at reducing the overall energy costs. The proposed DRL method is a centralized control approach, having a single-point and communication delay problem.

An innovative real-time autonomous energy management approach for residential multi-energy systems is proposed in Ref. [88]. It uses a DRL-based approach combined with a novel deep deterministic policy gradient (DDPG) method with a prioritized experience replay strategy. Through the proposed approach, the agent is able to receive accurate feedback concerning its impact on energy management decisions in the multi-energy systems, thus allowing for better discovery of more cost-effective control strategies by exploring the entire action domain. In addition, the state-of-the-art DDPG allows for performance improvements when compared to state-of-the-art DRL methods, traditional stochastic programming, and robust optimization.

As shown by these works, today's energy management models need to deal with multiple variables, including variable renewable generation, electricity market prices, and uncertain consumption. It becomes essential that such systems are able to incorporate and motivate the means to reach suitable flexibility from the consumer side. In this context, DR becomes a central topic of research.

3.2.3.2 Demand Response

A model-free DRL method for DR management of interruptible load is proposed in Ref. [43]. The proposed DRL approach is combined with a Dueling DQN structure, in order to optimize DR management of interruptible load under the time of use tariff and variable electricity consumption patterns. An automatic DR architecture provides the possibility of real-time application of DR. In addition, the problem at hand is addressed as a Markov decision process (MDP), that is, the state, action, and reward function are defined, to obtain the maximum long-term profit of the DR management through the analysis of cumulative reward. The proposed architecture, when compared to other traditional DQN, overcomes the noise and instability. Moreover, the proposed DRL realizes the goal of minimizing the peak load demand and operational costs on the premise of regulating voltage to the safe limit.

A novel multi-agent DRL for DR participation in discrete manufacturing systems is presented in Ref. [89], which aims at minimizing electricity costs and improving grid stability. To achieve this goal, the industrial manufacturing system is designed as a partially observable Markov game. Also, in order to obtain the optimal schedule, for energy consumption, of the different machines, a multi-agent DDPG algorithm is proposed.

The successful widespread application of DR is not only dependent on the quality of the models that are used to manage it, but also on the intrinsic flexibility of consumers and on the incentives that are provided so that consumers are motivated to act in a flexible way. Such incentives are often associated with electricity market prices, which makes it crucial that suitable models are proposed to predict market prices, propose novel efficient market mechanisms, and support players' actions in the market.

3.2.3.3 Electricity Market

In Ref. [44] an innovative DRL with a replay mechanism for an event-driven local energy market is proposed. It aims at facilitating energy trading between prosumers at the distribution level. It uses a deep Q-learning algorithm for local energy trading modified from a DQN, to facilitate decision-making in the energy system as well as promote participation among prosumers. Furthermore, the trading process of a prosumer is modeled as an MDP to consider the volatile market and physical conditions.

Also regarding electricity markets, Xu et al. [90] proposes a DRL approach using a DDPG to solve an MDP for the optimal bidding in a wholesale electricity market and the energy price charged in the retailed electricity market for a load-serving entity. The MDP is formulated for the joint bidding and pricing problem, with continuous state and action spaces where the energy price and bid are two actions that share a common goal. Furthermore, a DNN is applied in order to reduce the significant number of state transitions that the DDPG algorithm requires, thus reducing application costs. The DNN learns from historical data, in order to model the wholesale electricity market and end-use customers for dynamic bid and price responses, respectively. Results have shown that the

Table 3.5 Summary of other existing work for decision-making problems concerning DL approaches.

Article	Objective[a)]	Method[b)]	Application[c)]
[93]	EC	DRL, DQN	Micro grid
[94]	CS	DRL, Q-learning	Smart grid
[95]	ED	DRL, Q-learning	ESU, DGU
[96]	ED, KT	DRL, Q-learning	Smart grid
[97]	EO	DRL, DQN, GDPG	Smart grid
[98]	OC	DRL, Q-learning	Wind turbine
[99]	OC	DRL, A3C	HVAC

a) EC: edge computing, CS: cyber security, ED: economic dispatch, KT: knowledge transfer, EO: energy optimization, OC: operational control.
b) GDPG: Gibbs deep policy gradient, A3C: asynchronous advantage actor critic.
c) ESU: energy storage units, DGU: distributed generation units.

proposed methodology makes more profit using the bidding and pricing policies, however, aggregate energy consumptions are reduced significantly.

3.2.3.4 Other Decision-Making Works

Further relevant decision-making related works can be found in the literature, for instance, the work, Diao et al. [91] proposes a novel framework, Grid Mind, for autonomous grid operational control using DRL. It proposes an intelligent agent that learns voltage control policies through interactions with offline simulations and is capable of adapting to new changes in load/generation variations and topological changes. To achieve this, a DQN is employed.

For optimal HVAC and window control strategy, Chen et al. [92] presents an approach using DRL to maximize natural ventilation. The proposed DRL utilizes a model-free Q-learning algorithm that focuses on minimizing both energy consumption and thermal discomfort by controlling HVAC and window systems. It takes into consideration temperature, humidity, solar radiation, and wind speed, both for indoor and outdoor environments. The proposed solution, when compared to a rule-based heuristic control strategy, outperforms it in all aspects, from lower energy consumption to lower discomfort degree hours and fewer high humidity hours.

Other noteworthy works in the field, relating to decision-making problems in DL, are represented in Table 3.5.

3.3 Accomplishments, Limitations, and Challenges

Although still a rather recent technology, DL is already bringing significant benefits to PES. DL forecasting accuracy has become one of the biggest accomplishments in the solution of PES-related problems. DL models provide forecasting capabilities that greatly outperform all simpler models and traditional ML techniques. LSTMs are at the forefront of this development, however many of them are hitting a wall in accuracy. Therefore, in recent years, new LSTMs combined with other methods and/or architectures are being proposed, such as the works described in Refs. [34, 38, 45, 54]. In addition, other approaches that aim at outperforming LSTMs were presented by Refs.

[33, 38, 53]. Furthermore, there is also a variety of new approaches being proposed for power quality disturbance and FD, that aim at detecting and classifying multiple combinations of disturbances or faults simultaneously, while retaining high levels of accuracy. These works are presented in Refs. [39, 66–68, 74]. Also noteworthy, are the novel automatic feature engineering methods in Refs. [41, 79]. Finally, the increasing usage of DRL for energy management has allowed real-time control systems to considerably minimize energy costs and consumption in buildings. Moreover, they take into consideration the usage of renewable energies to further reduce costs and outperform traditional optimization models with ease, such as in Refs. [42, 43, 86–88]. The good performance of DL models in the PES domain relies mainly on the ability to deal with large datasets, being able to extract relations between different data features and temporal patterns. Such a process is difficult, if not impossible when using simpler models. The complexity of DL approaches enables learning much more complex tasks and connections, being therefore a hugely promising approach for the future, as sustained by the already achieved results.

On the backside of the reviewed works, some papers do not focus on comparing their work to other state-of-the-art DL approaches; instead, they compare the proposed models to more conventional ML algorithms or traditional optimization algorithms. The problem with this approach is that DL relies heavily on performance to showcase its relevance, while aspects such as the potential application in other fields, the explainability, applicability to small datasets, etc., should be considered secondary. Therefore, it is unfair to compare two algorithms using different architectures based only on performance. These papers should focus more on comparing, for example, a DBN with a CNN for classification performance, and not with an SVM, as done in Refs. [55, 59, 68], as it is known that application requirements, e.g. size of the data set, are very different. Another shortcoming is the lack of real-world deployment or case studies, mainly in DRL methods. Works such as Refs. [43, 44, 86, 87, 89, 90, 92], etc. only showcase their results in a simulated environment, while no research is done for when the proposed solution is deployed in a real-world environment, where uncertainty and errors in data are common. Such assessment could improve the paper's reputation by comparing the results from the simulated and real-world environments. Moreover, in PES there is a big gap in the number of papers related to regression and classification problems. The majority of DL models in PES literature are focused on forecasting and consumption prediction methods, while a minority of research is done on the classification of power quality disturbances, FD, or other classification problems. This difference is felt when comparing the proposed works. For example, regression papers normally present new architectures or combinations of DNNs such as in Refs. [33–35, 38, 45, 53]; while classification works typically only explore different applications using already existing DL architectures, for instance in Refs. [40, 73, 81, 82], not innovating in DL techniques.

The extraction of explanations from DL algorithms is one of the biggest challenges in the upcoming years. The main problem of DL is that, while it shines by having the best performance among other ML techniques, DNNs represent "*black boxes*" and cannot explain why they came up with a number, classification, or decision [100]. Achieving explainable AI in DNNs allows for EMSs to provide the reasons for their actions. It would allow answering questions such as why was the air conditioner turned off, why does the forecast indicate that there will be an increase in electricity prices, or why is this power quality disturbance classified as flicker. As a result, trust in these systems would increase among consumers, leading to a possible rise in the usage of EMSs in residential spaces and workplaces. In addition, these systems could be used by field experts to verify if the DNNs are following the correct logic and associations. Another challenge is the incorporation of human expertise in DNNs. Integrating prior knowledge into DL systems is not as easy as it seems, primarily because DNNs represent knowledge through correlations between features, rather than abstractions such as in quantified statements [101]. Some works have started to tackle this problem, such as in Ref. [102], however with little progress. Achieving success in this area would allow for expert knowledge, for example, heuristic or physics laws, to be embedded into EMSs using DL. Subsequently, performance should increase as well as stability, as it would be able to deal more easily with problems outside of the training dataset. The integration of explainable AI with embedded human expertise would allow the management of knowledge in EMSs, providing experts and consumers better control of their "*black boxes*."

3.4 Conclusions

In this chapter, the current state-of-the-art in DL literature for intelligent PES has been reviewed. An introduction is made to contextualize the goals of the chapter and the evolution of DL since it first emerged. Afterward, the most recent and promising works in regression, classification, and decision-making problems using DL are presented and explored. These works demonstrate an ever-increasing trend of research in this area that aims at employing new models for particular domains, which outperform the more general conventional ML methods.

New approaches and methods for DL forecasting have become one of the biggest accomplishments, with many new methods outperforming traditional models and even the first architectures of LSTMs, according to various metrics. The detection and classification of multiple power quality disturbances and system faults, simultaneously, have also become a hot topic, as they retain levels of accuracy on par with single disturbance/fault classifiers. Novel automatic feature engineering methods are also being proposed, which aim at improving performance in PES-related problems. Furthermore, a wide variety of innovative DRL approaches are being proposed for energy management, taking into account renewable energy availability, energy storage, heating and power units, water heaters, and much more.

In the upcoming years, explainable AI and embedded human expertise in EMSs are going to be essential to increase trust among system users. However, there are still questions that need to be answered such as how far is humanity open to having fully AI-controlled homes and workplaces, as well as intelligent automatic energy trading and purchase between homes and industries.

References

1 Bhattarai, B.P., Paudyal, S., Luo, Y. et al. (2019). Big data analytics in smart grids: state-of-the-art, challenges, opportunities, and future directions. *IET Smart Grid* 2 (2): 141–154. https://doi.org/10.1049/IET-STG.2018.0261.

2 Gantz, B.J., Reinsel, D., and Shadows, B.D. (2012). Big data, bigger digital shadows, and biggest growth in the far east executive summary: a universe of opportunities and challenges. *IDC iView: IDC Analyze the Future* 2007: 1–16.

3 Gill, S.K., Nguyen, P., and Koren, G. (2009). Adherence and tolerability of iron-containing prenatal multivitamins in pregnant women with pre-existing gastrointestinal conditions. *Journal of Obstetrics and Gynaecology* 29 (7): 594–598. https://doi.org/10.1080/01443610903114527.

4 Manyica, J., Chui, M., Brown, B. et al. (2011). *Big Data: The Next Frontier for Innovation, Competition, and Productivity*. McKinsey Global Institute (Issue May). https://www.mckinsey.com/business-functions/mckinsey-digital/our-insights/big-data-the-next-frontier-for-innovation.

5 Lecun, Y., Bengio, Y., and Hinton, G. (2015). Deep learning. *Nature* 521 (7553): 436–444. https://doi.org/10.1038/nature14539.

6 Jones, N. (2014). Computer science: the learning machines. *Nature* 505 (7482): 146–148. https://doi.org/10.1038/505146a.

7 Efrati, A. (2013). *How "Deep Learning" Works at Apple, Beyond – The Information*. http://www.theinformation.com/articles/how-deep-learning-works-at-apple-beyond.

8 Hossain, M.S. and Mahmood, H. (2020). Short-term photovoltaic power forecasting using an LSTM neural network and synthetic weather forecast. *IEEE Access* 8: 172524–172533. https://doi.org/10.1109/ACCESS.2020.3024901.

9 Abrishambaf, O., Lezama, F., Faria, P., and Vale, Z. (2019). Towards transactive energy systems: an analysis on current trends. *Energy Strategy Reviews* 26: 100418. https://doi.org/10.1016/j.esr.2019.100418.

10 Vural, G. (2020). Renewable and non-renewable energy-growth nexus: a panel data application for the selected Sub-Saharan African countries. *Resources Policy* 65: 101568. https://doi.org/10.1016/j.resourpol.2019.101568.

11 Runge, J. and Zmeureanu, R. (2021). A review of deep learning techniques for forecasting energy use in buildings. *Energies* 14 (3): 608. https://doi.org/10.3390/en14030608.

12 McCulloch, W.S. and Pitts, W. (1943). A logical calculus of the ideas immanent in nervous activity. *The Bulletin of Mathematical Biophysics* 5 (4): 115–133. https://doi.org/10.1007/BF02478259.

13 Shaw, G.L. (1986). Donald Hebb: the organization of behavior. In: *Brain Theory*, 231–233. https://doi.org/10.1007/978-3-642-70911-1_15.

14 Rosenblatt, F. (1958). The perceptron: a probabilistic model for information storage and organization in the brain. *Psychological Review* 65 (6): 386–408. https://doi.org/10.1037/H0042519.

15 Hopfield, J.J. (1982). Neural networks and physical systems with emergent collective computational abilities. *Proceedings of the National Academy of Sciences of the United States of America* 79 (8): 2554–2558. https://doi.org/10.1073/PNAS.79.8.2554.

16 Rumelhart, D.E., Hinton, G.E., and Williams, R.J. (1986). Learning representations by back-propagating errors. *Nature* 323 (6088): 533–536. https://doi.org/10.1038/323533a0.

17 Ackley, D.H., Hinton, G.E., and Sejnowski, T.J. (1985). A learning algorithm for Boltzmann machines. *Cognitive Science* 9 (1): 147–169. https://doi.org/10.1016/S0364-0213(85)80012-4.

18 Hinton, G.E., Osindero, S., and Teh, Y.-W. (2006). A fast learning algorithm for deep belief nets. *Neural Computation* 18 (7): 1527–1554. https://doi.org/10.1162/NECO.2006.18.7.1527.

19 Bengio, Y. and LeCun, Y. (2007). Scaling learning algorithms towards AI. In: *Large-Scale Kernel Machines* (ed. L. Bottou, O. Chapelle, D. DeCoste, and J. Weston), 182–192. MIT Press. https://nyuscholars.nyu.edu/en/publications/scaling-learning-algorithms-towards-ai.

20 Vincent, P., Larochelle, H., Bengio, Y., and Manzagol, P.A. (2008). Extracting and composing robust features with denoising autoencoders. In: *Proceedings of the 25th International Conference on Machine Learning*, 1096–1103. https://doi.org/10.1145/1390156.1390294.

21 Collobert, R. and Weston, J. (2008). A unified architecture for natural language processing: deep neural networks with multitask learning. In: *Proceedings of the 25th International Conference on Machine Learning*, 160–167. https://doi.org/10.1145/1390156.

22 Le Roux, N. and Bengio, Y. (2008). Representational power of restricted Boltzmann machines and deep belief networks. *Neural Computation* 20 (6): 1631–1649. https://doi.org/10.1162/NECO.2008.04-07-510.

23 Salakhutdinov, R. and Murray, I. (2008). On the quantitative analysis of deep belief networks. In: *ICML '08: The 25th Annual International Conference on Machine Learning held in conjunction with the 2007 International Conference on Inductive Logic Programming*, Helsinki Finland (5–9 July 2008). New York, NY, United States: Association for Computing Machinery. https://doi.org/10.1145/1390156.

24 Jarrett, K., Kavukcuoglu, K., Ranzato, M., and LeCun, Y. (2009). What is the best multi-stage architecture for object recognition? In: *Proceedings of the IEEE International Conference on Computer Vision*, 2146–2153. https://doi.org/10.1109/ICCV.2009.5459469.

25 Lee, H., Grosse, R., Ranganath, R., and Ng, A.Y. (2009). Convolutional deep belief networks for scalable unsupervised learning of hierarchical representations. In: *ICML '09: The 26th Annual International Conference on Machine Learning held in conjunction with the 2007 International Conference on Inductive Logic Programming*, Montreal, Quebec, Canada (14–18 June 2009). New York, NY, United States: Association for Computing Machinery, pp. 609–616. https://doi.org/10.1145/1553374.1553453.

26 Vincent, P., Larochelle, H., Lajoie, I. et al. (2010). Stacked denoising autoencoders: learning useful representations in a deep network with a local denoising criterion. *Journal of Machine Learning Research* 11: 3371–3408.

27 Katsigiannis, Y.A., Tsikalakis, A.G., Georgilakis, P.S., and Hatziargyriou, N.D. (2006). Improved wind power forecasting using a combined neuro-fuzzy and artificial neural network model. In: *Lecture Notes in Computer Science (Including Subseries Lecture Notes in Artificial Intelligence and Lecture Notes in Bioinformatics)*, vol. 3955. LNAI, *SETN '06: Proceedings of the 4th Helenic conference on Advances in Artificial Intelligence*, Heraklion Greece (18–20 May 2006). Berlin, Heidelberg: Springer-Verlag, pp. 105–115. https://doi.org/10.1007/11752912_13.

28 Al Mamun, M. and Nagasaka, K. (2005). Artificial neural networks applied to long-term electricity demand forecasting. In: *HIS '04: Proceedings of the Fourth International Conference on Hybrid Intelligent Systems*, Kitakyushu, Japan (5–8 December 2004). NW Washington, DC, United States: IEEE Computer Society, pp. 204–209. https://doi.org/10.1109/ichis.2004.27.

29 Bonanno, F., Capizzi, G., and Tina, G. (2009). Long-term energy performance forecasting of integrated generation systems by recurrent neural networks. In: *ICCEP 2009 : International Conference on Clean Electrical Power*, Capri, Italy (9–11 June 2009). NW Washington, DC, United States: IEEE Computer Society, pp. 673–678. https://doi.org/10.1109/ICCEP.2009.5211956.

30 Mellit, A. and Shaari, S. (2009). Recurrent neural network-based forecasting of the daily electricity generation of a Photovoltaic power system. *Ecological Vehicle and Renewable Energy (EVER)*, Monaco (26–29 March 2009).

31 Jiang, H. and Tan, Z. (2012). Load forecasting in demand response. In: *2012 Asia-Pacific Power and Energy Engineering Conference (APPEEC 2012)*, Shanghai, China (27–29 March 2012). NW Washington, DC, United States: IEEE Computer Society. https://doi.org/10.1109/APPEEC.2012.6307716.

32 Pinto, T., Morais, H., and Corchado, J.M. (2019). Adaptive entropy-based learning with dynamic artificial neural network. *Neurocomputing* 338: 432–440. https://doi.org/10.1016/j.neucom.2018.09.092.

33 Abdel-Basset, M., Hawash, H., Chakrabortty, R.K., and Ryan, M. (2021). PV-Net: an innovative deep learning approach for efficient forecasting of short-term photovoltaic energy production. *Journal of Cleaner Production* 303: 127037. https://doi.org/10.1016/j.jclepro.2021.127037.

34 Marszałek, A. and Burczynski, T. (2021). Forecasting day-ahead spot electricity prices using deep neural networks with attention mechanism. *Journal of Smart Environments and Green Computing* 21–31. https://doi.org/10.20517/jsegc.2021.02.

35 Fioretto, F., Mak, T.W.K., and Van Hentenryck, P. (2020). Predicting AC optimal power flows: combining deep learning and Lagrangian dual methods. *Proceedings of the AAAI Conference on Artificial Intelligence* 34 (01): 630–637. https://doi.org/10.1609/aaai.v34i01.5403.

36 Li, C., Ding, Z., Zhao, D. et al. (2017). Building energy consumption prediction: an extreme deep learning approach. *Energies* 10 (10): 1525. https://doi.org/10.3390/en10101525.

37 Xu, C. and Chen, H. (2020). Abnormal energy consumption detection for GSHP system based on ensemble deep learning and statistical modeling method. *International Journal of Refrigeration* 114: 106–117. https://doi.org/10.1016/j.ijrefrig.2020.02.035.

38 Wang, H.Z., Li, G.Q., Wang, G.B. et al. (2017). Deep learning based ensemble approach for probabilistic wind power forecasting. *Applied Energy* 188: 56–70. https://doi.org/10.1016/j.apenergy.2016.11.111.

39 Gonzalez-Abreu, A.-D., Delgado-Prieto, M., Osornio-Rios, R.-A. et al. (2021). A novel deep learning-based diagnosis method applied to power quality disturbances. *Energies* 14 (10): 2839. https://doi.org/10.3390/en14102839.

40 Lee, K.P., Wu, B.H., and Peng, S.L. (2019). Deep-learning-based fault detection and diagnosis of air-handling units. *Building and Environment* 157: 24–33. https://doi.org/10.1016/j.buildenv.2019.04.029.

41 Fan, C., Sun, Y., Zhao, Y. et al. (2019). Deep learning-based feature engineering methods for improved building energy prediction. *Applied Energy* 240: 35–45. https://doi.org/10.1016/j.apenergy.2019.02.052.

42 Sanaye, S. and Sarrafi, A. (2021). A novel energy management method based on Deep Q Network algorithm for low operating cost of an integrated hybrid system. *Energy Reports* 7: 2647–2663. https://doi.org/10.1016/j.egyr.2021.04.055.

43 Wang, B., Li, Y., Ming, W., and Wang, S. (2020). Deep reinforcement learning method for demand response management of interruptible load. *IEEE Transactions on Smart Grid* 11 (4): 3146–3155. https://doi.org/10.1109/TSG.2020.2967430.

44 Chen, T. and Su, W. (2018). Local energy trading behavior modeling with deep reinforcement learning. *IEEE Access* 6: 62806–62814. https://doi.org/10.1109/ACCESS.2018.2876652.

45 Luo, X., Zhang, D., and Zhu, X. (2021). Deep learning based forecasting of photovoltaic power generation by incorporating domain knowledge. *Energy* 225: 120240. https://doi.org/10.1016/j.energy.2021.120240.

46 Jebli, I., Belouadha, F.Z., Kabbaj, M.I., and Tilioua, A. (2021). Deep learning based models for solar energy prediction. *Advances in Science, Technology and Engineering Systems* 6 (1): 349–355. https://doi.org/10.25046/aj060140.

47 Li, P., Zhou, K., Lu, X., and Yang, S. (2020). A hybrid deep learning model for short-term PV power forecasting. *Applied Energy* 259: 114216. https://doi.org/10.1016/j.apenergy.2019.114216.

48 Gensler, A., Henze, J., Sick, B., and Raabe, N. (2017). Deep learning for solar power forecasting – an approach using autoencoder and LSTM neural networks. In: *2016 IEEE International Conference on Systems, Man, and Cybernetics (SMC)*, Budapest, Hungary (9–12 October 2016). NW Washington, DC, United States: IEEE Computer Society, pp. 2858–2865. https://doi.org/10.1109/SMC.2016.7844673.

49 Sogabe, T., Ichikawa, H., Sakamoto, K. et al. (2016). Optimization of decentralized renewable energy system by weather forecasting and deep machine learning techniques. In: *IEEE PES*

Innovative Smart Grid Technologies Conference Europe, 1014–1018. https://doi.org/10.1109/ISGT-Asia.2016.7796524.

50 Weng, G., Pei, C., Ren, J. et al. (2021). Modeling and forecasting of wind power output of urban regional energy internet based on deep learning. *Journal of Physics: Conference Series* 1732 (1): 12190. https://doi.org/10.1088/1742-6596/1732/1/012190.

51 Nam, K.J., Hwangbo, S., and Yoo, C.K. (2020). A deep learning-based forecasting model for renewable energy scenarios to guide sustainable energy policy: a case study of Korea. *Renewable and Sustainable Energy Reviews* 122: 109725. https://doi.org/10.1016/j.rser.2020.109725.

52 Mocanu, E., Nguyen, P.H., Gibescu, M., and Kling, W.L. (2016). Deep learning for estimating building energy consumption. *Sustainable Energy, Grids and Networks* 6: 91–99. https://doi.org/10.1016/j.segan.2016.02.005.

53 Abdel-Basset, M., Hawash, H., Chakrabortty, R.K., and Ryan, M. (2021). Energy-net: a deep learning approach for smart energy management in IoT-based smart cities. *IEEE Internet of Things Journal* https://doi.org/10.1109/JIOT.2021.3063677.

54 Berriel, R.F., Lopes, A.T., Rodrigues, A. et al. (2017). Monthly energy consumption forecast: a deep learning approach. In: *2017 International Joint Conference on Neural Networks (IJCNN)*, Anchorage, Alaska (14–19 May 2017). NW Washington, DC, United States: IEEE Computer Society, pp. 4283–4290. https://doi.org/10.1109/IJCNN.2017.7966398.

55 Hwang, J.K., Yun, G.Y., Lee, S. et al. (2020). Using deep learning approaches with variable selection process to predict the energy performance of a heating and cooling system. *Renewable Energy* 149: 1227–1245. https://doi.org/10.1016/j.renene.2019.10.113.

56 Brusaferri, A., Matteucci, M., Portolani, P., and Vitali, A. (2019). Bayesian deep learning based method for probabilistic forecast of day-ahead electricity prices. *Applied Energy* 250: 1158–1175. https://doi.org/10.1016/j.apenergy.2019.05.068.

57 Zhang, R., Li, G., and Ma, Z. (2020). A deep learning based hybrid framework for day-ahead electricity price forecasting. *IEEE Access* 8: 143423–143436. https://doi.org/10.1109/ACCESS.2020.3014241.

58 Huang, T.E., Guo, Q., Sun, H. et al. (2019). A deep learning approach for power system knowledge discovery based on multitask learning. *IET Generation, Transmission and Distribution* 13 (5): 733–740. https://doi.org/10.1049/iet-gtd.2018.5078.

59 Wen, L., Zhou, K., Yang, S., and Lu, X. (2019). Optimal load dispatch of community microgrid with deep learning based solar power and load forecasting. *Energy* 171: 1053–1065. https://doi.org/10.1016/j.energy.2019.01.075.

60 Soleimanzade, M.A. and Sadrzadeh, M. (2021). Deep learning-based energy management of a hybrid photovoltaic-reverse osmosis-pressure retarded osmosis system. *Applied Energy* 293: 116959. https://doi.org/10.1016/j.apenergy.2021.116959.

61 Pramono, S.H., Rohmatillah, M., Maulana, E. et al. (2019). Deep learning-based short-term load forecasting for supporting demand response program in hybrid energy system. *Energies* 12 (17): 3359. https://doi.org/10.3390/en12173359.

62 Wu, N. and Wang, H. (2018). Deep learning adaptive dynamic programming for real time energy management and control strategy of micro-grid. *Journal of Cleaner Production* 204: 1169–1177. https://doi.org/10.1016/j.jclepro.2018.09.052.

63 Huang, H., Yang, Y., Ding, Z. et al. (2020). Deep learning-based sum data rate and energy efficiency optimization for MIMO-NOMA systems. *IEEE Transactions on Wireless Communications* 19 (8): 5373–5388. https://doi.org/10.1109/TWC.2020.2992786.

64 Bugbee, B., Bush, B.W., Gruchalla, K. et al. (2019). Enabling immersive engagement in energy system models with deep learning. *Statistical Analysis and Data Mining* 12 (4): 325–337. https://doi.org/10.1002/sam.11419.

65 Deng, Y., Wang, L., Jia, H. et al. (2019). A sequence-to-sequence deep learning architecture based on bidirectional GRU for type recognition and time location of combined power quality disturbance. *IEEE Transactions on Industrial Informatics* 15 (8): 4481–4493. https://doi.org/10.1109/tii.2019.2895054.

66 Mohan, N., Soman, K.P., and Vinayakumar, R. (2018). Deep power: deep learning architectures for power quality disturbances classification. In: *2017 International Conference on Technological Advancements in Power and Energy (TAP Energy)*, Kollam, Kerala, India (21–23 December 2017). NW Washington, DC, United States: IEEE Computer Society, pp. 1–6. https://doi.org/10.1109/TAPENERGY.2017.8397249.

67 Shi, X., Yang, H., Xu, Z. et al. (2019). An independent component analysis classification for complex power quality disturbances with sparse auto encoder features. *IEEE Access* 7: 20961–20966. https://doi.org/10.1109/ACCESS.2019.2898211.

68 Li, C.M., Li, Z.X., Jia, N. et al. (2018). Classification of power-quality disturbances using deep belief network. In: *2018 International Conference on Wavelet Analysis and Pattern Recognition (ICWAPR)*, Chengdu, China (15–18 July 2018). NW Washington, DC, United States: IEEE Computer Society, pp. 231–237. https://doi.org/10.1109/ICWAPR.2018.8521311.

69 Sindi, H., Nour, M., Rawa, M. et al. (2021). A novel hybrid deep learning approach including combination of 1D power signals and 2D signal images for power quality disturbance classification. *Expert Systems with Applications* 174: 114785. https://doi.org/10.1016/J.ESWA.2021.114785.

70 Wang, W., Yin, H., Chen, C. et al. (2020). Frequency disturbance event detection based on synchrophasors and deep learning. *IEEE Transactions on Smart Grid* 11 (4): 3593–3605. https://doi.org/10.1109/TSG.2020.2971909.

71 Qiu, W., Tang, Q., Liu, J., and Yao, W. (2020). An automatic identification framework for complex power quality disturbances based on multifusion convolutional neural network. *IEEE Transactions on Industrial Informatics* 16 (5): 3233–3241. https://doi.org/10.1109/TII.2019.2920689.

72 Gonzalez-Abreu, A.D., Saucedo-Dorantes, J.J., Osornio-Rios, R.A. et al. (2020). Deep learning based condition monitoring approach applied to power quality. In: *2020 25th IEEE International Conference on Emerging Technologies and Factory Automation (ETFA)*, Vienna, Austria (8–11 September 2020). NW Washington, DC, United States: IEEE Computer Society, pp. 1427–1430. https://doi.org/10.1109/ETFA46521.2020.9212076.

73 Cheng, F., Wang, J., Qu, L., and Qiao, W. (2018). Rotor-current-based fault diagnosis for DFIG wind turbine drivetrain gearboxes using frequency analysis and a deep classifier. *IEEE Transactions on Industry Applications* 54 (2): 1062–1071. https://doi.org/10.1109/TIA.2017.2773426.

74 Zhang, Z., Li, S., Xiao, Y., and Yang, Y. (2019). Intelligent simultaneous fault diagnosis for solid oxide fuel cell system based on deep learning. *Applied Energy* 233, 234: 930–942. https://doi.org/10.1016/j.apenergy.2018.10.113.

75 Mohammadi, H.G., Arshad, R., Rautmare, S. et al. (2020). Deepwind: an accurate wind turbine condition monitoring framework via deep learning on embedded platforms. In: *2020 25th IEEE International Conference on Emerging Technologies and Factory Automation (ETFA)*, Vienna, Austria (8–11 September 2020). NW Washington, DC, United States: IEEE Computer Society, pp. 1431–1434. https://doi.org/10.1109/ETFA46521.2020.9211880.

76 Belagoune, S., Bali, N., Bakdi, A. et al. (2021). Deep learning through LSTM classification and regression for transmission line fault detection, diagnosis and location in large-scale multi-machine power systems. *Measurement* 177: 109330. https://doi.org/10.1016/J.MEASUREMENT.2021.109330.

77 Afrasiabi, S., Afrasiabi, M., Parang, B., and Mohammadi, M. (2020). Designing a composite deep learning based differential protection scheme of power transformers. *Applied Soft Computing* 87: 105975. https://doi.org/10.1016/J.ASOC.2019.105975.

78 Zhang, Y., Mei, W., Dong, G. et al. (2018). A cable fault recognition method based on a deep belief network. *Computers and Electrical Engineering* 71: 452–464. https://doi.org/10.1016/J.COMPELECENG.2018.07.043.

79 Wang, S. and Chen, H. (2019). A novel deep learning method for the classification of power quality disturbances using deep convolutional neural network. *Applied Energy* 235: 1126–1140. https://doi.org/10.1016/j.apenergy.2018.09.160.

80 Xuan, W., Shouxiang, W., Qianyu, Z. et al. (2021). A multi-energy load prediction model based on deep multi-task learning and ensemble approach for regional integrated energy systems. *International Journal of Electrical Power & Energy Systems* 126: 106583. https://doi.org/10.1016/j.ijepes.2020.106583.

81 Elsisi, M., Tran, M.Q., Mahmoud, K. et al. (2021). Deep learning-based industry 4.0 and internet of things towards effective energy management for smart buildings. *Sensors (Switzerland)* 21 (4): 1–19. https://doi.org/10.3390/s21041038.

82 Balouji, E., Gu, I.Y.H., Bollen, M.H.J. et al. (2018). A LSTM-based deep learning method with application to voltage dip classification. In: *2018 18th International Conference on Harmonics and Quality of Power (ICHQP)*, Ljubljana, Slovenia (13–16 May 2018). NW Washington, DC, United States: IEEE Computer Society, pp. 1–5. https://doi.org/10.1109/ICHQP.2018.8378893.

83 Li, S., Han, Y., Yao, X. et al. (2019). Electricity theft detection in power grids with deep learning and random forests. *Journal of Electrical and Computer Engineering* 2019: https://doi.org/10.1155/2019/4136874.

84 Lu, X., Zhou, Y., Wang, Z. et al. (2019). Knowledge embedded semi-supervised deep learning for detecting non-technical losses in the smart grid. *Energies* 12 (18): 3452. https://doi.org/10.3390/en12183452.

85 Feng, C., Mehmani, A., and Zhang, J. (2020). Deep learning-based real-time building occupancy detection using AMI data. *IEEE Transactions on Smart Grid* 11 (5): 4490–4501. https://doi.org/10.1109/TSG.2020.2982351.

86 Kolodziejczyk, W., Zoltowska, I., and Cichosz, P. (2021). Real-time energy purchase optimization for a storage-integrated photovoltaic system by deep reinforcement learning. *Control Engineering Practice* 106: 104598. https://doi.org/10.1016/j.conengprac.2020.104598.

87 Zhang, G., Hu, W., Cao, D. et al. (2021). Data-driven optimal energy management for a wind-solar-diesel-battery-reverse osmosis hybrid energy system using a deep reinforcement learning approach. *Energy Conversion and Management* 227: 113608. https://doi.org/10.1016/j.enconman.2020.113608.

88 Ye, Y., Ye, Y., Qiu, D. et al. (2020). Model-free real-time autonomous control for a residential multi-energy system using deep reinforcement learning. *IEEE Transactions on Smart Grid* 11 (4): 3068–3082. https://doi.org/10.1109/TSG.2020.2976771.

89 Lu, R., Li, Y.C., Li, Y. et al. (2020). Multi-agent deep reinforcement learning based demand response for discrete manufacturing systems energy management. *Applied Energy* 276: 115473. https://doi.org/10.1016/j.apenergy.2020.115473.

90 Xu, H., Sun, H., Nikovski, D. et al. (2019). Deep reinforcement learning for joint bidding and pricing of load serving entity. *IEEE Transactions on Smart Grid* 10 (6): 6366–6375. https://doi.org/10.1109/TSG.2019.2903756.

91 Diao, R., Wang, Z., Shi, D. et al. (2019). Autonomous voltage control for grid operation using deep reinforcement learning. In: *2019 IEEE Power & Energy Society General Meeting (PESGM)*, Atlanta, Georgia, United States of America (4–8 August 2019). NW Washington, DC, United States: IEEE Computer Society. https://doi.org/10.1109/PESGM40551.2019.8973924.

92 Chen, Y., Norford, L.K., Samuelson, H.W., and Malkawi, A. (2018). Optimal control of HVAC and window systems for natural ventilation through reinforcement learning. *Energy and Buildings* 169: 195–205. https://doi.org/10.1016/J.ENBUILD.2018.03.051.

93 Munir, M.S., Abedin, S.F., Tran, N.H., and Hong, C.S. (2019). When edge computing meets microgrid: a deep reinforcement learning approach. *IEEE Internet of Things Journal* 6 (5): 7360–7374. https://doi.org/10.1109/JIOT.2019.2899673.

94 Ni, Z. and Paul, S. (2019). A multistage game in smart grid security: a reinforcement learning solution. *IEEE Transactions on Neural Networks and Learning Systems* 30 (9): 2684–2695. https://doi.org/10.1109/TNNLS.2018.2885530.

95 Liu, W., Zhuang, P., Liang, H. et al. (2018). Distributed economic dispatch in microgrids based on cooperative reinforcement learning. *IEEE Transactions on Neural Networks and Learning Systems* 29 (6): 2192–2203. https://doi.org/10.1109/TNNLS.2018.2801880.

96 Han, C., Yang, B., Bao, T. et al. (2017). Bacteria foraging reinforcement learning for risk-based economic dispatch via knowledge transfer. *Energies* 10 (5): 638. https://doi.org/10.3390/EN10050638.

97 Sogabe, T., Malla, D.B., Takayama, S. et al. (2018). Smart grid optimization by deep reinforcement learning over discrete and continuous action space. In: *2018 IEEE 7th World Conference on Photovoltaic Energy Conversion (WCPEC) (A Joint Conference of 45th IEEE PVSC, 28th PVSEC & 34th EU PVSEC)*, Waikoloa, Hawaii, United States of America (10–15 June 2018). NW Washington, DC, United States: IEEE Computer Society, pp. 3794–3796. https://doi.org/10.1109/PVSC.2018.8547862.

98 Saenz-Aguirre, A., Zulueta, E., Fernandez-Gamiz, U. et al. (2019). Artificial neural network based reinforcement learning for wind turbine yaw control. *Energies* 12 (3): 436. https://doi.org/10.3390/EN12030436.

99 Zhang, Z., Chong, A., Pan, Y. et al. (2018). A deep reinforcement learning approach to using whole building energy model for HVAC optimal control. In: *2018 Building Performance Analysis Conference and SimBuild* co-organized by ASHRAE and IBPSA-USA, Chicago, Illinois, United States of America (28 September 2018). NW Washington, DC, United States: IEEE Computer Society, pp. 675–682. https://www.researchgate.net/publication/326711617.

100 Holzinger, A. (2018). From machine learning to explainable AI. In: *2018 World Symposium on Digital Intelligence for Systems and Machines (DISA)*, Košice, Slovakia (23–25 August 2018). NW Washington, DC, United States: IEEE Computer Society, pp. 55–66. https://doi.org/10.1109/DISA.2018.8490530.

101 Marcus, G. (2018). Deep learning: a critical appraisal. CoRR. https://arxiv.org/abs/1801.00631v1.

102 Mitsuhara, M., Fukui, H., Sakashita, Y. et al. (2019). Embedding human knowledge into deep neural network via attention map. In: *Proceedings of the 16th International Joint Conference on Computer Vision, Imaging and Computer Graphics Theory and Applications*, Volume 5 VISAPP: VISAPP, Online Streaming, 2021 (8–10 February 2021), pp. 626–636. https://arxiv.org/abs/1905.03540v4.

Part II

Clustering

4

Data Mining Techniques Applied to Power Systems

Sérgio Ramos[1], João Soares[1], Zahra Forouzandeh[1], and Zita Vale[2]

[1] GECAD – Research Group on Intelligent Engineering and Computing for Advanced Innovation and Development, Polytechnic of Porto, School of Engineering (ISEP), LASI – Intelligent Systems Associate Laboratory, Porto, Portugal
[2] Polytechnic of Porto, School of Engineering (ISEP), LASI – Intelligent Systems Associate Laboratory, Porto, Portugal

Acronyms

ANN	artificial neuronal network
CAISO	California Independent System Operator
CDI	clustering dispersion indicator
D	Dunn index
DB	Davies–Bouldin index
DM	data mining
FCM	fuzzy C-means
KDD	knowledge discovery in a database
KM	K-Means
LMP	locational marginal prices
MAE	mean absolute error
MIA	mean index adequacy
MLP	multi-layer perceptron
MV	medium voltage
NH	normalized Hubert statistic index
OMIE	Iberian Energy Market Operator
PS	point symmetric index
S	Silhouette statistic
TOU	time-of-use
XB	Xie and Beni clustering validity index

4.1 Introduction

Companies can improve their efficiency using data-driven decision-making, but only a tiny part of the world's data is processed and used. As a result, companies all over the world are losing massive prospects because they cannot get the data they need, and when they need it [1]. That is where database extraction can be useful and important to support companies in retrieving data and making the best use of it to assure better business decisions.

Intelligent Data Mining and Analysis in Power and Energy Systems: Models and Applications for Smarter Efficient Power Systems, First Edition. Edited by Zita Vale, Tiago Pinto, Michael Negnevitsky, and Ganesh Kumar Venayagamoorthy.

The study and analysis of huge amounts of data stored by companies over time, to discover knowledge contained in that same data, remain a difficult problem to be solved [2]. In addition to the growing increase in data accumulation, new problems have also emerged, which have given rise to new opportunities in power system field [3]. The importance of information is, of course, reinforced, because, in direct proportion to the amount of data, there is a portion of knowledge that can be obtained to support the decision-making.

The traditional data access, without the possibility of discovering patterns and trends, nor inference some relations and rules among them, greatly limits the extraction of knowledge that may exist in the stored data. In contrast, data mining (DM), through the use of specific algorithms, tries to discover understandable patterns contained in the data, as well as to attribute a relationship among them [4].

DM can be considered as a part of the KDD process. According to Fayyad et al. [5]: "Data Mining means the application of algorithms for extracting patterns from the data without the additional steps of the KDD process, such as the inclusion of previous knowledge and the correct interpretation of the results." Thus, KDD has already been defined as a process that uses DM techniques to extract and identify knowledge from databases. On the other hand, DM has been termed as a step in the entire KDD process, since DM algorithms require data pre-processing and subsequent interpretation. Figure 4.1 graphically depicts the several steps that belong to the KDD process.

The KDD refers to the entire process of knowledge discovery, while DM comprises merely the step referring to the application of algorithms for extracting patterns from the initial data [6].

In the real world, data tends to be inconsistent, incomplete, and/or wrong. Indeed, some questions need to be answered: "- What happens when data are not correct?", or, "- The knowledge extracted from databases can be reliable?". The answer to these questions is related to the initial database, i.e. the main obstacles to knowledge discovery are the use of poor data. Therefore, researchers must assure that the initial database is reliable and of good quality.

4.1.1 Data Selection

After defining the domain on which the knowledge discovery process is to be carried out, the data selection set is performed. For instance, in the case of electrical load profiles, it can be specified the voltage level, electrical energy consumption, type of activity, period of analysis, and others parameters.

The data volume definition for analysis is very important and, sometimes, statistical models (statistical inference) are required to define the data volume, whereby a representative sample of the population under study, it is possible to apply methods that allow estimates and make conclusions about that study population.

Another important issue about databases is the sampling frequency. Samples too long concerning to temporal axis in a system that varies very rapidly tend to be very incomplete and there is a loss of

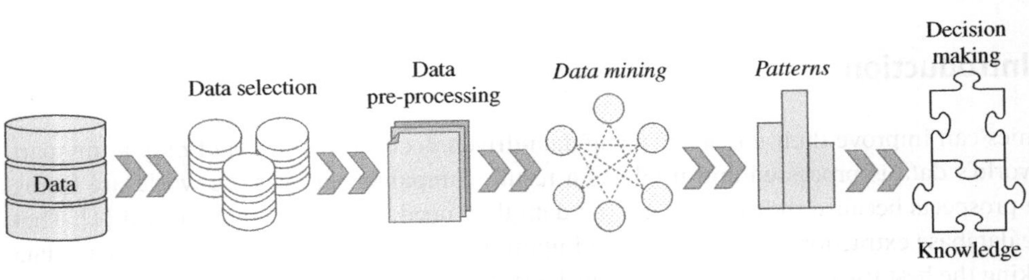

Figure 4.1 Knowledge discovery in database process.

information. On the other hand, with very small samples for systems that vary little, samples tend to be repetitive. Thus, it is important to ensure that should be chosen a sampling rate according to the system dynamics.

4.1.2 Data Pre-processing

This step is justified since the databases are highly susceptible to contain data with noise, missing records, or incomplete and inconsistent data due to their size, which is usually high, and maybe gathered from multiple and heterogeneous sources.

In this pre-processing step, the operation of data "cleansing" is performed, i.e. it is identified and removed extraneous or inconsistent data. To perform this task, it is important to know the study field so that there is a clear understanding of the data to be analyzed, as well as identifying and estimating missing data, converting non-numeric data to numeric values, disposing of data in disrepair, units' conversion, etc. There are several data pre-processing techniques, as described in Ref. [7], such as:

- **Data cleaning:** can be applied to remove noise and correct inconsistencies in data;
- **Data integration:** merges data from multiple sources into a coherent data store, such as a data warehouse;
- **Data transformations:** such as normalization. Normalization may improve the accuracy and efficiency of DM algorithms involving distance measurements;
- **Data reduction:** can reduce the data size by aggregating, eliminating redundant features, or clustering.

It is in this phase that the data are transformed into their final form to be processed by DM algorithms. Typically, it is estimated that the data preparation phase takes 70–80% of the whole effort to develop a DM study. Figure 4.2 depicts the operations most frequently applied during the data cleansing phase.

In this step, it is necessary to deal with corrupted and duplicated data, and an ordering operation is required to organize data in their locations (e.g. tables for further analysis and handling).

4.1.3 Data Mining

The knowledge discovery from a database focuses on the study and application of theories and algorithms that take models and relationships from data. DM is a variety of data analysis techniques applied to the problem of extracting knowledge contained and stored in large data volumes that, apparently, have no meaning.

DM algorithms are used to find trends and interesting relationships in the data, such as regression, neuronal networks, and clustering. Usually, a pattern representation is generated using classification rule sets, decision trees, association rules, regression, and other techniques.

For instance, according to Santos and Azevedo [2], most DM methods are based on concepts of machine learning, pattern recognition, statistics, artificial intelligence, and graphic models. Thus, it is at this stage that algorithms are applied to discover patterns in the data, involving the selection of methods and techniques that best fit the fulfillment of the established goals.

Indeed, Han et al. [7] refer that DM is the process of discovering interesting patterns and knowledge from large amounts of data, and the data sources can include databases, data warehouses, the Web, other information repositories, or data that are streamed into the system dynamically. Examples of DM algorithms are regression, classification, and neural networks, clustering.

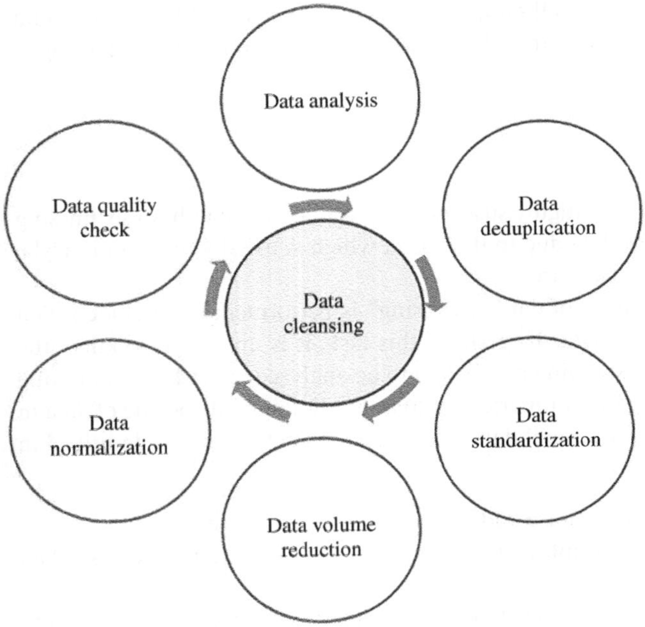

Figure 4.2 Frequently operations in data cleansing phase.

The definition of the models and parameters to be applied is sometimes difficult to choose, as these must be compatible with the objectives of the KDD process. When exploring data and discovering patterns, a representation is normally performed, such as classification rules, decision trees, association, or regression rules.

4.1.4 Analysis and Interpretation

This step corresponds to the analysis of the obtained results from the DM algorithms application. The identified patterns in this phase are interpreted as knowledge, which can be used to support decision-making. It is at this stage that the participation of experts is important to solving specific technical issues that may influence the analysis.

Thus, once the model is obtained, and validated, with the experts' help, it is proved that the knowledge and consequent found conclusions are valid and sufficiently satisfactory. It also includes the verification and resolution of possible conflicts that may arise from the new knowledge obtained with the previous knowledge.

The represented KDD process does not always assume a linear path, since in any of the stages, the need to return to one of the previous stages and restart the entire knowledge discovery process can always be identified.

Data analysis is, therefore, crucial for obtaining knowledge and decision-making, having an exploratory (hypothesis formulation and decision making) or confirmatory (model validation) character. In this context, Section 4.2 introduces the concept and notion of DM, resulting from the entire process of data exploration, which aims to identify interesting patterns and relationships in electricity consumption data.

4.2 Data Mining Techniques

The traditional concept of work, which involved large amounts of papers and reports more or less complicated and takes a decision, in most cases, based on inaccurate information, is gradually being replaced by the modern concept supported in the use of information technologies.

The exploration and analysis of large amounts of data in an automatic or semi-automatic way, for the discovery of patterns using the study of theories and algorithms, constitute the so-called Data Mining (DM) techniques. Knowledge discovery systems are usually complex systems, so they require rigor in their development process.

As presented in Figure 4.3, DM represents the task of finding new knowledge, generally unpredictable, based on a dataset previously collected and properly prepared for this purpose. Therefore, the term DM refers to the discovery phase of the KDD process and clusters all the theories and techniques of the different areas of research that deal with the knowledge extraction from a database.

Therefore, DM is the process of predicting outcomes by looking for anomalies, trends, and correlations in large data sets. This new information can be used to increase profits, cut costs, improve customer relationships, minimize risks, and more using a variety of strategies.

The application of DM techniques makes it possible to prove the assumption of transforming data into information and, later, into knowledge. This possibility makes DM techniques fundamental to the decision-making process. In fact, enables to discover insights and drive better opportunities.

Across a wide range of industries and disciplines, DM is at the heart of analytics efforts, therefore, DM is used by:

- Telecom, media and technology;
- Education;
- Banking;
- Insurance;
- Manufacturing;
- Retail, and;
- Power systems.

The DM process can be applied in two ways:

1. As a verification process, where the analyst suggests a hypothesis about the possible relationship among the data and tries to prove it using statistical and multidimensional techniques;
2. As a discovery process, in which no previous assumption is made, but the application of DM techniques, such as decision trees, clustering algorithms, and neural networks to discover interesting relationships in the data.

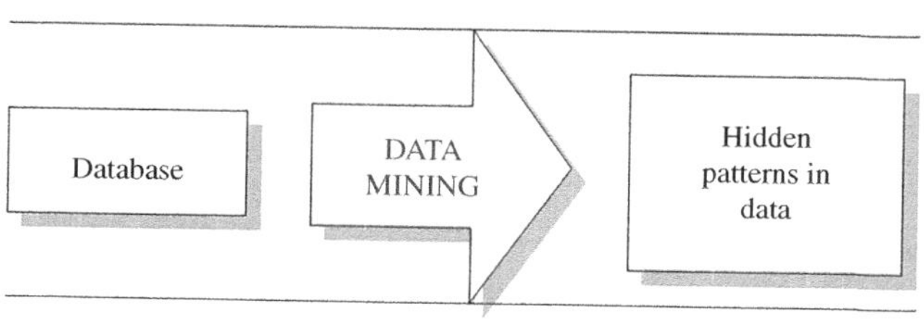

Figure 4.3 DM application: from data to knowledge.

The DM process involves the use of methods (algorithms) for determining patterns and relationships in the data. DM algorithms generally perform tasks for predicting unknown data and describing patterns. The main tasks of DM are:

- Classification;
- Estimation;
- Grouping by affinity or association;
- Clustering;
- Analysis of deviations.

Pattern recognition operations can be referred to as supervised and unsupervised operations [8]. In supervised operation, the vector of the characteristics and its "label" (correct answer) is supplied to the system. Thus, considering a pair of variables $\{X, W\}$, X constitutes a set of characteristics and W the class to which it belongs. In this way, before the DM technique execution, an identification of a specific objective is made.

In unsupervised operation, the algorithms assume that the class to which a given data collection belongs is not known "*a priori*." As do not require a target specified in advance, these operations analyze a collection of data and discover relationships among them, offering little meaning to those relations.

No technique solves all DM problems. Different methods serve different purposes, with each method providing advantages and disadvantages in its application. Thus, a certain familiarity with the different DM techniques is necessary, to facilitate the choice of one of them, according to the problems presented.

DM techniques come from different areas, such as statistics (descriptive statistics, decision trees, and clustering), association rules, and artificial intelligence (genetic algorithms, fuzzy sets, and neural networks). The most well-known and used DM techniques are:

- Decision trees;
- Regression;
- Neural networks;
- Genetic algorithms;
- Clustering algorithms;
- Nearest neighborhood algorithm;
- Fuzzy logic;
- Rough sets.

The techniques used in DM contain a certain degree of automatic search, that is, algorithms are applied that, without user's interference, elicit relations or models from the data. Each technique can be implemented using different algorithms.

It is important to emphasize that DM is not the end of the knowledge discovery activity, but it is only the beginning. It is to have trained analysts who know how to interact with systems to lead them to extract useful and relevant standards.

4.2.1 Clustering Algorithms

Clustering can be defined as the process of partitioning a large database into groups (or clusters) based on a concept of similarity or proximity among data [9]. Despite existing a wide variety of clustering algorithms, there is no single algorithm that can, by itself, discover all sorts of cluster shapes and structures among data.

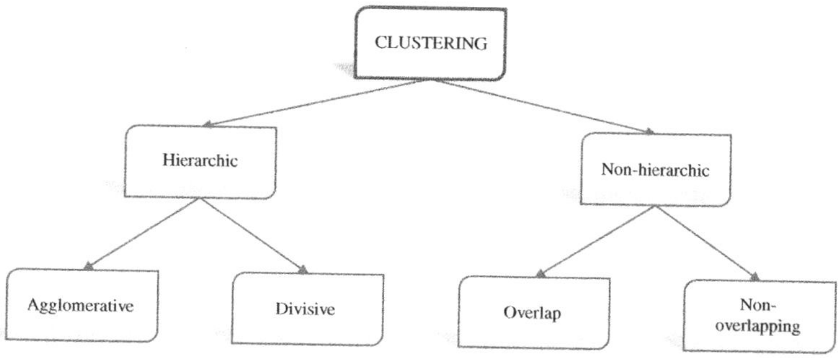

Figure 4.4 Taxonomy of the clustering methods.

Clustering approaches aim to divide a data set X composed of "n" data patterns $\{x_1, \ldots, x_n\}$ into "K" clusters $\{C_1, \ldots, C_K\}$, such that similar data patterns are placed in the same cluster and dissimilar data patterns are clustered into different groups.

Typically, an optimal clustering should be one whose partition minimizes the distances between the elements of the same group and maximizes the distance between the other groups. Since these distances can be defined in different ways, the definition of optimal grouping assumes many variations. Figure 4.4 illustrates the classic taxonomy of clustering methods, basically considering only a distinction between hierarchical and non-hierarchical algorithms. In turn, hierarchical algorithms are divided into agglomerative and divisive algorithms. Non-hierarchical algorithms are divided into overlapping and non-overlapping partition algorithms.

Thus, the most common clustering algorithms are divided into the following categories:

1. **Partitive algorithms:** This algorithm identifies "K" seed points $\overline{x}_k$ (centroids) for each cluster and iteratively updates these points to optimize some objective function. At each iteration, the most similar seed point is assigned to each object x_i.
   One of the best-known partitive algorithms is K-Means (KM) algorithm [10]. This algorithm attempts to minimize the within-cluster sum of squares (Eq. (4.1)), where $\|x_i - \overline{x}_i\|^2$ is the Euclidean distance between pattern x_i and its closest cluster centroid $\overline{x}_k$. The KM algorithm takes as parameter the desired number of clusters "K" and randomly chooses "K" data patterns as the initial centroids $\{\overline{x}_1, \ldots, \overline{x}_k\}$ of each cluster. Then, for each pattern $x_i \in X$, the KM algorithm finds the nearest centroid $\overline{x}_k$and assigns it to the corresponding cluster C_k, updating each centroid $\overline{x}_k$ as the mean vector of the corresponding cluster. This process is repeated until no pattern assignments change from iteration to iteration, indicating that the algorithm converged to a local minimum.

$$\left(\sum_{k=1}^{k} \sum_{x_i \in C_k} \|x_i - \overline{x}_k\|^2 \right) \tag{4.1}$$

2. **Hierarchical algorithms:** A hierarchical decomposition of a set of data objects is created using these algorithms.
   Based on how the hierarchical decomposition is created, a hierarchical algorithm can be agglomerative or divisive. In the agglomerative method, each object forms a separate cluster, which merges the objects, or clusters, that are close to one another in each iteration until all of the clusters are merged into one, or until a termination condition is met.
   On the other hand, the divisive method begins with all of the objects in the same cluster, and then divides each cluster into smaller clusters in each iteration until all of the objects are in one

cluster, or until a termination condition is met. Examples of agglomerative algorithms include single-link, average-link, and complete-link [11].

3. **Density-based algorithms:** High-density regions in space are considered clusters, while objects in low-density regions are considered outliers or noise by this algorithm. The general idea is that clusters will continue to expand as long as the density (number of objects) in the "neighborhood" exceeds a certain threshold [12].
4. **Grid-based algorithms:** By dividing each data feature into intervals, this algorithm quantizes the object space into a finite number of cells. A grid structure is formed by these cells. Clusters are created by locating contiguous cells with the minimum number of objects [13].
5. **Spectral algorithms:** To create a new representation of data, this approach uses the highest K eigenvalues. After that, a fast clustering algorithm, such as K-Means is used to cluster the new representation. Another approach, such as the Normalized Cut algorithm can be used, transforming the clustering algorithm into a weighted graph partitioning problem [14].
6. **Model-based algorithms:** This algorithm assumes a model for each cluster and finds the data that best fits the model. The clusters are located by creating a density function that represents the data points' spatial distribution [15].
7. **Fuzzy clustering:** In this clustering technic, a data point either belongs to a cluster or not and, that means that an object can be classified into more than one cluster. The uncertainty in the clustering task has led to the development of algorithms that employ fuzzy logic concepts. The fuzzy C-means (FCM), an extension of the classical C-Means algorithm for fuzzy applications, is a common fuzzy clustering algorithm [16]. FCM algorithm attempts to find the most characteristic point in each cluster (cluster center), as well as the grade of membership for each object in the clusters.

There is a vast range of clustering algorithms that it is possible to find out in the literature [2, 17]. Here, is presented a short-list concerning some of them:

- K-Means algorithm;
- Two-step algorithm;
- Single-link;
- Average-link:
- Complete-link;
- Ward's-link;
- Normalized cut algorithm;
- PC-k-Means – pairwise constrained K-Means;
- MPC-K-Means – metric pairwise constrained K-Means;
- Fuzzy C-means;
- SOM – self organizing features maps (Kohonen network);
- WEACS – weighted evidence accumulation clustering using subsampling.

Despite the wide variety of clustering algorithms, none is successfully applied to all possible structures present in multidimensional data sets. For a database, not all clustering algorithms correctly identify the different existing patterns, since the algorithms adopt implicit assumptions regarding the shape of the groups, based on the similarity measures and the used grouping criteria. This explains the high number of existing clustering algorithms, each of which, due to its specificity, presents better or worse performance in a given data set with a specific distribution.

Clustering analysis is one of the major tasks presented in a variety of fields of study, and it is unsupervised learning in pattern recognition that aims to find and extract groups in the underlying data, following a similarity (or dissimilarity) criterion. Thus, and based on that defined criterion, the data are grouped such that objects in the same cluster are more similar than objects in different clusters.

Some important issues concerning clustering data require to be addressed. For instance, it is really important to know the number of clusters that are presented in the data, and also to evaluate how good the data partition is.

4.2.2 Clustering Validity Indices

For answering the questions mentioned above, validity measures indices for clustering provide the formal mechanisms. Although there are several clustering validity measures, none consistently deliver the best results across all data sets.

When using data sets with, for example, two dimensions, the clustering validation results can be done using visual assessment, although such assessment is not always possible. When multidimensional data are grouped, validating the results using visual assessment becomes extremely difficult or even impossible. For this purpose, clustering validity measures are used, as a clustering strategy, to assist researchers in the clustering results validation, namely regarding partition quality, as well as the appropriate number of clusters.

The performance of a clustering algorithm may have different assessments depending on the used validation technique. To increase the confidence of the obtained results, it is possible to apply several measures, arriving at a compromise and consensus solution in its evaluation.

The purpose of these metric validation measures is to evaluate how a given clustering data corresponds, in fact, to the natural clustering of the data structure. Two main properties are followed in many of the clustering validity indices:

1. Dispersion, which evaluates how close the objects in a group are to other objects in the same group (a very common measure of dispersion is variance);
2. Separation, which assesses how far apart the groups are.

The use of data cluster validation indices, in which the user defines, "*a priori*," the minimum and a maximum number of "*k*" groups to be obtained, aims to support the choice of the best partition, as well as the optimal number of groups. Thus, for each "*k*" value, the results of these indices are checked and their tendency is verified (increasing, decreasing, or "knee analysis" of the curve, according to their criteria). Typically, the choice of the best partition, depending on the results of the indices with the increase of "*k*," follows two approaches. If the value of the indices does not assume an upward or downward trend (maximization or minimization criteria, respectively), then the objective will be to find the maximum or minimum value of the respective validity index. In the case of an upward/downward trend index, due to the increase in "*k*," the "*k*" value in which such variation occurs must be identified, usually, analyzing the most significant variation called "knee curve."

Figure 4.5 depicts an example concerning the evaluation of the clustering process, namely by employing the Normalized Hubert Statistic index, and Hubert Statistic index [18], to support the choice of the best number of clusters that better fit a certain database, where per each "*k*" (number of clusters), it is obtained the index value.

Table 4.1 presents some of the validation indices that can be used for the clustering analysis, which include:

The use of the described indices, as well as other indices existing in the literature, may support the researcher to have a criterion to evaluate the results from the clustering process, namely to

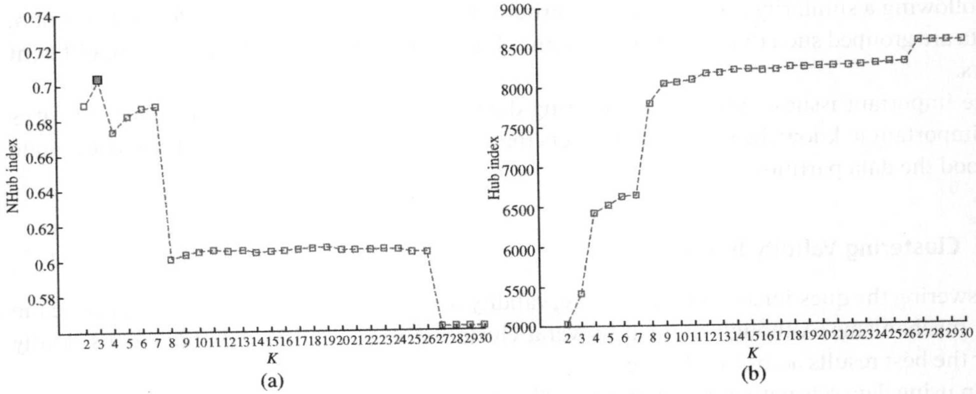

Figure 4.5 Example of two clustering validation indices results. (a) Normalized Hubert Statistic (Criterion: Maximization). (b) Hubert Statistic (Criterion: "knee" analysis).

Table 4.1 Clustering validity indices.

Abbreviation	Clustering validity indices	Optimization criteria	References
NH	Normalized Hubert Statistic	Maximization	[18]
H	Hubert Statistic	Knee analysis	[18]
PS	Point symmetry index	Minimization	[19]
D	Dunn index	Maximization	[20]
DB	Davies–Bouldin index	Minimization	[21]
SD	SD validity index	Minimization	[22]
S	Silhouette statistic	Maximization	[23]
I	Index I	Maximization	[24]
XB	XB cluster validity index	Minimization	[25]
SQE	Squared error index	Knee analysis	[26]
RMSSTD	Root-mean-square standard error	Knee analysis	[27]
RS	*R*-squared index	Knee analysis	[28]

support the decision for the best number of groups, and also the optimal number of clusters. It is important to highlight that this type of analysis often has the experience of experts in the study field, so the empirical analysis of the results is essential.

4.2.3 Classification Algorithms

The classification algorithms, unlike data clustering techniques, are supervised since the class of each object in the data set is known "*a priori*." The main goal of these algorithms is to learn a function or set of rules, referred to as a classifier, which enables a new (unobserved) object to be assigned to the correct class.

In classification problems, a set of pre-classified data points are given and the classification algorithm tries to discover a rule (supervised learning task), allowing the attribution of a new object (electrical consumer, for example) to a certain cluster, based on the rules generated by the classification model.

There are several types of algorithms for training classifiers, which can be organized based on the learning strategy:

- **Statistical models:** This model assumes that the classes of objects are generated according to some probabilistic distribution. The linear and quadratic discriminant analysis [28] are two examples.
- **Classification tree algorithms:** learn a tree structure in which each leaf node is labeled with a class and all non-leaf nodes are given a function. Starting with the root node, the data is split at each node by determining which function and splitting threshold best distinguishes classes in the corresponding subset of data. The C5.0 algorithm is an example of a classification tree that makes use of the inform function [29];
- **Artificial neural networks:** This algorithm intends to mathematically model the human brain. The multilayer perceptron is one of the most common types of the artificial neural network, and it is usually trained using the backpropagation algorithm [30];
- **Support vector machine:** This algorithm aims to find hyperplanes in a high-dimensional data space that divides the objects into groups. A new object is classified by identifying its position concerning the built hyperplanes [31];
- **Ensembles of classifiers:** Using a voting mechanism, this approach combines several classifiers to create a more robust classifier. Boosting algorithms are an example of a very successful class of this type of algorithms [32].

This task is multidisciplinary and belongs to several fields of study. Examples of classification operations are:

- The detection of bank fraud;
- Classify credit applications as low, medium, or high risk;
- Clarify fraudulent insurance claims;
- Identify the most favorable treatment for a patient, based on patient classes that respond well to a particular type of medical treatment;
- Classify the electricity consumption profile of a given customer in a typical load profile previously identified.

In order to predict the cluster of new electrical consumers a classification model should be implemented, as well as a new representation of the customers' typical load profiles, with simple and straightforward input data and classification rules. A rule-based modeling technique, such as a classification tree, is recommended for this step because the resulting model is easily understandable.

The definition of specific indices related to typical load profiles is required. These indices may be derived from the load profile curves capturing relevant information about the electricity customers' behavior. Table 4.2 presents some indices that can be used as input attributes in the classification model including load factor (f_1), the off-peak factor (f_2), the night impact coefficient (f_3), the lunch impact coefficient (f_4), and the modulation coefficient at off-peak hours (f_5), where P_{max} (or P_{min}) is the maximum (or minimum) power demand of the representative day and P_{av} is the average power demand [33, 34].

Figure 4.6 illustrates the classification model methodology that can be implemented to classify new electrical consumers.

Table 4.2 Normalized indices to characterize electricity customers' behavior.

Parameter	Definition	Acquisition period
Daily P_{av}/P_{max}	$f_1 = P_{av,day}/P_{max,day}$	1 day
Daily P_{min}/P_{max}	$f_2 = P_{min,day}/P_{max,day}$	1 day
Night impact	$f_3 = 1/3^*(P_{av,night}/P_{av,day})$	1 day (8 h night, from 11 pm to 6 am)
Lunch impact	$f_4 = 1/8^*(P_{av,lunch}/P_{av,day})$	1 day (3 h lunch, from 12 am to 3 pm)
Daily P_{min}/P_{av}	$f_5 = P_{min,day}/P_{av,day}$	1 day

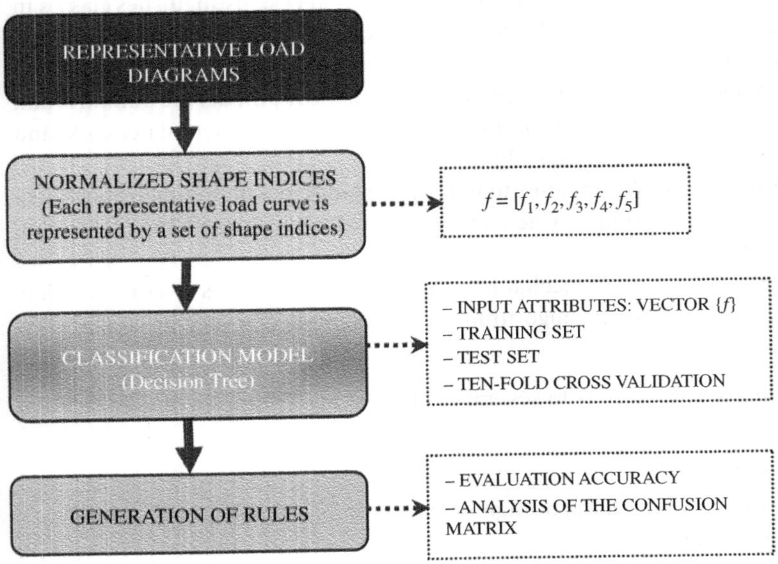

Figure 4.6 Electrical customers' classification model methodology.

Derived from the representative daily load consumption curves, the normalized shape indices belong to the [0, 1] range. The classification model generates a rule set to support the classification of new consumers.

4.3 Data Mining Techniques Applied to Power Systems

This section relies on DM-based methodologies applied to power systems. The characterization of electricity consumers is addressed using clustering techniques, and also by the implementation of a classification model. It is proposed the design of new electrical tariffs taking into consideration the behavior of the consumers' typical load profile. DM techniques are also used to characterize zonal prices in an electrical transmission network. Finally, forecasting techniques are explored to foreseen wind speed values.

4.3.1 Electrical Consumers Characterization

DM techniques can be applied to characterize electricity customers, that is, to support the identification of typical load profiles.

4.3.1.1 Typical Load Profile

The research work presented in Ref. [34] presents an electricity consumer characterization framework supported by knowledge discovery in database process. The main goal was to identify electricity consumers, medium voltage (MV) consumers, with the same consumption pattern. Figure 4.7 presents the proposed methodology to the electrical costumers' characterization.

The initial database, concerning MV consumers were used to identify clusters with the same consumption pattern. A huge number of clustering algorithms were used and their clustering performance compared and evaluated. Thus, by varying the number of clusters "k" between k_{min} and k_{max}, which are the minimum and maximum number of clusters for each clustering algorithm, $(k_{max} - k_{min} + 1)$ data partitions are created. Then, the quality of all generated data partitions was evaluated using "V" clustering validity indices, as shown in Figure 4.7. Equation (4.2), presents the data partitions produced by all clustering algorithms (N).

$$D_p = (k_{max} - k_{min} + 1) \times N \tag{4.2}$$

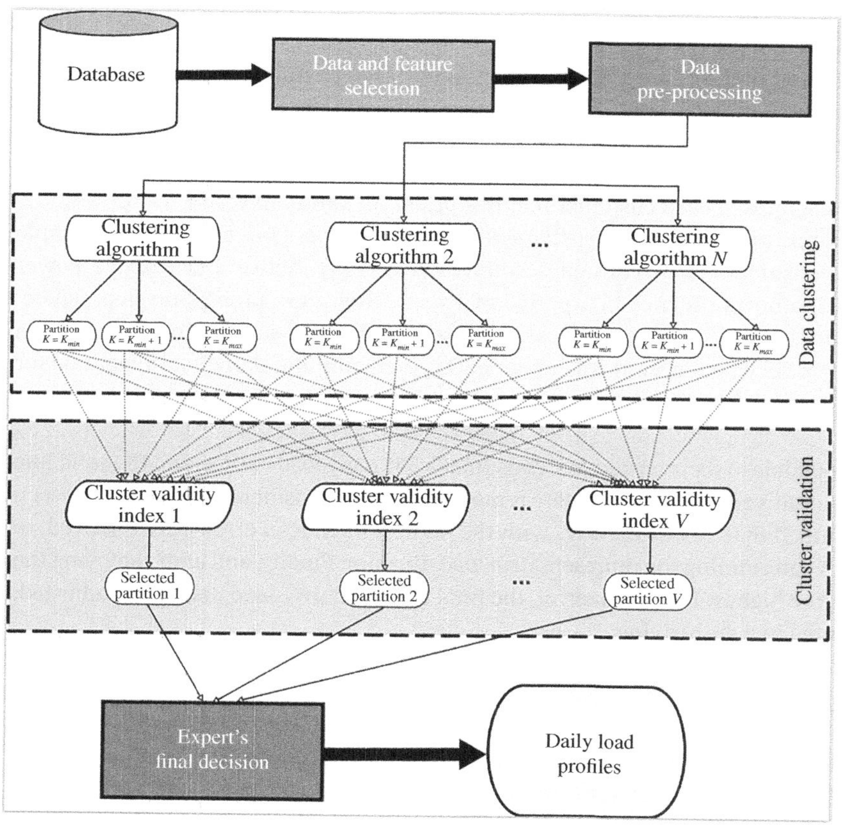

Figure 4.7 Electrical customers' characterization methodology. Source: Ramos et al. [34].

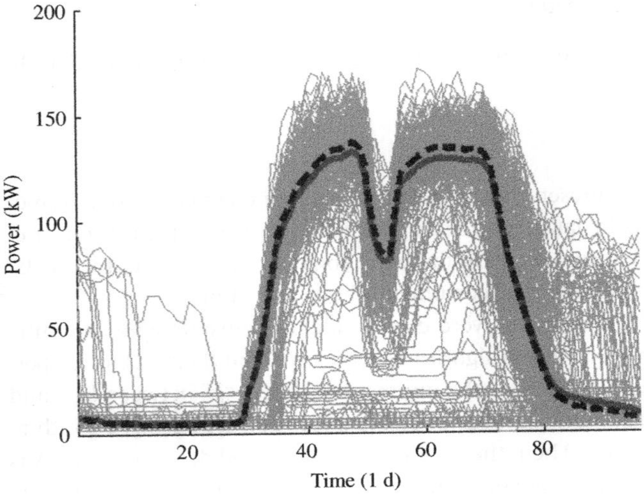

Figure 4.8 Representative load diagram with (gray curve) and without (dashed dark gray line) untypical.

Expert analysis is extremely valuable to support the evaluation of the clustering results. Thus, the expert has to select the partition that better fits his purpose from a limited subset of "*V*" data partitions (in case of no consensus among clustering validity indices in the identification of the best partition). Clustering findings must be closely analyzed. One possible approach is to choose a partition result (for the same number of derived classes) based on the largest number of indices that point to the same result [34].

The typical daily load profiles of the electrical consumers are obtained by the average of the representative load diagrams of each customer that belongs to the same cluster.

The case study concerning electricity customers characterization [34] used a data sample regarding 1022 MV consumers from Portuguese utility (E-REDES). Active and reactive power was recorded with 15 minutes cadence in a period of 1 year. After a pre-processing step, where, for instance, some duplicated data were removed (caused by daylight saving clock changes), a representative load profile consumption was obtained by averaging the daily load diagrams for each consumer. As illustrated in Figure 4.8, each consumer customer is represented by one typical load diagram.

In Figure 4.8, it is possible to see a certain MV consumer's 365 daily load curves, and the solid line represents the representative load diagram before removing the days considered untypical. For each customer, it was assumed that 10% of the days with the highest Euclidean error were removed, to minimize the error in determining the representative load diagram. Finally, and after removing the 10% of the days with the highest Euclidean error, the final representative load diagram is adjusted, in this case, represented in a dashed line.

As it is intended to compare the consumption pattern among consumers, the power consumption was normalized to the [0, 1] range, using as a normalization factor the peak power of each representative load curve. This way, it assures that the information concerning the initial load profile shape is maintained. Thus, each consumer is represented by a normalized representative daily load diagram [34]. Figure 4.9 illustrates an example of the use of the normalization factor to reduce the consumption pattern to the same scale, enabling the comparison of the loads' curves shapes without the effect of the power consumption quantity.

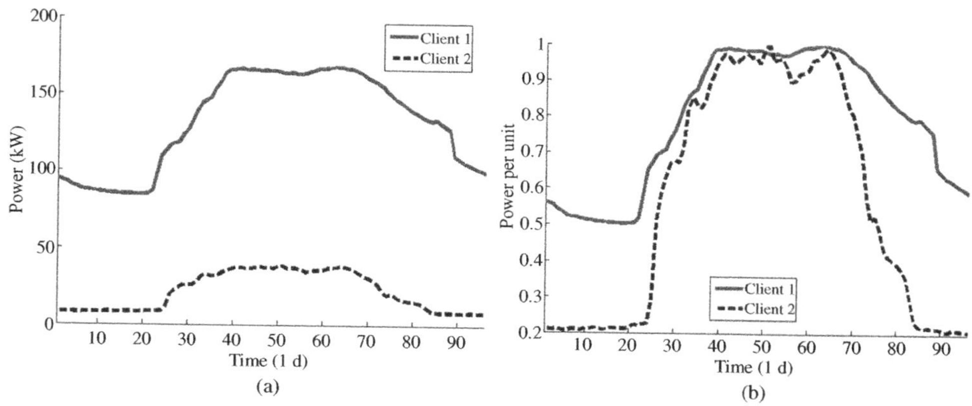

Figure 4.9 Representative load diagrams (b) after and (a) before normalization.

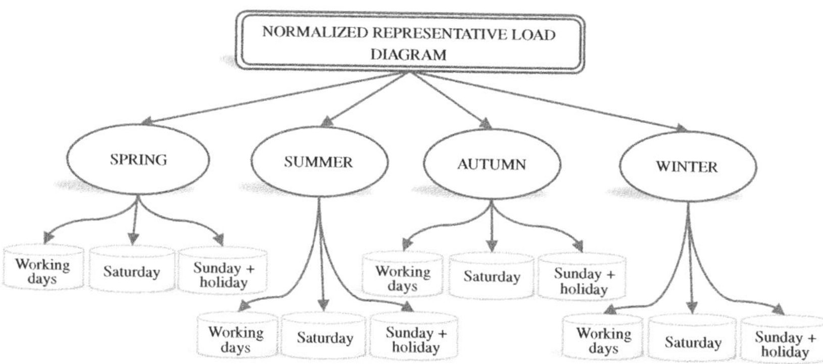

Figure 4.10 Division and reduction of database volume.

The normalized representative load diagram of all consumers was divided in several loading conditions. Figure 4.10 illustrates possible divisions that can be implemented.

In the proposed case study, Ramos et al. [34], the normalized representative diagrams were divided into three loading conditions: working days, Saturdays, and Sundays/Holidays, since then as MV consumers it is expected that during the working days the power consumption is different in comparison to weekend and special days.

In this case study, four clustering algorithms (K-Means, N-Cut, PC-K-Means, and MPC-K-Means) were used and their clustering performance was compared.

To assess the data partitions, eight validity indices from Table 4.1 were used, namely: Normalized Hubert statistic index (NH), Dunn index (D), Davies–Bouldin index (DB), the SD validity index, Silhouette statistic (S), index (I), Xie and Beni clustering validity index (XB), and Point symmetry index (PS). Iterative calculations were used to determine the minimum and a maximum number of clusters. For instance, it was imposed that the number of clusters "k" should range between 2 and 20. An index value within the defined range of "k" was produced for each clustering algorithm.

The results of the partitions selected by the validity indices are presented in Table 4.3 [34]. It is possible to verify that from the obtained data clustering, the K-mean algorithm performed better partitions because was selected as the best partition by three of the eight used clustering validity indices. Following the assessment of the partition results, the optimal number of groups was four, supported by five validity indices.

Table 4.3 Results of the partitions selected by the validity indices.

Cluster Range	Analysis of best result criteria	SD	PS	DB	XB	S	I	D	NH
2–20 Clusters	Clustering algorithm	KM	NC	KM	**KM**	**KM**	**KM**	MPCKM	PCKM
	Number of clusters	4	2	2	**3**	**3**	**3**	18	3
4–17 Clusters	Clustering algorithm	**KM**	**KM**	**KM**	PCKM	**KM**	**KM**	MPCKM	PCKM
	Number of clusters	**4**	**4**	**4**	5	**4**	**4**	15	5
5–14 Clusters	Clustering algorithm	KM	MPCKM	**KM**	PCKM	KM	**KM**	PCKM	PCKM
	Number of clusters	8	6	**6**	5	7	**6**	12	5

Source: Ramos et al. [34]/with permission of Elsevier.

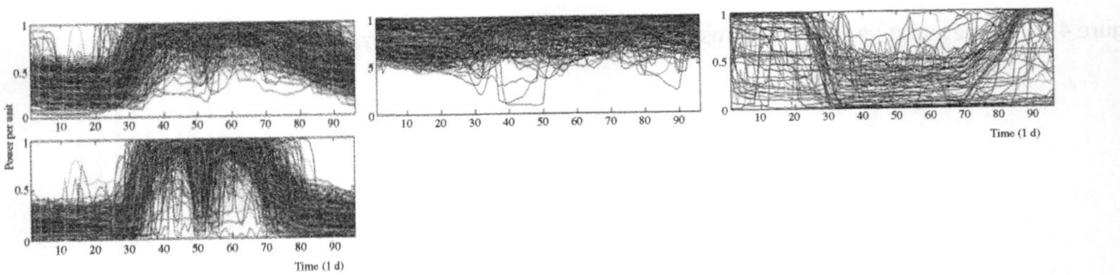

Figure 4.11 K-Means algorithm results for $k = 4$ (annual working days). Source: Ramos et al. [34]/with permission of Elsevier.

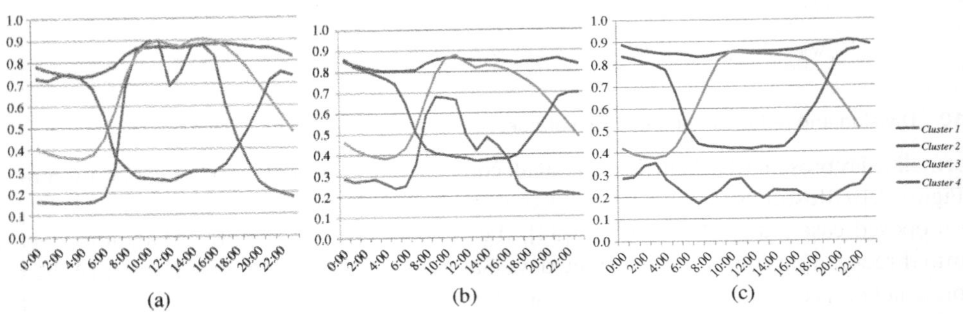

Figure 4.12 Cluster representative load diagrams using normalized shape indices. (a) Working days, (b) Saturdays, and (c) Sundays/Holidays. Source: Ramos et al. [34]/with permission of Elsevier.

Figure 4.11 presents the obtained four clusters, in this example for the annual working days.

After the choice of the best data partition, as well as the best number of clusters, the normalized daily load curves were used to obtain the cluster representatives' diagrams. Figure 4.12, depicts the obtained representative diagram for each cluster, for the proposed three loading conditions: working days, Saturday, and Sundays/Holidays.

Each representative load curve that represents a cluster also represents a customer group with the same consumption pattern.

4.3.2 Electrical Consumers Characterization – Classification

In order to classify new electrical consumers in one of the obtained clusters, a classification model is implemented. The classification of a new consumer is a supervised operation and basically consists of building a model that can be applied to unclassified objects, aiming to categorize them into classes, assigning them to one of the predefined groups.

The relevant information concerning the electrical customers' consumption pattern was obtained using the indices presented in Table 4.2, and directly derived from the representative daily load diagrams, including [34]:

- **Load factor (f_1):** The linearity of a load curve is expressed by this index. If the index value is high, the diagram tends to be straight, and if it is low, the consumption curve is not uniform, may have accentuated peaks and troughs;
- **Off-peak factor (f_2):** The sags in the load diagram are captured by this index. A high value indicates that the difference between the lowest and highest peak is small. A small value implies a bigger difference between the load diagram's minimum and maximum values;
- **Night impact factor (f_3):** This index can be used as a criterion for assessing the effect of evening electricity use (11:00 pm to 6 am). The higher this index's value, the more energy was consumed during that time;
- **Lunch impact factor (f_4):** A criterion for determining the effect of electricity use during lunch hours (12:00 am to 3 pm). The higher the value of this measure, the more energy was consumed during that period.
- **Modulation coefficient at off-peak hours (f_5):** Similar to index f_2, however, based on the difference between the minimum value and the average consumption for the period in analysis. The greater the difference between the minimum and average consumption in the period in analysis, the lower the value of this measure.

To achieve intelligible rules, these shape indices were used as input attributes in the classification model. For the classification model it was performed a rule-based modeling technique, the C5.0 classification algorithm (decision tree).

To obtain the classification indices for each defined cluster, a decision tree was performed. This algorithm may be implemented in MATLAB or using other approaches. Figure 4.13 presents a classification model (C5.0 algorithm) using the IBM SPSS-Modeler software.

Based on the initial data set (clusters indices), the classification model is trained, obtaining, for the different load regimes in analysis, the clusters associated with the defined shape indices. Typically, the training class consists of 2/3 of the input data and 1/3 of the data are used for testing the model. In this way, the classification model generates a decision tree, as well as the associated set of rules (Table 4.5), indicating the overall accuracy (94.8% for annual working days and 95.1% for annual weekend days) [34].

Table 4.4 reports the obtained of the shape indices derived from the normalized daily load profile diagrams, in this example for the annual working days' load condition (Figure 4.12a).

Figure 4.14 illustrates 4 MV consumers (client no. 70, 200, 451, and 921) from each obtained cluster (Figure 4.12a), as well the obtained shape indices values that describes the load diagram pattern of each MV customer.

Table 4.5 presents the obtained rule sets from the performing of the decision tree algorithm (C5.0 algorithm), in this example related to annual working days data sets. The obtained rules are simple and understandable.

Figure 4.15 illustrates an example of a new MV consumer classification. Figure 4.15a shows the typical daily load consumption of that customer. Table 4.6 reports the shape parameters of the consumers derived from the indices presented in Table 4.2, as well as the classification rule that was used to classify this client, based on the obtained rules from the classification model (Table 4.5).

Table 4.4 Classification indices results for the obtained clusters – annual working days.

Clusters	f_1	f_2	f_3	f_4	f_5
1	0.513	0.174	0.119	0.217	0.339
2	0.938	0.834	0.307	0.132	0.889
3	0.732	0.395	0.200	0.168	0.539
4	0.656	0.339	0.476	0.072	0.517

Source: Ramos et al. [34]/with permission of Elsevier.

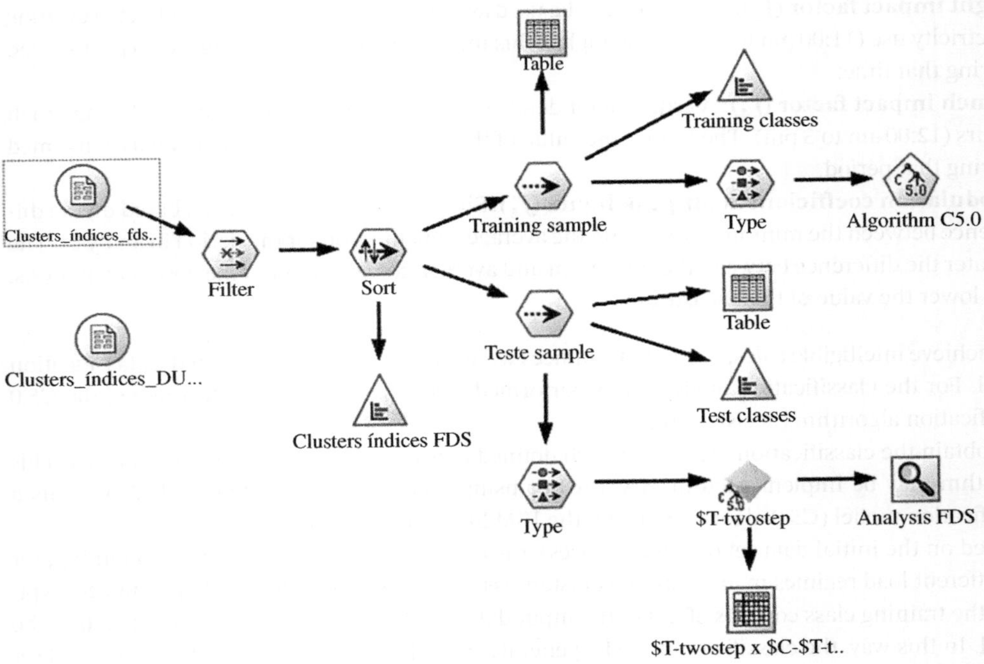

Figure 4.13 Classification model using a decision tree.

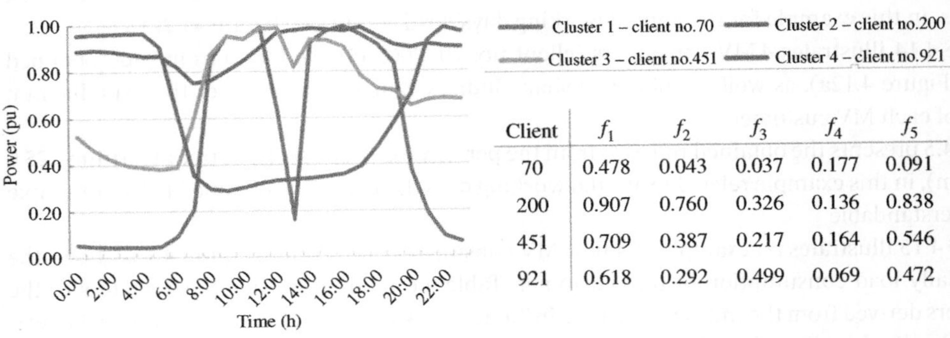

Client	f_1	f_2	f_3	f_4	f_5
70	0.478	0.043	0.037	0.177	0.091
200	0.907	0.760	0.326	0.136	0.838
451	0.709	0.387	0.217	0.164	0.546
921	0.618	0.292	0.499	0.069	0.472

Figure 4.14 Normalized shape indices of 4 MV costumers. Source: Ramos et al. [34]/with permission of Elsevier.

Table 4.5 Rule set of the classification model – annual working days.

Rules	Cluster
If $f_1 \leq 0.681 \wedge f_3 \leq 0.277 \wedge f_4 \leq 0.154 \wedge f_1 \leq 0.461$	Cluster – 4
If $f_1 \leq 0.681 \wedge f_3 \leq 0.277 \wedge f_4 \leq 0.154 \wedge f_1 > 0.461 \wedge f_1 \leq 0.640 \wedge f_1 \leq 0.579$	Cluster – 1
If $f_1 \leq 0.681 \wedge f_3 \leq 0.277 \wedge f_4 \leq 0.154 \wedge f_1 > 0.461 \wedge f_1 \leq 0.640 \wedge f_1 > 0.579 \wedge f_5 \leq 0.556$	Cluster – 3
...then	Cluster – 1
If $f_1 \leq 0.681 \wedge f_3 \leq 0.277 \wedge f_4 \leq 0.154 \wedge f_1 > 0.461 \wedge f_1 > 0.640$	Cluster – 3
If $f_1 \leq 0.681 \wedge f_3 \leq 0.277 \wedge f_4 > 0.154 \wedge f_1 \leq 0.364 \wedge f_1 \leq 0.306$	Cluster – 4
...then	Cluster – 1
If $f_1 \leq 0.681 \wedge f_3 \leq 0.277 \wedge f_4 > 0.154 \wedge f_1 > 0.364$	Cluster – 1
If $f_1 \leq 0.681 \wedge f_3 > 0.277 \wedge f_3 \leq 0.352 \wedge f_2 > 0.244$	Cluster – 4
If $f_1 \leq 0.681 \wedge f_3 > 0.277 \wedge f_3 \leq 0.352 \wedge f_2 > 0.244 \wedge f_4 \leq 0.115$	Cluster – 2
...then	Cluster – 3
If $f_1 \leq 0.681 \wedge f_3 > 0.277 \wedge f_3 > 0.352$	Cluster – 2
If $f_1 > 0.681 \wedge f_4 \leq 0.097$	Cluster – 2
If $f_1 > 0.681 \wedge f_4 > 0.097 \wedge f_4 \leq 0.167$	Cluster – 3
If $f_1 > 0.681 \wedge f_4 > 0.097 \wedge f_4 > 0.167 \wedge f_3 \leq 0.204 \wedge f_1 \leq 0.706$	Cluster – 1
...then	Cluster – 3
If $f_1 > 0.681 \wedge f_4 > 0.097 \wedge f_4 > 0.167 \wedge f_3 > 0.204$	Cluster – 3

Source: Adapted from Ramos et al. [34]/with permission of Elsevier.

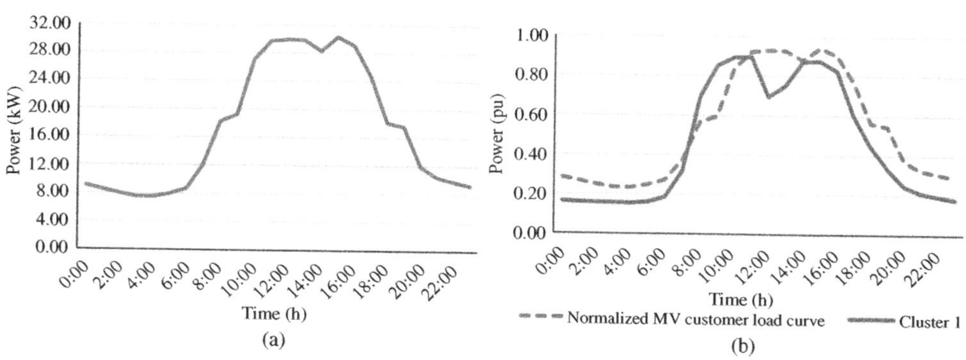

Figure 4.15 Example of a new electricity customer classification. (a) Typical daily load profile of a real MV customer. (b) Normalized load profiles: Real MV customer and Cluster 1. Source: Ramos et al. [34]/with permission of Elsevier.

Figure 4.15b illustrates the normalized MV customer load curve and the class that was assigned to it. It could be verified the similarity between the two diagrams indicating that the customer was well classified. Related work conducted by the authors can also be found in Ref. [35–37].

4.3.3 Conclusions

This section presented a methodology for the characterization of electricity consumers. DM techniques (clustering and classification) have been assessed to identify similar consumption patterns

Table 4.6 Rule set classification for a real MV consumer.

Consumer parameters – derived from Table 4.2	
$f_1 = 0.567; f_2 = 0.247; f_3 = 0.161; f_4 = 0.215; f_5 = 0.435;$	
Classification rule – Table 4.4	
If $f_1 \leq 0.681 \wedge f_3 \leq 0.277 \wedge f_4 > 0.154 \wedge f_1 > 0.364$	Cluster – 1

Source: Ramos et al. [34]/with permission of Elsevier.

among a set of electricity consumers. The knowledge about the typical load profile can benefit both utilities and customers, and this knowledge can strengthen the electric power supplier–consumer agreements.

4.4 Electrical Tariffs Design Based on Data Mining Techniques

Based on the use of DM techniques and the consequent characterization of a certain set of electricity consumers, the aggregators can propose and design a suitable tariff scheme for each consumer cluster.

4.4.1 Electrical Tariffs Design

The proposed methodology for the definition of the several tariff structures takes into account the typical daily load profile of electricity consumers, the value of the electricity price formed in the market, and also the expected aggregator's profits [38].

The designed tariffs were defined for each representative load diagram, with the distinction of the season of the year (winter and summer, for example) working and weekend days. This issue is important since electricity consumption can vary from season to season, weekdays, and type of consumer. This methodology gives to the aggregator the possibility to assess the impact of the proposed tariff and the consequent response (in terms of consumption) by its customers.

Thus, Figure 4.16 presents the proposed methodology to support aggregator to design their electricity tariffs, including:

- Knowledge of the electricity characterization (load profiling);
- Market clearing price value;
- Aggregator profit;
- Minimization of an objective function (Eq. (4.3)).

Based on the knowledge of the electricity price formed in the market (or its forecast based on historical data), the typical consumption profile resulting from the clustering analysis process, the profit expected by the aggregator, and the aggregator's strategy, a final tariff is proposed, which must be properly validated, namely, by comparison with the tariffs of the other competing market entities or with regulated tariffs (by governmental entities).

In the proposed methodology for defining tariff structures, suitable for each typical load profile, the duration of the hourly periods was defined for each cluster, considering Eq. (4.3) [38].

$$f = \sqrt{\sum_{p} \sum_{i \in p} \left(\left(P_{(i,c)} + \alpha \frac{C_{(i,AGG)}}{C_{avg(AGG)}} \right) - P_{(p,c)} \right)^2} \tag{4.3}$$

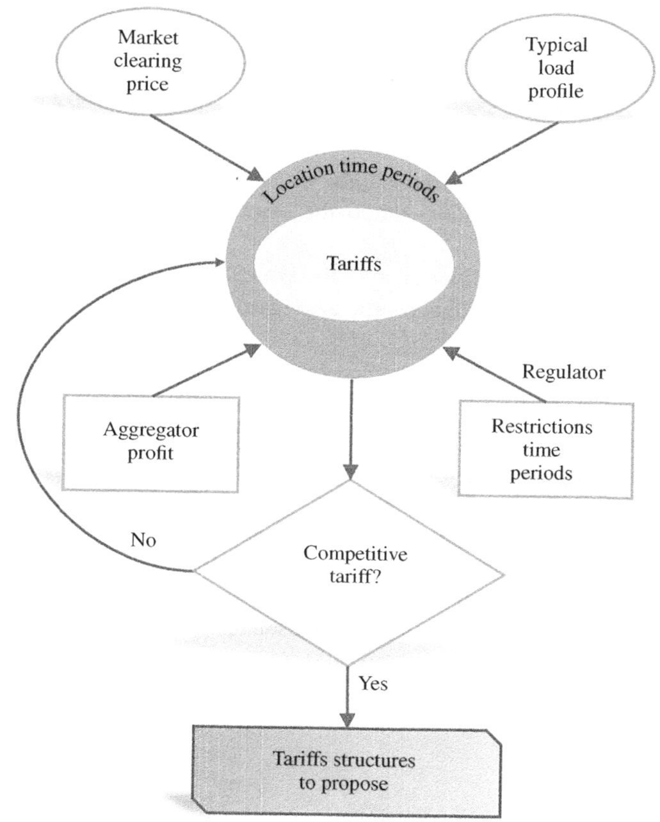

Figure 4.16 Electrical tariffs design methodology based on data mining techniques.

where:

f	–	Objective function to be minimized. It depends on the time of the tariff period "p"
i	–	30-minute intervals in the tariff period "p"
P	–	Tariff for peak hours (h_p), half-peak hours (h_{hp}), normal off-peak hours (h_{nop}), and super off-peak hours (h_{sop})
$P_{(i,c)}$	–	Average electricity demand, in the interval (i) of 30 minutes, of year "t," for the typical consumption profile "c"
$P_{(p,c)}$	–	Average electricity demand, in tariff period "p," of year "t," for the typical load profile "c"
$C_{(i,AGG)}$	–	Estimation of the electricity purchasing cost from the aggregator (AGG) in period "i"
$C_{avg(i,AGG)}$	–	Estimation of the average of the electricity purchasing cost from the aggregator (AGG), in year "t"
α	–	Impact factor of the electricity price

In the definition of tariffs proposals, aggregators may consider the consumption profile of each customer (or groups of customers – clusters) in intervals of 30 minutes, seeking to minimize the difference between that consumption and the value of the average demanded power ($P_{(p,c)}$), in periods (P), for each typical consumption profile (c).

The coefficient shown by Eq. (4.4) represents the coefficient of the energy acquisition cost in the time interval (i), compared with the average energy acquisition cost by the aggregator.

$$\frac{C_{(i,AGG)}}{C_{avg(AGG)}} \tag{4.4}$$

This coefficient enables the aggregator to signal the hours when occurs the highest energy cost ($C_{(i,AGG)} > C_{avg(AGG)}$), and to benefit the hours when the electricity cost is lowest ($C_{(i,AGG)} < C_{avg(AGG)}$).

The factor (α) allows aggregators to attach greater importance to the typical load consumption profile or the cost of energy acquisition. If (α) is equal to zero, the aggregator only considers the consumers' typical load profile.

The location of the hourly periods for each cluster is performed by minimizing the objective function (Eq. (4.4)) reflecting the distance of the load diagram representative of the cluster with a load diagram obtained by calculating the average powers of each of the defined hourly periods (peak, half-peak, normal off-peak and super off-peak hours).

Figures 4.17 and 4.18 show two examples of the obtained tariff structures taking into account the knowledge of the typical consumption profile of the cluster 1 and 2 from Figure 4.12a, as well as the price formed in the market (winter – working days) [38]. The daily market prices were used from the Iberian Energy Market Operator (OMIE) at www.omie.es. Here, the electricity price value was used as an example. In practice, a model of price forecasting can be used to anticipate the tariff structure proposal.

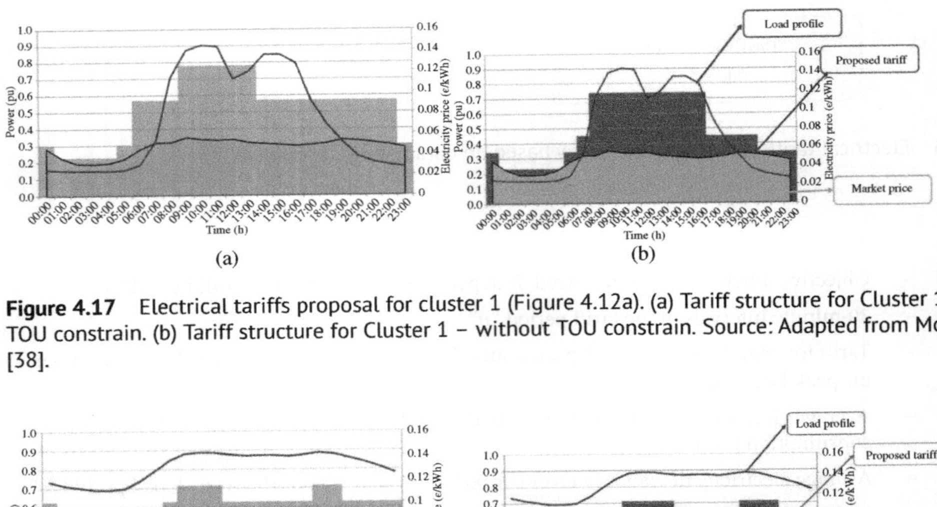

Figure 4.17 Electrical tariffs proposal for cluster 1 (Figure 4.12a). (a) Tariff structure for Cluster 1 – with TOU constrain. (b) Tariff structure for Cluster 1 – without TOU constrain. Source: Adapted from Morais et al. [38].

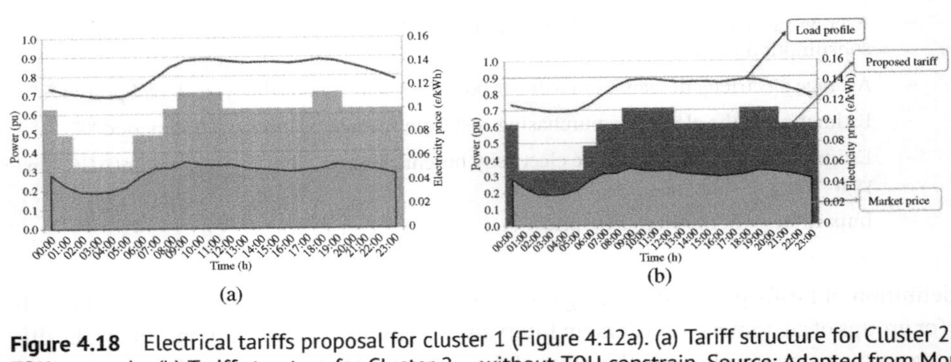

Figure 4.18 Electrical tariffs proposal for cluster 1 (Figure 4.12a). (a) Tariff structure for Cluster 2 – with TOU constrain. (b) Tariff structure for Cluster 2 – without TOU constrain. Source: Adapted from Morais et al. [38].

Despites the existence of electricity market liberalization, national energy regulator entity is responsible to publish annual regulated tariffs. Depending on the type of consumer (low, medium, or high voltage customer), those tariffs could have different prices depending on the time-of-use (TOU). Usually, for MV consumers there are four different prices for each day.

Thus, each figure reports two approaches results: Figures 4.17a and 4.18a present the proposed electricity tariff for cluster 1 considering that the aggregator has to offer the same number of daily TOU tariff as the energy regulator, and Figures 4.17b and 4.18b do not consider that time scheme constrains.

The results show that for both cases the aggregator proposes the same number of hourly tariff prices as the national energy regulator (four daily electricity tariffs: peak, half-peak, normal off-peak, and super off-peak hours).

4.4.2 Conclusions

Within the scope of the electricity markets, there is "total freedom" in the formulation of tariff schemes to be proposed to electricity consumers. It is desirable to determine transparent, fair tariffs. Thus, taking into account the value of the electricity price formed in the market and the customers typical load profile consumption, an optimization method was proposed that minimizes the distance between the typical consumption profile (in time intervals of 15, 30, or 60 minutes) and the value of the demanded average power in TOU periods.

In the mathematical formulation, it was considered an electrical energy price impact factor (α) that enables the aggregator to consider greater importance to the typical consumption profile or the cost of energy acquisition, depending on whether the factor is equal to zero or greater than zero. In the presented case study, a profit margin of 100% for aggregators was considered [38].

4.5 Data Mining Contributions to Characterize Zonal Prices

In the electrical sector, mainly in liberalized markets, the transmission and distribution power network remains a non-competitive business, due to the large infrastructure investments required, having one entity responsible for all exploration. However, this sector has yet to establish the best business model for transmission and distribution power networks in the scope of a competitive environment.

In this section, it is proposed the application of DM contributions to characterize zonal prices from real power transmission networks.

4.5.1 Zonal Prices Characterization

LMP reflects electrical and economic values at nodes (buses) or in areas of a certain transmission power network, providing economic indicator signals to business agents, as a result of bidding competition from the electricity market. The application of LMP in electrical transmission lines can be used to define zonal boundaries and whether or not a zone should be merged with another or separated into new zones.

LMP represents the marginal price variation with time and location in a certain transmission power grid. The LMP consist of three components, as presented in Eq. (4.5):

$$LMP_i = LMP^{energy} + LMP_i^{loss} + LMP_i^{cong} \tag{4.5}$$

where LMP_i is the locational marginal price at bus "i" (€/MWh), LMP^{energy} represents the system's marginal energy price (€/MWh), LMP_i^{loss} is the marginal losses price at bus "i" (€/MWh), and LMP_i^{cong} the marginal congestion price at bus "i" (€/MWh).

In short, LMP energy corresponds to the energy price and is the same for all buses. The LMP loss component reflects the costs of losses, which can be positive or negative, depending on the power flow away (from or toward the reference bus). The LMP congestion component accounts for the costs of congestion measured between a given node and a reference node, which can be positive or negative, depending on whether serving additional demand increases or reduces congestion, respectively.

In the electricity market environment, important pricing signals can be given by LMP to participants because the effects of transmission losses and binding constraints are embedded in it, providing valuable information of each location node or area.

Usually, before the calculation of the LMP for location "i" and time "t," an economic dispatch optimization problem is first considered [39], as follows:

$$\text{Min} \sum_i c_i \times Pg_i \tag{4.6}$$

Subject to:

$$\sum_i Pg_i = \sum_i Pd_i(\lambda) \tag{4.7}$$

$$\begin{aligned} Pl^{min} &\leq |P_{i-k}| = |Pl|(\mu_l) \\ &= \left| \sum_l A_{i,l}(Pg_i - Pd_i) \right| \\ &\leq Pl^{max}, \forall_l \end{aligned}$$

$$Pg_i^{min} \leq Pg_i \leq Pg_i^{max} \ \forall_i \left(\gamma_i^{min},\ \gamma_i^{max}\right) \tag{4.8}$$

where c_i (€/MWh) represents the energy bid price of generator "i," λ is the shadow prices associated with equality constraints, μ_l is the shadow prices associated with the transmission constraint for line "l," and $\left(\gamma_i^{min},\ \gamma_i^{max}\right)$ is the shadow prices associated with the generation constraints. Pg_i is the output of generator "i" (MW), Pd_i is the demand level at bus "i" (MW), and Pl is the power flow in line "l" (MW). Pl^{min} and Pl^{max} are the minimum and maximum limits of power flow in line "l," due to stability and thermal constraints. $A_{i,l}$ is the power flow's sensitivity on line "l" due to injection at bus "i" [39].

A DM-based methodology is proposed to support the characterization of the zonal prices in real power transmission networks. The main goal is to use clustering techniques to group the buses of a certain transmission power network, taking into consideration the information of the LMP values of each bus, following a similarity criterion of the LMP value. Two different clustering algorithms were used, compared and evaluated (two-step clustering algorithm and K-Means clustering algorithm). To test the proposed methodology, an LMP database from the California Independent System Operator (CAISO) was used to identify distinguish economic zones [39].

Figure 4.19 illustrates the proposed methodology, which based on the application of DM techniques follows the knowledge discovery in databases process.

To evaluate the quality of partition as well as the optimal number of clusters two clustering validity indices were used, the mean index adequacy (MIA) and clustering dispersion indicator (CDI),

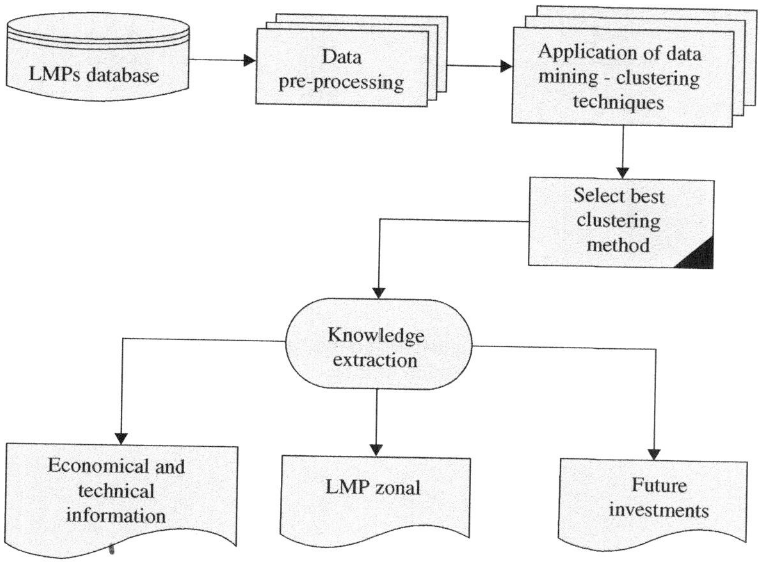

Figure 4.19 Data-mining-based methodology to characterize locational marginal pricing. Source: Ferreira et al. [39]/with permission of IEEE.

as described in [33].

$$MIA = \sqrt{\frac{1}{K}\sum_{k=1}^{K} d^2(r^{(k)}, C^{(k)})} \tag{4.9}$$

$$CDI = \frac{\sqrt{\frac{1}{K}\sum_{k=1}^{K}\frac{1}{2 \cdot n^{(k)}}\sum_{n=1}^{n^{(k)}} d^2(l^{(m)}, C^{(k)})}}{\sqrt{\frac{1}{2K}\sum_{k=1}^{K} d^2(r^{(k)}, R)}} \tag{4.10}$$

where K is the total number of clusters, and each class is formed by a subset $C^{(k)}$ of diagrams, $r^{(k)}$ is a pattern assigned to cluster "k," and R is the set of the class representative diagrams.

The lower the value obtained from the indices validity, the better the partition quality. These indices evaluation can also support the decision to define the optimal number of clusters.

Based on real data concerning CAISO Market Oasis site (www.oasis.caiso.com), the transmission network in the analysis is formed by 3458 buses, and the LMP (total LMP, LMP energy, LMP loss, and LMP congest) for each bus was recorded with a cadence of one hour during one year.

After a pre-processing step (121, 168, 320 records were analyzed), the mentioned clustering algorithms were applied. The algorithm that produced the smallest MIA and CDI index values prevailed in terms of partition performance. Indeed, the smallest value of MIA indicates more compact clusters [39].

Figure 4.20 depicts the results related to the clustering performance.

The K-Means clustering algorithm generated a better partition based on total LMP and LMP congestion, as shown in Figure 4.20. The two-step algorithm generated a better partition for LMP loss, but the difference is not significant. Thus, the K-Means algorithm was chosen for the clustering

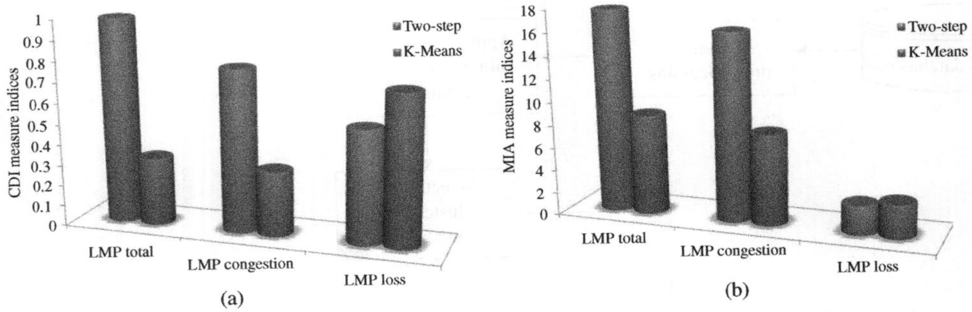

Figure 4.20 Clustering algorithms evaluation. (a) Clustering dispersion indicator and (b) mean index adequacy performance. Source: Ferreira et al. [39]/with permission of IEEE.

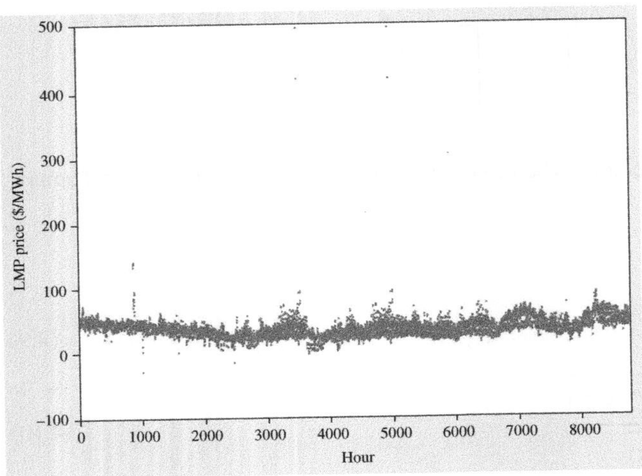

Figure 4.21 LMP energy value per hour and for each bus. Source: Ferreira et al. [39]/with permission of IEEE.

based on these findings. Related to the optimal number of clusters, and based on these two indices, the results have pointed to 30 clusters [39].

Figure 4.21 gives the LMP energy diagram that resulted from the CAISO transmission power network. Each point indicates the LMP energy value per hour and for each node.

Figure 4.22 shows the CAISO LMP loss clustering results. Each curve represents a typical annual LMP loss diagram for a certain set of transmission buses. Several clusters have a negative LMP loss component (namely shadows 2, 3, 5, 7), meaning that serving additional demand reduces transmission losses. On the opposite, the clusters located in (1, 4, 6, 8) serving additional demand increases transmission losses, which corresponds to a positive loss component. Based on the obtained results, the best location for the distributed generation to reduce this LMP component can be identified.

The LMP congestion diagram is shown in Figure 4.5. This LMP component has as well positive and negative values that represent what happens, system-wide, when the reference bus serves an additional increment of demand at a specific location. Knowing the effect of LMP congestion in a power transmission network can help in the identification of the areas where additional investment, such as extending or improving existing power transmission lines, is desirable.

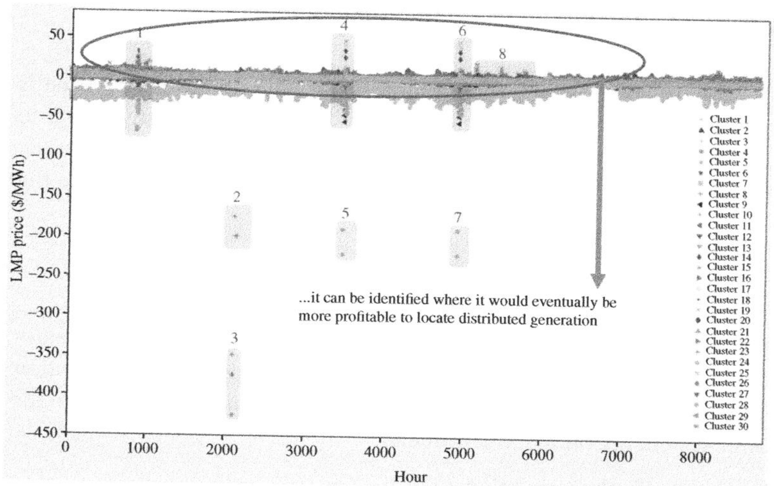

Figure 4.22 Typical annual LMP loss diagram for a certain set of transmission buses. Source: Ferreira et al. [39]/with permission of IEEE.

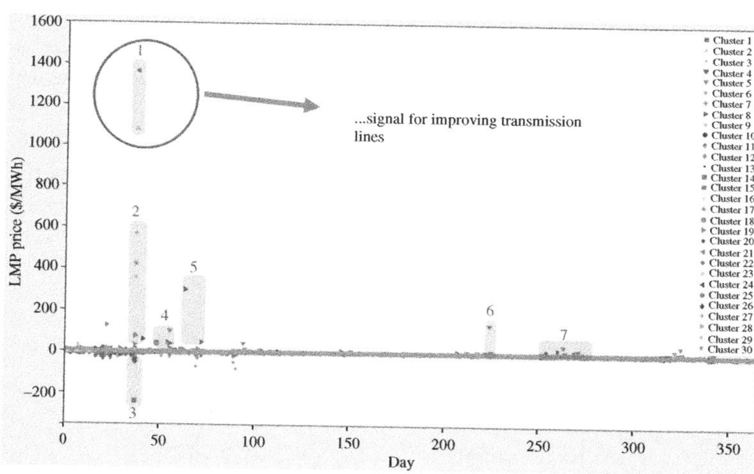

Figure 4.23 Typical annual LMP congest. Source: Ferreira et al. [39]/with permission of IEEE.

In fact, distributed generation can help to decongest some parts of the power network. In this case (based on Figure 4.23), it should be preferably installed near the nodes that belong to the clusters included in the zones: 1, 2, 4, 5, 6, and 7 [39].

4.5.2 Conclusions

Based on the proposed DM methodology, the knowledge extracted from the LMP databases may enable the ISO to formulate relevant decisions related to power transmission network planning and investment in electricity generation.

4.6 Data Mining-Based Methodology for Wind Forecasting

This section proposes a DM-based methodology for wind forecasting. The proposed approach is based on the use of DM techniques applied to a real database, including a case study related to three years of recorded data. The main goal is to use an artificial neuronal network to predict the speed of wind using different input objects, such as the number of previous hours to foresee the wind speed.

4.6.1 Wind Forecasting

The introduction of environmental policies justifies the increase of renewable energy in many countries around the world, and wind power generation represents nowadays a significant penetration share on electric generation in several European countries. The investment in wind energy entails a great challenge for the power system operator, namely due to:

- The variability and uncertainty in weather conditions;
- The scheduling dispatch, based on wind power generation, may have a significant portion of uncertainty.

Wind forecasting may contribute to introduce improvements in the efficiency of the power system operator. To forecast wind speed, DM techniques were applied to a historical database. The next step is to analyze and preprocess the data, which includes detecting irregular and missing objects as well as proper file formatting so that the multi-layer perceptron (MLP) artificial neuronal network (ANN) models can use them properly [40].

Figure 4.24 presents the proposed implemented methodology. In the pre-processing step, all values from the real database concerning wind speed, wind direction, and temperature were properly checked and the lack of value was filled. As the initial database had a record interval cadence of

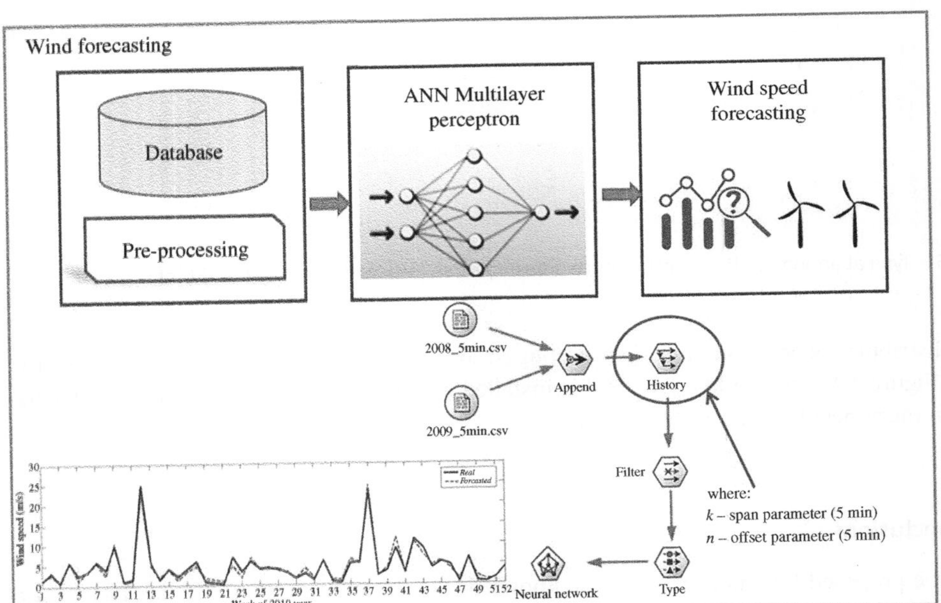

Figure 4.24 Wind forecasting proposed methodology. Source: Adapted from Ramos et al. [40].

Table 4.7 Estimated accuracy of the ANN simulations model.

Simulation	Attributes	Offset (5 min)	Span (5 min)	Previous hours	*MAE*	Accuracy (%)
1	Speed, direction, temperature	288	288	24	1.977	92.46
2	Speed, direction, temperature	288	120	10	2.066	92.39
3	Speed	1	288	24	0.840	97.11
4	Speed	1	120	10	0.823	97.17
5	Speed	1	60	5	0.818	97.61
6	Speed, direction	1	60	5	0.843	97.18
7	Speed, direction	1	60	5	0.846	97.19

Source: Adapted from Ramos et al. [40].

one minute, the database was adequate for five minutes interval cadence (data reduction volume). Then, an adequate file extension was made to be used by the neuronal net model.

An ANN (Multilayer Perceptron) was used to predict the wind speed values. Two different software were performed, MATLAB and IBM SPSS-Modeler (see an example of the implemented stream in Figure 4.24). Finally, the results were evaluated, and for that, it was used the mean absolute error (MAE), accordingly Eq. (4.11), where "x_i" represents the prediction value, "y_i" represents the actual (true) value, and "n" is the number of observations.

In this approach it was considered wind speed at times "t" and "k" the offset time value that represents the last previous record (concerning the current record value) that will be used as an input attribute by the ANN. Thus, "$k + 1$" reflects the number of records that the ANN can use as input attributes. In this case, in addition to other basic inputs such as temperature and wind direction, two input attributes relating to "k" and "$k + 1$" were used [40].

Several simulations with different input attribute values were performed to determine which model performs better based on the information on the previous state of the speed wind. As a result, the offset and span parameters proved to be crucial in the prediction model. Thus, based on the specification of these two parameters (span and offset), the efficiency of the forecasting model was evaluated.

Table 4.7 presents the several simulation that were performed for different values of span and offset parameters.

Figure 4.25 illustrates the obtained results from the forecasting model related to simulation 1 and 2, which uses as input attribute of the ANN the wind speed, wind direction, and temperature values. The used offset and span parameters were both 24 hours for simulation 1, and offset = 24 hours and span = 10 hours for simulation 2.

In simulations 3 and 4 was used, as input attribute, the wind speed value relating the previous 24 and 10 hours, respectively.

Figure 4.26 depicts the forecasting results from simulation 5, which uses the wind speed as input attribute and the previous five hours (span = 5 hours).

Figure 4.27 reports the forecasting results from simulations 5, 6, and 7. Simulation 6 uses as input attributes to the ANN the wind speed, wind direction, and temperature (using the previous five hours). Simulation 7 uses as input attributes the wind speed and wind direction, and also a span of five hours.

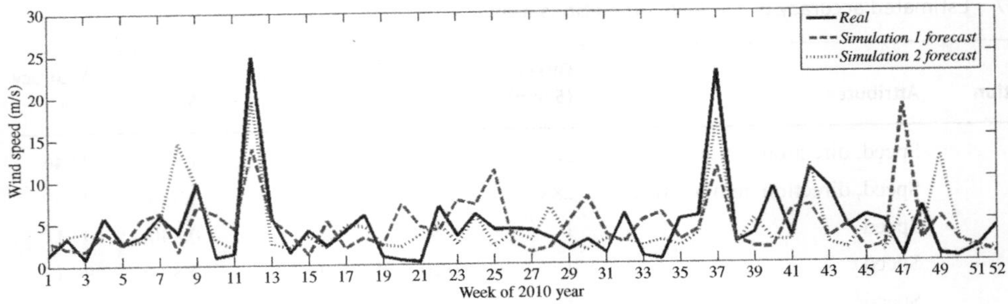

Figure 4.25 Wind speed forecasting – simulations 1 and 2 (Table 4.7). Source: Ramos et al. [40]/ with permission of IEEE.

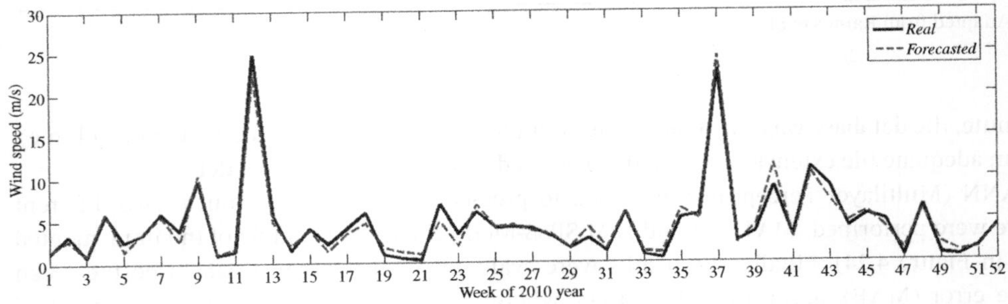

Figure 4.26 Wind speed forecasting – simulation 5 (Table 4.7). Source: Ramos et al. [40]/with permission of IEEE.

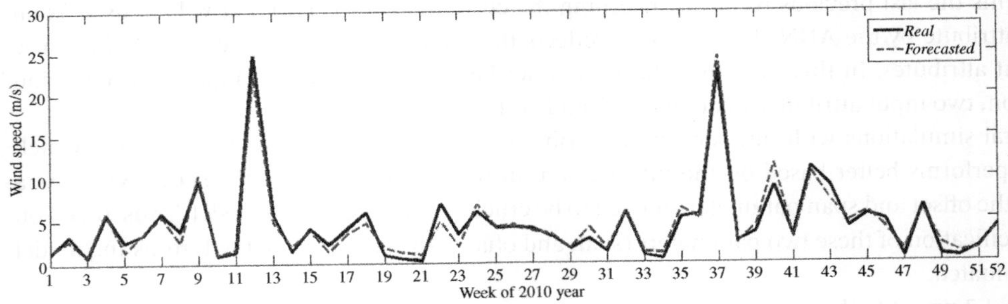

Figure 4.27 Wind speed forecasting – simulations 5, 6, and 7 (Table 4.7). Source: Ramos et al. [40]/ with permission of IEEE.

The results pointed that information containing wind speed values from a few hours before the actual estimated hour (in this case with a period of five hours – simulation 5, Table 4.7) had better estimated accuracy rather than including information starting from the same hour but of the previous day [40].

4.6.2 Conclusions

Based on the KDD process, a DM-based methodology was implemented to support wind speed forecasting. A real database concerning wind speed, wind direction, and temperature were used to predict wind speed. A pre-processing step was required to prepare data to be used by a DM algorithm (an ANN). The accuracy of the obtained forecasting results was analyzed, taking into consideration several scenario simulations.

4.7 Final Remarks

This chapter presented some examples of DM techniques applied to power systems, based on the process of knowledge discovery in databases.

In electricity market environment the consumers' characterization, and the knowledge of the daily typical load profile is essential to formulate personalized electricity tariffs, accordingly customers' consumption.

Electricity aggregators can also take some advantages from consumers' characterization to formulate electricity tariffs, taking into account the prices formed in the market, as well as the knowledge of the electricity consumers' behavior

In some markets the electricity prices may vary from geographical areas, depending on the energy transmission, loss, and congestion (Locational Marginal Prices – LMP). DM techniques may play an important rule by analyzing the buses with the same LMP characteristics and give some signal to improve the electricity transmission lines or locate the best buses to invest in new production units.

Finally, DM techniques were used to support the wind speed forecasting. The prediction of the wind speed in power systems, mainly in wind power generation, is extremely important to predict the electricity generation for the next hours or day, given a certain control margin to the electrical system operator.

References

1 Liu, Y., Zhang, D., and Gooi, H.B. (2021). Data-driven decision-making strategies for electricity retailers: deep reinforcement learning approach. *CSEE Journal of Power and Energy Systems* 7: 358–367. https://doi.org/10.17775/CSEEJPES.2019.02510.

2 Santos, M.F. and Azevedo, C.S. (2005). *Data Mining – Descoberta de conhecimento em bases de dados*. FCA – Editora de Informática, Lda *Repositorium*.

3 Soares, J., Pinto, T., Lezama, F., and Morais, H. (2018). Survey on complex optimization and simulation for the new power systems paradigm. *Complexity* https://doi.org/10.1155/2018/2340628.

4 Mahmud, M.S., Huang, J.Z., Salloum, S. et al. (2020). A survey of data partitioning and sampling methods to support big data analysis. *Big Data Mining and Analytics* 3: 85–101. https://doi.org/10.26599/BDMA.2019.9020015.

5 Fayyad, U., Piatetsky-Shapiro, G., and Smyth, P. (1996). From data mining to knowledge discovery: an overview. In: *Advances in Knowledge Discovery and Data Mining*, MIT Press, 1–34.

6 Goebel, M. and Gruenwald, L. (1999). A survey of data mining and knowledge discovery software tools. *ACM SIGKDD Explorations Newsletter* 1: 20–33. https://doi.org/10.1145/846170.846172.

7 Han, J., Kamber, M., and Pei, J. (2012). *Data Mining: Concepts and Techniques*, Morgan Kaufmann Series in Data Management Systems, 3e. Morgan Kaufmann https://doi.org/10.1016/C2009-0-61819-5.

8 Witten, I.H., Hall, M.A., Frank, E., and Pal, C.J. (2017). *Data Mining – Practical Machine Learning Tools and Techniques*, 4e. Morgan Kaufmann https://doi.org/10.1016/C2015-0-02071-8.

9 Aggarwal, C.C. and Reddy, C.K. (2013). *Data Clustering: Algorithms and Applications*, Chapman & Hall/CRC Data Mining and Knowledge Discovery Series. Chapman and Hall/CRC.

10 MacQueen, J.B. (1967). Some methods for classification and analysis of multivariate observations. In: *Proceedings of 5th Berkeley Symposium on Mathematical Statistics and Probability* (ed. L.M.L. Cam, and J. Neyman), vol. 1, University of California Press, Berkeley, USA, pp. 281–297.

11 Jain, K.A. and Dubes, C.R. (1988). *Algorithms for Clustering Data*, Prentice-Hall Advanced Reference Series. Prentice Hall.

12 Ester, M., Kriegel, H.-P., Sander, J., and Xu, X. (1996). A density-based algorithm for discovering clusters in large spatial databases with noise. In: *Proceedings of the Second International Conference on Knowledge Discovery and Data Mining* (ed. E. Simoudis, J. Han, and U. Fayyad), AAAI Press, pp. 226–231.

13 Grabusts, P. and Borisov, A. (2002). Using grid-clustering methods in data classification. In: *Proceedings of the International Conference on Parallel Computing in Electrical Engineering*. Warsaw, Poland: IEEE https://doi.org/10.1109/PCEE.2002.1115319.

14 Shi, J. and Malik, J. (2000). Normalized cuts and image segmentation. *IEEE Transactions on Pattern Analysis and Machine Intelligence (PAMI)* 22: 888–905. https://doi.org/10.1109/34.868688.

15 Fraley, C. and Raftery, A.E. (2002). Model-based clustering, discriminant analysis, and density estimation. *Journal of the American Statistical Association* 97: 611–663. https://doi.org/10.1198/016214502760047131.

16 Bezdeck, J.C., Ehrlich, R., and Full, W. (1984). FCM: fuzzy C-means algorithm. *Computers and Geosciences* 10: 191–203. https://doi.org/10.1016/0098-3004(84)90020-7.

17 G. Gan, C. Ma and J. Wu. *Data Clustering Theory, Algorithms and Applications*. ASA-SIAM Series on Statistics and Applied Mathematics, 2007. https://doi.org/10.1137/1.9780898718348.

18 Hubert, L.J. and Schultz, J. (1975). Quadratic assignment as a general data-analysis strategy. *British Journal of Mathematical and Statistical Psychology* 29: 190–241. https://doi.org/10.1111/j.2044-8317.1976.tb00714.x.

19 Chou, C.H., Su, M.C., and Lai, E. (2004). A new cluster validity measure and its application to image compression. *Pattern Analysis and Applications* 7: 205–220. https://doi.org/10.1007/s10044-004-0218-1.

20 Dunnt, J.C. (1974). Well separated clusters and optimal fuzzy partitions. *Journal of Cybernetics* 4: 95–104. https://doi.org/10.1080/01969727408546059.

21 Davies, D.L. and Bouldin, D.W. (1979). A cluster separation measure. *IEEE Transactions on Pattern Analysis and Machine Intelligence* 1: 224–227. https://doi.org/10.1109/TPAMI.1979.4766909.

22 Halkidi, M., Batistakis, Y., and Vazirgiannis, M. (2001). Clustering algorithms and validity measures. In: *Proceedings 13th International Conference on Scientific and Statistical Database Management*. https://doi.org/10.1109/SSDM.2001.938534.

23 Kaufman, L. and Rousseeuw, P.J. (1990). *Finding Groups in Data: An Introduction to Cluster Analysis*. Wiley Series in Probability and Statistics, New York: John Wiley & Sons, Inc. https://doi.org/10.1002/9780470316801.

24 Maulik, U. and Bandyopadhyay, S. (2002). Performance evaluation of some clustering algorithms and validity indices. *IEEE Transactions on Pattern Analysis and Machine Intelligence* 24: 1650–1654. https://doi.org/10.1109/TPAMI.2002.1114856.

25 Xie, X.L. and Beni, G. (1991). A validity measure for fuzzy clustering. *IEEE Transactions on Pattern Analysis and Machine Intelligence* 13: 841–847. https://doi.org/10.1109/34.85677.

26 Krazanowski, W.J. and Lai, Y.T. (1988). A criterion for determining the number of groups in a data set using sum of squares clustering. *Biometrics* 44: 23–34. https://doi.org/10.2307/2531893.

27 Sharma, S.C. (1996). *Applied Multivariate Techniques*. Wiley https://doi.org/10.1177/0019556119960301.

28 G. J. Mclachlan. *Discriminant Analysis and Statistical Pattern Recognition. Wiley Series in Probability and Statistics*, 1992. https://doi.org/10.1002/0471725293.

29 Patil, N., Lathi, R., and Chitre, V. (2012). Comparison of C5.0 & CART classification algorithms using pruning technique. *International Journal of Engineering Research & Technology* 1: 1–5.

30 Rumelhart, D.E. and McClelland, J.L. (1987). *Parallel Distributed Processing: Explorations in the Microstructure of Cognition: Foundations*, vol. 1, 318–362. MIT Press Direct https://doi.org/10.7551/mitpress/5236.001.0001.

31 Cristianini, N. and Shawe-Taylor, J. (2000). *An Introduction to Support Vector Machines: And Other Kernel-Based Learning Methods*. Cambridge University Press https://doi.org/10.1017/CBO9780511801389.001.

32 Freund, Y. and Schapire, R.E. (1997). A decision-theoretic generalization of online learning and an application to boosting. *Journal of Computer and System Sciences* 55: 119–139. https://doi.org/10.1006/jcss.1997.1504.

33 Chicco, G., Napoli, R., Postolache, P. et al. (2003). Customer characterization options for improving the tariff offer. *IEEE Transactions on Power Systems* 18: 381–387. https://doi.org/10.1109/TPWRS.2002.807085.

34 Ramos, S., Duarte, J., Duarte, J., and Vale, Z. (2015). A data-mining-based methodology to support MV electricity customers' characterization. *Energy and Buildings* 91: 16–25. https://doi.org/10.1016/j.enbuild.2015.01.035.

35 Ramos, S. and Vale, Z. (2008a). Data mining techniques application in power distribution utilities. In: *IEEE/PES Transmission and Distribution Conference and Exposition*, 1–8. https://doi.org/10.1109/TDC.2008.4517229.

36 Ramos, S. and Vale, Z. (2008b). Data mining techniques to support the classification of MV electricity customers. In: *IEEE Power and Energy Society General Meeting - Conversion and Delivery of Electrical Energy in the 21st Century*, 1–7. https://doi.org/10.1109/PES.2008.4596669.

37 Ramos, S., Duarte, J.M., Duarte, F.J. et al. (2013). A data mining framework for electric load profiling. In: *IEEE PES Conference on Innovative Smart Grid Technologies (ISGT Latin America)*, 1–6. https://doi.org/10.1109/ISGT-LA.2013.6554489.

38 Morais, H., Vale, Z., Faria, P., and Ramos, S. (2013). Defining electricity tariffs using the knowledge about the consumers profiles in ELECON project. In: *IEEE PES Conference on Innovative Smart Grid Technologies (ISGT Latin America)*, 1–7. https://doi.org/10.1109/ISGT-LA.2013.6554488.

39 Ferreira, J., Ramos, S., Vale, Z., and Soares, J. (2011). A data-mining-based methodology for transmission expansion planning. *IEEE Intelligent Systems* 26: 28–37. https://doi.org/10.1109/MIS.2011.4.

40 Ramos, S., Soares, J., Vale, Z., and Morais, H. (2011). A data-mining based methodology for wind forecasting. In: *16th International Conference on Intelligent System Applications to Power Systems*, 1–6. https://doi.org/10.1109/ISAP.2011.6082223.

5

Synchrophasor Data Analytics for Anomaly and Event Detection, Classification, and Localization

Sajan K. Sadanandan[1], Arman Ahmed[2], Shikhar Pandey[3], and Anurag K. Srivastava[4]

[1]Smart Grid Integration, R&D Center, Dubai Electricity & Water Authority (DEWA), Dubai, UAE
[2]Smart Grid Demonstration and Research Investigation Lab, Washington State University, Pullman, WA, USA
[3]Smart Grid and Technology, ComEd, Oakbrook Terrace, IL, USA
[4]Smart Grid Resiliency and Analytics Lab, West Virginia University, Morgantown, WV, USA

5.1 Introduction

Software tools in the control center help in monitoring the power system using real-time measurements from various sensors like Intelligent Electronic Devices (IED), Remote Terminal Unit (RTU), and other measurement and sensing devices in the network. Applications utilizing these measurements provide decision support for the corrective actions for the reliable operation of the power system. However, if the measurements are incorrect, the applications can provide false results affecting the regular operation of the grid. The push to automate and digitize the electric grid has led to the widespread installation of Phasor Measurement Units (PMUs) for improved real-time wide-area system monitoring and control with its ability to provide high-resolution time-stamped measurements. Nevertheless, transforming large volumes of high-resolution PMU measurements into actionable insights remains challenging. A central challenge is creating flexible and scalable online anomaly detection in PMU data streams. PMU data can hold multiple types of anomalies arising in the physical system or the cyber system (measurements and communication networks). These anomalies as they seem inevitable and a result of synchrophasor architecture the onus to make use for grid's situational awareness with noisy measurement data and bad data (BD) lies upon researchers and application developers. The synchrophasor system has the potential to enable several applications that can enhance monitoring and control of the power systems.

The dynamic behavior of the power grid (e.g. change in load, generation, distributed energy resources (DERs) switching, network, controls) causes the definition of data anomalies to vary in time and requires real-time assessment and online training. A fundamental challenge is to enable online (i.e. real-time) training of machine or deep learning algorithms for anomaly detection over streaming PMU data. While machine and deep learning are often desirable to manage data streams, training a deep learning algorithm over streaming PMU data is nontrivial due to changes in data statistics caused by the dynamic nature of streaming data. This chapter will discuss PMUNET: a novel device-level deep learning-based data-driven algorithm for streaming PMU data, using online learning and multivariate data-drift (DD) detection algorithm.

Moreover, with an increasing number of extreme events, grid components, and complexity, more alarms are being observed in the power grid control centers. Operators in the control center need to monitor and analyze these alarms to take suitable control actions, if required, to ensure the system's

Intelligent Data Mining and Analysis in Power and Energy Systems: Models and Applications for Smarter Efficient Power Systems, First Edition. Edited by Zita Vale, Tiago Pinto, Michael Negnevitsky, and Ganesh Kumar Venayagamoorthy.

reliability, stability, security, and resiliency. Although existing alarm and event processing tools help in monitoring and decision making, synchrophasor data along with the topology and component location information can be used in detecting, classifying, and locating the event, which is the focus of this work. In this chapter, PMU's data quality issue is also addressed before using data for event analysis. The algorithms discussed in this chapter include statistical, clustering, and Maximum Likelihood Criterion (MLE)-based anomaly detection. Density-based spatial clustering of applications with noise (DBSCAN) is used for event detection, and physics-based rule and decision tree for event classification. Further, topology information, statistical techniques, and graph search algorithms are used for event localization.

5.2 Synchrophasor Data Quality Issues and Challenges

Synchrophasor data obtained from the PMU provides a multitude of information that has been promising technology enabling the grid operator to perform enhanced power system monitoring, protection, diagnosis, and control [1–3]. The conventional power system uses supervisory control and data acquisition (SCADA) for monitoring and control. The data obtained by the SCADA from the meters, RTUs, IEDs, merging terminal units (MTUs) provide data at a rate that is within a time window that is typically a few seconds long. But PMUs can provide time-synchronized wide-area measurements at a data rate as high as 60 samples per second compared to two to four seconds per record in the case of SCADA data. Figure 5.1 shows PMUs' high-resolution data makes the invisible event by SCADA data visible [4]. Additionally, PMU data contains time-synchronized current and voltage phasors, frequency, and rate of change of frequency (ROCOF). In contrast, SCADA data is analog measurements of root mean square (RMS) voltages, currents, real and reactive power, making PMU data suitable for wide-area monitoring and control application.

Various applications based on time-synchronized PMU measurements have been developed and are currently used by utilities for power system monitoring and control. Some of the synchrophasor-based monitoring applications used in the power grid control room are linear state estimation [5, 6], island detection [7], disturbance monitor [8], voltage stability analysis [9], power system restoration [10], load modeling [11], parameter estimation, event detection [12, 13].

PMU data are prone to anomalies that can either be due to network or device error (bad data) or due to any power system events such as fault, load or generation tripping, switching operation,

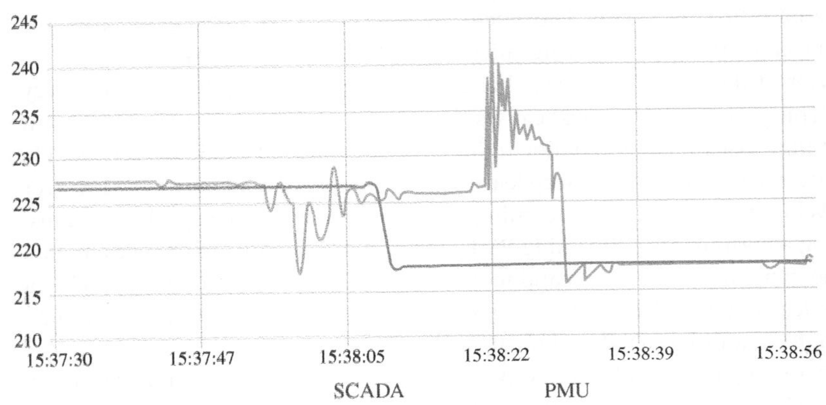

Figure 5.1 Synchrophasor Technology vs. SCADA data.

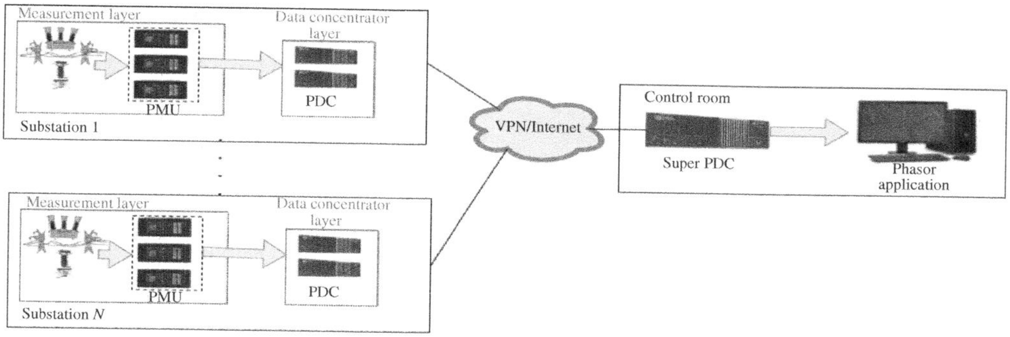

Figure 5.2 Synchrophasor data flow from phase conductor to phasor application.

etc. Bad data in the PMU measurement can be induced anywhere between the phase conductor to which PMU via potential transformer (PT) and current transformer (CT) is connected and phasor application in the control room. A detailed explanation of the different types of errors is explained in the following subsection. If this abnormal data is fed to the phasor application, it might lead to a misleading output and lead to wrong control action by operators in the control room.

5.2.1 PMU Data Flow: Data Quality Issues

The complete data processing of PMU data before it reached for PMU application is shown in Figure 5.2. The instantaneous voltage and current measurement obtained by CT and PT are fed to PMU via instrumentation cables; then, PMU will use these measurements to calculate voltage and current phasor, frequency, and ROCOF with a time tab obtained from the GPS antenna. Further, data from multiple PMUs are aggregated in phasor data concentrators (PDCs) and super-PDC via communication lines to control centers where it is received for applications. More details are accessible in the North American Synchrophasor Initiative (NASPI), PMU Applications Requirements Task Force (PARTF) Framework [14].

The following are the potential areas from where data quality issues can arise in PMU data:

- **At or within PMU**: The most common cause of error in PMU data is due to CT and PT bias, filters, and algorithms within PMU, or incorrect or missing timestamps associated with measurement.
- **At any intermediate data aggregator**: All the data from Multiple PMUs are aggregated at PDCs at a different level. This can cause an error due to PDCs, historians, databases, or archives of the data, misalignment or erroneous compression of data, duplicate data, or it can be due to late deliveries of data to PDC.
- **Within communications network**: Communication network is the backbone structure to send the PMU dataset from its measuring point to the control room where it is used by different phasor applications as shown in Figure 5.2. Data quality issues in PMU data can be caused in communication networks due to loss of communications network nodes, excess latency, or data corruption introducing erroneous values.
- **At the point of use by the application**: Data quality of PMU can be affected due to data storage issues or not enough historical data for the application's use. This may lead to degraded results by the phasor applications.

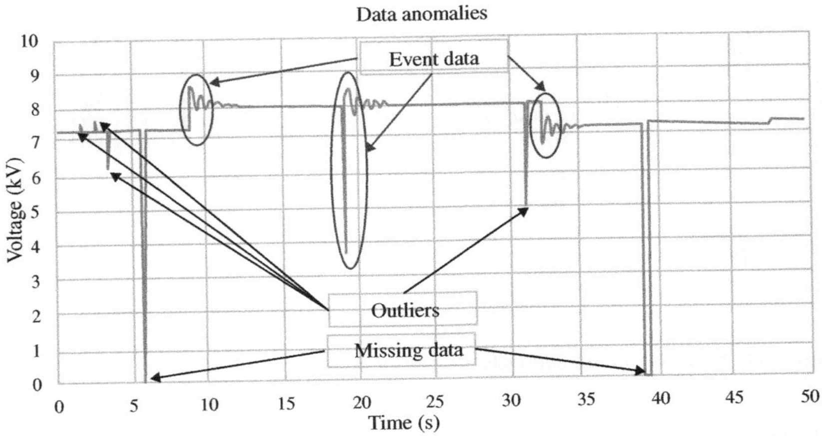

Figure 5.3 Synchrophasor data snapshot with anomalies.

The issues mentioned above can corrupt the PMU data and make it unsuited for PMU applications. Therefore data preprocess and anomaly detection is a must before it is fed to phasor applications. PMU data may have different types of anomalies. Figure 5.3 shows a snapshot of PMU data with outliers, missing data, and event data.

5.2.2 PMU Data Anomalies

The previous layer's anomalies can contaminate available PMU data in the application layer, which will affect the performance of phasor applications output. Distinguishing between the source of the anomaly is challenging.

Figure 5.4 summarizes the anomalies and their origin in the PMU data. In common, it is hard to differentiate between bad data anomaly and event-based anomaly. This leads to confusion between the source of abnormalities, including but not limited to noise and bad data detection (BDD) and event, sensor problems, data poisoning due to cyberattacks, and physical events. To address these issues we present two different methods for anomaly and event detection classification and localization in this chapter. The first method used is a novel device-level unsupervised deep learning-based data-driven approach for event detection, localization, and classification over streaming PMU data, using online learning and multivariate DD detection algorithm. Secondly, DBSCAN for event detection and physics-based rule and decision tree for event classification is presented.

5.3 ML-Based Anomaly Detection, Classification, and Localization (ADCL) Over Data Drifting Multivariate Synchrophasor Data Streams

This section introduces a machine learning (ML)-based anomaly detection, localization, and classification using a multivariate PMU data stream. This algorithm is taken from our previous work [15], where we proposed a device-level ML-based framework called PMUNET. The developed framework has the following features: (i) active learning considering possible concept drift in PMU data stream, (ii) Stochastic Gradient Descent (SGD) with context-based learning rate schedule to scale the learning of autoencoder, (iii) a stacked deep autoencoder with z-score. The main objective of this algorithm is to address the challenges in the existing ML solutions, which, when

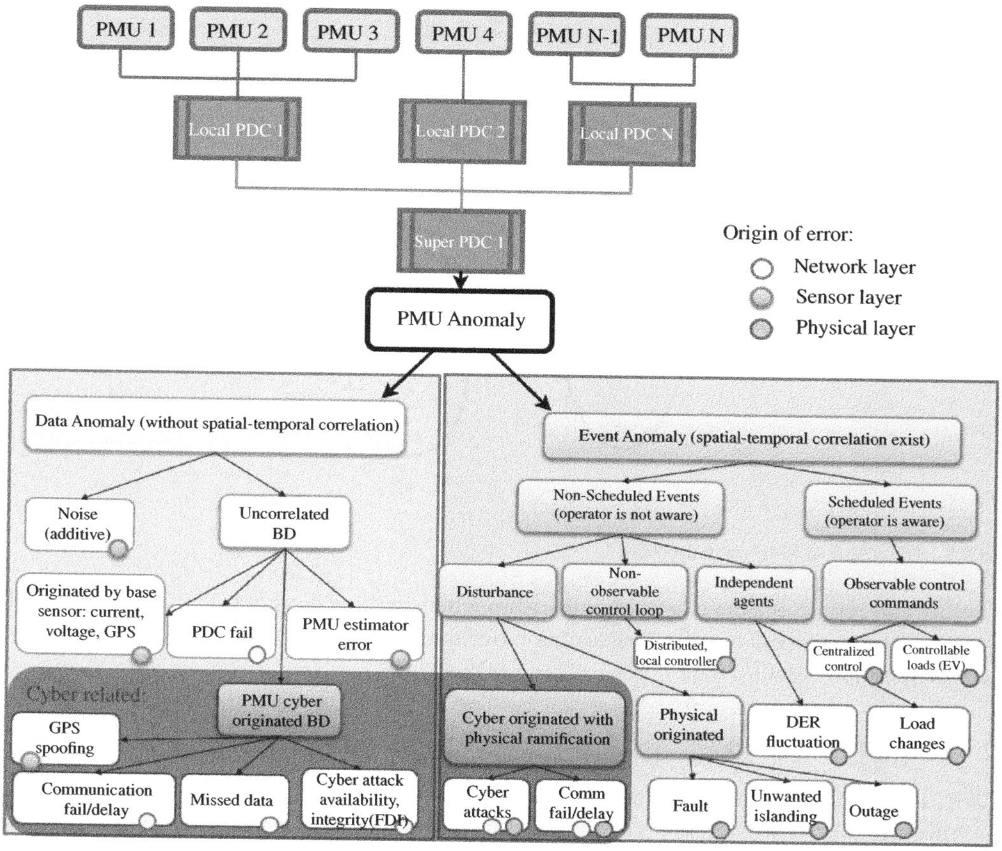

Figure 5.4 Possible causes for synchrophasor data anomalies.

trained offline, don't consider the dynamic behavior of the power system in real-time, resultant in a low-performance model. In the following subsection, we will discuss the definition of data drift in PMU data and PMUNET architecture for anomaly detection, classification, and localization (ADCL).

5.3.1 Data Drift in Synchrophasor Measurements

Data drift is defined as a phenomenon in which the data statistical characteristic changes over time due to dynamic changes in the system [16]. The data drift concept is prevalent in the power system as it experiences various dynamic changes depending on different operations. Since we are using PMU data to perform data analytic techniques to detect anomalies and the performance of these techniques will degrade due to the statistical data changes in the system data. An example of such a phenomenon is shown in Figure 5.5, where streaming PMU data statistics and its characteristics arbitrarily change beyond the threshold over time due to different operations in the power grid.

Usually, in ML-based anomaly detection techniques, the learning is done offline, and the retaining is done when the performance of the method degrades a certain threshold. On the other hand, retraining the process continuously irrespective of data drift will be very costly and increase the computational burden on the system. Therefore, it is beneficial to detect the data drift in PMU

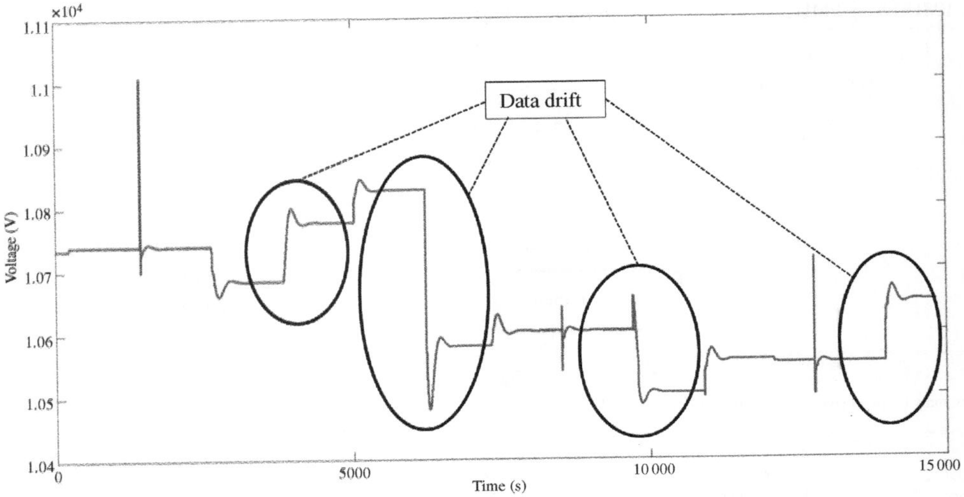

Figure 5.5 Data drift in PMU measurement.

data online and retrain the deep learning algorithm whenever it is detected. This work proposed the PMUNET framework, a novel deep learning-based data-driven approach to enable ADCL over concept-drifting PMU data.

5.3.2 PMUNET Framework

PMUNET is a device-level anomaly detection approach over a PMU data stream. Figure 5.6 shows the architecture overview diagram for PMUNET. The input to PMUNET is PMU data stream consists of three-phase voltage and current phasor measurement data point. In the proposed approach, the PMU data stream is processing in batches B_i with data points W (e.g. 5K PMU measurement data points) as (B_1, W), (B_2, W), …, (B_i, W), …(i refers to the batch number) for online training. Once the PMUNET is training with the first batch of PMU data stream, it will perform anomaly detection every second and check for data drift for every batch. In this work, we also leverage synchronous SGD [17] with a learning rate scheduler for faster training of the deep learning algorithm in an online fashion.

The functional modules in the PMUNET framework as Figure 5.6 are explained as follows:

5.3.2.1 Data Pre-Processing (DPP) Module

The data pre-processing (DPP) module is used to preprocess every batch of PMU data (B_i, W) by scaling the data values to mean "0" and standard deviation of "1." Then these normalized data points are used to generate the Resilient Distributed Datasets (RDDs) input and output features. RDD [18] are unchanging elements that are used in trigger for doing operations on data set in a distributed manner.

5.3.2.2 Data-Drift (DD) Module

The DD module is used to detect the data drift in the PMU data stream from the second PMU batch. The drift detection algorithm proposed in this work is as explained in Algorithm 5.1.

In this module, the PMU data stream batch (B_i, W) is first divided into two sub batches $b_1, b_2 \in (B_i, W)$. b_1 consists of $d_1, d_2, \dots, d_{W/2}$ and b_2 consists of $d_{(W/2)+1}, \dots, d_W$ PMU measurements, where

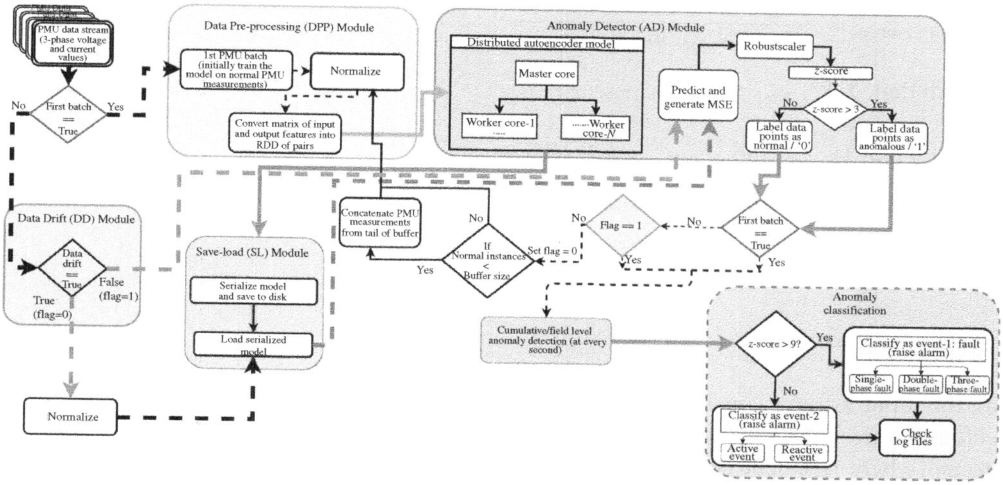

Figure 5.6 PMUNET active leaning framework for ADCL.

Algorithm 5.1 Data drift algorithm

Input: PMU Data Stream batch (B_i, W)
Output: True (if drift detected) else False
Initialization **foreach** $(B_i, W) \in PMU\ Data\ Stream$ **do**
  $b_1, b_2 \leftarrow (B_i, W)/2$

  foreach $f \in F$ **do**
    Evaluate $H_{1:W/2}$ and $H_{(W/2)+1:W}$ for f
    foreach $q \in (H_{1:W/2}, H_{(W/2)+1:W})$ **do**
      Evaluate ${R^q}_{1:W/2}, {R^q}_{(W/2)+1:W}$
    end
    Evaluate $L^f_{b_1,b_2}$ for f
  end
  $L_{max} = \arg\max_{1 \le f \le F}(L^f_{b_1,b_2})$
  $Th = -log_2(\alpha)$
  if $L_{max} > Th$ **then**
    Return True
    # Data Drift Detected
  else
    Return False
    # No Data Drift Detected
  end
end

---

d_i is the PMU measurement data point at timestamp i. The histograms with q bins for b_1, b_2 are $H_{1:W/2}$ and $H_{(W/2)+1:W}$ respectively.

$$ {R^q}_{1:W/2} = \frac{N_{PMU} \text{ in bin } q \text{ of } H_{1:W/2}}{\text{Total } N_{PMU} \text{ in batch } b_1} \tag{5.1} $$

$$R^q{}_{(W/2)+1:W} = \frac{N_{PMU} \text{ in bin } q \text{ of } H_{(W/2)+1:W}}{\text{Total } N_{PMU} \text{ in batch } b_2} \tag{5.2}$$

$$\text{such that,} \sum_q R^q{}_{1:W/2} = \sum_q R^q{}_{(W/2)+1:W} = 1 \tag{5.3}$$

$$\text{and } \forall \text{ q in } H_{1:W/2} \text{ and } H_{(W/2)+1:W} \neq \text{NULL} \tag{5.4}$$

Hence, the log-likelihood ratio (LLR) can be defined as:

$$L_{b_1,b_2} = \sum_{q=1}^{q} R^q{}_{(W/2)+1:W} \log_2\left(\frac{R^q{}_{(W/2)+1:W}}{R^q{}_{1:W/2}}\right) \tag{5.5}$$

$$L_{max} = \underset{1 \leq f \leq F}{\arg\max}(L^f_{b_1,b_2}) \tag{5.6}$$

For all the input features considered in PMUNET, LLR is calculated for both sub-batch. f is the individual feature ($f \in F$), where F represents the set of all features. If L_{max} exceeds th, a data drift is detected; where Th can be defined as:

$$Th = -text \log_2(\alpha) \tag{5.7}$$

In most of the literature $\alpha = 0.05$ [19] is used, here alpha (significance level) is the region of rejection, which is the probability of false detection.

5.3.2.3 Save-Load (SL) Module

Save-load (SL) module has two tasks to perform. The first task is to save the model whenever the deep autoencoder algorithm learns new weights into memory. Secondly, it is used to load the model whenever the prediction is requested over PMU data points.

5.3.3 Anomaly Detector (AD) Module

Anomaly detector (AD) module mainly consists of four main components:

- **Distributed autoencoder model (DAM)**: The input/output features obtained from RDD are fed to DAM and used to train the autoencoder. The autoencoder has two parts: "Encoder" and "Decoder." At every timestamp "t" the encoder encodes the input vector feature x^t into a low dimension interpretation, using an function (Eq. (5.8)).

  $$f(x) = \sigma(M_w * x^t + b) \tag{5.8}$$

  The decoder then decodes the low dimension input vector as $\hat{x}$, by applying the function (Eq. (5.9)).

  $$g(z) = \sigma(M_w * z + b) \tag{5.9}$$

  Where b is bias vector, M_w is the weight matrix, $\sigma(g)$ is the sigmoid activation function. The deep autoencoder aims to minimize the reconstruction error between x and $\hat{x}$ using a loss function as follows:

  $$\text{MSE} = \frac{1}{N}\sum_{i=1}^{n}(x_i - \hat{x}_i)^2 \tag{5.10}$$

- **Predict and generate MSE module (PGM)**: This component is used to predict the reconstruction output of the autoencoder and generate the error vector and attach it with respective batches.

- **Robustscaler**: This component is used to scale the error vector obtained from PGM. In scenarios where outliers are present in the dataset and scaling of data to mean "0" and standard deviation of "1," the scaled data gets skewed toward the outliers. In such scenarios scaling using median and interquartile ranges always provide good results (i.e. it eliminates the outlier effect during scaling).
- **z-score**: The z-score value indicates whether the PMU data has an anomaly or not. It uses the three-sigma rule [20] for labeling a PMU data point as an anomaly. z-score is calculated as shown in Eq. (5.11).

$$z\text{-score} = \frac{\text{MSE} - \mu}{\sigma} \tag{5.11}$$

where μ represents the mean and σ represents the standard deviation of the window, and N is the size of the window. The threshold of 3 is taken for z-score, if for any data point the z-score is greater than 3 is marked as Anomaly.

5.3.3.1 Anomaly Classification

Once the anomaly is detected based on the z-score, the anomaly classification process is done. The anomaly is broadly classified into two categories:

- **Event 1**: All the anomalies caused by faults, i.e. "line to ground, double line to ground, and 3-phase fault," fall under the "Event-1" category. The classification is done based on its z-score. If the z-score value is greater than 9 then the anomaly is marked as "Event-1." Then for further sub-classification, the z-score of each phase is computed, and based on the value if one or two or all phases exceed the threshold, we can find the type of fault.
- **Event 2**: All the anomalies caused by events, such as capacitor bank operation, DER switching, load switching, load or generation decrease and increase, fall under the "Event-2" category. If the z-score value is between 3 and 9, then the anomaly is marked as "Event-2." Then for further sub-classification, the z-score of voltage phasor and current phasor is computed separately. Based on the value, it is classified as active events or reactive events. if the z-score values of voltage measurements exceed 3 standard deviations, then the event is labeled as a reactive event on the other hand if the z-score value of the current phasor exceeds the threshold, then the event is labeled as an active event.

The threshold values are decided based on experiments (discussed in the result section). As a next step, the event can be further validated and identified using network data and log files.

5.3.3.2 Anomaly Localization

Based on the physics rule, the maximum deviation will be observed by the measurement of PMU, which is closer to the anomaly location. In this work, localization is done using the z-score value obtained from the framework. The PMU closed to the place of anomaly will have the largest, and the respective PMU will be labeled as anomalous (1) others as normal (0). Hence during localization, only that particular PMU is selected whose measurement is labeled anomalous "1" by PMUNET.

5.3.4 Distributed Deep Autoencoder Learning

In this work, the configuration parameter of SGD is optimally selected using distributed training principle by parallelizing the learning of autoencoder for enabling fast anomaly detection. To show the effectiveness of using distributed training in selecting the parameter, we have shown the performance of two configurations described as follows:

- **Baseline configuration**: In baseline configuration, out of the shelf (TensorFlow) SGD was trained over 15 epochs on a batch size of 1 using "mean squared error" as its loss function.
- **Optimal configuration**: In the optimal configuration, a tweaked version of SGD (momentum of 0.8 and initialized learning rate of 0.1 decaying after every epoch) was trained over 15 epochs with a batch size of 128 using "mean squared error" as its loss function.

5.4 Synchrophasor Data Anomaly and Event Detection, Localization, and Classification (SyncAED)

Synchrophasor data anomaly and event detection, localization and classification (SyncAED) method performs two main tasks. In the first task, SyncAED performs DPP by detecting and flagging or mitigating the anomalies present in the synchrophasor measurements. The second task uses this processed data for event detection, localization, and classification. In this work, we used statistical and clustering methods as base detectors and unsupervised ML algorithm to develop an ensemble technique for anomaly detection [21, 22]. Further, we use prony analysis [23] to differentiate the anomalies as bad data and event data to enhance the developed technique. The second step in the proposed framework is for event detection, classification, and localization. We have used a data-driven cluster change method for event detection, physics-based rules for event classification, and a graph-based approach for event localization [24].

5.4.1 Synchrophasor Data Anomaly Detection (SyncAD)

PMU data can have multiple types of anomalies, such as outlier or missing data. In this work, we have developed an algorithm to flag the anomalies (bad data) as shown in Figure 5.7. This algorithm uses multiple statistical and clustering-based base detectors namely, linear regression

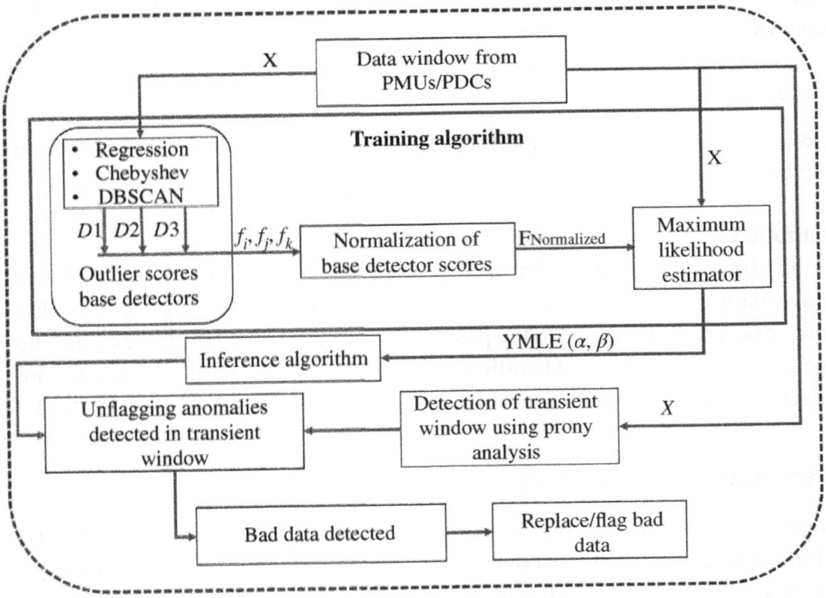

Figure 5.7 Data flow architecture for anomaly detection.

[25], Chebyshev [26], and DBSCAN [27]. This algorithm can add more base detectors depending on the user's choice. All the base detectors work independently on the data and flag the outliers; after that ensemble-based learning is used to combine the output of all the base detectors to make the final decision on outliers in PMU data. Since an event can also create dynamics in the PMU data stream, which can be labeled as bad data by the ensemble algorithm, therefore we have used prony analysis in our work. The prony analysis will differentiate the outliers caused by transients in the power system and unflag that data points and use this preprocessed data for the event detection method.

5.4.1.1 Base Detectors

1. **Linear regression detector**: A regression-based detector fits a linear model on PMU data in the window by minimizing the sum of squared residuals. An upper and lower threshold is defined based on the distance of data points from the regression line. Regression line is represented as follows:

$$\begin{aligned} &\text{Regression} = ax + b \\ &\text{Th}_H = ax + b + k * \text{RMS} \\ &\text{Th}_L = ax + b - k * \text{RMS} \end{aligned} \tag{5.12}$$

$$\text{RMS} = \sqrt{\frac{1}{N-1}\sum_{i=1}^{N}[x(i) - \text{Regression}(i)]} \tag{5.13}$$

   where a is the slope, b is the intercept of the regression line (the value of regression when $x = 0$), and x is the closest point on the regression line from the actual data point. Data points lying outside the low thresholds (TH_L) or high thresholds (TH_H) are labeled anomalous data. RMS is the root mean square of the difference between the regression line and data point.
2. **Chebyshev base detector**: The Chebyshev-based method is used to detect anomalies in the data when probability distribution is unknown. According to Chebyshev, inequality for any probability distribution of no more than $1/k^2$ of total samples can be k standard deviations away from the mean. It is a 2 two-step process in which a strict threshold of k_1 is chosen for the first step, and then a relaxed threshold k_2 is chosen for second step to filter out the outliers. Chebyshev method is based on the equation below:

$$\Pr(|X - \mu| \le k\sigma) \le (1 - 1/k^2) \tag{5.14}$$

3. **DBSCAN base detector**: The third base detector used for anomaly detection is DBSCAN-based detector. DBSCAN has two hyperparameters ϵ and *minPts*. DBSCAN clusters data points lying within ϵ distance within one cluster and considers other points as anomalies.

5.4.1.2 Ensemble Method

The ensemble method uses three base detectors: linear regression base detector, Chebyshev base detector, and DBSCAN base detector. The ensemble-based approach then combines the results using unsupervised ML to compute an anomaly score for PMU data points.

5.4.1.3 Prony-Based Transient Window Estimation

Prony analysis is a well-studied tool in the power community to discover the transient window [23]. Prony analysis has been utilized to capture the dynamic window in the PMU measurement data stream in this work. For the Prony analysis, we have considered the voltage magnitude since it has smaller oscillations due to PMU measurement uncertainty compared to other types of PMUs' measurements for disturbance detection. Consider the vector $\tilde{V}[j] = [V_1[j], \ldots, V_r[j]]^T$ where $V_i[j]$ shows the voltage magnitude measured by ith PMU at sample j. The Hankel matrix (H) is constructed as follows for a data window with length k. A 2.5-second window (this size window selection provides a trade-off between speed and accuracy) of voltage measurement data is used. The total number of samples for a 2.5-second window having a PMU reporting rate of 120 frames per second is $J = 300$.

$$H = \begin{bmatrix} \tilde{V}[0] & \tilde{V}[1] & \ldots & \tilde{V}[\frac{J}{2}-1] \\ \tilde{V}[1] & \tilde{V}[2] & \ldots & \tilde{V}[\frac{J}{2}] \\ . & . & . & . \\ . & . & . & . \\ \tilde{V}[\frac{J}{2}-1] & \tilde{V}[\frac{J}{2}] & \cdots & \tilde{V}[J-1] \end{bmatrix} \quad (5.15)$$

where $\tilde{V}[i]$ demonstrates the ith sample of the window. The rank of the Hankel matrix can be utilized to detect transient in the window and distinguish it with measurement noises and bad data. It can be approximated using eigenvalue decomposition. In steady-state, the Hankel matrix of the window is of rank one. It has one dominant singular value, and the rest of the singular values relate to the measurement noises, and bad data are far smaller. If the ratio between the dominant and the second singular values exceeds a pre-set limit, the Prony analysis will recognize that the window carries a dynamic event. The pre-set limit is set as −4.27 by tuning it using the known events obtained from the real-time digital simulator (RTDS) simulation and industry data for transmission PMUs.

5.4.2 Event Detection, Classification, and Localization

The event detection, classification, and localization architecture are shown in Figure 5.8. The change point detection in voltage, current, real power, reactive power, and frequency measurements are used to detect an event. Once the event is detected and classified using the physic rule

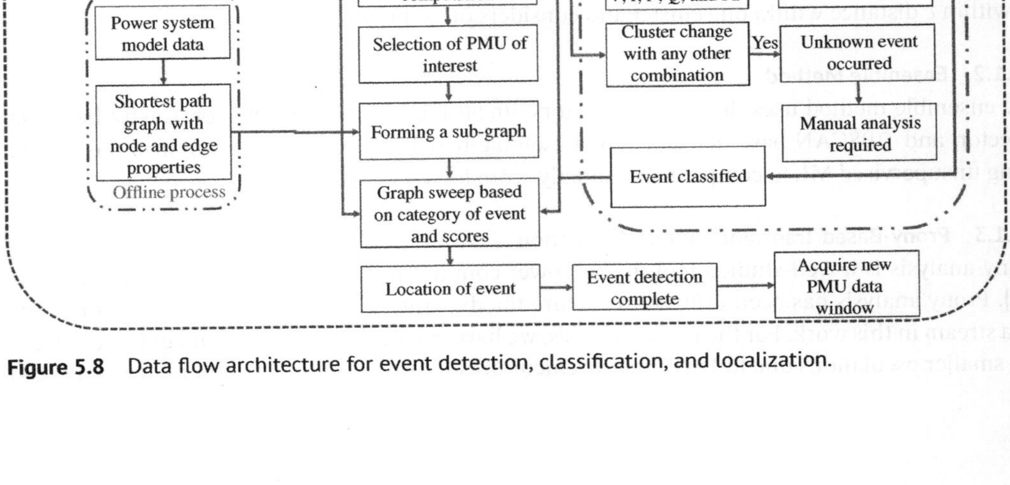

Figure 5.8 Data flow architecture for event detection, classification, and localization.

or decision tree, the top two PMUs are chosen based on statistical and normalized scores. Finally, using the top PMUs, a subgraph is formed, and a graph sweep algorithm is used to detect the event's location.

PMU data is first checked for bad data or missing data using the ensemble method. Once the anomalies are detected, the mitigation is done using the average of the following and preceding data points. After the data is cleaned using the synchrophasor data anomaly detection (SyncAD), the voltage phasor and current phasor measurements from PMU are used to compute active and reactive power flows as follows:

$$P_{ij} = 3 * |V_{LL}^{i}| * |I_{ij}| * \cos(V_{angle} - I_{angle}) \tag{5.16}$$

$$Q_{ij} = 3 * |V_{LL}^{i}| * |I_{ij}| * sin(V_{angle} - I_{angle}) \tag{5.17}$$

Where P_{ij} and Q_{ij} are three-phase active and reactive power flow (PU) between node i and j, I_{ij} is the line i and j current (PU), V_{LL}^{i} is the line to line voltage in (PU) and V_{angle} and I_{angle} are voltage and current angles. After computing, P_{ij} and Q_{ij}, measurements voltage, current, real power, reactive power, and frequency are fed to the DBSCAN algorithm for further processing.

5.4.2.1 Event Detection

The PMU measurements from all PMUs (V, I, P, Q, F) are fed to the DBSCAN algorithm for event detection. Once an event occurs, the operating points change, and it will create multiple clusters in one or more measurements. Our interest is to detect the boundary point instance of the first cluster, i.e. cluster change point. If we observe a cluster change in any of the measurements, the window is considered an event window.

5.4.2.2 Event Classification

After an event is detected, a physics-based rule or decision tree for event classification is used as explained below:

1. **Active power event**: If cluster changes are observed in I and P, and no changes in V and Q, it is classified as an active power event.
2. **Reactive power event**: If cluster changes are observed in V and Q, and no changes in I, P, and F, it is classified as a reactive power event.
3. **Fault event**: If cluster changes are observed in all the measurements V, I, P, Q, and F, it is a fault event.

This active, reactive, and fault event classification is done as shown in Figure 5.8. If it doesn't follow any of the rules mentioned above, it is marked as an unknown event, and a manual analysis is required to determine the root cause of that event.

5.4.2.3 Event Localization

After event classification, we perform event localization based on two steps. The first step is to score the PMUs based on the event window. Once the topmost impacted PMU, we localize the event using the graph method. We used five statistical parameters for scoring the PMUs for the event window. Table 5.1 shows the formulation of all the statistical parameters used to score the PMUs for the event window.

In Table 5.1, "DP" is the data point in the PMU data stream, i defines the ith data point in the window of length "WL." A window length of 30 data points is considered in this work (15 data

Table 5.1 Statistical parameter formulation for scoring PMUs.

S.No.	Statistical parameters	Formulation
1	Shannon entropy (SE)	$SE= -\sum_{i=1}^{WL} DP_i^2 \log(DP_i^2)$
2	Standard deviation (SD)	$SD= \sqrt{\frac{1}{WL}\sum_{i=1}^{WL}(DP_i-\mu)^2}$
3	Range	$Range= \mid \max(DP_i) - \min(DP_i) \mid$
4	Mean difference (MD)	$MD= \mid \mu_{Cluster1} - \mu_{Cluster2} \mid$
5	Crest Factor (CF)	$CF= \frac{\mid \max(DP_i) \mid}{\frac{1}{WL}\sum_{i=1}^{WL}(DP_i)^2}$

points before the event and 15 points after the event). Then finally, all these statistical parameters are combined to compute a final Normalized score for each PMUs as per Eq. (5.18).

$$NS = (SE + CF) * (SD + Range + MD) \tag{5.18}$$

Algorithm 5.2 Sub graph and graph sweep

Input: *Shortest Path Graph* : S_G; *Event Type*
Output: Sub Graph, Event Location

```
Compute Normalized Score [eq. 5.12] for each PMUs
Form Sub Graph
if Number of PMU > 1 then
    Select top 2 PMUs based on NS
    Radius = Shortest Distance(PMU^Bus1, PMU^Bus2)
    Sub Graph = Buses from PMU^Bus1, PMU^Bus2 within Radius
else
    Number of PMU = 1
    Sub Graph = Adjacent Buses from PMU^Bus1
end if
Graph Sweep
Total Bus = Buses(Sub Graph)
for each bus n ∈ Total Bus do
    if Event Type = Reactive then
        Location = Reactive Power Capability Bus
    else if Event Type = Active then
        Location = Active Power Capability Bus
    else if Event Type = Fault then
        Location = PMU^Bus with highest NS
    end if
    Event Location = Location
end for
return Sub Graph, Event Location
```

The next step is to select the top PMUs based on the scores. Certain events are detected by several PMUs as their system-level impact is higher. A fault event would be seen by multiple PMUs, and a transformer tap change or a load change would be sensed locally by PMUs at those buses, at adjacent buses, or PMUs placed close to the event bus in terms of electrical distance. We use

Dijkstra's algorithm [28] to find the shortest path and store it offline for a particular system. The process of the selection of a subgraph is explained in Algorithm 5.2. It has to be noted here that a subgraph formation can be an offline process. Based on the PMUs of interest, the subgraph can be chosen for graph sweep analysis and updated only when the network changes.

5.5 Test-Bed and Test Cases

5.5.1 Cyber-Power Test-Bed Architecture

Figure 5.9 shows the multi-layered testbed architecture used in this work to generate the data to test the algorithms. The testbed setup consists of four layers. The first layer is the power system layer, where the power system modeling is done on the RTDS. The second layer is the sensor and actuator layer that includes all the measurement devices and aggregators (PMU and PDCs). The third layer consists of the communication layer, where the communication network is modeled using NS3. And the last layer is the application layer, where the proposed algorithms are running to provide cleaned data to other synchrophasor applications so that the operator can have a better situational awareness or take control actions to operate the power system in a resilient and reliable manner.

5.5.1.1 Test Case

The IEEE-14 bus system was modeled in RTDS, and various events were simulated. Figure 5.10 shows the one-line diagram of IEEE 14 bus system with 5 PMUs installed on Buses 2, 3, 4, 5, and 9 to monitor voltage phasors and the current phasors of lines 2-3, 3-4, 4-2, 5-1, and 9-10, respectively. Different events such as switching cap bank, DER, generators or load, changing generation or loading, faults, etc. The periodicity of the PMU data stream is set to 60 samples per seconds for all the PMUs.

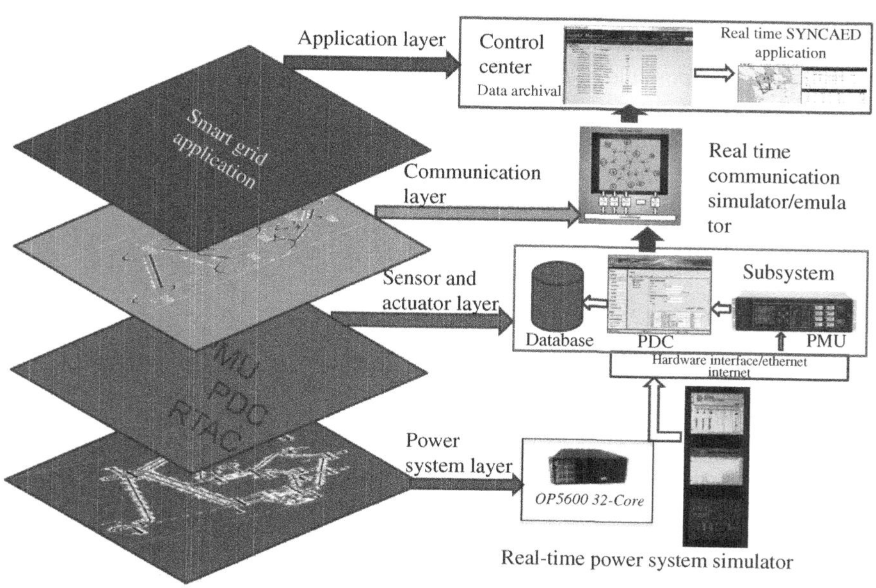

Figure 5.9 Cyber-power test-bed architecture. Source: Schweitzer Engineering Laboratories.

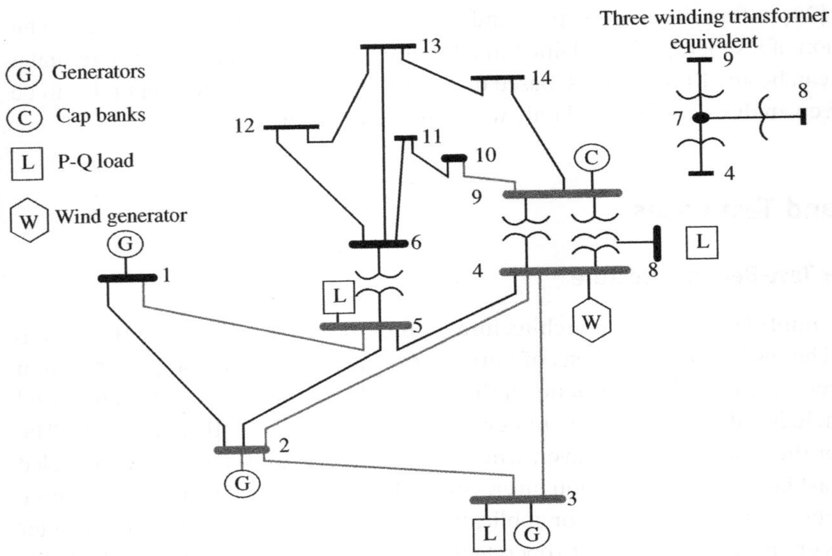

Figure 5.10 IEEE-14 bus system test cases using RTDS.

5.6 Results and Discussion

5.6.1 Simulation Results for PMUNET

5.6.1.1 Performance Evaluation Metrics

To evaluate the PMUNET framework, two types of metrics: accuracy measures and learning time, has been used in this chapter, which can be defined as follows:

- **Accuracy measures**: This Include Accuracy, False Positive Rate (FPR), and False Negative Rate (FNR). These measures are intersection between the actual values and the predicted values, which are evaluated using four possible below situations:
  True positives (TP): Positive instances correctly classified.
  False positives (FP): Negative instances are classified as positive.
  True negatives (TN): Negative instances correctly classified as non-positive.
  False negatives (FN): Positive instances are classified as negative.
  The formula for accuracy can be defined as:

$$\text{Accuracy} = \frac{\text{TP} + \text{TN}}{\text{TP} + \text{FP} + \text{TN} + \text{FN}} \tag{5.19}$$

  The formula for false-negative rate can be defined as:

$$\text{FNR} = \frac{\text{FN}}{\text{FN} + \text{TP}} \tag{5.20}$$

  The formula for false-positive rate can be defined as:

$$\text{FPR} = \frac{\text{FP}}{\text{FP} + \text{TN}} \tag{5.21}$$

- **Learning time**: It is defined as the time (seconds) taken for training the deep learning algorithm. The aim is to train the deep learning algorithm as fast as possible (with the given hardware constraints). The proposed approach was tested on an intel core i7-7700HQ quad-core processor.

Table 5.2 PMUNET simulation result with distributed deep autoencoder learning.

SGD configuration	Number of threads	FNR	FPR	Accuracy	Learning time (S)
Baseline configuration	2	2.05	1.53	97.94	24.63
	4	1.68	2.71	97.94	20.58
	6	2.65	1.78	97.95	14.58
	8	2.63	1.72	97.94	13.26
Optimal configuration	2	1.58	2.71	97.96	2.7
	4	2.53	1.55	97.94	2.5
	6	1.65	2.72	97.92	2.4
	8	1.63	2.72	97.57	2.1

5.6.1.2 Experimental Analysis

In this subsection, experiments are performed in data drift settings in which data statistics are varying over time, making the PMU data stream evolve due to the change in dynamics of the simulation testbed. Table 5.2 demonstrates the effectiveness of using distributed training principle by parallelizing the learning of autoencoder for optimally selecting the configuration parameters of SGD.

The results show that there is improvement in the performance of PMUNET with the optimal configuration of SGD has been improved in terms of accuracy and learning rate when compared with the base configuration. The PMUNET learning time decreased to 2.1 seconds from 13.26 seconds by selecting the optimal configuration of SGD when trained on eight CPU threads.

PMUNET trained on 8 CPU threads using optimal configuration achieves the best performance for SGD. This configuration is further validated with different scenarios simulated using a testbed on the IEEE-14 bus system. The PMUNET once detects the anomaly based on the z-score threshold. It will further classify and localize the anomaly. In the classification part, the anomaly is classified into two types: "Event-1" ("line to ground, double line to ground and 3-phase fault") or "Event-2" ("capacitor switching, DER switching, load or generator switching and decrease or increase"). Tables 5.3 and 5.4 show the simulation results of PMUNET for anomaly detection & localization and anomaly classification, respectively.

As shown in Table 5.3 on IEEE 14 bus system, the PMUNET was able to detect anomaly at Bus 2 (z-score of 12.54) and Bus 3 (z-score of 12.65) with an accuracy of 97.82%, FPR of 1.66%, and FNR of 2.35% corresponding to the actual simulated scenario. And as shown in Table 5.4, the same anomaly was classified as Event-1, labeled as fault event. Similarly, other events were also localized and classified accurately as the simulated scenario.

Table 5.5 showed the result for the test system by the PMUNET when it detected data drift in the PMU data stream. In this work, the threshold has been set to detect data drift over streaming PMU data. As observed from Table 5.5, in IEEE 14 bus system, data drift was seen in "cap bank out" with a L_{max} value of 0.69, "load switched" with a L_{max} value of 0.66, "generation trip" with a L_{max} value of 0.65. It was observed that when multiple events are occurring in the system such as "Cap bank Out and Load Switched In," the system dynamics change on a significant scale as L_{max} value of 0.94 was observed.

Table 5.3 Simulation result for PMUNET for event detection and localization.

Scenario number	Actual event	Actual location	Accuracy	FRP	FNR	Detected location PMU at bus (*z*-score value)
1	Cap bank closed	9	97.92	1.58	2.35	Bus 9 (3.76)
2	3-Phase fault	Line 2-3	97.11	1.87	2.89	Bus 2 (12.54) and Bus 3 (12.65)
3	Load increased	3	97.35	2.22	2.67	Bus 3 (3.22)
4	Generation increased	2	97.36	1.98	1.52	Bus 2 (4.76)
5	Load removed	6	97.58	2.39	1.91	Bus 5 (4.22)
6	Generation removed	3	97.63	2.39	1.91	Bus 3 (4.16)
7	Load decreased	3	97.57	1.59	2.78	Bus 3 (3.89)
8	1-Phase to ground fault	Line 4-3	97.29	1.86	2.73	Bus 4 (12.14) and Bus 3 (12.35)
9	Generation decreased	2	97.84	2.54	1.37	Bus 2 (5.82)
10	Load decreased	4	96.97	3.13	1.81	Bus 4 (3.35)
11	Load switched ON	6	97.45	1.79	2.96	Bus 5 (3.45)
12	Cap bank removed/ load removed	Bus 9/ Bus 6	97.66/ 97.36	2.857/ 2.64	1.58/ 1.94	Bus 9 (3.68) /Bus 5 (3.65)
13	3-Phase fault	Line 2-3	97.51	1.53	2.87	Bus 2 (12.84) and Bus 3 (12.95)
14	Load decreased	9	97.89	1.89	1.95	Bus 9 (3.65)
15	DER trip	Bus 4	97.51	1.53	2.87	Bus 4 (5.86)

5.6.2 Simulation Results for SyncAED

5.6.2.1 Anomaly Detection

In this section, we demonstrate the performance of the anomaly detection technique discussed in this chapter. The simulated IEEE-14 bus system data is used for anomaly detection by manually inserting anomalies into the clean simulated data. The results are shown with two evaluation metrics precision and recall. The recall is defined as the percentage of correctly identified bad data to actual inserted bad data. Precision is expressed as the percentage of inserted bad data that were correctly identified over the total number of times data were identified as bad data by the algorithm. Table 5.6 shows the performance of SynAED for anomaly detection. Each row on this table represents the 5 PMUs installed in the IEEE-14 bus system. The results show that the algorithm can detect the inserted bad data given by recall is above 98% in most cases and precisely detect it, i.e. without false detection, is around 95%.

5.6.2.2 Event Detection and Classification Using Clustering and Decision Tree

The simulation results for the IEEE-14 bus system are shown in Table 5.7 and Table 5.8 for event detection and localization, and characterization of events using clustering and decision tree,

Table 5.4 Simulation result for PMUNET for event classification.

Scenario number	Actual event	Actual label	Classified label
1	Cap bank closed	Reactive event	Event-1 (Reactive event)
2	3-Phase fault	Fault	Event-1 (Fault)
3	Load increased	Active Event	Event-2 (Active event)
4	Generation increased	Reactive event	Event-2 (Reactive event)
5	Load removed	Active event	Event-2 (Active event)
6	Generation removed	Active event	Event-2 (Active event)
7	Load decreased	Active event	Event-2 (Active event)
8	1-Phase to ground fault	Fault	Event-1 (Fault)
9	Generation decreased	Active event	Event-2 (Active event)
10	Load decreased	Active event	Event-2 (Active event)
11	Load switched ON	Active event	Event-2 (Active event)
12	Cap bank removed/ Load removed	Reactive event/ Active event	Event-2 (Reactive event)/ Event-2 (Active event)
13	3-Phase fault	Fault	Event-1 (Fault)
14	Load decreased	Active event	Event-2 (Active event)
15	DER trip	Active event	Event-2 (Active event)

Table 5.5 Data drift scenarios for IEEE 14 bus system.

Scenarios	L_{max}
Cap bank out	0.69
Load switched out	0.66
Generation trip	0.65
Cap bank out/ Load switched in	0.94

Table 5.6 Performance of anomaly detector.

S.No.	Precision	Recall
1	0.956	0.9842
2	0.9478	0.9842
3	0.9436	0.9836
4	0.9866	0.9864
5	0.9468	0.9842

Table 5.7 Simulation result for SyncAED for event detection and localization.

Scenario number	Time (S)	Actual event	Actual location	Detected location PMU 1 (normalized score)	Detected location PMU 2 (normalized score)
1	56	Cap bank closed	9	9 (0.22)	4(0.15)
2	76	3-Phase fault	Line 2-3	2(35.56)	3(25.32)
3	96	Load increased	3	3(2.11)	—
4	116	Generation increased	2	2(2.73)	3(1.96)
5	136	Load removed	6	5(0.90)	—
6	156	Generation removed	3	3(2.06)	4(1.95)
7	176	Load decreased	3	3(0.22)	—
8	196	1-Phase to ground fault	Line 4-3	4(26.35)	3(22.98)
9	216	Generation decreased	2	2(0.67)	1(0.47)
10	236	Load decreased	4	4(0.20)	5(0.16)
11	256	Load switched ON	6	5	9
12	276	Cap bank removed/Load removed	Bus 9/Bus 6	Bus 9(0.22)/—	Bus 4(0.17)/—
13	296	3-Phase fault	Line 2-3	2(35.56)	3(25.32)
14	316	Load decreased	9	9(0.17)	—
15	336	DER trip	4	4(0.39)	3(0.27)

Table 5.8 Simulation result for SyncAED for event classification.

Scenario number	Actual event	Actual label	Classified label
1	Cap bank closed	Reactive event	Reactive power event
2	3-Phase fault	Fault	Fault event
3	Load increased	Active event	Active power event
4	Generation increased	Reactive event	Reactive power 0event
5	Load removed	Active event	Active power event
6	Generation removed	Active event	Active power event
7	Load decreased	Active event	Active power event
8	1-Phase to ground fault	Fault	Fault event
9	Generation decreased	Active event	Active power event
10	Load decreased	Active event	Active power event
11	Load switched ON	Active event	Active power event
12	Cap bank removed/ Load removed	Reactive event/ Active event	Reactive power event/—
13	3-Phase fault	Fault	Fault event
14	Load decreased	Active event	Active power event
15	DER trip	Active event	Active power event

respectively. It is clear from Tables 5.7 and 5.8 that all the events were detected successfully, except for an active load drop at Bus 6 at 276 seconds. The reason for no detection is that multiple events happened simultaneously, and the cap bank switching was dominant compared to the load event, and also, the bus with cap bank has PMU connected to it. Normalized scores for PMUs that detected a particular event are computed, and the top two PMUs are chosen, and a subgraph is formed. The subgraphs can also be computed and stored offline to make the process fast. The formation of subgraphs can be changed according to the power system model.

Once the subgraph is formed and the type of event is known, a Graph Sweep algorithm is run, which queries each bus for its capability. In Table 5.7, at 56 seconds, PMUs at Bus 9 and 4 detected the events, but the score of Bus 9 was higher, and only Bus 9 has the potential of causing a reactive power event. Therefore, it can be certainly determined that the cap bank at Bus 9 was operated.

5.7 Summary

This chapter provides a comprehensive assessment and effects of possible data anomalies and events data observed in PMUs. In this work, we classified anomalies into two parts, namely bad data and events. We presented two different methods for anomaly and event detection, classification, and localization. The first method used is a device-level unsupervised deep learning-based data-driven approach for event detection, localization, and classification over streaming PMU data, using online learning and multivariate DD detection algorithm. In the second method, we presented Synchrophasor Anomaly and Event Detection (SyncAED), an Ensemble-based anomaly detection technique to detect the bad data and treat event data as good data in the first phase and then in the second phase of SyncAED, detection, classification, and localization of event using PMU data and applying a suite of mathematical and statistical methods comprise of DBSCAN-based cluster change, physics-based rule and decision tree, Shannon entropy, crest factor, and graph theory. The simulation results prove the effectiveness of the algorithm to detect, classify, and locate the anomaly and events. The tool will be most valuable if it can be incorporated with the existing Energy Management System (EMS) tools, and real-time topology changes are updated to develop an updated subgraph to provide the exact root cause of the events and provide better decision support to operators.

Acknowledgments

The authors would like to acknowledge financial support from the National Science Foundation (NSF) award 1840192 and Pacific Northwest National Lab (PNNL) Distinguished Graduate Research fellowship Program (DGRP) for this work. The authors would also like to thank Dr. Amirkhosro Vosughi and Dr. Yinghui Wu for technical support.

References

1 De La Ree, J., Centeno, V., Thorp, J.S., and Phadke, A.G. (2010). Synchronized phasor measurement applications in power systems. *IEEE Transactions on Smart Grid* 1 (1): 20–27. https://doi.org/10.1109/TSG.2010.2044815.

2 Pandey, S., Srivastava, A.K., Ishchenko, D., and Cintuglu, M. (2020). Application-aware adaptive reporting of synchrophasor measurements. *2020 IEEE Power Energy Society General Meeting (PESGM)*, pp. 1–5. https://doi.org/10.1109/PESGM41954.2020.9281524.

3 Kundu, P. and Pradhan, A. (2016). Enhanced protection security using system integrity protection scheme (SIPS). *2016 IEEE Power and Energy Society General Meeting (PESGM)*, p. 1. https://doi.org/10.1109/PESGM.2016.7741137.

4 Evangelos, F. (2020). Let's talk about synchrophasors, PMUs & applications. https://www.naspi.org/sites/default/files/2020-03/20200320_naspi_webinar_farantatos.pdf (accessed 25 April 2021).

5 Fernandes, E.R., Ghiocel, S.G., Chow, J.H. et al. (2017). Application of a phasor-only state estimator to a large power system using real PMU data. *IEEE Transactions on Power Systems* 32 (1): 411–420. https://doi.org/10.1109/TPWRS.2016.2546947.

6 NASPI-LSE (2019). Real time applications using linear state estimation technology, NASPI. https://www.naspi.org/sites/default/files/2019-10/01_FOA1492_RealTimeApplications__Nayak_20191030.pdf (accessed 25 April 2021).

7 NASPI-ID (2015). Using synchrophasor data during system islanding events and blackstart restoration, NASPI control room solutions task team paper. https://www.naspi.org/sites/default/files/reference_documents/10.pdf?fileID=1537 (accessed 25 April 2021).

8 NASPI-DM (2019). Using synchrophasor data to determine disturbance location, NASPI control room solutions task team paper. https://naspi.org/sites/default/files/reference_documents/naspi_crstt_disturbance_detection_final_20190220.pdf (accessed 25 April 2021).

9 Zhang, H., Ning, J.A., Vaiman, M., and Vaiman, M. (2019). Implementing the real-time voltage stability ananlysis tool on a large scale system model for IROL assessment in the Western interconnection. *2019 IEEE Power Energy Society General Meeting (PESGM)*, pp. 1–5. https://doi.org/10.1109/PESGM40551.2019.8974011.

10 Sarmadi, S.A.N., Dobakhshari, A.S., Azizi, S., and Ranjbar, A.M. (2011). A sectionalizing method in power system restoration based on WAMS. *IEEE Transactions on Smart Grid* 2 (1): 190–197. https://doi.org/10.1109/TSG.2011.2105510.

11 Tushar, Pandey, S., Srivastava, A.K. et al. (2018). Online estimation of steady-state load models considering data anomalies. *IEEE Transactions on Industry Applications* 54 (1): 712–721. https://doi.org/10.1109/TIA.2017.2753719.

12 Kundu, P. and Pradhan, A. (2017). Real-time event identification using synchrophasor data from selected buses. *IET Generation, Transmission and Distribution*. https://doi.org/10.1049/iet-gtd.2017.1100.

13 Dahal, O.P., Brahma, S.M., and Cao, H. (2014). Comprehensive clustering of disturbance events recorded by phasor measurement units. *IEEE Transactions on Power Delivery* 29 (3): 1390–1397. https://doi.org/10.1109/TPWRD.2013.2285097.

14 NASPI-PARTF (2017). PMU data quality: a framework for the attributes of PMU data quality and quality impacts to synchrophasor applications 2017. https://www.naspi.org/sites/default/files/reference_documents/PARTF_WhitePaper_20170314_Final_PNNL_26313.pdff (accessed 25 April 2021).

15 Ahmed, A., Sajan, K.S., Srivastava, A., and Wu, Y. (2021). Anomaly detection, localization and classification using drifting synchrophasor data streams. *IEEE Transactions on Smart Grid* 12 (4): 3570–3580. https://doi.org/10.1109/TSG.2021.3054375.

16 Saurav, S., Malhotra, P., Tv, V. et al. (2018). Online anomaly detection with concept drift adaptation using recurrent neural networks. *Proceedings of the ACM India Joint International Conference on Data Science and Management of Data*, pp. 78–87.

17 Bertsekas, D.P. and Tsitsiklis, J.N. (1989). *Parallel and Distributed Computation: Numerical Methods*, vol. 23. Englewood Cliffs, NJ: Prentice Hall.

18 Zaharia, M., Xin, R.S., Wendell, P. et al. (2016). Apache spark: a unified engine for big data processing. *Communications of the ACM* 59 (11): 56–65.

19 Emmert-Streib, F. and Dehmer, M. (2019). Understanding statistical hypothesis testing: the logic of statistical inference. *Machine Learning and Knowledge Extraction* 1 (3): 945–961.

20 El-naggar, M.F. and Shetaya, A.A. (2016). Wide area angle stability assessment tool based on statistics standard score technique. *2016 IEEE 16th International Conference on Environment and Electrical Engineering (EEEIC)*, pp. 1–6. https://doi.org/10.1109/EEEIC.2016.7555623.

21 Banerjee, P., Pandey, S., Zhou, M. et al. (2017). Data mining based anomaly detection in PMU measurements and event detection. *North American Synchrophasor Initiative (NASPI)*, Gaithersburg, MD, pp. 1–24. https://doi.org/05.1109/JSIS.2017.7914501.

22 Zhou, M., Wang, Y., Srivastava, A.K. et al. (2018). Ensemble based algorithm for synchrophasor data anomaly detection. *IEEE Transactions on Smart Grid* 10 (3): 2979–2988. https://doi.org/10.1109/TSG.2018.2816027.

23 Hauer, J.F., Demeure, C.J., and Scharf, L.L. (1990). Initial results in Prony analysis of power system response signals. *IEEE Transactions on Power Systems* 5 (1): 80–89. https://doi.org/10.1109/59.49090.

24 Pandey, S., Srivastava, A.K., and Amidan, B.G. (2020). A real time event detection, classification and localization using synchrophasor data. *IEEE Transactions on Power Systems* 35 (6): 4421–4431. https://doi.org/10.1109/TPWRS.2020.2986019.

25 Sedgwick, P. (2013). Simple linear regression. *BMJ* 346: f2340.

26 Amidan, B.G., Ferryman, T.A., and Cooley, S.K. (2005). Data outlier detection using the Chebyshev theorem. *2005 IEEE Aerospace Conference*, pp. 3814–3819. IEEE.

27 Ester, M., Kriegel, H.-P., Sander, J., and Xu, X. (1996). A density-based algorithm for discovering clusters in large spatial databases with noise. *KDD*, Volume 96, pp. 226–231.

28 Fuhao, Z. and Jiping, L. (2009). An algorithm of shortest path based on Dijkstra for huge data. *2009 6th International Conference on Fuzzy Systems and Knowledge Discovery*, Volume 4, pp. 244–247. https://doi.org/10.1109/FSKD.2009.848.

6

Clustering Methods for the Profiling of Electricity Consumers Owning Energy Storage System

Cátia Silva[1], Pedro Faria[1], Juan M. Corchado[2], and Zita Vale[1]

[1]GECAD – Research Group on Intelligent Engineering and Computing for Advanced Innovation and Development, LASI – Intelligent Systems Associate Laboratory, Polytechnic of Porto, Porto, Portugal
[2]BISITE Research Centre, University of Salamanca, Salamanca, Spain

Acronyms

ASW	Average silhouette width
CLARA	Clustering large applications
DG	Distributed generation
DR	Demand response
DSO	Distribution system operator
ESS	Energy storage systems
NSP	Non-supplied power
OF	Objective function
PAM	Partitioning around medoids
PF	Power flow
WSS	Within-cluster sum of square

6.1 Introduction

In the current paradigm, small consumers have indirect influence regarding the transactions in the energy market. With the smart grids concept, namely, to avoid the usage of non-renewable generation, new players and resources will be added to the energy system [1]. Distributed generation technologies, such as wind and solar, will be the main source in the future. However, their volatile behavior will impact the system's reliability and security [2]. To avoid this problem, the consumers must provide flexibility, introducing a new paradigm where the generation no longer follows demand [3].

The demand response (DR) concept brings the empowerment of the consumers and increases their influence in the operations [4]. DR allows active players to receive and react to signals and change their consumption accordingly. These signals can be price-based or incentive-based and are triggered considering the system balance achievement. For instance, at peak hours, DR event participants receive a signal to reduce their consumption when the generation is low. Those who comply are rewarded, and continuous participation is remunerated [5–7].

Intelligent Data Mining and Analysis in Power and Energy Systems: Models and Applications for Smarter Efficient Power Systems, First Edition. Edited by Zita Vale, Tiago Pinto, Michael Negnevitsky, and Ganesh Kumar Venayagamoorthy.

However, the implementation of these programs in the real market is still very scarce. Efforts are being made to provide all the means necessary to be successful. In Europe, for instance, regarding policy and rule creation, Directive 2019/944 was created [8] and recasts Directive 2012/27/EU on common rules used in the internal electricity market. This market started in 1999 in the European Union and was progressively presented in competitive and cross-border electricity markets. Directive 2003/54/EC and Directive 2009/72/EC have contributed to the creation of the internal electricity market as it is currently. Several communications by the committee were announced after Directive 2012/27/EU, mainly in 2015. One of which, on 24 February, puts citizens at the center of concerns, taking ownership of the energy transition, taking advantage of new technologies to reduce costs by actively participating in the market, with the most vulnerable consumers being protected. Also, in the same line of thought, on 15 July, it was mentioned that retail market should serve consumers better, notably by improving the links between the wholesale and retail markets – allowing all consumers to participate fully in the energy transition to participate in the overall reduction of energy consumption by providing efficient solutions. Also, in the same communication, the need to adapt the rules applied to production due to the transition from power generation in large installations to decentralized installations by changing the roles in the existing market is recalled. This change results in more flexible markets and fully integrates all market players, including renewable energy producers, new energy service providers, energy storage, and flexible demand. The importance of active consumers and their core role is well referenced throughout these Directives, as mentioned previously. Moreover, to achieve the targets with effectiveness, innovation must be encouraged, and the flexibility from the consumers well compensated. After implementing the DR and fully functional energy market, the possibility of adding renewable energy to the actual grid can become a reality taking a new step to decarbonize the system [9].

The lack of knowledge from these new players adds complexity to the system management [10]. Therefore, to supervise the new communities created, a new entity was designed – the aggregators. By becoming responsible for coordinating all the small resources associated with the community, the aggregator will participate in wholesale electricity markets as an intermediary in the transactions with the Independent System Operator [11]. With this approach, the active consumers will take a step closer to the energy market without jeopardizing the system's functioning.

The authors propose a methodology to ease the complex task of managing such uncertain resources in the current chapter. The methodology comprises several steps: scenario definition, optimal scheduling, aggregation, remuneration, and classification. The method may serve as a planning approach with all the aggregate resources, being capable of dealing with distributed generation (DG), prosumers, energy storage systems (ESS), and participants in DR events. But it can also be used in the case of real-time operation, resorting to the last phase, classification. With this, the authors worked to find a solution that contemplates the same phases, but quickly.

The study performed will be crucial for the aggregation phase, where a clustering method is used to find patterns in the consumers' resources profiles. The core idea is to gather similar ones and recompense them according to their actual performance – incentivizing fairness between all players. This way, four different clustering methods were tested and evaluated according to this context. Some of them are sensitive to the number of clusters used as input, so three different methods used to find the optimal one will also be compared.

The chapter is organized into six sections. Section 6.1 serves as an introduction to the topic addressed. Section 6.2 details the methodology proposed by the authors. Section 6.3 presents the several methods to be used in this study. Section 6.4 presents the case study chosen to prove the feasibility of the methodology which the results are further presented in Section 6.5. Finally, Section 6.6 presents the conclusions from the present study.

6.2 Methodology Definition

To better define an effective business model, able to adapt to the future changes in the energy market the authors consider that there is an important question that should be analyzed: since the active consumers are now such an important player in the future market, providing flexibility when generation is not able to suppress all the demand requests, how can the entity manager motivate their participation? Also, predict and reduce the uncertainty of their behaviors and the resources associated? Because it is already known that these volatile resources will impact the system's reliability and security if not well managed.

In this way, the authors proposed a multifaceted methodology for different approaches (both planning and operation) that support the decisions from the aggregator perspective. The previous works developed by the authors were able to gather several steps considered crucial to define the proper methodology to manage an active community with different types of resources. Figure 6.1 specifies the five main phases, further explored after scenario definition, optimal

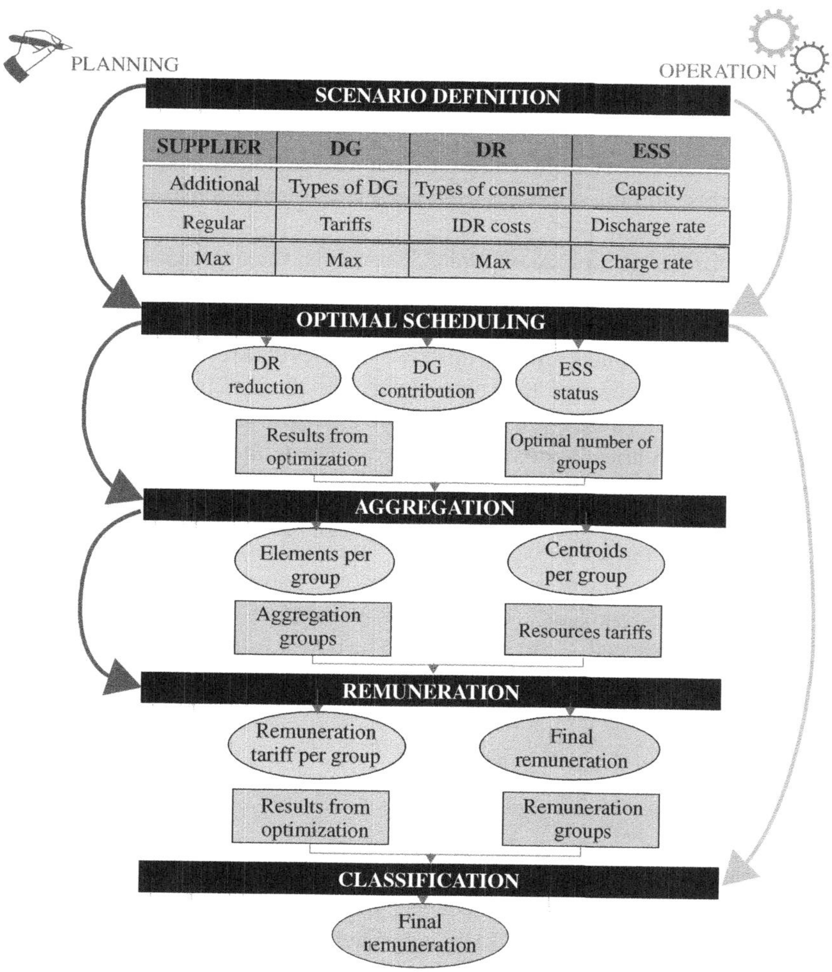

Figure 6.1 Proposed methodology.

scheduling, aggregation, remuneration, and classification. One of the key factors introduced, highly important for the continuous participation in the authors' opinion, is the fair compensation of the participants in DR events wherever requested. Although the DR contract cannot always imply a guaranteed reduction from the demand side, the participation is voluntary. Even with the appliance control, the active consumer can withdraw this permission at any time.

Being a continuity from previous works [12], The authors define the model according to a real scenario. Starting with Distributed System Operator (DSO) performing the load forecast for the community – in a higher or lower time range, in other words, weekly or real-time approach. After, DSO performs a power flow (PF) and the respective analysis, using the PF service and looking for any issues in the considered network. If none is detected, the scheduling is performed normally. On the other hand, if any violation is detected, the DSO must request a load reduction to each community manager, the aggregators. This entity will then trigger a DR event using the proposed methodology. It must be highlighted that aggregator may or may not have the load forecast information, so both operation and planning approaches can be useful.

Firstly, as entity manager from the active community, the aggregator must gather all the essential information to serve as input for the proposed methodology and do a pre-treatment. This solution has already proved to be robust since it can easily be used for a small database and one that reaches thousands of resources [2]. Remember, however, that there may be a hardware limitation when applying in these cases. Furthermore, regarding the time to process the information, the authors advise gathering all the information in the planning case to ease a day-ahead or hour-ahead configuration. So, the input database must collect useful contextual knowledge from the external supplier, the participants in the DR events, and the resources associated with them, such as the DG found in the active community or ESS installed.

Having the database ready for the next phase, Optimal Scheduling, the aggregator collects the current information for the DR participants, the DG contribution, and the ESS status. The active consumers must provide information on the available demand flexibility and the willingness to participate in this context. External suppliers are only introduced if the resources in the active community cannot suppress the demand. However, the optimization objective function must minimize the operation costs, so the priority must be given to the resources within the community, always considering the fair remuneration. A linear approach is employed and defined in Eq. (6.1). Let c be the number of consumers, t be the period, g the DG unit, s be the external supplier, and b be the number of ESS. The power from DG unit g on period t, in kW, is represented by $P_{RES(g,t)}$ with the respective cost $C_{RES(g,t)}$. The Power from DR active consumer c on period t, in kW, is represented by $P_{DR(c,t)}$ with the respective cost $C_{DR(c,t)}$. The non-supplied power (NSP) on period t, in kW, is represented by $P_{NSP(t)}$ with the respective cost $C_{NSP(t)}$. The power from supplier s on period t, in kW, is represented by $P_{Supplier(s,t)}$ with the respective cost $C_{Supplier(s,t)}$. The power from ESS b on period t, in kW, is represented by $P_{ESSd(b,t)}$ with the respective cost $C_{ESSd(b,t)}$.

$$\begin{aligned}\text{Min. OF} = &\sum_{p=1}^{P}\left[P_{RES(g,t)}C_{RES(g,t)}\right] + \sum_{c=1}^{C}\left[P_{DR(c,t)}C_{DR(c,t)}\right] + P_{NSP(t)}C_{NSP(t)} \\ &+\sum_{s=1}^{S}\left[P_{Supplier(s,t)}C_{Supplier(s,t)}\right] + \sum_{b=1}^{B}\left[P_{ESSd(b,t)}C_{ESSd(b,t)}\right] \\ &c,t,g,s,b \in \mathbb{Z} \,:\, c,t,g,s,b > 0\end{aligned} \tag{6.1}$$

The objective function (OF) represented in Eq. (6.1) is subjected to several constraints. To maintain the system reliability and achieve the network power balance – stability between consumption and generation, Eq. (6.2) was defined. The sum of the difference between the initial load, the ESS charge, and the requested reduction should match the value from the total generation within DG units, external suppliers, and ESS discharge.

$$\sum_{c=1}^{C}\left[P^{initial}_{(c,t)} + P_{EVc(v,t)} + P_{ESSc(b,t)} - P_{DR(c,t)}\right] = \sum_{g=1}^{G}[P_{RES(g,t)}]$$
$$+\sum_{s=1}^{S}[P_{Supplier(s,t)}] + \sum_{v=1}^{V}[P_{EVd(v,t)}] + \sum_{b=1}^{B}[P_{ESSd(b,t)}] + P_{NSP(t)}$$
$$c,t,g,s \in \mathbb{Z} : c,t,g,s > 0 \tag{6.2}$$

The OF is restricted by the maximum contribution requested from each active consumer to a DR event ($P^{Max}_{DR\ (c,t)}$), according to the context in which the event is triggered. To define boundaries for the DG units' contribution, the aggregator to control the upper ($P^{Max}_{RES\ (g,t)}$) and lower limits ($P^{Min}_{RES\ (g,t)}$) and the total value of generation provided from each different technology found in the active community ($P^{Total}_{RES\ (t)}$). In the case where the ESS and DG units cannot suppress the demand, other constraints are introduced and represent the external suppliers' constraints, restricting the maximum capacity ($P^{Max}_{Supplier\ (s,t)}$) and the total amount of generation provided from this source ($P^{Total}_{Supplier\ (t)}$).

Regarding ESS, firstly the limits for the operation capacity is added, both max ($E^{stormax}_{(s,t)}$) and ($E^{stormin}_{(s,t)}$). After, the ESS charge ($P^{chmax}_{(s,t)}, X^{essch}_{(s,t)}, X^{essch}_{(s,t)}$) and ESS discharge ($P^{dchmax}_{(s,t)}, X^{essdch}_{(s,t)}, X^{essdch}_{(s,t)}$), respectively. Each of these equations includes one binary variable, which guarantees the impossibility of charging and discharging during the same period t ($X^{essdch}_{(s,t)} + X^{essch}_{(s,t)}$). Finally, to maintain the power balance within the ESS – the previous state of what was charged and discharged.

In the meantime, the optimal number of clusters to be used in the clustering method is also defined to serve as input to the third phase of the methodology: aggregation, along with the optimization results, as shown in Figure 6.1. The mentioned phase may be used with several intermediate goals and for different resources (load consumption or ESS status to find similar profiles, load reduction to understand the behavior in a certain context, and many more), but the main goal is to always achieve the fairness between participating resources – granting that they are compensated according to their contribution. The aggregator will be more confident in other decisions by finding resource groups with similar characteristics.

The remuneration phase will allow to the aggregator to define the proper compensation for each participant and, with that, increase the probability of continuous participation so, this phase is one of the most important ones in the author's opinion. All the work done in the previous phases culminate in this final step, guarantying the active community's continuous participation in the management: the Remuneration phase. The compensation tariff must be well defined to serve as a good incentive and always according to the context and the active consumer characteristics.

A new phase is added for the operation stage concerning the planning approach – the Classification phase. As already mentioned earlier, it will be necessary to find a feasible solution promptly. Thus, it is proposed to use classification methods to assist this task. With the output from the Optimal Scheduling phase, as shown in Figure 6.1, the aggregator can use any classification method to predict, for instance, which remuneration group should assign each resource skipping tasks from other phases.

This proposal presents several contributions to successfully implement DR in real market, aiding the aggregator in the complex task of managing this active communities. In this way, the proposed methodology highlights:

- Active consumers will be the future center of the energy market. Therefore, the new business models should study their behavior, the associated resources (prosumers), and ways to motivate their participation to maintain the system reliable and secure. In addition, the authors believe that unique characteristics should be further studied to develop better solutions.
- Definition of resource groups according to their similarity. For example, for both participation selection and remuneration definition, the individual approach may not be the quickest one in a real-time approach. In this way, assembly resources with similar behaviors in the same group may lead to more precise and fair decisions.
- When aggregating the active consumers, and since some clustering methods may be input sensitive, the introduction of methods to find the optimal number of clusters is discussed so the aggregator can determine the best number of groups to be implemented.
- A contextual approach aids the reduction prediction of trustworthy consumers. Furthermore, it introduced information from different contexts in the model definition, for instance, period, day or even season, and historical. Finally, performances add variability to the methodology, thus increasing its interest in dealing with the new and complex players in the energy market.

The present study's focus will be the aggregation phase detailing all the steps and comparing well-known methods to find the optimal number of clusters and several Clustering algorithms.

Twenty prosumers in an active community can participate in DR events following a Time-of-use (TOU) tariff as incentive signals for DR events by the aggregator. This DR program is the most widely used strategy in academic research and practical projects [13]. The case study presented TOU tariff is divided into three different periods: peak, intermediate, and off-peak, as can be seen in Table 6.1 and according to the Portuguese legislation. The table also presents the parameter W, representing the DR weight according to the periods.

The preferred DG technologies used are the Photovoltaic panels. Still, all of them also have Energy Storage Systems (ESS) installed, and the main appliances used for DR events are the dishwasher, air conditioning, and the water heater. The information gathered regarding the mentioned community resources is on a 15-minute interval, resulting in 96 periods throughout the day. Figure 6.2 was then created to visualize better the expected community resource consumption, production, and DR flexibility from each main type of appliance.

Since the ESS focus on the present study, the idea is to optimize their behavior according to the remaining resources. With the results, such as their status throughout the day, the charge and discharge curve are withdrawn from the Optimal Scheduling step for further analysis and discussion. After only the status will be analyzed in the aggregation phase.

Table 6.1 Active community characterization.

Transactions	Peak	Intermediate	Off-Peak
Sell (m.u./kWh)	0.1659	0.1659	0.1659
Buy (m.u./kWh)	0.3326	0.1681	0.0930
Periods	10 a.m. -p.m. 7 p.m. - 9p.m.	8 a.m. – 10 a.m. 1 p.m. – 7 p.m. 9 p.m. – 10 p.m.	10 p.m. – 8 a.m.
DR weight	0.000	0.2000	0.4000

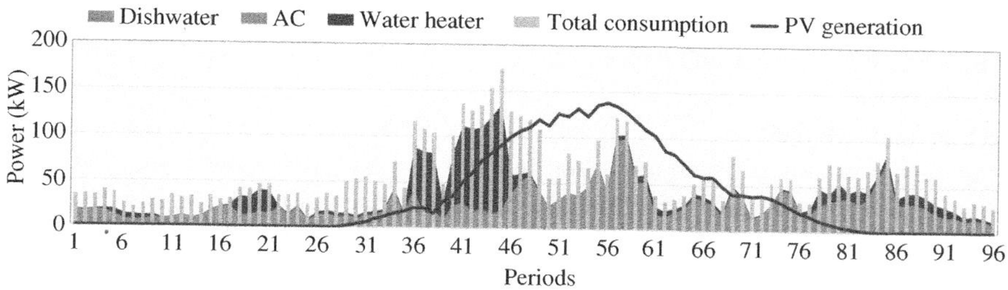

Figure 6.2 Discriminated consumption for the active community: DR flexibility, total consumption, and PV generation.

6.3 Clustering of Consumers with ESS

A comparison between several methods results will be analyzed and discussed in a planning perspective further to improve the aggregation phase from the proposed methodology. The results from using the problem database described previously applied to the proposed methodology will be used as input for this phase, namely the ESS dataset. The focus is on ESS status to find similar profiles between these resources. The dataset to be used is represented in Figure 6.3. which presents the state of charge for each consumer throughout the day. The authors assumed that the minimum should be 1.5 kW to all the ESS in the active community.

The authors use this type of study to answer questions like: which ones should be activated in a certain context, according to previous experiences? Instead of calling all the resources in the community, by knowing their expected behavior, there may be some more trustworthy than others for the DR event.

In this way, the current section will be divided into two main topics: discovering the optimal number of clusters and the group definition according to several clustering methods for the value found.

6.3.1 Optimal Number of Clusters

Three different methods will be compared: average silhouette, elbow, and gap statistic methods. The first two are direct methods, and the last one is statistical testing. Adding the definition of

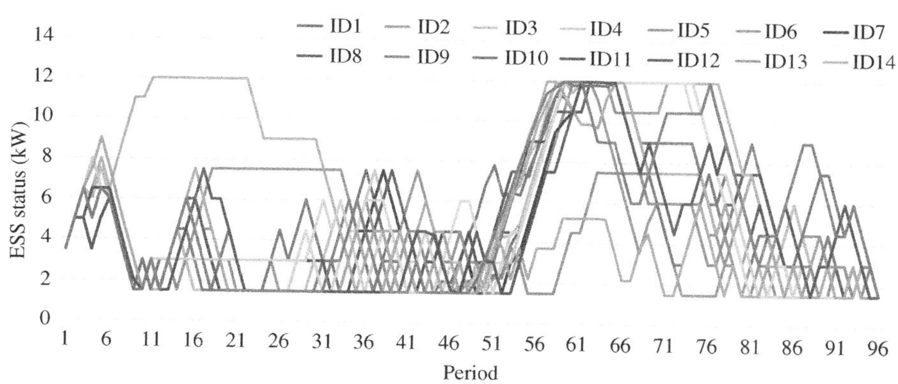

Figure 6.3 Optimal scheduling results: state of charge.

each method, the direct ones consist of optimizing a criterion. The second one, statistical methods, consists of relating evidence against a null hypothesis.

6.3.1.1 Average Silhouette Method

Kaufman and Rousseeuw proposed the first method in 1990 [14], and it is a direct method. The criterion used for this method, the average silhouette, determine the clustering quality. In other words, the k with higher average silhouette width (ASW) is considered the optimal cluster in a certain range, as can be seen in Eq. (6.3).

$$\overline{S}(C,d) = \frac{1}{n}\sum_{i=1}^{n}\frac{b(i)-a(i)}{max\{a(i),b(i)\}} \tag{6.3}$$

Let $a(i)$ be the average distance between the points in the cluster to which it was assigned, and $b(i)$ be the average distance of the nearest cluster to which it was not assigned. Considering that clusters are meant to be homogeneous and well separated, the larger the silhouette score and of ASW, the better the clustering quality is. Unfortunately, t ASW cannot be calculated for $k=1$ because this method cannot directly be used to decide whether the dataset is homogeneous and whether there should be any clustering at all [15]. The results from the selected dataset can be seen in Figure 6.4. All ASW curves were drawn to ease the visualization of the optimal number of clusters found, resorting to this method. Data treatment was done previously, and all the data was standardized.

In the selected range, from $k=2$ to $k=10$, the maximum value of ASW can be found in $k=2$. In this way, according to the silhouette method, dividing the ESS into two different groups is the best approach. It should be highlighted that all the following clustering methods tested obtained the same value for an optimal number of clusters regarding the silhouette method and the remaining.

6.3.1.2 Elbow Method

The following method is one of the oldest ones since it traced to Robert L. Thorndike in 1953 [14] and, like silhouette method, it is a direct method. So, starting the number of clusters at 2 and incrementing this value with step 1, the total within-cluster sum of square (WSS) is calculated for each of them according to Eq. (6.4) [16].

$$WSS = \sum_{i=1}^{k}\sum_{xj\varepsilon Si} x_j - \mu_i^{\,2} \tag{6.4}$$

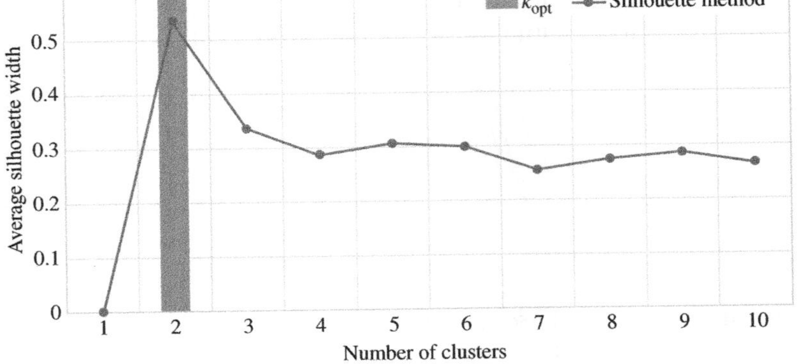

Figure 6.4 Average silhouette method results for the dataset selected – ESS status.

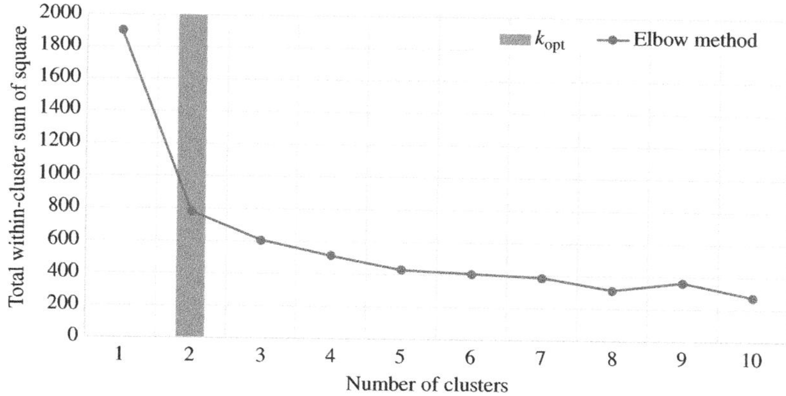

Figure 6.5 Elbow method results for the dataset selected – ESS status.

Let x be the observations, S be the clusters, and μi be the mean of observations in cluster Si. Since the sum of squares is the squared Euclidean distance, the minimum WSS will be generated by choosing the nearest mean. In this way, Figure 6.5 represents the results found. For better visualization, these results are plotted by the analyst. The value where the cost drastically reduces and then hits the plateau is where the optimal number of clusters was found.

In other words, by plotting the results of the study set, it will be possible to find an "elbow" considered an indicator of the appropriate number of clusters – but it can lead to misreading. In this case, for the same range as in the average silhouette method, the optimal number of clusters is $k = 2$.

6.3.1.3 Gap Statistic Method

Finally, the Statistical Testing Method chosen was Gap statistic published by R. Tibshirani, G. Walther, and T. Hastie in 2001 [14]. The algorithm steps in this method intend to compare the total intra-cluster variation for a range of k clusters with their expected values under null reference distribution of the data – in other words, with no apparent grouping. So, to find the optimal number of clusters, the gap statistic value must be maximum. Figure 6.6 shows the results from this method applied to the selected dataset in the case study. The results show a different optimal number of clusters. In the following sub-section, the values will be compared.

6.3.2 Clustering Methods

Many authors use clustering methods within the current literature on consumer profiling modeling [17–19]. Being an unsupervised learning method, from the machine learning perspective, this technique of dividing data into different object groups with similar characteristics – the so-called clusters – can find hidden patterns in the same information. In other words, this algorithm can be used to explore existing relations in the patterns set and organize the objects into homogeneous groups. So, unlike the supervised learning methods, such as Classification, no data labeling is provided for a-priori differentiation.

In this way, to measure the density of connection between the objects within each cluster, intra-connectivity must be used so that the higher the value, the more certainty the analyst can have. On the other hand, the concept of inter-connectivity measures the degree of connectivity between different clusters. Therefore, this value will be important to be low, meaning that each

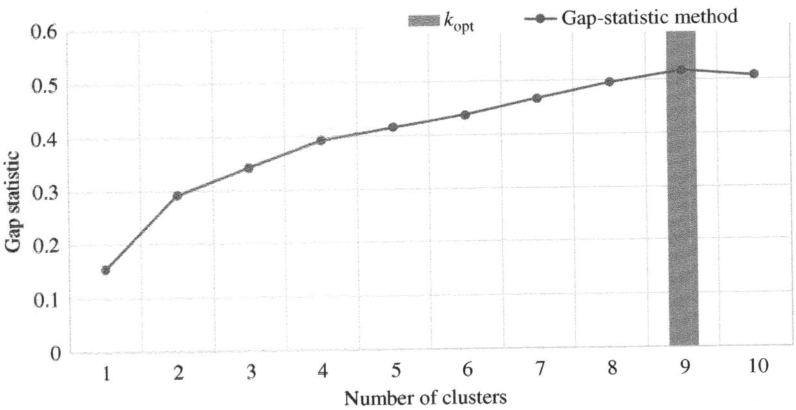

Figure 6.6 Gap statistic method results for the dataset selected – ESS status.

cluster is individually disparate from the rest. To further understand the several types of clustering methods, a comparison is made in this section and a detailed explanation of each one. For this study were selected four of the most-well know methods.

6.3.2.1 Partitional Clustering

Firstly, the following methods distribute the data into non-overlapping subsets, guaranteeing that all data belongs to a cluster. As mentioned, the number of *k* clusters to be generated must first be defined by the analyst – as done before. Partitioning clustering includes several known methods, such as *k*-means or *k*-medoids clustering.

k-means is the most common and used unsupervised machine learning algorithm for partitioning according to [20]. Between many adaptations done throughout the years, the one defined by Hartigan-Wong in 1979 [21] treats the total variation within a cluster as the sum of the squares of Euclidean distance between a point and the center of the cluster, assigning the point to the nearest *k* cluster. Iteratively, each cluster is represented by a new centroid, which corresponds to the average of the points assigned to the *k* cluster in question. However, the main problems from this method are the sensitivity to noise and outliers, so the previous data treatment is crucial to the successful implementation [22].

Resorting to the ESS dataset, regarding the status throughout the day from each ESS in the active community, Figure 6.7 represents the comparison between the optimal numbers found with the three methods used in the previous sub-section.

In Figure 6.7a, the centroid values to all the periods are represented and can be seen for Group 0 and Group 1. The first one gathered 14 ESS profiles, and the remaining were assigned to Group 1. From around period 51, Group 0 gathered the ESS with power higher value, achieving almost 12 kW – the maximum capacity from these resources. On the other hand, group 1 consumers did not achieve the 10 kW. These consumers were probably the ones called to participate, being unable to charge until the maximum level.

Moving to Figure 6.7b, it became harder to visually understand a pattern between groups and why the algorithm gathered these consumers. Table 6.2 is introduced to aid with more information regarding the results obtained.

By analyzing this table, in $k = 9$, Group 0, Group 1, and Group 5 gather more ESS within a cluster. These follow a similar pattern to the groups in Figure 6.7a. Furthermore, it can also be observed that Group 2, Group 4, Group 6, and Group 8 are composed of only one object. Taking a closer look,

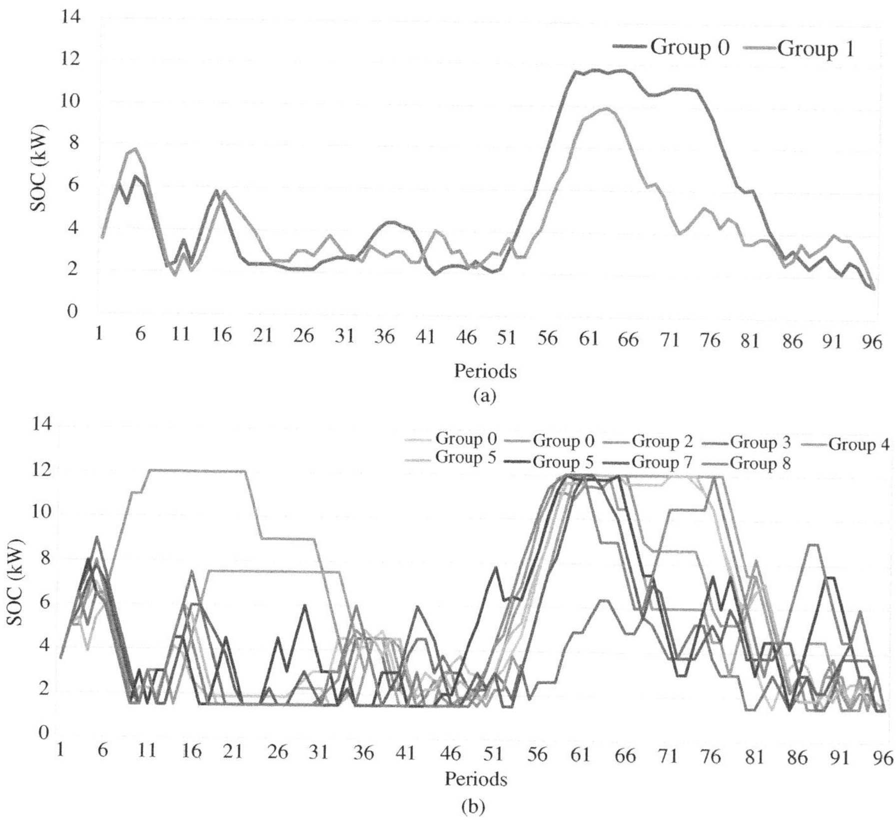

Figure 6.7 Partitional clustering: k-means results for the dataset selected for (a) $k_{\text{opt}} = 2$; (b) $k_{\text{opt}} = 9$.

these ESS profiles are very different from the remaining, for instance, from period 6 to period 36. Their dissimilarity from the other objects can impact the motivation from the active community due to the fairness matter. For example, suppose the aggregator opt with the Figure 6.7a solution. In that case, the remaining groups contributed to the market's electricity transactions. In contrast, these groups – for instance, the ESS assigned on Group 2, Group 4 in the mentioned period in Figure 6.7b, these resources were able to charge until they reached max capacity and received the same remuneration as if they contributed.

Moving on to k-medoids, two algorithms can be identified: partitioning around medoids (PAM) and clustering large applications (CLARA). The second is an extension of the first to deal with datasets with thousands of observations, where typically PAM is unsuccessful. This algorithm is search-based for objects, medoids, representing a cluster. After each medoid exchange with each non-medoid and only if there is an improvement over a criterion (minimization of the sum of the dissimilarities of all objects relative to the nearest medoid), this non-medoid will represent the group. Iteratively, all the non-medoids are tested until found achieve the criterion value.

The PAM method was also tested, and the results can be seen in Figure 6.8. In Figure 6.8a, both groups have a similar behavior overall. However, differences can be seen around period 67 until period 80 – Group 1 objects may have participated in the market transactions since the ESS status reduces almost 1/3 of their capacity in certain periods. Similar behavior was seen in Figure 6.7a for the same group.

Table 6.2 Partitional clustering – ESS per group (*k*-means).

k-means	$k = 2$	$k = 9$
1	0	0
2	0	5
3	0	0
4	0	1
5	1	4
6	1	7
7	0	1
8	0	5
9	0	1
10	1	6
11	0	5
12	0	5
13	1	3
14	0	2
15	0	0
16	0	0
17	1	3
18	0	1
19	1	7
20	0	8

Also, the difference between these curves in the PAM method results and those shown in Figure 6.7a, where the lines are curvy.

In Figure 6.7a, *k*-means assigned to Group 0 a total of 14 objects, while in Figure 6.8a, the dataset was divided more evenly: Group 0 with 11 elements and Group 1 with nine elements. In this way, the centroids found within these groups result in a distinct curve in Figure 6.7a – remembering that *k*-means centers are created resorting to the mean. However, the PAM method to find the cluster centers, medoids, is different, so the curves are more like those seen in Figure 6.3.

Moving to Figure 6.8b, the first highlight when compared with Figure 6.7b are the periods mentioned earlier (period 6 to period 36). There are no longer the two groups with values closer to the maximum capacity. However, with the help from Table 6.3, there are still two groups with only one object. By taking a closer look and analyzing the Groups in Figure 6.8b, Group 2 was assigned consumer ID 12, and Group 4 was assigned consumer ID 11.

Regarding the results from *k*-means in Figure 6.7b, Group 2 was assigned consumer ID 14, and Group 4 was assigned consumer ID 5. Compared with Figure 6.3, these were the ones with higher values in the mentioned periods – so their group should be different from those with a higher contribution in the market transactions.

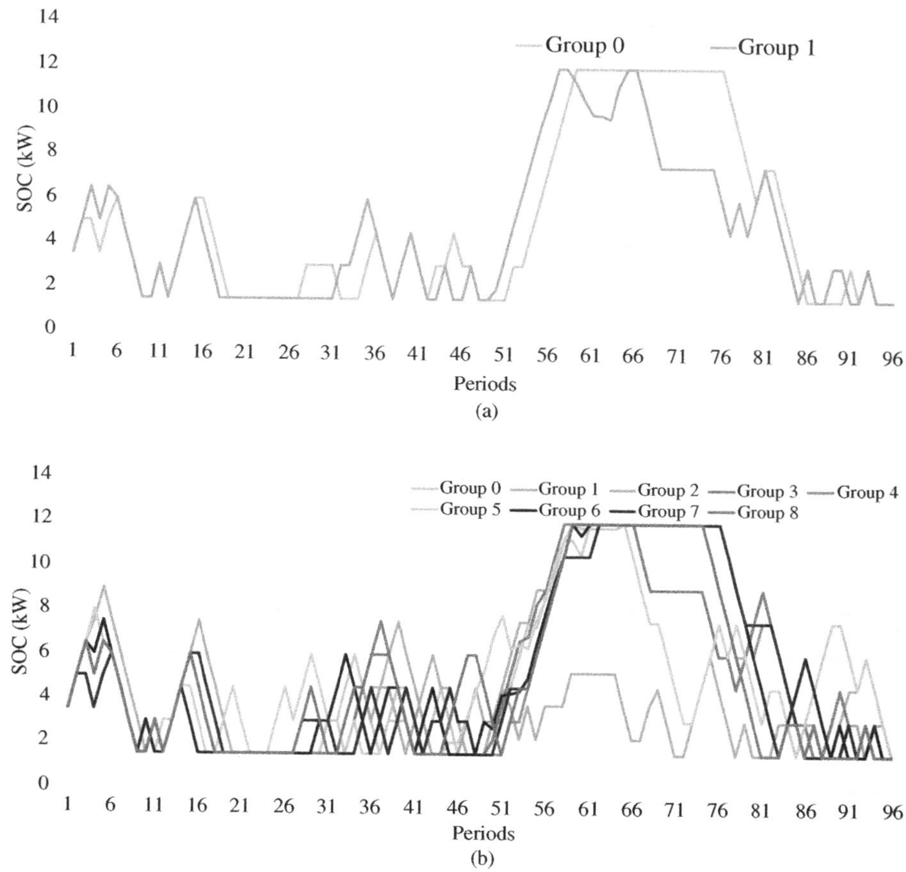

Figure 6.8 Partitional clustering: PAM results for the dataset selected for (a) $k_{opt} = 2$; (b) $k_{opt} = 9$.

6.3.2.2 Fuzzy Clustering

This algorithm has been used in many areas: cluster analysis, pattern recognition, image processing, and so forth. *c*-means is the most used within fuzzy clustering methods according to the literature, [23, 24]. Using fuzzy theory divides data points into a set of fuzzy clusters according to a certain partitioning criterion.

The core idea behind this concept is that objects within the same cluster have high values of similarity and minimal values of similarity among the different clusters [23]. To do this, it is considered a membership degree to which an element can belong to a given cluster, being this degree between 0 and 1. With this, objects near the center of the cluster may have a degree greater than the points at the cluster's edges. Furthermore, due to this membership degree, the final center coordinates will always be affected by choice of initial center coordinates – stochastic algorithm. In this way, selecting the right center will have a high impact on the results [24].

The results for $k = 2$ and $k = 9$ using the ESS status dataset can be seen in Figure 6.9. Starting with Figure 6.9a and comparing with the outcomes from the partitional clustering methods, the results from *c*-means were more like the *k*-means method in Figure 6.7a. Again, the calculation method uses the mean. Here, Group 0 also had 14 elements and assigned six elements to Group 1. The most

Table 6.3 Partitional clustering – ESS per group (PAM).

PAM	$k = 2$	$k = 9$
1	0	6
2	0	8
3	0	6
4	1	3
5	1	5
6	1	1
7	1	3
8	0	7
9	1	0
10	1	5
11	0	4
12	0	2
13	1	1
14	0	7
15	0	6
16	0	8
17	1	1
18	1	0
19	1	0
20	1	6

notorious difference, although small, can be seen between period 50 and period 60, where the curve from Group 1 is closer to Group 0.

Regarding Figure 6.9b, the groups can again be well spotted when confronting the previous results, but the consumer ID 14 from Figure 6.3. has not a single group assigned to him like Figure 6.7a.

In this way, Table 6.4 is added to understand how many elements each group has. According to the values shown, the groups with a single element are Group 2, Group 3, and Group 7.

In this way, for $k = 9$, Group 2 was assigned consumer ID 2, Group 3 was assigned consumer ID 19, and Group 7 was assigned consumer ID 16. None of the consumers previously mentioned in the single groups were also attributed in this method. It is important to highlight that consumer IDs 5 and 14 were assigned to the same group in this method – Group 1.

6.3.2.3 Hierarchical Clustering

Unlike other algorithms, the number of groups to be defined is not required a priori. As the name suggests, this method defines a hierarchical structure according to a proximity matrix. This structure is usually presented in dendrograms or binary trees. Consequently, if the analyst wants any other specific number of clusters, just cut the dendrogram at the desired level. Hierarchical clustering is subdivided into two main groups: agglomerative clustering and divisive clustering.

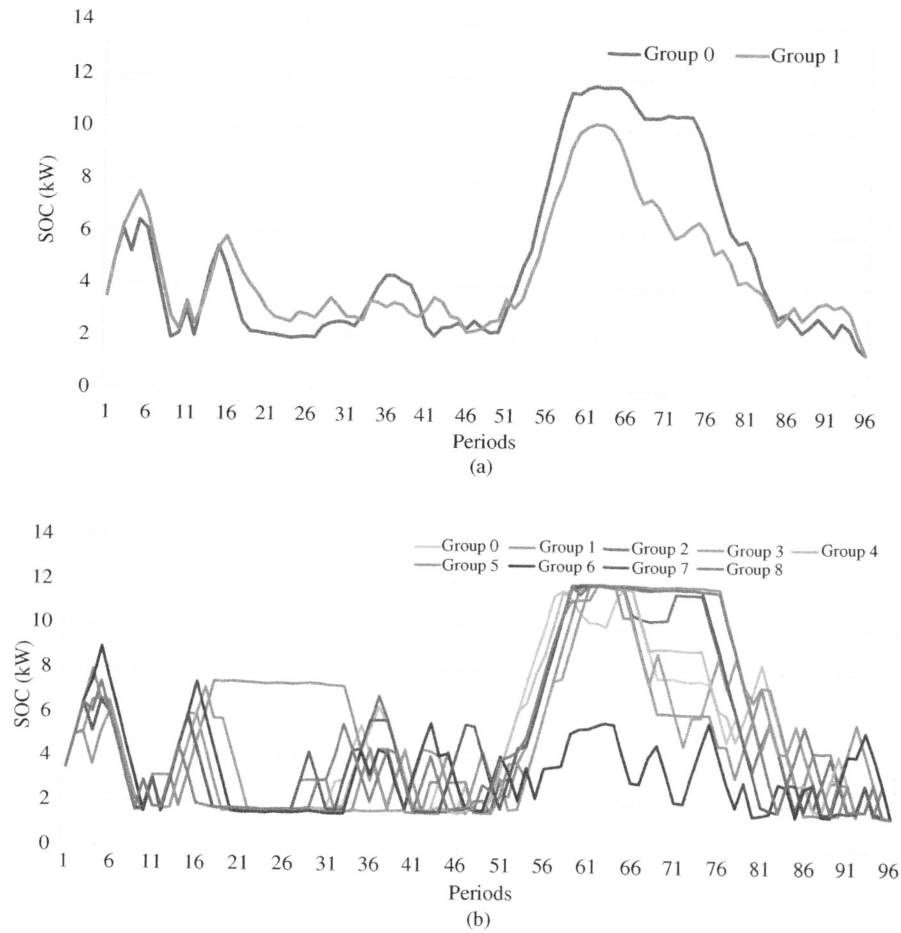

Figure 6.9 Fuzzy clustering: *c*-means results for the dataset selected for (a) $k_{opt} = 2$; (b) $k_{opt} = 9$.

Firstly, agglomerative clustering starts by gathering all objects in the database as a singleton cluster or a "leaf." After, clusters are successfully created until each object is in a single cluster, or "root." For divisive clustering, the opposite logic applies – as a top-down approach. In this way, it is considered that agglomerative clustering is a good option to identify smaller clusters. On the other hand, divisive clustering is the preferred choice for recognizing larger clusters.

Considering this assumption, Figure 6.10 represents the results from applying the hierarchical clustering method chosen to the dataset selected. Both values of the optimal number of clusters can be seen in the figure by cutting the dendrogram in $k = 2$ and $k = 9$.

However, to ease the understanding of the results, Table 6.5 was added to indicate the group for each consumer, in $k = 2$ and $k = 9$.

For the first scenario, where $k = 2$, there were 14 elements in Group 0 and 6 elements in Group 1, the same result as in Figure 6.7a and b. Regarding the scenario where $k = 9$, Group 0 gathered consumer ID 10 and consumer ID 19; to Group 1 was assigned consumer ID 2, consumer ID 4, consumer ID 7, consumer ID 9, and consumer ID 18; Group 2 has a single element (consumer ID 20); Group 3 has two elements, consumer ID 13 and consumer ID 17; Group 4 has four elements

Table 6.4 Fuzzy clustering – Number of ESS per group (c-means).

c-means	k = 2	k = 9
1	0	8
2	0	2
3	0	8
4	0	4
5	1	1
6	1	6
7	0	4
8	0	5
9	0	4
10	1	0
11	0	5
12	0	5
13	1	6
14	0	1
15	0	8
16	0	7
17	1	6
18	0	0
19	1	3
20	0	8

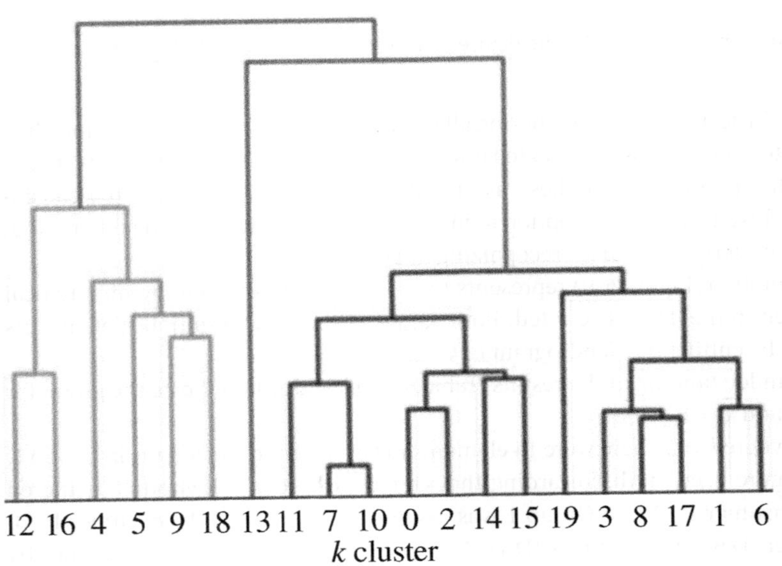

Figure 6.10 Hierarchical clustering: agglomerative results for the dataset selected.

Table 6.5 Hierarchical clustering: group attribution.

Consumer ID	$k = 2$	$k = 9$
1	0	4
2	0	1
3	0	4
4	0	1
5	1	5
6	1	7
7	0	1
8	0	8
9	0	1
10	1	0
11	0	8
12	0	8
13	1	3
14	0	6
15	0	4
16	0	4
17	1	3
18	0	1
19	1	0
20	0	2

being consumer ID 1, consumer ID 3, consumer ID 15 and consumer ID 16; Group 5, Group 6 and Group 7 are single elements groups gathering consumer ID 5, consumer ID 14, and consumer ID 6 respectively; finally, Group 8 has the remaining consumers. Again, consumer IDs 5 and 14 are highlighted with a single group.

6.4 Conclusion

A methodology to optimally manage an active community from the aggregator perspective was presented throughout this chapter. In the future, in smart grids, the consumers will become the core player mainly due to the load flexibility provided by these resources to suppress the distributed generation resources volatile behavior. However, these new players knowledge is still very scarce regarding the market transactions. Dependence on their behavior to guarantee the reliability and security of the entire system can be faulty. In this way, the authors believe in understanding the consumers, how they react in the event context, their behaviors, and new ways to motivate their participation in the market transactions to reduce the uncertainty.

Clustering methods are widely used in different areas such as cluster analysis, pattern recognition, image processing, among others. In the present chapter, the authors intend to use these algorithms to find patterns and use consumer profiling to find the proper motivation for demand response events, with information not only from the load consumption but also the remaining resources that can be associated with this new player, for instance, distributed generation or energy storage systems. The core idea is to find, within the dataset, objects with similar characteristics. So, the number of groups to be formed is crucial for the method's success.

Three of the most well-known methods were compared: elbow, silhouette, and gap-statistic. From the results comparison, the direct methods obtain a different value from the statistical method from the study. Both elbow and silhouette methods considered that the optimal number of clusters for the dataset should be 2. In the other hand, the Gap-statistic bet that the proper number of clusters should be 9.

So, the authors went further and tested both approaches in the second comparison: different types of clustering methods. Partitional, fuzzy, and hierarchical clustering methods were confronted for both "optimal" numbers of clusters found. From the authors' perspective and after the analysis of all the methods performed, the solution $k = 2$ might be very reductive for a 96 periods perspective. In this way, one of the unique contributions from the study is a new perspective from what should be included in further works regarding the number of groups, since might be essential in this perspective. Consumers with different behaviors were assigned to the same group to receive the same remuneration. So, fairness, in this case, may have been jeopardized, impacting one of the may goals of the proposal. From the aggregator viewpoint, this information is highly impactful in order to modify the motivation approach on the DR event participants.

Although the perspective presented was on a planning approach, the aggregator might benefit from performing the aggregation phase each period and understanding the complex behavior in a deeper form of contextualization. To ease the task, the authors suggest using hierarchical clustering since several k can be visualized.

Acknowledgments

This work has received funding from the European Union's Horizon 2020 research and innovation program under project DOMINOES (grant agreement No 771066), from FEDER Funds through COMPETE program, and from National Funds through (FCT) under the projects UIDB/00760/2020, MAS-SOCIETY (PTDC/EEI-EEE/28954/2017), CEECIND/02887/2017, and SFRH/BD/144200/2019.

References

1 Osório, G.J., Shafie-khah, M., Lotfi, M. et al. (2019). Demand-side management of smart distribution grids incorporating renewable energy sources. *Energies* 12 (1): 143. https://doi.org/10.3390/en12010143.

2 Silva, C., Faria, P., and Vale, Z. (2019). Demand response and distributed generation remuneration approach considering planning and operation stages. *Energies* 12 (14): 2721. https://doi.org/10.3390/en12142722.

3 de São José, D., Faria, P., and Vale, Z. (2021). Smart energy community: a systematic review with metanalysis. *Energy Strategy Reviews* 36 (July): 100678. https://doi.org/10.1016/j.esr.2021.100678.

4 Faria, P. and Vale, Z. (2021). Distributed energy resource scheduling with focus on demand response complex contracts. *Journal of Modern Power Systems and Clean Energy* 9 (5): 1172–1182. https://doi.org/10.35833/MPCE.2020.000317.

5 Gomes, L., Vale, Z.A., and Corchado, J.M. (2020). Multi-agent microgrid management system for single-board computers: a case study on peer-to-peer energy trading. *IEEE Access* 8: 64169–64183. https://doi.org/10.1109/ACCESS.2020.2985254.

6 Morais, H., Faria, P., and Vale, Z. (2014). Demand response design and use based on network locational marginal prices. *International Journal of Electrical Power & Energy Systems* 61 (October): 180–191. https://doi.org/10.1016/j.ijepes.2014.03.024.

7 Vale, Z., Morais, H., Ramos, S. et al. (2011). Using data mining techniques to support DR programs definition in smart grids. In: *2011 IEEE Power and Energy Society General Meeting*, 1–8. IEEE https://doi.org/10.1109/PES.2011.6039081.

8 European Parliament, and Council of the EU (2019). Directive (EU) 2019/944 on common rules for the internal market for electricity and amending directive 2012/27/EU. *Official Journal of the European Union* 158: 18. http://eur-lex.europa.eu/pri/en/oj/dat/2003/l_285/l_28520031101en00330037.pdf.

9 Abrishambaf, O., Lezama, F., Faria, P., and Vale, Z. (2019). Towards transactive energy systems: an analysis on current trends. *Energy Strategy Reviews* 26 (November): 100418. https://doi.org/10.1016/j.esr.2019.100418.

10 Zheng, S., Sun, Y., Li, B. et al. (2021). Incentive-based integrated demand response for multiple energy carriers under complex uncertainties and double coupling effects. *Applied Energy* 283 (June 2020): 116254. https://doi.org/10.1016/j.apenergy.2020.116254.

11 Okur, Ö., Voulis, N., Heijnen, P., and Lukszo, Z. (2019). Aggregator-mediated demand response: minimizing imbalances caused by uncertainty of solar generation. *Applied Energy* 247 (October 2018): 426–437. https://doi.org/10.1016/j.apenergy.2019.04.035.

12 Silva, C., Faria, P., and Vale, Z. (2019). Multi-period observation clustering for tariff definition in a weekly basis remuneration of demand response. *Energies* 12 (7): 1248. https://doi.org/10.3390/en12071248.

13 Jin, H., Li, Z., Sun, H. et al. (2019). Coordination on industrial load control and climate control in manufacturing industry under TOU prices. *IEEE Transactions on Smart Grid* 10 (1): 139–152. https://doi.org/10.1109/TSG.2017.2733578.

14 Subbalakshmi, C., Rama Krishna, G., Rao, S.K.M., and Rao, P.V. (2015). A method to find optimum number of clusters based on fuzzy silhouette on dynamic data set. *Procedia Computer Science* 46: 346–353. https://doi.org/10.1016/j.procs.2015.02.030.

15 Batool, F. and Hennig, C. (2021). Clustering with the average silhouette width. *Computational Statistics & Data Analysis* 158 (June): 107190. https://doi.org/10.1016/j.csda.2021.107190.

16 Kashani, A.G. and Graettinger, A.J. (2015). Cluster-based roof covering damage detection in ground-based lidar data. *Automation in Construction* 58 (October): 19–27. https://doi.org/10.1016/j.autcon.2015.07.007.

17 Granell, R., Axon, C.J., and Wallom, D.C.H. (2015). Impacts of raw data temporal resolution using selected clustering methods on residential electricity load profiles. *IEEE Transactions on Power Systems* 30 (6): 3217–3224. https://doi.org/10.1109/TPWRS.2014.2377213.

18 Granell, R., Axon, C.J., and Wallom, D.C.H. (2015). Clustering disaggregated load profiles using a dirichlet process mixture model. *Energy Conversion and Management* 92 (March): 507–516. https://doi.org/10.1016/j.enconman.2014.12.080.

19 Sun, L., Zhou, K., and Yang, S. (2020). An ensemble clustering based framework for household load profiling and driven factors identification. *Sustainable Cities and Society* 53 (February): 101958. https://doi.org/10.1016/j.scs.2019.101958.

20 Singh, A., Yadav, A., and Rana, A. (2013). K-Means with three different distance metrics. *International Journal of Computer Applications* 67 (10): Citeseer.

21 Hartigan, J.A. and Wong, M.A. (1979). Algorithm AS 136: A K-means clustering algorithm. *Applied Statistics* 28 (1): 100. https://doi.org/10.2307/2346830.

22 Sinaga, K.P. and Yang, M.-S. (2020). Unsupervised K-means clustering algorithm. *IEEE Access* 8: 80716–80727. https://doi.org/10.1109/ACCESS.2020.2988796.

23 Wan, R., Xiong, N., Qinghui, H. et al. (2019). Similarity-aware data aggregation using fuzzy c-means approach for wireless sensor networks. *EURASIP Journal on Wireless Communications and Networking* 2019 (1): 59. https://doi.org/10.1186/s13638-019-1374-8.

24 Wu, S., Pang, Y., Shao, S., and Jiang, K. (2018). Advanced fuzzy C-means algorithm based on local density and distance. *Journal of Shanghai Jiaotong University (Science)* 23 (5): 636–642. https://doi.org/10.1007/s12204-018-1993-y.

Part III

Classification

7

A Novel Framework for NTL Detection in Electric Distribution Systems

Chia-Chi Chu[1], Nelson Fabian Avila[2], Gerardo Figueroa[3], and Wen-Kai Lu[4]

[1]*Department of Electrical Engineering, College of Electrical Engineering and Computer Science, National Tsing Hua University, HsinChu, Taiwan, R.O.C.*
[2]*Operations Support Engineering, Electric Reliability Council of Texas (ERCOT), Austin, TX, USA*
[3]*Sentiance NV, Antwerpen, Belgium*
[4]*Department of Information Management, College of Management, National Taiwan University, Taipei, Taiwan, R.O.C.*

Acronyms

AdaBoost	Adaptive boosting
AMI	Advanced metering infrastructure
AUC	Area under curve
DT	Decision tree
MID	Meter identification number
MCC	Matthews correlation coefficient
MODWPT	Maximal overlap discrete wavelet-packet transform
NTL	Non-technical loss
ROC	Receiver operating characteristic
ROS	Random over-sampling
RUS	Random under-sampling
RusBoost	Random under-sampling boosting
XGBoost	Extreme gradient boosting
SVM	Support vector machine

7.1 Introduction

Over the past few years, electric utilities around the world have raised concerns about the negative financial impact of non-technical losses (NTLs) in electric distribution systems [1–4]. Generally speaking, NTLs refer to a type of loss generally involving electricity theft, errors in meter readings, or defective meters, and they account for a significant part of total losses in most countries, particularly in developing countries [3–6]. The current practice for detecting NTLs in most countries is rather empirical. Particular patterns in the collected power consumption data are manually analyzed, and on-site inspections are conducted on suspicious consumers for verification. These traditional methods require lots of additional resources. Recently, owing to the widespread deployment of smart meters, we can efficiently collect an immense amount of fine-grained consumption data, enabling new methods like artificial intelligence and machine

Intelligent Data Mining and Analysis in Power and Energy Systems: Models and Applications for Smarter Efficient Power Systems, First Edition. Edited by Zita Vale, Tiago Pinto, Michael Negnevitsky, and Ganesh Kumar Venayagamoorthy.

learning techniques to detect fraudulent consumers. Therefore, developing an automatic NTL detection system by exploring massive smart meter data has become an emerging issue in various electric distribution companies [4].

Recently, existing solutions for NTL detection began to emerge. However, these solutions are time-consuming and high cost for building up the domain know-how of power systems and collecting enough anomaly data points. Theoretically speaking, the task of NTL detection can be treated as a special type of anomaly detection, and as such, it involves a high data imbalance between anomalous and regular cases [7, 9]. More explicitly, the number of fraudulent electricity consumers is remarkably smaller than that of non-fraudulent ones. The problem of class imbalance is of particular concern in supervised machine learning [10–14], which is a widely used methodology throughout the literature to detect NTLs. Most supervised machine learning algorithms experience reduced performance when the classes in the training data are highly imbalanced. The reason is that only a few fraudulent examples (positive class) are learned compared to non-fraudulent examples (negative class). This under-representation of the positive class heavily affects the classification algorithm, a problem that should be addressed in the NTL detection process. Unfortunately, previous research has paid little attention to address the class imbalance problem in the application of supervised machine learning algorithms for NTL detection. To review recent progress in the NTL detection for imbalance data, please refer to [15].

7.1.1 State-of-the-Art

During the past two decades, various approaches have been studied for NTL detection. In the recent survey paper in [17], existing techniques for NTL detection are be categorized into (i) data-oriented, (ii) network-oriented, and (iii) hybrid techniques. In data-oriented approaches, only consumption profiles from meter readings are investigated by data analytics and machine learning technology. Other data profile can also merge with consumption profile to increase accuracy in predicting NTL. These additional data profile may contain environmental and temperature features.

Furthermore, some interest has developed by utilizing power network data to identify potential NTL. Some efforts have also been made to develop hybrid techniques by combining consumption and network data. Thus, better performance can be expected. For brevity, we will only focus on data-oriented approaches in this chapter.

Data-oriented approaches can be categorized into the following three categories: (i) statistical approaches, (ii) expert systems, and (iii) data mining and machine learning techniques [4, 8, 9, 16–18]. Statistical approaches employing state estimation and analysis of variance have been studied to identify consumers suspected of having metering problems [18]. On the other hand, expert systems are constructed based on expert knowledge and typically involve handcrafted rules in their decision-making process. Lastly, data mining and machine learning methods attempt to learn patterns from the data to find meaningful results. All these approaches require an in-depth analysis of consumer load profiles using collected time series data [19]. Patterns in load profiles can be statistically analyzed to detect fraudulent behavior or used as special features in supervised or unsupervised machine learning algorithms, which are the focus of this chapter.

A significant amount of research has been devoted to developing expert systems for NTL detection. Some of these works implement fuzzy rules [7, 20], consumer categorization [21], and Boolean logic [7]. Other types of methods that can be included in this category are those that perform statistical analysis of the load profiles of electricity consumers. For instance, the variability in consumer consumption patterns was analyzed in [22] to detect fraudulent behavior, and the correlation between electricity consumption and time was proposed in [23] to identify irregular customers. Of these works, only work studied in [7] presented an adequate evaluation metric for NTL detection: the area under the receiver operating characteristic (ROC) curve, also known as area under the curve (AUC) [26].

Data mining and machine learning approaches have also addressed consumer load profiles. In Nizar et al. [27], load profiles of electricity consumers were utilized in a composite data mining framework to automatically detect NTLs, applying clustering and classification. However, their performance was evaluated using the accuracy metric, which is not suitable for evaluating performance in highly imbalanced datasets such as the data used for NTL detection. In the past, several supervised machine learning algorithms have been applied for NTL detection, including support vector machines (SVMs) [5, 7, 28, 31], decision trees [22, 23], neural networks [29–31], and Bayesian networks [23].

More recently, some works have integrated data mining and machine learning techniques with expert systems for the detection of NTLs. For example, a methodology was proposed in [24] that utilized the output of a text mining module and data mining module to generate the rules for an expert system. Their system leveraged a wide variety of data sources and information obtained from an electricity company in Spain, rather than just using consumer consumption data. Similarly, a knowledge-based system was developed in [25] that generated rules created from expert knowledge with rules obtained from statistical techniques, text mining, and neural networks to enhance the system's efficiency and efficacy.

To the best of our knowledge, only [28, 29, 31] attempted to counteract the effects of imbalanced data in their classification methods. Muniz proposed a system composed of neural networks ensembles to detect irregular consumers in low-voltage networks [29]. The training and validation datasets were equalized to diminish the imbalance between irregular and normal consumer data. A different approach was taken in [28] to outweigh this problem, encompassing one-class classifiers, cost-sensitive learning, and the ensemble of diverse classifiers. In both of these works, proper performance metrics were not considered to evaluate performance. In addition, the selected attributes used in their methods were obtained by aggregating the power data, and this process can lead to severely reduced performance if the dataset contains a significant amount of noise and outliers.

The use of accurate performance measures for NTL detection methods was addressed by [7]. The authors claim that the AUC is the most suitable metric. Their methodology, however, does not attempt to counteract the imbalance of data in the framework. Our preliminary work [31] approached the imbalance problem by applying data sampling on three different classifiers. Furthermore, we analyzed the performance of our methodology using seven different metrics. In addition, more advanced signal processing techniques were utilized to extract the most important features from the load profiles for detecting irregular consumers accurately.

7.1.2 Proposed Framework

So far, no previous research on NTL detection has aimed at extracting the most important patterns from consumer load profiles using state-of-the-art signal processing techniques, correcting the effects of imbalanced data in the classification task by applying new advanced machine learning strategies, and simultaneously reporting the effectiveness of the methodology with adequate performance measures. To this end, and inspired by our previous work [31, 32], a framework for NTL detection is proposed in this chapter to addresses the challenges mentioned above. The proposed framework can be divided into four main stages, namely:

(1) data acquisition and pre-processing,
(2) feature extraction and selection,
(3) classification using imbalanced-data strategies, and
(4) performance evaluation.

In summary, the proposed framework for automatic NTL detection will explore supervised machine learning techniques and offers the following contributions:

(1) Since collected smart meters data have noise and outliers, this paper implements a comprehensive data pre-processing stage to lessen the noise and remove outliers from the dataset.
(2) Previous works conventionally use the daily average power consumption values over a month as features of the machine learning method [5, 20, 27, 31, 33]. In contrast, we have developed a maximal overlap discrete wavelet-packet transform (MODWPT) as the main mechanism for feature extraction and selection. The idea is to decompose the load profile in coefficients for various frequency levels using low- and high-pass filters. Next, crucial information in the load profile of each customer is gathered by:
   (i) computing the standard deviation of the wavelet coefficients at each frequency band, and
   (ii) selecting the coefficients with a higher absolute magnitude of each consumer.
(3) We employed a novel technique, called Random Under-Sampling Boosting (RusBoost), for the NTL detection problem [34]. RusBoost is recognized as a state-of-the-art supervised machine learning method for imbalanced data applications. Furthermore, since RusBoost offers superior robustness to the challenge introduced by imbalanced data, better performance can be ensured when applied to the NTL detection problem.
(4) Several meaningful metrics evaluate the performance of the proposed approach for imbalanced binary classification problems.

The proposed framework for NTL detection is based on our previous work [32]. The remainder organization of this chapter is listed as follows. Section 7.2 presents the data acquisition procedure and the pre-processing stage. Section 7.3 introduces the proposed feature selection mechanism using the MODWPT. The RusBoost technique for NTL detection will be explained in Section 7.4. Various evaluation metrics are presented in Section 7.5. Experimental results are reported in Section 7.6. Finally, conclusions and future work are given in Section 7.7.

7.2 Data Acquisition and Pre-Processing

7.2.1 Data Acquisition

The utilized data are based on historical load consumption from different industrial and commercial consumers in the electric distribution system in Honduras, obtained from the Advanced Metering Infrastructure (AMI) installed in the most populated cities around the country, including Tegucigalpa and San Pedro Sula. The dataset consisted of $N = 881$ days of electricity consumption data from 2014 to 2016 for 2271 selected consumers. Since data samples represent each consumer's daily accumulated power consumption, the proposed framework can detect NTL daily after aggregating the 24-hour power consumption data. The dataset consists of $N = 881$ days of electricity consumption data from 2014 to 2016, corresponding to the top consumers in Honduras' industrial and commercial sectors. Every customer c has an assigned meter identification number (MID). The original dataset was composed of two parts:

(1) The accumulated active and reactive power consumption for each consumer c until day t, expressed as P_t^c and Q_t^c, respectively, the currently active and reactive power of consumer c, expressed as P_{now}^c and Q_{now}^c, respectively, and the time and date of reading t.
(2) A list of verified cases of fraud identified by performing multiple on-site inspections. Of the attributes included in the consumption data, only P_t^c was retained in our methodology. Since P_t^c has accumulated values, the actual power consumed by consumer c on day t can be calculated as $\Delta P_t^c = P_t^c - P_{t-1}^c$.

In brief, power samples P_t^c of each consumer are represented in the vector form as $S_c = \left[P_1^c, P_2^c, \ldots, P_N^c\right]^T$ and normalized as $\grave{S}_c = S_c/|S_c|$.

7.2.2 Pre-Processing

The pre-processing step is divided into the following steps:

(1) **Consumer selection**: Smart meters in Honduras were installed at different times during the analyzed time interval. As a result, some electricity consumers have only a few months of historical power consumption data. Therefore, at the expense of reducing the number of examples when training the supervised machine learning classifier [35], to create a concise dataset with a similar number of power measurements for each consumer, some clients were filtered from the original dataset. These consumers that are removed from the dataset are those with less than 600 days of consumption data.
(2) **Correct missing data**: If consumer c has a series of $m < 7$ consecutive days of missing samples P_t^c, the missing values are filled using linear interpolation on existing data points. If $m \geq 7$, the missing values are replaced with zeros.
(3) **Outlier detection**: The detection of outliers in the data is essential for the better performance of machine learning algorithms. In our framework, outliers are detected using a smoothing spline function [36] on the power samples of each consumer c. The required steps are described as follows:
   a. For each consumer c, a smoothing spline function with smoothing parameter $\gamma \in [0,1]$, denoted as $F_\gamma(t)$ is fitted to the consumption samples P_t^c. The spline function $F_\gamma(t)$ can be found by minimizing (7.1) overall samples in P_t^c. Parameter γ determines the level of smoothness of the spline. For example, small values of $\gamma(\gamma \to 0)$ produce quasi straight-line fits to P_t^c, determined by a least-square solution to the data samples. On the other hand, large values of $\gamma(\gamma \to 1)$ produce more smooth curves, such as cubic splines [37]. The appropriate selection of γ varies from one dataset to another, and it normally depends on the level of separation of the data points. In this sense, experimentation and knowledge of the spacing in the data samples often require selecting the smoothing parameter. For a comprehensive understanding of smoothing splines and efficient solutions to (7.1), please refer to Chapter 2 in [37].

$$\gamma \sum_{t=1}^{N} \left(F_\gamma(t) - P_t^c\right)^2 + (1-\gamma) \int \left(\frac{d^2 F_\gamma}{dt^2}\right)^2 dt \tag{7.1}$$

   b. Subsequently, P_t^c is detected as an outlier if $\left|F_\gamma(t) - P_t^c\right| < k\sigma(\grave{S}_c)$, where $\sigma(\grave{S}_c)$ is the standard deviation of the elements in $\sigma(\grave{S}_c)$ and k is a positive constant. More precisely, P_t^c is identified as an outlier if located at a distance greater than k standard deviations from the smoothing spline.

7.3 Feature Extraction

7.3.1 Overview

Temporal features are widely considered in the feature extraction of NTL detection. For example, electricity consumption data have two peaks in the morning and the afternoon each day. Capturing the consumption values at these two peaks characterizes households' patterns, and these features can be abstracted into a simpler model which uses a smaller number of features to reflect these two characteristic peaks. Recently, the so-called "V-shape" model has been proposed to construct features on load profiles [38]. However, in these approaches, a labeled dataset is still needed to handle abnormal behaviors of thefts. To tackle this difficulty, some statistic characteristics, such as Pearson correlation coefficient (PCC) and maximum information coefficient (MIC), have also been proposed for feature extraction [39].

Analytical signal decomposition techniques, such as Fast Fourier Transform (FFT), Wavelet decomposition, have also been applied for feature extraction in several areas of science and engineering [40]. One of the most salient advantages of this approach is that abrupt transitions in the input signal result in an increased absolute value of the obtained wavelet coefficients. Consequently, important features of the input signal can be captured from the magnitude and the standard deviation of these wavelet coefficients. In the following, a brief review of MODWPT, which is the core technique of our feature selection approach, will be presented.

7.3.2 MODWPT

The MODWPT is a time-invariant transformation that decomposes an input signal into coefficients. These coefficients can capture features of the original signal at various frequency sub-bands. Let the original signal for consumer c be represented in vector form as $u_0^0 = \grave{S}_c$. Now, given a pair of low-pass and high-pass decomposition filters g_d and h_d, respectively, the MODWPT coefficients at bands $2z$ and $2z+1$ and level j, denoted as u_j^{2z} and u_j^{2z+1} respectively, are determined in the following manner [43]:

$$u_j^{2z}(k) = \frac{1}{\sqrt{2}} \sum_{n=-\infty}^{\infty} g_d(n) u_{j-1}^{z}(k-n) \tag{7.2}$$

$$u_j^{2z+1}(k) = \frac{1}{\sqrt{2}} \sum_{n=-\infty}^{\infty} h_d(n) u_{j-1}^{z}(k-n) \tag{7.3}$$

Filters h_d and g_d are referred to as wavelet and scaling decomposition filters, respectively. These two filters must satisfy the finite energy signal, the admissibility, and the zero-mean properties [41]. Under this framework, both the wavelets and the scaling filters are numerically derived from Daubechies wavelet [42]. It is important to mention that, after the signal decomposition process, the original signal can be reconstructed using reversed reconstruction filters g_r and h_r by following an identical process as in (7.2) and (7.3).

7.3.3 Feature Extraction Mechanism

Figure 7.1 shows the proposed MODWPT decomposition, where the input signal $\grave{S}_c$ is decomposed according to (7.2) and (7.3) from $j = 0$ to $j = 3$. For the NTL detection problem, both the absolute magnitude and the standard deviation of the obtained wavelet coefficients provide relevant

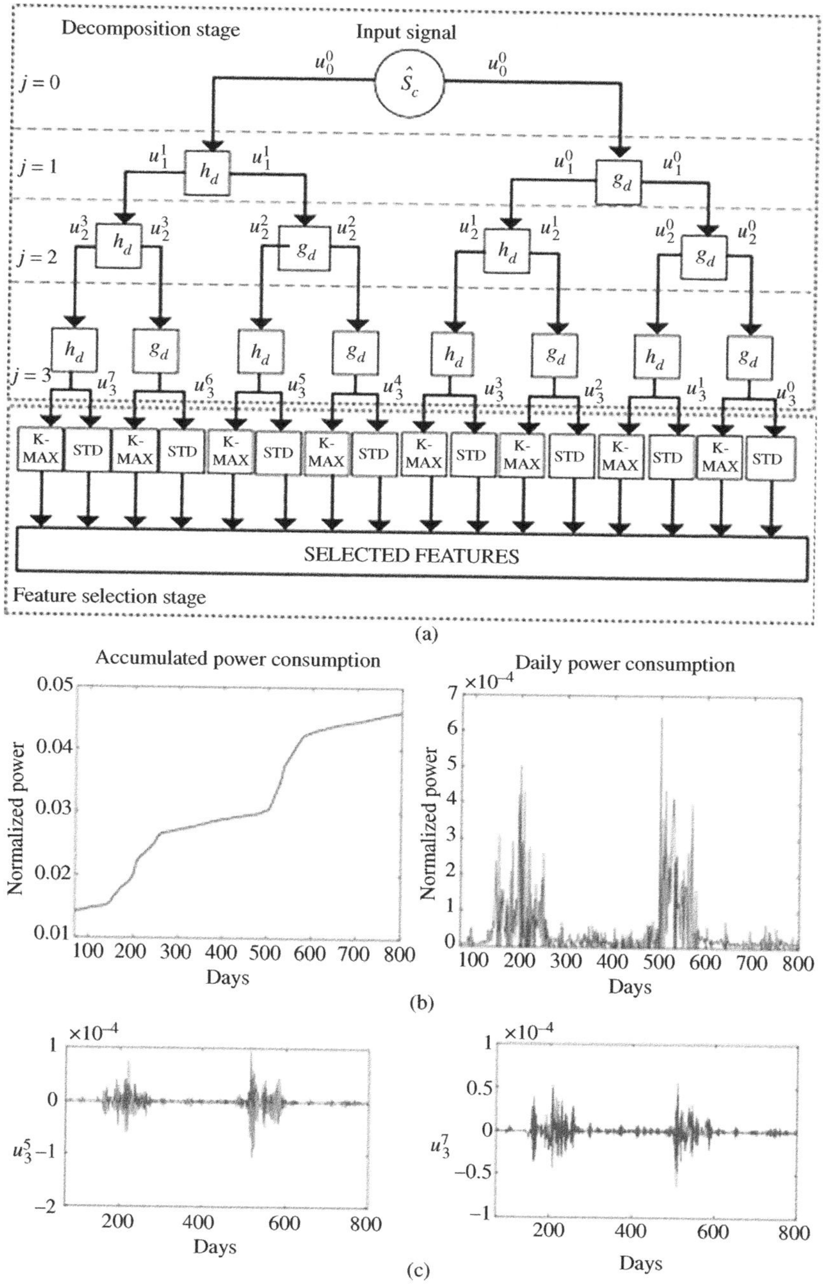

Figure 7.1 (a) Diagram of the proposed feature selection technique for three decomposition levels ($l = 3$). (b) Load profile of a fraudulent electricity consumer. (c) Wavelet decomposition example of the fraudulent consumer in (b). Source: Avila et al. [32]/with permission of IEEE.

information about important changes in the load profile of each consumer. For illustrational purposes, Figure 7.1b shows the load profile of a fraudulent consumer. This load profile has been decomposed according to Figure 7.1a, and two out of the eight wavelet signals at $j = 3$ are shown in Figure 7.1c. As observed in Figure 7.1c, sudden transitions in the load profile result in increased absolute magnitude and standard deviation of the wavelet coefficients at $j = 3$. Therefore, our framework selects the k-greatest wavelet coefficients and computes the standard deviation of each generated signal u_3^z. In general, if $j = l$ at the last decomposition level, we select $f = (k+1)\times 2^l$ features, including the $k \times 2^l$ greatest wavelet coefficients and the 2^l standard deviations. In Figure 1(a) for instance, our framework selects $f = 8\,K + 8$ features, including 8 K greatest coefficients and 8 standard deviations at $j = 3$. The feature selection process is possible by using the K–MAX and STD blocks shown in Figure 7.1a. Next, these selected features are used as examples $\boldsymbol{x}_i \in \boldsymbol{R}^{1\times f}$ and assigned a class y_i. Lastly, each pair $(\boldsymbol{x}_i, y_i)$ is appended in the training or testing sets of a robust supervised machine learning technique for NTL detection, which will be described later.

7.4 Classification Strategies

Since a high data imbalance between anomalous and regular cases in this NTL detection problem, classification strategies for imbalanced data will be provided in this section. These techniques will be explored in the NTL detection problem in the coming sections.

7.4.1 Random Under-Sampling (RUS) and Random Over-Sampling (ROS) Techniques

Data sampling is one of the most common strategies to alleviate the class imbalance problem [44, 45]. The basic idea behind these approaches is to diminish the imbalance ratio by either duplicating examples of the minority class (oversampling), or removing examples from the majority class (under-sampling) [46]. For example, in [31], the NTL detection problem was addressed in part using a random oversampling (ROS) technique, where copies of the minority class were included in the training set at different proportions to counteract the imbalance property of the dataset. This technique, however, increases the time to build a classifier, since the size of the training set is significantly increased. In addition, since the training set contains several copies of the minority class, the model is prone to overfitting [46].

Unlike ROS, random under-sampling (RUS) techniques reduce the computational burden by decreasing the size of the training set, and it is unlikely to overfit the data because of the lack of repetition of examples. Recently, researchers developed a hybrid framework, referred to as RusBoost [34], which makes use of the simple RUS technique and the powerful adaptive boosting (AdaBoost) algorithm for class imbalance problems [47]. This framework has been recognized in the literature as state-of-the-art technology to counteract the problems introduced by imbalanced data. Both the conventional AdaBoost and the modified RusBoost algorithms will be briefly reviewed in this section.

7.4.2 Adaptive Boosting Algorithm

The Adaptive Boosting (AdaBoost) algorithm was formally introduced in [47] as a meticulous classification technique to solve highly non-linear problems. A pseudo-code of the algorithm is shown in Algorithm 1 for a non-linear binary classification problem with training set $\boldsymbol{S} = (\boldsymbol{x}_1, y_1)$,

$(\boldsymbol{x}_2, y_2), \ldots, (\boldsymbol{x}_m, y_m)$, where $\boldsymbol{x}_i \in \boldsymbol{R}^{1\times f}$ and $y_i \in -1, 1$. AdaBoost aims to increase the accuracy of a given classifier h_n, also called a weak learner, by creating an ensemble of h_ns and repeatedly training each h_n in a series of iterations $n = 1, 2, \ldots N$. At each iteration n, the weak learner h_n is trained using a subset $\boldsymbol{S}'$ of the given training set $\boldsymbol{S}$. Elements $(\boldsymbol{x}_i, y_i) \in \boldsymbol{S}'$ are selected according to a probability distribution D_n, such that incorrectly classified examples at the current iteration n have a higher probability D_{n+1} to be selected in the next iteration $n + 1$. This process is possible by utilizing the special error formula shown in lines 7 and 8 of Algorithm 7.1, and the probability updates given in lines 10 and 12. Lastly, as shown in line 16, a weighted vote of all the h_n s is used to assign a class $y_j = H(\boldsymbol{x}_j)$ to the un-labelled examples $\boldsymbol{x}_j$.

Algorithm 7.1 Adaptive Boosting Algorithm

1: Given: $\boldsymbol{S} = (\boldsymbol{x}_1, y_1), (\boldsymbol{x}_2, y_2), \ldots, (\boldsymbol{x}_m, y_m)$
2: where $\boldsymbol{x}_i \in \boldsymbol{R}^{1\times f}$ and $y_i \in -1, 1$
3: Initialize $D_1(\mathrm{i}) = 1/m$, $\forall\, \mathrm{i} \in \{1, \ldots, m\}$
4: **for** $n = 1 \ldots N$ **do**
5: Create $\boldsymbol{S}'$ by randomly selecting from $\boldsymbol{S}$ with probability $D_n(\mathrm{i})$;
6: Train h_n using $\boldsymbol{S}'$ as training set;
7: Calculate the error: $\epsilon_n = \sum_{i:h_n(x_i)\neq y_i} D_n(i), \forall \boldsymbol{x}_i \in \boldsymbol{S}$
8: Let $\alpha_n = log_e \frac{1-\epsilon_n}{\epsilon_n}$;
9: **if** $h_n(\boldsymbol{x}_i) == y_i$ **then**
10: $D_{n+1}(i) = D_n(i)e^{-\alpha n}$;
11: **else**
12: $D_{n+1}(i) = D_n(i)e^{\alpha n}$;
13: **end if**
14: $D_{n+1} = normalize(D_{n+1})$;
15: **end for**
16: Output: $y_j = H(\boldsymbol{x}_j) = sign\left(\sum_{n=1}^{N} \alpha_n h_n(\boldsymbol{x}_j)\right)$;

7.4.3 Random Under-Sampling Boosting Algorithm

The Random Under-Sampling Boosting (RusBoost) algorithm combines the powerful features of RUS and AdaBoost. A pseudo-code of this technique is provided in Algorithm 7.2. The functionality of RusBoost is derived from the AdaBoost algorithm, with an additional RUS step required at each iteration n on the selected subset of examples $\boldsymbol{S}'$. This RUS step improves the performance of the conventional AdaBoost algorithm when dealing with the problems introduced by imbalanced data. As can be observed in line 7 of Algorithm 2, the RUS process randomly removes elements in $\boldsymbol{S}'$ to create a $\boldsymbol{S}'_{RUS} \in \boldsymbol{S}'$ until $p\%$ of the selected examples in $\boldsymbol{S}'_{RUS}$ belong to the majority class ($y_i = -1$), where $p\%$ is a user-defined variable. The appropriate imbalance ratio $p\%$ should select through experimental designs in the dataset. The RusBoost method aims to alleviate the class imbalance problem when training each weak learner h_n by decreasing the imbalance ratio of the dataset used for training h_n. Similar to AdaBoost, a class $y_j = H(\boldsymbol{x}_j)$ is assigned to the unlabeled examples $\boldsymbol{x}_j$ using a weighted voting scheme on all weak learners h_n. In our framework, the RusBoost technique uses an ensemble of 150 weak learners, where each weak learner is a simple classification tree.

Algorithm 7.2 Random Under-Sampling Boosting Algorithm

Given an imbalanced dataset $\boldsymbol{S}$ with minority class $y_i = 1$
$\boldsymbol{S} = (\boldsymbol{x}_1, y_1), (\boldsymbol{x}_2, y_2), \ldots, (\boldsymbol{x}_m, y_m)$
where $\boldsymbol{x}_i \in \boldsymbol{R}^{1\times \mathrm{f}}$ and $y_i \in -1, 1$
Initialize $D_1(\mathrm{i}){=}1/m$, $\forall\, \mathrm{i} \in \{1, \ldots, m\}$
for $n = 1 \ldots N$ **do**
Create $\boldsymbol{S}'$ by randomly selecting from $\boldsymbol{S}$ with probability $D_n(i)$;
Create $\boldsymbol{S}'_{RUS} \in \boldsymbol{S}'$ by performing RUS of the majority class until $p\%$ of the elements in $\boldsymbol{S}'_{RUS}$ belong to $y_i = -$yelong tol$p\%$ belong to $y_\mathrm{i} = 1$;
Train h_n. using $\boldsymbol{S}'_{RUS}$ as training set;
Calculate the error: $\epsilon_n = \sum\limits_{i:h_n(\boldsymbol{x}_i)\neq y_i} D_n(i), \forall \boldsymbol{x}_i \in \boldsymbol{S}$
Let $\alpha_n = \log_e \frac{1-\epsilon_n}{\epsilon_n} \alpha$;
if $h_n(\boldsymbol{x}_i) == y_i$ **then**
$D_{n+1}(i) = D_n(i)e^{-\alpha n}$;
else
$D_{n+1}(i) = D_n(i)e^{\alpha n}$;
end if
$D_{n+1} = \text{normalize}(D_{n+1})$;
end for
Output: $y_j = H(\boldsymbol{x}_j) = sign\left(\sum_{n=1}^{N} \alpha_n h_n(\boldsymbol{x}_j)\right)$;

7.5 Evaluation

In the context of applying classification algorithms for automatic NTL detection, true positives (T^+) and true negatives (T^-) correspond to correctly classified fraud and no-fraud cases, respectively, false positives (F^+) indicate no-fraud cases incorrectly marked as fraud, and false negatives (F^-) are those fraud cases that were undetected. The objective of NTL detection is to reduce false positives and increase true positives, thus avoiding costly inspections and finding as many fraud cases as possible [7]. Seven different performance metrics for classification problems, given by the following expressions, are reported in this work:

(i) Accuracy:

$$\text{Accuracy} = \frac{T^+ + T^-}{T^+ + T^- + F^+ + \mathrm{F}^-} \tag{7.4}$$

(ii) Precision:

$$\text{Precision} = \frac{\boldsymbol{T}^+}{\boldsymbol{T}^+ + \boldsymbol{F}^+} \tag{7.5}$$

(iii) Recall:

$$\text{Recall} = \frac{\boldsymbol{T}^+}{\boldsymbol{T}^+ + \mathbf{F}^-} \tag{7.6}$$

(iv) F_β-score:

$$F_\beta = (1 + \beta^2) \times \frac{\text{Precision} \times \text{Recall}}{\beta^2 \times \text{Precision} + \text{Recall}} \tag{7.7}$$

(v) Specificity:

$$\text{Specificity} = \frac{T^-}{T^- + F^+} \tag{7.8}$$

(vi) Area Under the Curve (AUC) of a Receiver Operating Characteristic Curve:

$$AUC = \frac{\text{Recall} + \text{Precision}}{2} \tag{7.9}$$

(vii) Matthews correlation coefficient (MCC):

$$MCC = \frac{a_1}{\sqrt{a_2 a_3 a_4 a_5}} \tag{7.10}$$

where $a_1 = T^+ \times T^- - F^+ \times F^-$, $a_2 = T^+ + F^+$, $a_3 = T^+ + F^-$, $a_4 = T^-\ F^+$, and $a_5 = T^- + F^-$.

Despite being used extensively in the literature, the accuracy and recall have been proven to be inadequate performance measures on imbalanced datasets [16]. Likewise, the precision, F_1-score, and specificity are not reliable metrics when considered in isolation [16]. Nevertheless, we will use these metrics to appreciate their meaning in the context of NTL detection and better comprehend the effects of the proposed RusBoost technique. The F_β score in assigns weights to the precision and the recall where the value of β is larger than 0. Here, the F_β-score is calculated using $\beta = 0.5$, as false positives may potentially lead to incurring greater costs than false negatives. Another accurate metric suitable for imbalanced data is the AUC [7, 26], given in Eq. (7.9). The AUC is a measure between 0 and 1, where any value greater than 0.5 indicates better performance than random guessing. The last metric is the Matthews correlation coefficient (MCC) given in and suggested in [16]. This measure considers all four outcomes of a confusion matrix, suggesting a meaningful assessment of the quality of binary classification in imbalanced data. The MCC is a value between −1 (total disagreement) and +1 (total agreement), and is essentially a correlation between the observed and predicted classifications, where 0 is the result expected from random classification.

7.6 Experiments

The proposed NTL detection framework was implemented in the MATLAB 2017 environment. The dataset consisted of 881 days of power measurements for 2271 selected consumers. Out of all samples, 2153 consumers were marked as non-fraudulent ($y_i = -1$) and the remaining 118 consumers were marked as fraudulent ($y_i = 1$). Thus, the number of examples of the positive class is roughly 18 times larger than that of the negative class.

7.6.1 Outlier Detection Using Smoothing Splines

(1) **Parameter selection**: The selection of parameters in the outlier detection stage requires some knowledge of the dataset. Conventionally, γ and K are selected according to the spacing between data samples. In the utilized power consumption data, samples tend to have a nearly uniform separation daily. For this particular application, small values of the smoothing parameter γ are the most suitable to detect samples far away from the majority of the data points. Similarly, the selection of K depends on the average spacing between points. For datasets with closely separated samples, one would select small values of K to identify remotely located samples. Conversely, due to their high standard deviation, datasets with highly spread data

samples require higher values of K to detect anomalous data points. Tuning these parameters is an important procedure in the pre-processing stage. Failure to find appropriate values may lead to undetected outliers in input data of the machine learning technique, negatively affecting the performance of the classification algorithm. Our framework determined that the presence of undetected outliers could affect the Area under the Curve metric by up to 13.2%.

(2) **Outlier detection mechanism**: The application of the smoothing spline technique to outlier detection for a non-fraudulent consumer is depicted in Figure 7.2a, where the normalized accumulated power samples of the selected consumer $\grave{S}_c$ are depicted in dark gray and the corresponding spline function $F_\gamma(t)$, is shown in gray. The outlier detection parameters for this consumer have been selected as $\gamma = 10^{-3}$ and $k = 0.5$. As can be observed in Figure 7.2a, after fitting the smoothing spline into the power samples, three outliers were detected, and their values were removed from the time-series data of the consumer. This process was applied to all the

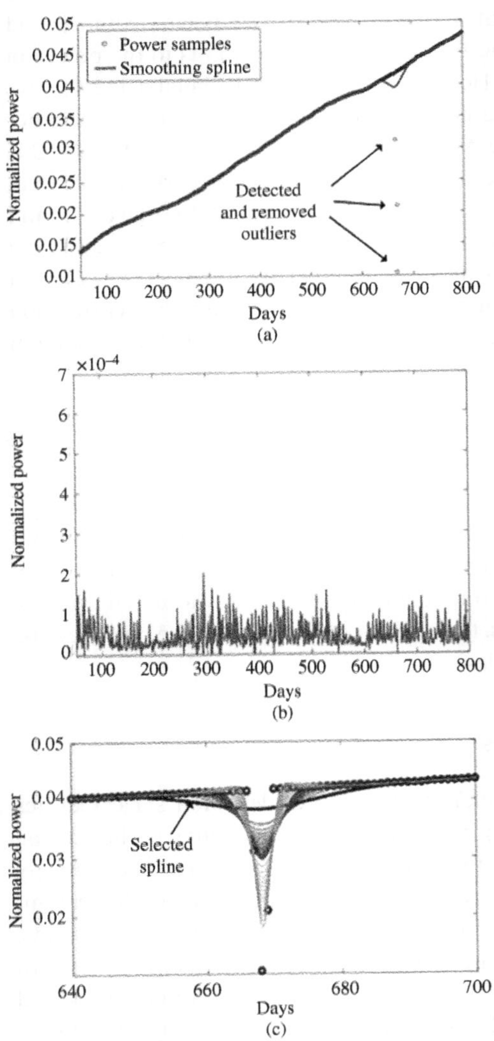

Figure 7.2 (a) Outlier detection and removal using smoothing splines. (b) Calculation of the daily consumption ΔP_t^c after detecting and removing outliers. (c) Zoomed-in view of (a) varying the smoothing parameters. Source: Avila et al. [32]/ with permission of IEEE.

existing consumers in the obtained dataset. After removing outliers, the daily power consumption of the consumer can be estimated from the accumulated power curve as $\Delta P_t^c = P_{t+1}^c - P_t^c$, as depicted in Figure 7.2b. Aiming to illustrate the effect of the smoothing parameter γ in the spline interpolation process, Figure 7.2c depicts a zoomed-in view of (a) for different values of γ. (c) shows the selected spline in black. The gray-colored lines are spline functions with $10^{-3} < \gamma < 10^{-1}$, the light gray-colored lines are splines with $10^{-1} < \gamma < 8 \times 10^{-1}$. As it can be observed in Figure 7.2c, small values of γ (i.e. $\gamma \ll 1$) are the most suitable to detect outliers in the NTL dataset, which is because small values of γ will produce quasilinear splines, which seem to create a good fit for daily accumulated power samples.

7.6.2 MODWPT-Based Signal Decomposition

After the pre-processing stage, the cleansed and consistent time series data of each consumer were decomposed into wavelet coefficients using the MODWPT (Figure 7.3), according to Eqs (7.2)–(7.3). illustrates the decomposition approach for the non-fraudulent consumer whose consumption data is depicted in. For this particular case, the load profile has been decomposed according to Figure 7.1a, of which two out of the eight obtained signals $u_3^{0-7}$are shown. Comparing the wavelet coefficients in Figures 7.1c and 7.3, it can be concluded that the magnitude of the wavelet coefficients of the fraudulent consumer is significantly larger than that of the non-fraudulent one, which is because the power consumption data of fraudulent consumers contain abrupt transitions, which occur at the time of the fraud incident. These transitions are rarely encountered on non-fraudulent consumers. Consequently, the magnitude and standard deviation of the obtained wavelet coefficients provide us with important insights into the power consumption pattern of each consumer. As indicated previously, these features are selected at the last decomposition level according to the absolute magnitude of the wavelet coefficients, using the $K - \text{MAX}$ and STD blocks in Figure 7.1a.

7.6.3 RusBoost NTL Detection Technique

The RusBoost technique shown in Algorithm 7.2 can now be applied to the NTL detection problem. To find the most appropriate hyperparameter values of the wavelet decomposition stage, the feature selection stage, and the RUsBoost algorithm, an exhaustive searching procedure known as grid search was conducted by varying the number of decomposition levels L, the number

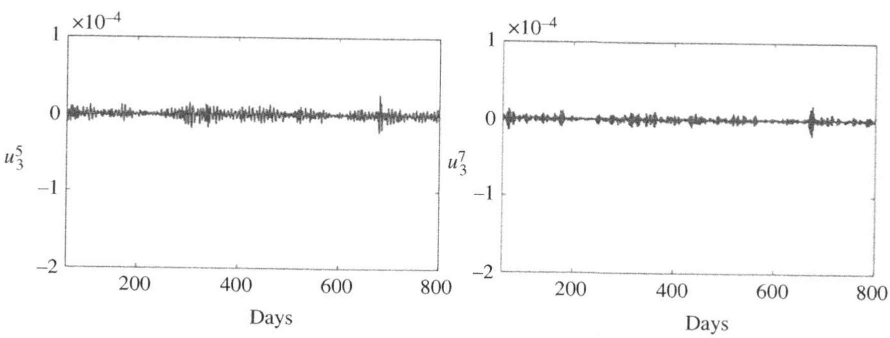

Figure 7.3 MODWPT decomposition of a non-fraudulent consumer. Source: Avila et al. [32]/ with permission of IEEE.

of selected features K (obtained by the K – MAX block in Figure 7.1), and the imbalanced ratio p. The grid search algorithm was conducted using the k-fold cross-validation technique to avoid model overfitting [48].

1) *RusBoost evaluation using a grid search parameter selection approach:* The performance of the NTL detection algorithm, referred to as Q, was evaluated with different values of L, K, and $p\%$. The initial imbalance ratio of the dataset can be calculated as $p_0\% = 2271/(2271+118)\times 100\% = 94.8\%$. Due to the high imbalance characteristic of the dataset, the classification performance must be evaluated using the AUC and the MCC given in Eqs. (7.9) and (7.10) respectively. Since both metrics are important when dealing with imbalanced datasets, the classification performance was assessed by taking the average of these two metrics, $Q = (AUC + MCC)/2$. As previously indicated, the main purpose of RUS is to diminish the imbalance ratio by randomly under-sampling the majority class in the dataset (Figure 7.4). With this in mind, the RUS procedure, shown in Line 6 of Algorithm 7.2, was executed starting from zero under-sampling, $p\% \approx p_0\%$ until the number of samples of each class was equalized, $p\% = 50\%$. It is important to point out that $p\% = 50\%$ should be the maximum allowed RUS. A value of $p\%$ less than 50% indicates a turnaround of the classes in the dataset. In other words, the majority class would become the minority class after applying RUS in the dataset. The turnaround would prevent the classifier from having enough samples of the majority class in the training stage. Moreover, since most of the samples in the testing set belong to the majority class, a reduced performance would be perceived in the testing stage. The performance Q when varying L, K, and $p\%$ can be observed in the heatmaps depicted in Heatmaps in aim to help us find the most appropriate parameters when the parameter space is chosen as $K \in [2, 10]$, $L \in [2, 6]$, and $p\% \in [50\%, 94.8\%]$. discloses the following:
   - Due to the large initial value of $p\% = p_0\%$, the ideal performance of the classification algorithm (Q) is achieved when the majority class is significantly under-sampled (say $p\% < 65\%$). When the majority class is under-sampled to create a training dataset of $50\% < p\% < 65\%$, the classification performance Q is significantly better than when no under-sampling is achieved (i.e. AdaBoost).
   - All utilized performance metrics are listed in Table 7.1. These metrics were obtained using the values of $p\%$, K, and L that resulted in the highest Q value of each heatmap depicted in Figure 7.4. As observed in Table 7.1, the number of decomposition levels seems to play an important role in the performance of the RusBoost algorithm. Table 7.1 reveals that when $p\% \geq 56\%$, five decomposition levels ($L = 5$) achieved the best classification performance. When $p\% \approx 62\%$, either five or four decomposition levels can be utilized, as they have nearly identical maximum Q value (see for $p\% = 62\%$ and $L \in \{4, 5\}$).
   - The highest value for the Area Under the Curve and the Matthews correlation coefficient occur when $(p\%, K, L)$ equal (65%, 7, 5) and (50%, 5, 2), respectively. Thus, the user can select the most appropriate parameters based on the priority given to each of these two metrics. In this paper, since we have assigned equal importance to both metrics, the selected parameters were (50%, 5, 2), with performance metrics given in the last row of Table 7.1.

7.6.4 Comparison with Existing Approaches

The proposed NTL detection approach was compared with existing supervised machine learning frameworks proposed in the literature, including Linear and Non-Linear SVM [5, 31] and Artificial Neural Networks [29, 30]. The techniques used for comparison were conducted using: (i) the average daily power consumption over a month as features [5] along with the oversampling

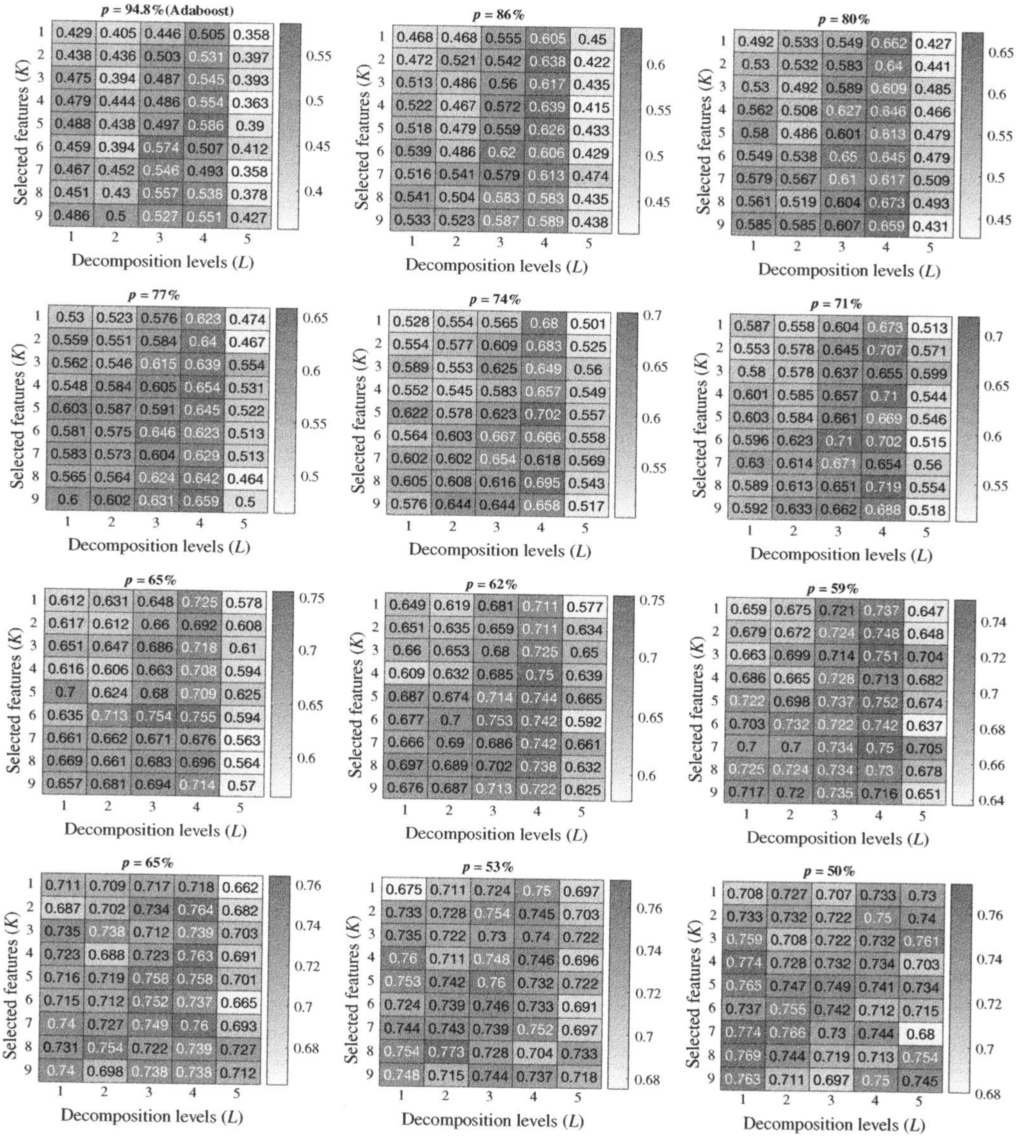

Figure 7.4 Performance evaluation $Q = (AUC + MCC)/2$ under different values of L, K, and p%. Source: Avila et al. [32]/with permission of IEEE.

method in [31] as the imbalance strategy, and (ii) the proposed pre-processing techniques long with the MODWPT-based feature extraction mechanism. With this in mind, the results obtained for the most relevant performance metrics are shown in Tables 7.2 and 7.3. By comparing the AUC and MCC in Table 7.1 with the results in Tables 7.2 and 7.3, we can observe that the proposed approach, which includes a robust pre-processing stage, a feature selection stage using the MODWPT, and the classification stage using the RusBoost algorithm, outperforms the three other methods.

Table 7.1 Performance metrics for the highest value of Q obtained in the heatmaps of Figure 7.4.

P%	*L*	*K*	*MCC*	*AUC*	F_β-score	Precision	Recall	Speciality	Accuracy
94.8	5	6	0.3893	0.7825	0.3351	0.2900	0.9000	0.6650	0.6957
86	5	5	0.4693	0.8096	0.4332	0.3892	0.8167	0.8025	0.8043
80	5	9	0.5207	0.8250	0.4942	0.4520	0.8000	0.8500	0.8435
77	5	10	0.5223	0.7958	0.5372	0.5159	0.7167	0.8750	0.8543
74	5	6	0.5771	0.8275	0.5791	0.5500	0.7500	0.9050	0.8848
71	5	9	0.6133	0.8246	0.6395	0.6250	0.7167	0.9325	0.9043
65	5	7	0.6796	0.8312	0.7327	0.7438	0.7000	0.9625	0.9283
62	4	7	0.6801	0.8263	0.7385	0.7738	0.7000	0.9525	0.9196
59	5	6	0.6909	0.8125	0.7743	0.8198	0.6500	0.9750	0.9326
56	5	3	0.7115	0.8162	0.8013	0.8524	0.6500	0.9825	0.9391
53	3	9	0.7400	0.8058	0.8572	0.9500	0.6167	0.9950	0.9457
50	2	5	0.7300	0.8187	0.8263	0.8864	0.6500	0.9875	0.9435

Table 7.2 Performance comparison without the pre-processing stage and using the average daily power consumption over a month.

Method	*MCC*	*AUC*	Recall	Speciality	Accuracy
Linear SVM	0.08	0.58	0.35	0.82	0.800
Nonlinear SVM	0.15	0.62	0.33	0.91	0.891
ANN	0.13	0.65	0.55	0.74	0.734

Table 7.3 Performance comparison with pre-processing and using the proposed MODWPT-based feature extraction.

Method	*MCC*	*AUC*	Recall	Speciality	Accuracy
Linear SVM	0.70	0.77	0.54	1.00	0.93
Nonlinear SVM	0.70	0.76	0.53	1.00	0.93
ANN	0.60	0.73	0.47	0.99	0.91

7.7 Conclusion

We have proposed a novel method for NTL detection in electric distribution systems. The proposed framework comprises a rigorous outlier detection technique using smoothing spline functions, a maximal overlap discrete wavelet packet transform for feature extraction and selection, and a RusBoost algorithm for classification of NTLs. In addition, the proposed technique has been evaluated using various metrics for binary classification problems. Experiments were conducted using smart meter data collected from 2271 consumers of the electric distribution system in Honduras. Experimental results show that the proposed framework has a Matthews correlation coefficient, area under the curve, F_β-score, precision, recall, specificity, and accuracy of approximately 0.7300, 0.8187, 0.8263, 0.8864, 0.6500, 0.9875, and 0.9435, respectively. Numerical comparisons with previously proposed supervised machine learning techniques show that the proposed framework has a significantly better performance when detecting NTLs.

In our future work, integrating advanced AI-based methods into NTL detection still keeps going [49]. Recently, extreme gradient boosting (Boost), an implementation of the gradient boosting decision tree (DT) algorithm, has also been applied to NTL detection [50, 51]. Furthermore, XGBoost can execute parallel and distributed computing to speed up the process of learning, enabling quicker model exploration. Some elaborations of XGBoost to handle the imbalanced data have also been proposed in [4]. We will continue our work in this direction.

On the other hand, due to the technical limitations of metering and privacy considerations of customers, the large-scale and real-time collection of residential load data remains a challenge. More recently, the concept of generative adversarial networks (GANs) in the area of deep neural networks (DNNs) has been proposed to produce synthetic residential loads as an alternative solution [4]. Generally speaking, the GAN model consists of one generator network that outputs synthetic load profiles and one discriminator network that can differentiate the real or faked load data. Due to the superior learning ability of DNNs, some hidden features of load patterns can be captured entirely. Thus, the accuracy of the NTL detection will be increasing significantly. Our current objective is to integrate GANs into the proposed framework for NTL detection. Hopefully, we will report this result soon.

References

1 Depuru, S.S., Wang, L., and Devabhaktuni, V. (2011). Smart meters for power grid: challenges, issues, advantages and status. *Renewable and Sustainable Energy Reviews* 15 (6): 2736–2742.

2 Viegas, J.L., Esteves, P.R., Melício, R. et al. (2017). Solutions for detection of non-technical losses in the electricity grid: a review. *Renewable and Sustainable Energy Reviews* 80: 1256–1268.

3 Ahmad, T., Chen, H., Wang, J., and Guo, Y. (2018). Review of various modeling techniques for the detection of electricity theft in smart grid environment. *Renewable and Sustainable Energy Reviews* 82: 2916–2933.

4 Wang, Y., Chen, Q., and Kang, C. (2020). *Smart Meter Data Analytics: Electricity Consumer Behavior Modeling, Aggregation, and Forecasting*. Springer Nature Singapore Pte Ltd.

5 Nagi, J., Yap, K.S., Ahmed, S.K., and Mohamad, M. (2010). Non-technical loss detection for metered customers in power utility using support vector machines. *IEEE Transactions on Power Delivery* 25 (2): 1162–1171.

6 Bihl, T.J. and Zobaa, A.F. (2018). Data-mining methods for electricity theft detection. In: *Big Data Analytics in Future Power Systems* (ed. Zobaa, A.F. and Trevor J.B.), 107–124. CRC Press.

7 Glauner, P.O. Boechat, A., Dolberg, L., State, R., Bettinger, F., Rangoni, Y., and Duarte, D. (2016). Large-scale detection of non-technical losses in imbalanced data sets. In: *Proceedings of the 7th IEEE Conference Innovative Smart Grid Technologies*, 1–5.

8 Ghori, K.M., Abbasi, R.A., Awais, M. et al. (2020). Performance analysis of different types of machine learning classifiers for non-technical loss detection. *IEEE Access* 8: 16033–16048.

9 Lin, X.-X., Lin, P., and Yeh, E.-H. (2020). Anomaly detection/prediction for the internet of things: state of the art and the future. *IEEE Network* 25 (1): 212–218.

10 Chawla, N.V., Bowyer, K.V., Hall, L.O., and Kegelmeyer, W.P. (2002). Smote: synthetic minority over-sampling technique. *Journal of Artificial Intelligence Research* 16: 321–357.

11 Tang, Y., Zhang, Y.Q., Chawla, N.V., and Krasser, S. (2008). SVMs modeling for highly imbalanced classification. *IEEE Transactions on Systems, Man, and Cybernetics, Part B (Cybernetics)* 39 (1): 281–288.

12 López, V., Fernández, A., García, S. et al. (2013). An insight into classification with imbalanced data: empirical results and current trends on using data intrinsic characteristics. *Information Sciences* 250: 113–141.

13 Thabtah, F., Hammoud, S., Kamalov, F., and Gonsalves, A. (2020). Data imbalance in classification: experimental evaluation. *Information Sciences* 513: 429–441.

14 Zhu, L., Lu, C., Dong, Z.Y., and Hong, C. (2017). Imbalance learning machine-based power system short-term voltage stability assessment. *IEEE Transactions on Industrial Informatics* 13 (5): 2533–2543.

15 Ghori, K.M., Awais, M., Khattak, A.S. et al. (2021). Treating class imbalance in non-technical loss detection: an exploratory analysis of a real dataset. *IEEE Access* 9: 98928–98938.

16 Glauner, P., Meira, J.A., Valtchev, P. et al. (2017). The challenge of non-technical loss detection using artificial intelligence: a survey. *International Journal of Computational Intelligence Systems* 10 (1): 760–775.

17 Messinis, G.M. and Hatziargyriou, N.D. (2018). Review of non-technical loss detection methods. *Electric Power Systems Research* 158: 250–266.

18 Huang, S.-C., Lo, Y.-L., and Lu, C.-N. (2013). Non-technical loss detection using state estimation and analysis of variance. *IEEE Transactions on Power Systems* 28 (3): 2959–2966.

19 Gerbec, D., Gasperi, S., Smon, I., and Gubina, F. (2005). Allocation of the load profiles to consumers using probabilistic neural networks. *IEEE Transactions on Power Systems* 20 (2): 548–555.

20 Nagi, J., Yap, K.S., Ahmed, S.K. et al. (2011). Improving SVM-based non-technical loss detection in power utility using the fuzzy inference system. *IEEE Transactions on Power Delivery* 26 (2): 1284–1285.

21 Nagi, J., Yap, K.S., Nagi, F., Tiong, S.K., Koh, S.P., and Ahmed, S.K. (2010). NTL detection of electricity theft and abnormalities for large power consumers in TNB Malaysia. In: *Proceedings of the IEEE Student Conference on Research and Development* (SCOReD), 202–206.

22 Léon, C., Biscarri, F., Monedero, I. et al. (2011). Variability and trend-based generalized rule induction model to NTL detection in power companies. *IEEE Transactions on Power Systems* 26 (4): 1798–1807.

23 Monedero, I., Biscarri, F., Léon, C. et al. (2012). Detection of frauds and other non-technical losses in a power utility using Pearson coefficient, Bayesian networks and decision trees. *International Journal of Electrical Power & Energy Systems* 34 (1): 90–98.

24 Léon, C., Biscarri, F., Monedero, I. et al. (2011). Integrated expert system applied to the analysis of non-technical losses in power utilities. *Expert Systems with Applications* 38 (8): 10274–10285.

25 Guerrero, J., Léon, C., Monedero, I. et al. (2014). Improving knowledge-based systems with statistical techniques, text mining, and neural networks for non-technical loss detection. *Knowledge-Based Systems* 71: 376–388.

26 Fawcett, T. (2005). An introduction to ROC analysis. *Pattern Recognition Letters* 27: 861–874.

27 Nizar, A., Dong, Z., Jalaluddin, M., and Raffles, M. (2006). Load profiling method in detecting non-technical loss activities in a power utility. In: *Proceedings of the IEEE International Power Energy Conference*, 82–87.

28 Di Martino, M., Decia, F., Molinelli, J., and Fernández, A. (2012). Improving electric fraud detection using class imbalance strategies. *ICPRAM* 2: 135–141.

29 Muniz, C., Figueiredo, K., Vellasco, M., Chavez, G., and Pacheco, M. (2009). Irregularity detection on low tension electric installations by neural network ensembles. In: *Proceedings of the IEEE International Joint Conference on Neural Networks*, 2176–2182. IEEE.

30 Costa, B.C., Alberto, B.L., Portela, A.M. et al. (2013). Fraud detection in electric power distribution networks using an ann based knowledge-discovery process. *International Journal of Artificial Intelligence & Applications* 4 (6): 17.

31 Figueroa, G., Chen, Y.S., Avila, N.F., and Chu, C.C. (2017). Improved practices in machine learning algorithms for NTL detection with imbalanced data. In: *Proceedings of the IEEE Power & Energy Society General Meeting*, July, 1–5. IEEE.

32 Avila, N.F., Figueroa, G., and Chu, C.C. (2018). NTL detection in electric distribution systems using the maximal overlap discrete wavelet-packet transform and random undersampling boosting. *IEEE Transactions on Power Systems* 33 (6): 7171–7180.

33 Nagi, J., Yap, K.S., Tiong, S.K. et al. (2010). Non-technical loss detection for metered customers in power utility using support vector machines. *IEEE Transactions on Power Delivery* 25 (2): 1162–1171.

34 Seiffert, C., Khoshgoftaar, T., Van Hulse, J., and Napolitano, A. (2010). Rusboost: a hybrid approach to alleviating class imbalance. *IEEE Transactions on Systems, Man, and Cybernetics - Part A: Systems and Humans* 40 (1): 185–197.

35 Raudys, S.J. and Jain, A.K. (1991). Small sample size effects in statistical pattern recognition: recommendations for practitioners. *IEEE Transactions on Pattern Analysis and Machine Intelligence* 13 (3): 252–264.

36 Craven, P. and Wahba, G. (1978). Smoothing noisy data with spline functions. *Numerische Mathematik* 31 (4): 377–403.

37 Green, P.J. and Silverman, B.W. (1993). *Nonparametric Regression and Generalized Linear Models: A Roughness Penalty Approach*. London, U.K.: Chapman & Hall.

38 Al-Otaibi, R., Jin, N., Wilcox, T., and Flach, P. (2016). Feature construction and calibration for clustering daily load curves from smart-meter data. *IEEE Transactions on Industrial Informatics* 12 (2): 645–654.

39 Zheng, K., Chen, Q., Wang, Y. et al. (2019). A novel combined data-driven approach for electricity theft detection. *IEEE Transactions on Industrial Informatics* 15 (3): 1809–1819.

40 Liu, H. and Motoda, H. (2001). *Feature Extraction, Construction and Selection: A Data Mining Perspective*. Norwell, MA, USA: Kluwer Academic Publishers.

41 Percival, D.B. and Walden, A.T. (2000). *Wavelet Methods for Time Series Analysis*, 1e, 206–253. Cambridge University Press.

42 Daubechies, I. (1988). Orthonormal bases of compactly supported wavelets. *Communications on Pure and Applied Mathematics* 41: 909–996.

43 Stephane, M. (2009). *A Wavelet Tour of Signal Processing*, 3e, 321–375. Elsevier.

44 Kubat, M. et al. (1997). Addressing the curse of imbalanced training sets: one-sided selection. In: *Proceedings of the 4th International Conference on Machine Learning*, Nashville, USA, vol. 97, 179–186.

45 Japkowicz, N. and Stephen, S. (2002). The class imbalance problem: a systematic study. *Intelligent Data Analysis* 6 (5): 429–449.

46 Weiss, G.M. (2004). Mining with rarity: a unifying framework. *SIGKDD Explorations* 6 (1): 7–19.

47 Freund, Y. and Schapire, R.E. (1997). A decision-theoretic generalization of on-line learning and an application to boosting. *Journal of Computer and System Sciences* 55 (1): 119–139.

48 Arlot, S. and Celisse, A. (2010). A survey of cross-validation procedures for model selection. *Statistics Survey* 4: 40–79. https://doi.org/10.1214/09-SS054.

49 Guerrero, J.I., Léon, C., Biscarri, F., Monedero, I., Biscarri, J., and Millán, R. (2010). Increasing the efficiency in non-technical losses detection in utility companies. In: *MELECON 2010–2010 15th Mediterranean Electrotechnical Conference*, 136–141. IEEE.

50 Yan, Z. and Wen, H. (2021). Electricity theft detection base on extreme gradient boosting in AMI. *IEEE Transactions on Instrumentation and Measurement* 70: 2504909.

51 Wang, C., Deng, C., and Wang, S. (2020). Imbalance-XGBoost: leveraging weighted and focal losses for binary label-imbalanced classification with XGBoost. *Pattern Recognition Letters* 136: 190–197.

8

Electricity Market Participation Profiles Classification for Decision Support in Market Negotiation

Tiago Pinto and Zita Vale

GECAD – Research Group on Intelligent Engineering and Computing for Advanced Innovation and Development, School of Engineering, Polytechnic of Porto, Porto, Portugal

Acronyms

AI	Artificial intelligence
ANN	Artificial neural network
P2P	Peer-to-peer
SVM	Support vector machine

8.1 Introduction

Electricity markets are evolving [1]. Consumers are becoming active players in power and energy systems, and increasing the capability of generating their own energy [2]. On the other hand, new players, such as aggregators [3], communities [4], and new types of loads and resources, such as electric vehicles [5], distributed storage, and demand flexibility are emerging. The new context is forcing the conventional electricity market environment to adapt to the new and upcoming reality [6].

The traditional market models are mostly designed to deal with a rather stable generation profile and consumption estimation. However, the new characteristics of the system require more dynamic and fair market models, able to accommodate the variability from most renewable-based generation, while boosting the potential of flexibility from the consumer side [7]. Among the several proposals for future electricity market evolution, peer-to-peer (P2P) trading is one of the most promising approaches, placing energy trading under direct responsibility of the consumer and small generator [8]. Direct, or bilateral negotiations, among small players, are therefore envisaged as one of the most adopted approaches for future energy trading.

Small players, however, are inexperienced in such a role, which demands for the emergence of new models able to support their actions in such trading process. Several artificial intelligence (AI) models are being proposed to deal with this problem, using, e.g. agent-based technology [9] and Bayesian models [10], but these are only the first steps towards the achievement of effective solutions.

Data mining and analysis have a huge potential to give very relevant contributions in this context, providing multiple models that can be useful to address this type of problem. This chapter illustrates how data mining approaches, namely clustering and classification models, can contribute to

Intelligent Data Mining and Analysis in Power and Energy Systems: Models and Applications for Smarter Efficient Power Systems, First Edition. Edited by Zita Vale, Tiago Pinto, Michael Negnevitsky, and Ganesh Kumar Venayagamoorthy.

the development of decision support approaches that envisage the improvement of players' results from bilateral negotiations.

The chapter is organized into five sections. Section 8.1 introduces the topic addressed in this chapter. Section 8.2 provides an overview of bilateral negotiations. Section 8.3 describes the methodology for players decision support in what regard market negotiations. Section 8.4 presents some illustrative results, showcasing the application of the clustering, classification, and automated adaptation models. Finally, Section 8.5 discusses the main conclusions from this chapter.

8.2 Bilateral Negotiation

Bilateral negotiation within electricity markets corresponds to direct negotiations made by two players, which may include energy prices and quantity, delivery time, and contract duration. Players of this market define contracts' terms and conditions independently from the market operator. Usually, players undertake bilateral contracts to protect themselves from market's fluctuation prices in the daily market. Furthermore, bilateral contracts are very flexible, since both parts can set the contract terms. Bilateral contracts can be very specific and include different types of contracts, the following being the most important ones:

- **Forward contracts**: In these contracts the price is agreed in a certain date, but the electricity is only delivered to the buyer later, from a few months to years;
- **Future contracts**: Like forward contracts, the electricity is delivered later in the future with a previously defined price. The difference is that future contracts are standard contracts, the quantity and quality of electricity are well defined, and all the terms of the contract are formalized;
- **Option contracts**: In this type of contracts, a player gets the option to buy or sell electricity in the future at an agreed price set when the contract is negotiated;
- **Contracts for difference**: In this contract, a player pays the difference between the markets' electricity price and the price defined by the established contract.
  Within these contracts there are three types of trading types [11]:
  - **Customized long-term contracts**: These contracts involve big amounts of electricity, hundreds, or thousands of Megawatts. The terms and conditions of these contracts do not need the intervention of the market operator. The operator only checks if there is the minimum quantity to make the transaction and also checks the security in the transmission of electricity.
    Advantages: Flexibility as terms and conditions are private.
    Disadvantages: Cost of the negotiation and not fulfillment of some counterparties;
  - **Trading "over the counter"**: Small amounts of electricity are involved. Usually, a standard profile is defined for the delivery of the electricity, for example, defining a certain amount of energy to be delivered in different times of the day.
    Advantages – Small cost;
  - **Electronic trading**: The negotiation is done in an electronic market. If there is a suitable offer for a proposal made, the software makes the deal. If there is not a suitable offer, the proposal remains in the market till a matching offer is made. This type of negotiations is usually performed near the closing of the market.
    Advantages: Allows an easy negotiation.

The negotiation of bilateral contracts can be divided into three steps [12]. In the first step, the players define their "*objectives, negotiation limits and initial offers.*" This is called the pre-negotiation step. Next, they try to reach an agreement, using their own strategies for the negotiation. In the final step, called implementation phase, if an agreement has been made, the contract is put in practice.

When seeking to reach advantageous agreements, it is essential that players are able to understand who they are negotiating with. While modelling an opponent in this type of negotiation, there are three main questions [13]:

- **Preference estimation**: What is the opponent trying to achieve?
- **Strategy prediction**: Which actions will he/she take, and when?
- **Opponent classification**: What is the profile of the opponent player and what can we do to against such profile to take as much benefit as possible?

In a negotiation, the knowledge on opponents expected behavior can be used to minimize costs, adapt strategies according to the opponent player, or reach agreements where both players win. There are four main attributes that need to be considered when trying to model another player [14]:

- **The acceptance strategy**: It regards whether a player will accept an agreement and can be determined through explicit knowledge of a certain player or by estimating a certain probability of acceptance;
- **Deadline**: It is the time till when a player must reach an agreement;
- **Preference profile**: It relates to the importance of different issues or criteria in negotiations for a player;
- **Bidding strategy**: It is determined by the whole negotiation strategy used by a player, defining how the negotiation and concession will occur. There are usually three main strategies used by the negotiating players [15]:
- **Concession making**: A negotiator reduces his profit to be able to make an agreement with the opposing player;
- **Competing**: A negotiator holds his selling/buying terms and negotiators and tries to persuade the other player to fulfill its terms;
- **Problem solving**: Negotiators work with each other in order to achieve a reasonable price for both. This strategy may be divided into two sub-strategies [16] which are named logrolling and nonspecific compensation. Logrolling consists in each player conceding in low-priority issues for themselves and high-priority for the other player. In nonspecific compensation, one of the players accomplishes his goals and pays the other for fulfilling his interests.
  Each strategy can be classified in one of three following types:
  - **Time dependent**: If a player has a certain deadline while negotiating, the strategy considers it, usually leading to a faster give in near the deadline;
  - **Resource dependent**: Similar to time dependent strategies, but in this case the factor taken into account is a certain limited resource, which affects the way of negotiating;
  - **Behavior dependent**: It depends on the behavior of the opponent player and tries to achieve the best result based on that behavior.

According to these types of strategies, some generic players' profiles can be considered [17]:

- **Determined**: Players do not change their bids through the negotiation;
- **Anxious**: Players change rapidly their initial prices after a short period of negotiation, in case they do not reach their goals;
- **Moderated**: Changes in prices are steady and are usually made in the middle of the negotiation period;
- **Gluttonous**: Players only make slight changes to their prices until the final stage of the negotiation. In the final stage, drastic changes are made in order to achieve an agreement.

There are, however, several other strategies for bilateral negotiations that have been developed. Some examples are the conceder, moderate conceder, linear, moderate hardheaded and hardheaded, as proposed by Awasthi et al. [16]. The basics of these strategies range, respectively, from a big yielding in prices in the beginning (conceder) till yielding only in the end (Hardheaded). Awasthi et al. [16] also propose a system which tries to figure out the opponents' strategy through the analysis of two consecutive offers. In case the opponent had a conceder strategy, they adopt a hardheaded strategy, if it was a hardheaded they would negotiate in a moderate conceder way. An algorithm for choosing the best strategy among a number of available strategies can be found in [18].

The study presented in [11] focused on forward contracts in the bilateral market and showed that among 7 different strategies the ones with better results were concession strategies. These strategies were low-priority concession making, meaning that a player concedes in low-priority aspects and energy dependent concession making, where concessions are made strategically along the negotiation according to the amount of energy traded and the period of the day.

In another work [19], which involved forward contracts' negotiation, a mathematical model for a concession strategy has been developed. The model, to be used by retailers, includes a concession factor which took into account the quantity of energy that was expected to sell in a certain period, making it possible to "*determine the concessions to be made throughout the negotiation.*"

Some strategies based on utility for sellers and buyers can be found in [20] and [9]. The work in [21] simulated the negotiation of contracts for difference and concluded that sellers can maximize their profit when they use a more risky attitude. Also, with this attitude sellers can make more profits with contracts for difference than with forward contracts.

A specific decision support system for bilateral contracts negotiation is proposed in [22] and demonstrated in [23]. This system comprehends two main phases: (i) pre-negotiation, in which the most suitable opponents are identified, and the negotiation planning is performed; and (ii) negotiation, which includes the actual learning process for automated bilateral negotiations. This model includes three main decision methods. Players using the pessimistic method negotiate to get the best possible profit considering the worst-case scenario while players using the optimistic method, which is riskier, aim at getting the highest possible gain. The "Most probable scenario" method tries to identify the scenario with higher possibility of occurrence and supports the decisions so that profits can be maximized according to that particular scenario.

8.3 Decision Support for Bilateral Negotiations

An efficient decision support for bilateral negotiations requires that specific machine learning models are proposed to enable deciding which negotiation tactics and strategies to apply against each opponent at each time, depending on the negotiation context and the characteristics of the negotiation – Figure 8.1.

In this work, a learning algorithm is proposed, which is characterized as a learning network, as it follows the same principles as a typical artificial neural network: several interconnected nodes with dynamic weights assigned to each link between nodes – Figure 8.2. Each node represents a different learning result, based on the analysis of different types of information:

- Analysis of data observed directly by the system (from current and previous negotiations);
- Learning from the experience of negotiations with similar players (similar players are identified through the models described in Section 8.3.1);
- Learning based on the history of negotiations in similar contexts.

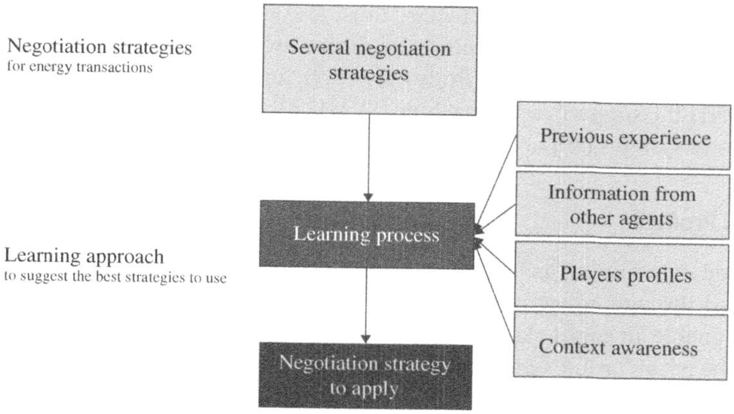

Figure 8.1 Learning process overview.

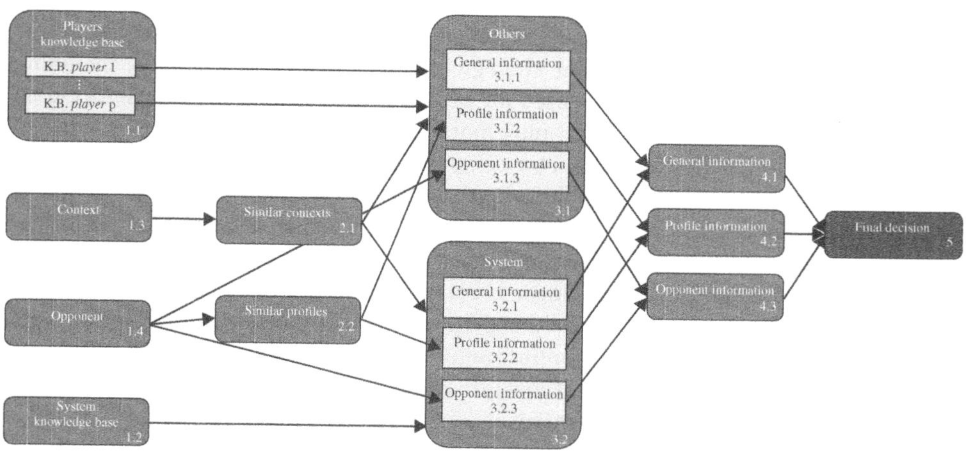

Figure 8.2 Learning network architecture.

The weights propagation between the nodes uses collaborative reinforcement learning, by using the results of different independent learning processes to reach a common solution [24]. The Bayesian theorem of probability is also experimented to propagate the learning network weights, by associating the output of each node with its probability of success [25]. The learning network provides as final output the strategy (or combination of strategies) to apply in each specific situation.

Another novel approach regards the association of expectation values with the agreements that are expected with each opponent (considering the targets for the multiple negotiation items). This expectation value is associated with a confidence/uncertainty value, which is adapted according to the forecasting/estimation error achieved in the prediction based on the opponent's previous agreements (according to the context). As the negotiation process progresses, both the expectation (target) values and the uncertainty values associated with each negotiation are updated (learned) according to the difference between the expected negotiation profile of the opponent and the actual actions that are verified. In this way, the priorities on reaching each agreement (in simultaneous negotiations with multiple opponents) and the expected target of each agreement are updated throughout the negotiation process, thus, also updating and improving the negotiation learning process itself. Multiple learning algorithms are experimented, namely reinforcement learning algorithms (*Q*-Learning [26], Roth-Erev [27]), and regression algorithms (artificial neural network [ANN] [28], support vector machine [SVM] [29], hidden Markov models, and hybrid neuro-fuzzy inference systems [30]).

8.3.1 Clustering of Players Profiles

Clustering is one the most used data mining techniques for finding similarities/patterns in data. In this work, clustering is used through *k-means* algorithm to easily group developed strategies into players' profiles and also enable an easier classification of data, as detailed in [31]. Thus, it is possible to identify which is the group profile of an opponent player using different strategies for each group, trying to use the best approach to deal with the different groups found. Clustering has the capability of congregating players into stereotype groups, so in a future negotiation an opponent is not put in a very restrict group, which could jeopardize the negotiation in case it was put in the wrong group. Clustering results are used to train classification models, as presented in Section 8.3.2, since the goal is to specify in which group, or negotiation profile, a certain strategy is.

For clustering analysis all time-dependent strategies presented in [15] are used, including mixed strategies. Three examples have been used for each strategy that has an increase/decrease factor. It was used the factor standard value plus more two with a higher and lower value to include cases where prices are increased (for a buyer) or decreased (for a seller) faster. Developed behavior-dependent strategies were not included in clustering because they do not follow a mathematical formula, not being possible to calculate fixed values, which is needed for clustering.

Data has been scaled, so that all values were put in a 0 to 1 scale using feature-scaling method as in (8.1), where z' is the obtained scaled value, z is the value to be scaled and min N and max N are the minimum and maximum values used during the negotiation. This method allows to scale data using price range's maximum and minimum value, so after scaling all minimum values are set to 0 and maximum values to 1. As the final goal is to classify proposals into a certain group and to correctly classify them, scaling data is essential.

$$z' = \frac{z - \min\ N}{\max\ N - \min\ N} \tag{8.1}$$

Clustering was performed using kml library of R software, this library is an implementation of *k*-means algorithm and it was specially developed to analyze longitudinal data, which is the current case. *K*-means is a hill-climbing algorithm capable of clustering. *K*-means generally follows the next steps:

1. User defines a k number of clusters to find.
2. The algorithm finds central points for each cluster found.
3. Each data point calculates the distance to the nearest central point, forming different groups of data points with a common nearest center.
4. Each center searches for the centroid of its group and takes that position.
5. First 4 steps are repeated until none center can change to another group.

After choosing a number of k clusters to find, Kml has seven different methods to assign points to clusters being the default one *k-means++*, full detail about these methods can be found in [32]. Kml library can calculate distance between points using different methods including Manhattan, Euclidean, and Minkowski. Default method is the Euclidean Distance with Gower Adjustment (capable of dealing with missing values). To calculate the distance between 2 points $\boldsymbol{y}_i$ and $\boldsymbol{y}_j$, if

y_i, y_j or both are missing variable w_{ijl} is 0, if none is missing w_{ijl} is 1. The distance is calculated as in (8.2).

$$Dist^E_{NA}(y_i, y_j) = \sqrt{\frac{t}{\sum_{l=1}^{t} w_{ijl}} \sum_{l=1}^{t} (y_{il} - y_{jl})^2 \cdot w_{ijl}} \tag{8.2}$$

Clustering results show then a quality criterion for a number of k clusters chosen. This criterion ranges from 0 to 1 and is used to understand if a certain number of clusters is better than another. Quality criteria is measured using five different criteria [32], namely Calinski and Harabasz, Calinski and Harabasz Kryszczuk variant, Ray and Turi, and Davies and Bouldin criterion. This is very helpful, as it is possible to undertake a decision based on different criteria. In the current case all criteria pointed to the same number of clusters.

8.3.2 Classification of New Players

Classification of proposals from sellers or buyers is one of the most crucial steps of this work. Through classification it is possible to put an opponent player in a class/group of negotiators, in this case one the clusters found in the previous step. Hence it is possible to use the best approach to negotiate against the other players' profile. In this work, classification is performed using ANNs and SVMs and it is implemented in *R* language. An opponent player is classified starting on the 3rd proposal and reclassified at each new proposal. This process makes it possible to calculate our next proposals during a negotiation. The training is performed considering the points from each cluster' trajectories.

8.3.2.1 Artificial Neural Networks

ANN training is performed using *nnet* library included in *R* software, according to Eq. (8.3), where $\boldsymbol{y}$ is the output, $\boldsymbol{w}$ corresponds to a weight, in this case the weight between neurons i and k, j and k and between i and j. Variable $\boldsymbol{\alpha}$ is the bias and $\boldsymbol{\emptyset_h}$ is the activation function of the hidden layer units, the logistic function expressed in (8.4).

$$y_k = \phi_o\left(\alpha_k + \sum_{i\to k} w_{ik}x_i + \sum_{j\to k} w_{jk}\phi_h\left(\alpha_j + \sum_{i\to j} w_{ij}x_i\right)\right) \tag{8.3}$$

$$f(h) = \frac{\exp(h)}{1+\exp(h)} \tag{8.4}$$

This ANN is a feed-forward neural network. It has an input layer that receives the values of the offers/proposals done by sellers or buyers. Thus, the number of neurons in this layer is the same as the number of proposals already made in a negotiation. The ANN was structured with a single hidden layer with eight neurons, as it was with this number that best results were achieved. Output layer has four neurons, as there are four possible groups where proposals can be classified. Figure 8.3 represents the topology used for a case where three proposals have been done.

8.3.2.2 Support Vector Machines

For training of SVM, e1071 library from *R* software has been used. The basic purpose of SVM is to define hyperplanes that can separate points from different classes. In this case, for formulation of the SVM model, *C*-Classification has been used along with a Radial basis kernel, as these usually have good performance. *C*-Classification is expressed in (8.5), where $\boldsymbol{e}$ is the unity vector, $\boldsymbol{C}$ is the upper bound, $\boldsymbol{\alpha}$ represents the weights, $\boldsymbol{Q}$ is an $\boldsymbol{l}$ by $\boldsymbol{l}$ positive semidefinite matrix defined

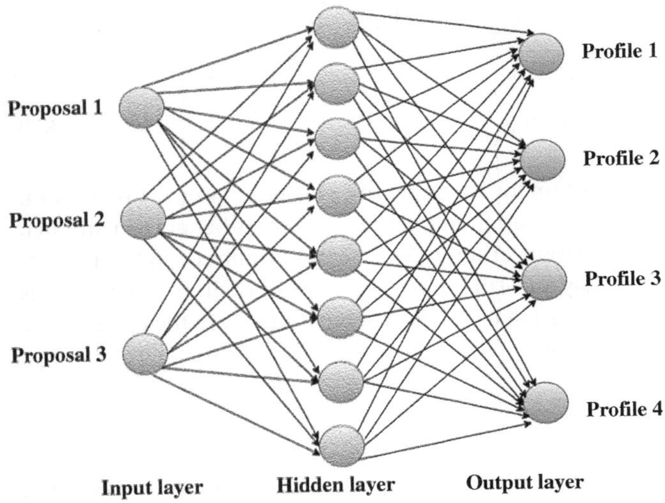

Figure 8.3 ANN topology.

in (8.6), $\boldsymbol{K}$ is the kernel, in this case the radial basis function expressed in (8.7). In (8.7) u and v represent two different points. Parameters $\boldsymbol{C}$ and γ must be set during training, which influences final classification results. The best values for $\boldsymbol{C}$ and γ have been chosen based on the analysis of charts produced by e1071 library, that make an estimation of the prediction error depending on the different values for $\boldsymbol{C}$ and γ.

$$\min_{\alpha} \frac{1}{2}\alpha^{\mathrm{T}}\boldsymbol{Q}\alpha - e^{\mathrm{T}}\alpha$$
$$\text{s.t.}\ 0 \leq \alpha_i \leq C,\ \ i = 1, \ldots, l,$$
$$\mathrm{y}^{\mathrm{T}}\alpha = 0, \tag{8.5}$$

$$Q_{ij} = y_i y_j * K \tag{8.6}$$

$$K(\boldsymbol{u}, \boldsymbol{v}) = \exp(-\gamma|\boldsymbol{u} - \boldsymbol{v}|^2) \tag{8.7}$$

8.4 Illustrative Results

This section describes a case study from a seller perspective, considering a negotiation against a buyer. In the clustering process, for determining the main buyer's profiles that could be found, a total of 15 strategies have been used. Nine basic strategies plus six strategies with different parameters that change the prices' change. These strategies are listed as follows [15]:

- **Anxious strategy**: players decrease prices in the beginning of the negotiation (with increment factor of 0.6, 0.8, and 1);
- **Gluttonous strategy**: price variations are only made near the end of a negotiation (with decrement factor of 0.8, 0.9, and 1);
- **Moderated strategy**: price variations are made in a steady way;
- **Determined strategy**: players only make slight changes to the initial price;
- Percentage decrease (3%, 4%, and 5%);
- Combination between Gluttonous (in the beginning of the negotiation) and Anxious (near the end of the negotiation);

- Combination between Gluttonous (in the beginning of the negotiation), Anxious (in the middle of the negotiation), and Gluttonous again (near the end of the negotiation);
- Combination between Anxious (in the beginning of the negotiation), Gluttonous (in the middle of the negotiation), and Anxious again (near the end of the negotiation);
- Gluttonous + Anxious + Gluttonous + Anxious + Gluttonous strategy.

Table 8.1 shows inputs for the clustering process considering a maximum of eight proposals during the negotiation. Prices shown were scaled using feature-scaling, 0 and 1 values correspond to minimum and maximum prices used during the negotiation, respectively.

Figure 8.4 shows a chart with clustering results for buyers. In x axis the number of clusters is presented, the y axis shows the quality criteria for each number of clusters. Nine basic strategies

Table 8.1 Clustering input values for all negotiation strategies.

No. of proposal Strategy	1	2	3	4	5	6	7	8
Anxious inc. fac. = 0.6	0	0.464	0.675	0.796	0.874	0.929	0.969	1
Gluttonous inc. fac. = 0.9	0	0.02	0.048	0.086	0.143	0.238	0.429	1
Moderated	0	0.143	0.286	0.426	0.571	0.714	0.857	1
Determined	0.5	0.5	0.5	0.5	0.5	0.5	0.5	0.5
Percentage inc. 4%	0	0.127	0.258	0.395	0.538	0.686	0.84	1
Gluttonous + Anxious	0	0.022	0.052	0.093	0.575	0.794	0.919	1
Gluttonous + Anxious + Gluttonous	0	0.024	0.056	0.595	0.841	0.981	0.989	1
Anxious + Gluttonous + Anxious	0	0.464	0.676	0.685	0.697	0.712	0.911	1
Gluttonous + Anxious + Gluttonous + Anxious + Gluttonous	0	0.022	0.524	0.752	0.762	0.775	0.994	1
Anxious inc. fac. = 1	0	0.571	0.762	0.857	0.914	0.952	0.98	1
Gluttonous inc. fac. = 1	0	0.020	0.048	0.086	0.143	0.238	0.429	1
Percentage inc. 5%	0	0.123	0.252	0.387	0.53	0.679	0.835	1
Anxious inc. fac. = 0.8	0	0.524	0.725	0.832	0.898	0.943	0.975	1
Gluttonous inc. fac. = 0.8	0	0.020	0.047	0.086	0.143	0.238	0.429	1
Percentage inc. 3%	0	0.130	0.265	0.403	0.546	0.693	0.844	1

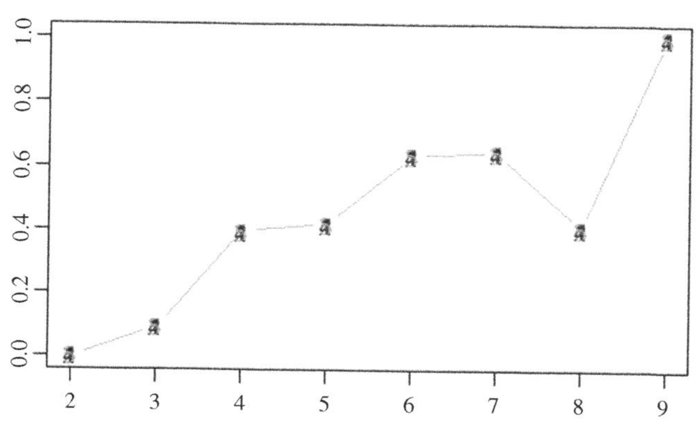

Figure 8.4 Clustering results using buyer strategies as input. Source: Pinto et al. [31]/with permission of IEEE.

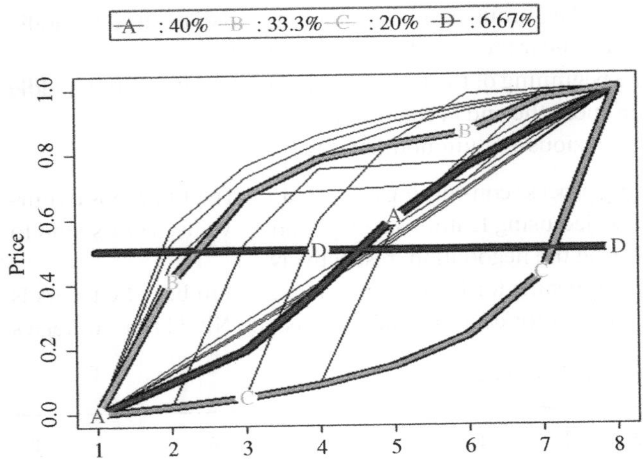

Figure 8.5 Trajectories for all the four profiles that buyers have, resulted from clustering process. Source: Pinto et al. [31]/with permission of IEEE.

were used for the clustering process, so choosing more than 5 clusters would not be reasonable, because it would not sufficiently group the strategies. Considering from 2 to 5 clusters, the best results are with 4 and 5 clusters. Analyzing clusters' trajectories, it was noticeable that if 5 clusters were chosen, two of them would have too much similar trajectories, so 4 was chosen as the best number of clusters, as all of the 4 clusters have distinct trajectories.

Figure 8.5 represents the trajectories for each of the 4 clusters and the percentage of strategies that each cluster covers and Table 8.2 details the cluster where each strategy profile is placed.

With the obtained results from the clustering process, it was possible to train the ANN and SVM models for classifying new players according to the profile of their negotiation behavior [13]. The training data used in the ANN and SVM with 8 proposals is represented in Table 8.3. Showing for each strategy its normalized values (using feature-scaling) and in which cluster it was placed.

After being able to classify an opponent player into a certain cluster, representing a group/profile of player, simulations have been done to assess which strategies had a better performance against each of the players' profiles. Table 8.4 shows the strategies to be used against profiles 1, 2, 3, and 4. These profiles are shown in Figure 8.3 with letters A, B, C, and D, respectively.

In the beginning of every negotiation, we always start using Gluttonous strategy, as it is the best one against most of the other strategies and only reduces the price slightly, until there is enough data to evaluate the opponent player. Two examples of negotiations are shown. Table 8.5 shows a negotiation against a buyer using a Gluttonous + Anxious strategy.

As one can see by Table 8.5, the first two proposals by the supported player used the Gluttonous strategy, then the classification tells us that the player's profile is 3, so Gluttonous + Anxious + Gluttonous strategy is adopted in proposals 3 and 4. After 5 proposals, with a sudden increase in price by the buyer, it becomes clear that the player's profile is not 3, but 1, so Gluttonous strategy is readopted. Table 8.6 shows the performance of different strategies against a Gluttonous + Anxious buyer.

Table 8.6 shows that Gluttonous strategy is the best against a Gluttonous + Anxious buyer, as the selling price is higher when negotiating with this strategy. Even though the strategy with a percentage decrease in prices has the same selling price, it is less constant than Gluttonous strategy and in other occasions has worse results, so Gluttonous is considered the best.

Table 8.2 Buyer's strategies and their respective cluster.

Strategy	Cluster
Anxious inc. fac. = 0.6	2
Gluttonous inc. fac. = 0.9	3
Moderated	1
Determined	4
Percentage inc. 4%	1
Gluttonous + Anxious	1
Gluttonous + Anxious + Gluttonous	1
Anxious + Gluttonous + Anxious	2
Gluttonous + Anxious + Gluttonous + Anxious + Gluttonous	2
Anxious inc. fac. = 1	2
Gluttonous inc. fac. = 1	3
Percentage inc. 5%	1
Anxious inc. fac. = 0.8	2
Gluttonous inc. fac. = 0.8	3
Percentage inc. 3%	1

Table 8.3 Training data used in the ANN and SVM with a maximum of 8 proposals.

No. of proposalStrategy	1	2	3	4	5	6	7	8	Cluster
Anxious inc. fac. = 0.6	0	0.464	0.675	0.796	0.874	0.929	0.969	1	2
Gluttonous inc. fac. = 0.9	0	0.02	0.048	0.086	0.143	0.238	0.429	1	3
Moderated	0	0.143	0.286	0.429	0.571	0.714	0.857	1	1
Determined	0.5	0.5	0.5	0.5	0.5	0.5	0.5	0.5	4
Percentage inc. 4%	0	0.127	0.258	0.395	0.538	0.686	0.84	1	1
Gluttonous + Anxious	0	0.022	0.052	0.093	0.575	0.794	0.919	1	1
Gluttonous + Anxious + Gluttonous	0	0.024	0.056	0.595	0.841	0.981	0.989	1	1
Anxious + Gluttonous + Anxious	0	0.464	0.676	0.685	0.697	0.714	0.911	1	2
Gluttonous + Anxious + Gluttonous + Anxious + Gluttonous	0	0.022	0.524	0.752	0.762	0.775	0.994	1	2
Anxious inc. fac. = 1	0	0.571	0.762	0.857	0.914	0.952	0.98	1	2
Gluttonous inc. fac. = 1	0	0.02	0.048	0.086	0.143	0.238	0.429	1	3
Percentage inc. 5%	0	0.123	0.252	0.387	0.529	0.679	0.835	1	1
Anxious inc. fac. = 0.8	0	0.524	0.725	0.832	0.898	0.943	0.975	1	2
Gluttonous inc. fac. = 0.8	0	0.02	0.048	0.086	0.143	0.238	0.429	1	3
Percentage inc. 3%	0	0.131	0.265	0.403	0.546	0.693	0.844	1	1

Table 8.4 Counter-strategies to use against buyers.

Cluster/Profile of buyer	Counter strategy to be used by seller
1	Gluttonous
2	Gluttonous
3	Gluttonous + Anxious + Gluttonous
4	Moderated

Table 8.5 Negotiation against buyer using Gluttonous + Anxious strategy.

	Our values (seller) min = 33 max = 53	Gluttonous + Anxious buyer (opponent) min = 30 max = 44	Strategy used for each proposal	Profile
1	53	30	Gluttonous	
2	52.679	30.225	Gluttonous	
3	52.257	30.525	Gluttonous + Anxious + Gluttonous	3
4	45.036	30.945	Gluttonous + Anxious + Gluttonous	3
5	44.494	35.841	Gluttonous	1
6	43.591	38.066	Gluttonous	1
7	41.786	39.338	Gluttonous	1
8	40.16	40.16	Gluttonous	1

Table 8.6 Strategies performance comparison against buyer using Gluttonous + Anxious strategy.

Seller max = 53. min = 33; Gluttonous + Anxious Buyer min = 30. max = 44			
Counter strategy	**Agreement achieved**	**Value of last offer**	**No. of last offer**
Anxious	Yes	38.07	8
Gluttonous	**Yes**	**40.16**	**8**
Moderated	Yes	39.33	8
Determined	No	43	8
Percentage dec.4%	Yes	40.16	8
Gluttonous + Anxious	Yes	39.66	8
Gluttonous + Anxious + Gluttonous	Yes	39.76	8
Anxious + Gluttonous + Anxious	Yes	39.34	8
Gluttonous + Anxious + Gluttonous + Anxious + Gluttonous	Yes	39.34	8
Behavior dependent	Yes	40.15	8

Table 8.7 Negotiation against buyer using Gluttonous strategy.

	Our values (seller) min = 33 max = 53	Gluttonous buyer (opponent) min = 30 max = 44	Strategy used for each proposal	Profile
1	53	30	Gluttonous	
2	52.679	30.225	Gluttonous	
3	52.257	30.525	Gluttonous + Anxious + Gluttonous	3
4	45.036	30.945	Gluttonous + Anxious + Gluttonous	3
5	41.753	31.575	Gluttonous + Anxious+ Gluttonous	3
6	41.566	32.625	Gluttonous + Anxiou + Gluttonous	3
7	41.303	34.725	Gluttonous + Anxiou + Gluttonous	3
8	41.025	41.025	Gluttonous + Anxious + Gluttonous	3

Table 8.8 Strategies performance comparison against buyer using Gluttonous strategy.

Seller max = 53. min = 33; Gluttonous buyer min = 30. max = 44			
Counter strategy	**Agreement achieved**	**Value of last offer**	**No. of last offer**
Anxious	Yes	36.85	8
Gluttonous	Yes	37.25	8
Moderated	Yes	35.5	8
Determined	No	43	8
Percentage dec. 4%	Yes	39.82	8
Gluttonous + Anxious	Yes	38.49	8
Gluttonous + Anxious + Gluttonous	**Yes**	**39.62**	**8**
Anx. + Glu. + Anx.	Yes	36.85	8

Table 8.7 shows a negotiation against a buyer with a Gluttonous strategy.

By Table 8.7 it is seen that the opponent player is always classified as having profile 3 (Gluttonous approach), thus, Gluttonous + Anxious + Gluttonous strategy is adopted. This strategy works well against Gluttonous player because it maintains prices in the end of the negotiation, while Gluttonous buyers increase their bids, enabling to obtain more profits. Table 8.8 shows a comparison of different strategies applied against a Gluttonous buyer.

Table 8.8 shows that Gluttonous + Anxious + Gluttonous strategy is the best against a Gluttonous buyer, showing that it obtained the highest selling price for the supported seller player.

8.5 Conclusion

This chapter addresses the topic of decision support of electricity market players for bilateral negotiations. Given the increasing need for reaching advantageous deals from negotiations and considering the increasing number of bilateral negotiations expected from the emergence of P2P transaction models, it becomes essential that models that are able to support players' actions in such negotiations are proposed. This chapter illustrates the application of data mining models to address this problem, namely clustering-based models to identify negotiation profiles and classification models, based on ANN and SVM to classify new players according to the identifies profiles. The illustrative results show that, using the proposed model, players are able to automatically adapt their negotiation approach, considering the expected behavior of opponents, in this way reaching more advantageous deals.

References

1 Pinto, T., Vale, Z., Widergren, S., and editors. (2021). *Local Electricity Markets*, 1e. Academic Press 384 p. https://www.elsevier.com/books/local-electricity-markets/pinto/978-0-12-820074-2. (accessed 1 June 2022).

2 Klein, M., Ziade, A., and De Vries, L. (2019). Aligning prosumers with the electricity wholesale market – the impact of time-varying price signals and fixed network charges on solar self-consumption. *Energy Policy* 134 (July).

3 Pinto, T., Wooldridge, M., and Vale, Z. (2021). Consumer flexibility aggregation using partition function games with non-transferable utility. *IEEE Access* 1.

4 de São, J.D., Faria, P., and Vale, Z. (2021). Smart energy community: a systematic review with metanalysis. *Energy Strategy Reviews* 36: 100678. https://www.sciencedirect.com/science/article/pii/S2211467X2100064X.

5 Tan, K.M., Ramachandaramurthy, V.K., and Yong, J.Y. (2016). Integration of electric vehicles in smart grid: a review on vehicle to grid technologies and optimization techniques. *Renewable and Sustainable Energy Reviews* 53: 720–732.

6 European Parliament and Council of the EU (2019). Directive (EU) 2019/944 on common rules for the internal market for electricity and amending directive 2012/27/EU. *Official Journal of the European Union* 18 (L 158).

7 Mengelkamp, E., Diesing, J., and Weinhardt, C. (2019). Tracing local energy markets: a literature review.

8 Sousa, T., Soares, T., Pinson, P. et al. (2019). Peer-to-peer and community-based markets: a comprehensive review. *Renewable and Sustainable Energy Reviews* 104: 367–378. https://www.sciencedirect.com/science/article/pii/S1364032119300462.

9 Chen, S. and Weiss, G. (2015). An approach to complex agent-based negotiations via effectively modeling unknown opponents. *Expert Systems with Applications* 42 (5): 2287–2304. http://www.sciencedirect.com/science/article/pii/S0957417414006800.

10 Zhang, J., Ren, F., and Zhang, M. (2015). Bayesian-based preference prediction in bilateral multi-issue negotiation between intelligent agents. *Knowledge-Based Systems* 84: 108–120. http://www.sciencedirect.com/science/article/pii/S0950705115001446.

11 Lopes, F., Algarvio, H., and Coelho, H. (2013). Bilateral contracting in multi-agent electricity markets: negotiation strategies and a case study. In: *2013 10th International Conference on the European Energy Market (EEM)*, 1–8. doi: 10.1109/EEM.2013.6607343.

12 Hermoza, J. and Reta, R. (2018). Implementation of bilateral contracts in an electricity wholesale market, through a hydrothermal simulator that considers the transmission network. In: *2018 IEEE 38th Central America and Panama Convention (CONCAPAN XXXVIII)*, 1–6. doi: 10.1109/CONCAPAN.2018.8596328.

13 Pinto, A., Pinto, T., Praça, I., and Vale, Z. (2019). Classification of local energy trading negotiation profiles using artificial neural networks. In: *2019 IEEE Power & Energy Society General Meeting (PESGM)*, 1–5. doi: 10.1109/PESGM40551.2019.8973600.

14 Essa, S.A.G., Dekker, H.C., and Groot, T.L.C.M. (2018). Your gain my pain? The effects of accounting information in uncertain negotiations. *Management Accounting Research* http://www.sciencedirect.com/science/article/pii/S1044500518300192.

15 Pinto, A., Pinto, T., Silva, F. et al. (2018). Automated combination of bilateral energy contracts negotiation tactics. In: *2018 IEEE Power & Energy Society General Meeting (PESGM)*, 1–5. doi: 10.1109/PESGM.2018.8586003.

16 Awasthi, S.K., Vij, S.R., Mukhopadhyay, D., and Agrawal, A.J. (2016). Multi-strategy based automated negotiation: BGP based architecture. In: *International Conference on Computing, Communication and Automation (ICCCA)*, 588–593. doi: 10.1109/CCAA.2016.7813789.

17 Vale, Z., Pinto, T., Morais, H. et al. (2011). VPP's multi-level negotiation in smart grids and competitive electricity markets. In: *IEEE Power and Energy Society General Meeting*, 1–8. doi: 10.1109/PES.2011.6039847.

18 Ilany, L. and Gal, Y. (2016). Algorithm Selection in Bilateral Negotiation. *Autonomous Agents and Multi-Agent Systems* 30 (4): 697–723. https://doi.org/10.1007/s10458-015-9302-8.

19 Algarvio, H., Lopes, F., and Santana, J. (2015). Multi-agent retail energy markets: bilateral contracting and coalitions of end-use customers. In: *2015 12th International Conference on the European Energy Market (EEM)*, 1–5. doi: 10.1109/EEM.2015.7216750.

20 Zhang, L., Song, H., Chen, X., and Hong, L. (2011). A simultaneous multi-issue negotiation through autonomous agents. *European Journal of Operational Research* 210: 95–105.

21 Sousa, F., Lopes, F., and Santana, J. (2015). Multi-agent electricity markets: a case study on contracts for difference. In: *2015 26th International Workshop on Database and Expert Systems Applications (DEXA)*, 86–90.

22 Pinto, T., Vale, Z., Praça, I. et al. (2015). Decision support for energy contracts negotiation with game theory and adaptive learning. *Energies* 8.

23 Silva, F., Pinto, T., and Vale, Z. (2019). Decision support system for opponents selection in electricity markets bilateral negotiations. In: *Proceedings of the 18th International Conference on Autonomous Agents and MultiAgent Systems*, 2384–2386. Richland, SC: International Foundation for Autonomous Agents and Multiagent Systems. (AAMAS '19).

24 He, M., Lu, D., Tian, J., and Zhang, G. (2021). Collaborative reinforcement learning based route planning for cloud content delivery networks. *IEEE Access* 9: 30868–30880.

25 Wu, H., Yuan, Y., Ma, K., and Fang, L. (2021). Estimate the probability distribution of imbalance-induced energy loss with minimal data. *IEEE Transactions on Power Apparatus and Systems* 36 (4): 3798–3801.

26 Pinto, T. and Vale, Z.A. (2020). Contextual Q-learning. In: *ECAI 2020 – 24th European Conference on Artificial Intelligence*, 2927–2928. Santiago de Compostela, Spain https://doi.org/10.3233/FAIA200457.

27 Vijaya, P.K. and KRM, C. (2020). New interactive agent based reinforcement learning approach towards smart generator bidding in electricity market with micro grid integration. *Applied Soft Computing* 97: 106762. https://www.sciencedirect.com/science/article/pii/S1568494620307006.

28 Ruan, L., Yan, Y., Guo, S. et al. (2020). ANN. *IEEE Transactions on Industrial Informatics* 16 (3): 1848–1857.

29 Jindal, A., Dua, A., Kaur, K. et al. (2016). Decision tree and SVM-based data analytics for theft detection in smart grid. *IEEE Transactions on Industrial Informatics* 12 (3): 1005–1016.

30 Dang, K.B., Burkhard, B., Windhorst, W., and Müller, F. (2019). Application of a hybrid neural-fuzzy inference system for mapping crop suitability areas and predicting rice yields. *Environmental Modelling & Software* 114: 166–180. http://www.sciencedirect.com/science/article/pii/S1364815218303761.

31 Pinto, Â., Pinto, T., Praça, I. et al. (2018). Clustering-based negotiation profiles definition for local energy transactions. In: *2018 IEEE International Conference on Communications, Control, and Computing Technologies for Smart Grids (SmartGridComm)*, 1–5. doi: 10.1109/SmartGridComm.2018.8587572.

32 Genolini, C., Alacoque, X., Sentenac, M., and Arnaud, C. (2015). kml and kml3d: R packages to cluster longitudinal data. *Journal of Statistical Software* 65(4 SE-Articles): 1–34. https://www.jstatsoft.org/index.php/jss/article/view/v065i04.

9

Socio-demographic, Economic, and Behavioral Analysis of Electric Vehicles

Rúben Barreto, Tiago Pinto, and Zita Vale

GECAD – Research Group on Intelligent Engineering and Computing for Advanced Innovation and Development, School of Engineering, Polytechnic of Porto, Porto, Portugal

Acronyms

ANN	artificial neural network
BMS	battery management system
CS	charging station
DM	data mining
EV	electric vehicle
FCM	fuzzy C-means
ICEV	internal combustion engine vehicle
IoT	internet of things
LSTM	long short-term memory
MSE	mean square error
RMSE	root mean square error
SoC	state of charge
SoH	state of health
V2G	vehicle-to-grid

9.1 Introduction

In the transport sector, the rapid integration of electric vehicles (EVs) seen today enables the emergence of new technologies and systems from which EVs can benefit, such as vehicle-to-grid (V2G) technologies [1]. Through V2G, EVs can provide ancillary services, such as voltage and frequency regulation [2], valley filling, and peak shaving [3], and even assist in the integration of renewable resources [4].

However, if not controlled, this high penetration of EVs has its consequences and may be responsible for overloading or degrading the performance of the grid itself. This can happen, for example, if several EVs are charging during peak hours of consumption [5]. Thus, implementing models that allow the control/management of EVs' charging behavior is crucial [6]. Similarly, to facilitate EV integration and preservation, the implementation of different models related to the parameters of EVs and batteries enables to assist EVs providing them with greater efficiency and safety. Regardless of the type of model developed specifically for a given problem/topic, the reliability and efficiency of the model in question must be tested, requiring large amounts of

Intelligent Data Mining and Analysis in Power and Energy Systems: Models and Applications for Smarter Efficient Power Systems, First Edition. Edited by Zita Vale, Tiago Pinto, Michael Negnevitsky, and Ganesh Kumar Venayagamoorthy.

data [7]. Thus, pre-processing the respective data and the subsequent application of data mining (DM) models are crucial steps. Depending on the context, DM models can be applied for different purposes, such as discovering patterns in data, classifying different data, predicting data, among many others [8].

Accordingly, this chapter provides an overview of the current and future status of EV integration, including a review of DM models and applications to solve EV-related problems. On the one hand, the chapter demonstrates how EV adoption is progressing in terms of the global market, economic and socio-demographic aspects, and EV policies. On the other hand, it highlights the different DM applications that are currently proposed to deal with EVs.

This chapter is divided into four sections. After Section 9.1, Section 9.2 provides a study on the progress of EVs expansion, economic and socio-demographic aspects, and culminates with some recommendations to policymakers are highlighted. Section 9.3 presents, through several examples, the impact that DM models have in the context of EVs regarding characteristics related to EVs, EV batteries, charging and user behavior and charging station (CS) planning. Finally, Section 9.4 presents the most relevant points extracted from this study.

9.2 Electric Vehicle Outlook

This section presents an overview of the current and future status of EVs penetration. Firstly, the most recent data on EV developments worldwide is presented. Next, the economic situation of EVs is presented, where the factors influencing consumers' opinion on EV purchasing are shown. Furthermore, socio-demographic aspects focusing on different regions of the world are discussed, as well as the related influence on the integration/expansion of EVs in the market.

9.2.1 Electric Mobility Market

Since the Paris Agreement, the reduction in greenhouse gas emissions has become a priority for several countries. Especially in the transport sector, initiatives are being taken in order to encourage the transition from internal combustion engine vehicles (ICEVs) to EVs, aiming to reach the 2030 and 2050 targets [9].

According to IEA [10, 11], one can see that the different countries have already started their transition to EVs. In 2019, the global EV stock reached 7.2 million, meaning an increase in about 40% compared to 2018. However, this increase was smaller than in previous years.

Figure 9.1 illustrates the respective global stock of EVs (battery EV and plug-in hybrid EV) between 2010 and 2019, where the predominance of China, Europe, and the USA can be seen. Similarly, Figure 9.2 shows, between 2013 and 2019, the amount of EVs sold as well as the market share in the most relevant countries and regions.

As seen from Figures 9.1 and 9.2, China has the highest number of EVs, in about 3.4 million (almost 50%), and also leads the sales market, selling about 1.11 million EVs in 2019. In second place is Europe, which had, in 2019, about 25% of the stock of EVs and sold more than 500 000 units. As for the USA, this was the third-largest market for EVs, which, in 2019, sold just over 300 000 units.

Considering these three largest EV market entities, only China and Europe reached a new peak in the market share of new EVs, as shown in Figure 9.2. In this follow-up, with respect to the European countries also illustrated in Figure 9.2, it is observed that in 2019, Norway and the Netherlands have higher EV market shares than the three largest EV market entities, with a share of 56% and 15%.

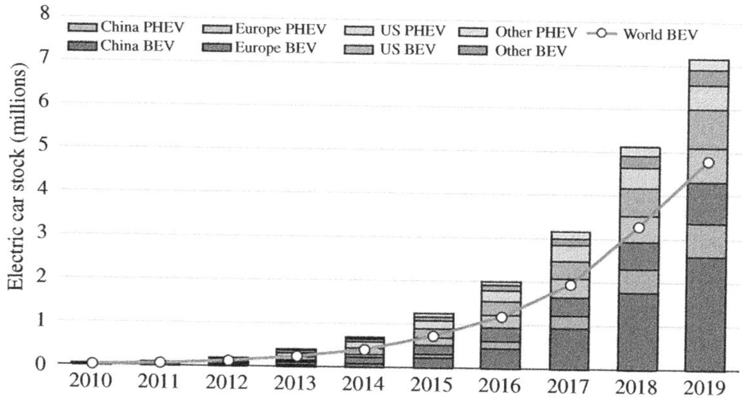

Figure 9.1 Global EV stock, 2010–2019. Source: IEA [10]/with permission of IEA.

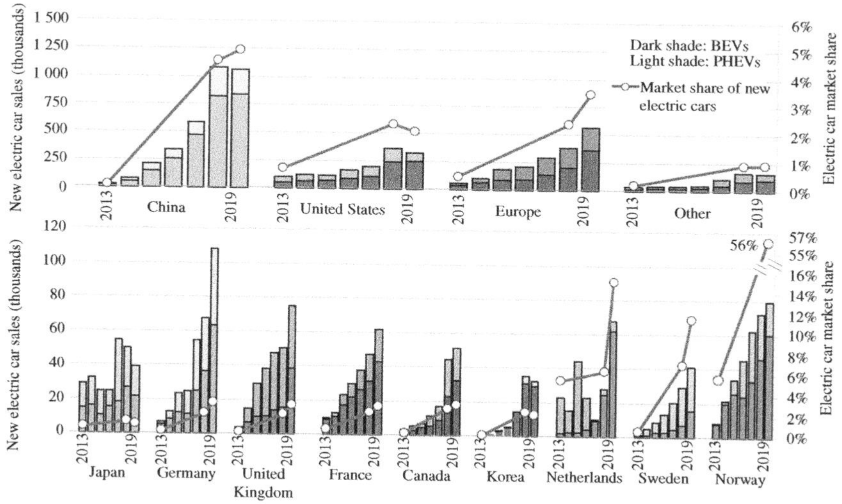

Figure 9.2 Passenger EV sales and market share in selected countries and regions, 2013–2019. Source: IEA [10]/with permission of IEA.

It is noteworthy that according to IEA [11], preliminary 2020 sales data shows that, even taking coronavirus disease 2019 (COVID-19) into account, global EV sales reach a new record high, exceeding 10 million EVs. This is indeed an astonishing increase, which is the equivalent of a 43% increase over 2019. China remains the entity with the most EVs, with around 4.5 million. However, the year-on-year increase in EVs from Europe was higher, reaching 3.2 million.

9.2.2 Economic Aspects

Although the worldwide share of EVs is increasing, we are still far from achieving the targets stipulated by international agreements. It is, thereby, essential to analyze and discuss different factors related to EVs that influence people's interest in purchasing/adopting EVs.

In Ref. [12] a study is presented to support EV companies in developing promotion strategies. In this study, only symbolic factors associated with EVs are considered, such as status, environmentalism, and innovation. This study concluded that the considered symbolic factors are the ones that have greater relevance in EVs adoption, and the influence of these symbols is moderated by culture and consumers' own identity. In this study, it was concluded that of the three mentioned symbols, environmentalism and innovation are the most important for consumers; hence, governments and educational institutions should value them in order to promote the popularity of EVs. In Ref. [13], a study conducted in two Chinese cities is presented, which concludes that the most sought-after characteristics of EVs are autonomy, reliability, and battery quality. The study of consumers' opinions regarding EVs features is also conducted in [14] The results showed that the main factors that influence the reluctance to purchase EVs are the price of the vehicle itself and of the batteries.

The work in Ref. [15] presents a study conducted in India on consumers' opinion regarding some factors of EVs such as price, maintenance, and safety. From the study, it was concluded that consumer opinion is constantly changing over time, making the process of understanding consumer opinion difficult. Similarly, it was found that regarding maintenance and price, consumers' opinions were negative. It is speculated that regarding maintenance, factors such as maintenance of EVs' batteries and uncertainty of EVs' lifespan are the leading causes of dissatisfaction. As for the price, dissatisfaction comes due to the fact that, compared to ICEVs, EVs are still quite expensive, thus hindering the adoption of this type of vehicle.

9.2.3 Socio-demographic Aspects

Consumers' interest in purchasing/adopting EVs is strongly influenced by the different characteristics of EVs and the impact they have on the environment. However, it is essential to highlight that the purchase/adoption of EVs is also significantly influenced by other factors, such as different people attributes like age, gender, education and occupation, and socio-demographic aspects. Thus, it is essential to analyze and discuss these factors.

Several kinds of literature have carried out the study on this topic, as seen in Ref. [16]. Based on the socio-demographic profile of 11 037 California EV users, this study presents a multi-year questionnaire to better understand EV users. This study, through Latent class cluster analysis, determines four heterogeneous groups of EV users. These groups correspond to households with different monetary conditions and ages. From the analysis of these groups, it was found that 49% are high-income households, 26% are old middle/high-income households, 20% are young middle/high-income households, and finally, 5% are middle-income renters. In this way, different entities such as car manufacturers and policymakers can benefit from this data, reaching a sense of how people's age and income influence the EV market.

Other studies are focusing more on the gender of the EV user, such as Refs. [17, 18]. In Ref. [18], to analyze the impact of EV penetration in Sweden, the study is based on a survey of early EV users. The results show that the typical EV user is male, well-educated, upper-middle-income and lives in places with low population density. Similarly, this study shows that both genders opt for EVs because of their cost-efficiency or the environment. Nickkar et al. [17] present a study recommending policy entities to encourage gender equity in the purchase and use of EVs. The results show that in the case of Maryland, there is a significant gender gap in EV ownership, and it is speculated that most households have registered their EVs are under male ownership. Furthermore, this study shows that most people who buy or rent EVs are well-informed, have a good income, and care about the environment.

Another example is Ref. [19], which concludes that there are gender differences even among the most educated EV users, with men opting more for EVs. Similarly, in this study, older people are the most interested in EVs, with older women with low incomes being the most aware of EVs. Thus, this study shows that potential EV buyers can be completely different people, helping decision-makers to make better business decisions, adapted to the different target groups.

9.2.4 Recommendations for Policymakers

In order to comply with the ambitious targets, several countries have already taken the initiative to switch to EVs. However, the broad shift towards EVs is a delicate process that has to be well organized and supported by policies that still need to be worked out to benefit all entities involved in this process, including car manufacturers and consumers.

Some examples of articles, such as Ref. [20], conclude that policies must be changed and improved to enable the broad adoption of EVs. This study was elaborated in Germany, and the results show that subsidies provided by incentive policies, even together with improved EV charging infrastructure, are not enough to make EVs the dominant consumer choice. This would require implementing more efficient policies, such as solid penalization of conventional cars, forcing car users to opt for EVs. In Ref. [21], unlike the previous study, it is proposed that the government ban the sale of ICEVs in public transport, taxis, and other public service areas. In addition, it also considers the perspective of EV companies, suggesting that a fund should be established to support industries/companies in the transition from ICEVs to EVs.

From the study proposed by Lee et al. [16], four types of households using EVs were defined, which differ by monetary conditions and age. The conclusions drawn from this study indicate that policymakers need to shape incentives more fairly according to the income of the respective households. Thus, lower-middle-income households (families which exist in more significant numbers in a society), might have more opportunities with more adequate subsidies, thus contributing to further integration of EVs. Other examples such as Refs. [22–24] also mention that the price difference between EVs and ICEVs is high, thus hindering the transition to EVs. Thus, both studies conclude that policymakers need to provide more adequate subsidies to bridge the purchase price gap between these two vehicle types, thus enabling better integration of EVs. The study presented in Ref. [22], also highlights the importance of creating campaigns to inform the interested public about the benefits of EVs' adoption regarding environmental aspects.

Finally, there are also studies, such as Ref. [25], which concluded that policymakers should form policies to encourage people to adopt used EVs or EVs for rent. This would allow several people who cannot afford to buy a new EV to use an EV daily, contributing to widespread EV adoption.

9.3 Data Mining Models for EVs

This section presents several examples of DM models and applications that address problems and studies related to EVs. The illustrated examples use DM models to contribute directly or indirectly to the problem at hand, i.e. DM models may provide a solution to the given problem itself or provide crucial basis data for other models.

9.3.1 Charging Behavior

One of the most problematic results from the high penetration of EVs is the associated randomness of their charging process, where many times, the charging coincides with the grid's peak hours, causing the latter to be debilitated or perform poorly [26]. Thus, it is urgent to find suitable ways to control this impact. In this context, from a grid operator's perspective, DM, together with models oriented to the charging behavior of EVs, can contribute to controlling the impact of high EV penetration.

Different kinds of works address this topic, e.g. in [27], a customer segmentation method based on the K-Means algorithm is presented as basis to classify different EV users. Through charging data from public CSs, this study initially extracts different critical variables associated with EV users, such as charge efficiency, quantity, and recency, to be used as input in the clustering algorithm. This way, after analyzing the resulting clustering, it is possible to perform the classification of the different EV users. Thus, with the implementation of this model, it is easier for CS operators to define more efficient strategies to manage the different EV users, contributing to the smooth operation of the respective grid. Another example that implements clustering models is Ref. [28]. This study explores the charging behavior of EV users, considering real EV charging data from three regions in China. This study uses two clustering algorithms, namely K-Means and Apriori. On the one hand, the K-Means algorithm is implemented to identify the different types of EV users, resulting in the identification of five classes. On the other hand, the Apriori algorithm is used to verify the impact that holidays and weather have on the EV users' charging behavior, and it is illustrated in the results that the influence is significant.

In Ref. [29], a study on EV charging behavior is also carried out, putting the main focus on optimizing the EV charging schedules. The methodology of this paper starts with a predictive algorithm to determine the EVs' charging behavior. The ensemble predicting algorithm is elaborated, which consists in the combination of three models, namely support vector regression, random forest, and diffusion-based kernel density estimator, in order to improve the prediction performance by decreasing the errors. Subsequently, the results of the ensemble predicting algorithm are applied to the EV charging scheduling optimization algorithm to minimize the charging cost. With this methodology, simulations were performed on real EV charging data and the results show a 4% and 27% reduction in costs and peak load, respectively.

V2G systems are also considered by some studies to analyze the charging behavior of EVs, as can be seen in Ref. [6], which has the goal of estimating the total load of an EV fleet with the slightest possible error. This study is based on real data from UK EV users and the presented methodology is divided into two steps. The first step consists of determining a specific set of EV users through the K-Means algorithm, which will serve as representatives. In this paper, the clustering algorithm considered several characteristics of EVs, such as the energy received from the grid (Grid-to-Vehicle), the energy supplied to the grid (V2G), and customer elasticity regarding electricity price charging frequency and demand. The second step consists of using the charging behavior data of these representatives in two mapping functions: a linear regression and a multi-layer perceptron, for estimating the total load of a fleet of EVs. The results show that the application of this methodology allows reducing the estimation error up to 12%.

9.3.2 EV User Behavior

From the EV users' perspective, the optimization of the EV charging schedule is crucial as it allows minimizing the energy costs. However, other factors such as comfort, safety, and driving efficiency have to be taken into account to better meet EV users' needs. These factors can be influenced by traffic, roads, weather, among others [30]. In this way, the possibility to predict the behaviors of EVs allows a better management and preparation for further mishaps [31].

In Ref. [30], an efficient routing model was proposed based on two DM techniques. The model itself allows minimizing the energy consumption of the EV or the travel time. In this study, considering several types of historical data from different streets and EVs, a clustering and classification model, respectively K-Means and Naïve Bayes, are implemented to obtain an efficient routing. The results show that the shortest route is not always the most efficient one due to external factors such as weather and road conditions. Another work is Ref. [32], which determines the driving pattern of EVs through clustering models. Considering real data from a fleet of 1463 EVs from China, this study presents a two-level methodology based on clustering models. In the first level, K-Means is used to identify the EVs' daily driving patterns, which considers only the daily driving data. In the second level, the results obtained in the previous level, namely the K-Medoids, are used to perform the recognition of multidimensional driving patterns based on the daily driving patterns. The experimental results show that the two-level method is efficient, being able to distinguish, by EV, several driving patterns.

The literature also offers studies that rely on internet of things (IoT) systems to improve road efficiency, e.g. [31]. This study, focusing on analyzing and monitoring EVs' behavior, implements a deep neural network based on Hadoop to determine EVs' paths. This model was compared with support vector machine and back propagation neural network, and the results obtained from simulations show that this model is faster and can improve the prediction accuracy by 9.76%. It is also highlighted that this model, through the travel characteristics of different EVs, can help traffic management.

There are also studies oriented towards aggregators, as seen in Ref. [33], which applies a rough artificial neural network (ANN) approach. The study aims to predict each EVs' travel behavior and subsequently predict the charging of the EVs present in the distribution grid, thus assisting the aggregator. Furthermore, compared to others, this approach preserves the correlation between the arrival and departure time and the trip duration. In order to prove the robustness and reliability of the proposed model, this study performed the comparison with conventional models, such as Monte Carlo simulation. The results show that the aggregator can benefit from this approach, where the financial loss is reduced by about 16 \$/EV per year.

9.3.3 Driving Range

Although EV user comfort is an essential factor, today, it is found that EV users may suffer range anxiety since EVs have a low driving range compared to ICEVs [34]. Thus, to mitigate this problem, it is crucial to develop autonomy prediction models that make it possible to consider all kinds of unfavorable factors, being these external factors, such as the environment surrounding the EV, and internal factors [35].

In Ref. [36], a strategy for predicting the range of EVs was proposed based on two models, the extreme gradient boosting regression tree and the light gradient boosting regression tree. This strategy's logic consists of learning the relationship of the range and other factors such as battery temperature, the accumulated output energy of the motor and battery, and different driving patterns. Furthermore, in this same strategy, an anchor model was included to handle the unbalanced dataset distribution used in this study. The use of the anchor model achieved a better performance than other benchmark models. Another work that applies regression models is Ref. [34], where robust non-linear regression models were proposed to estimate an EV's range accurately. This study is based on real data collected from an EV driving in Beijing, China. The considered models take into consideration different impact factors such as temperature, state of charge (SoC) and speed. In this study, the non-linearity between speed and driving distance per SoC was given special attention. In order to deal with these, quadratic and Weibull distribution were adopted to establish the non-linear models for estimating the driving distance per SoC. The results of this model indicated that the prediction has good accuracy.

Clustering models are also widely used to address driving range-related problems, such as [37, 38]. In Ref. [38], a model based on fuzzy C-means (FCM) was proposed to predict a lithium battery's respective state of function. In the context of EVs, the state of function is directly related to the respective SoC of the EV, so it can be indirectly considered as a parameter to perform the prediction of the driving range. This study implements FCM to optimize the respective state of function prediction model, considering the State of Health (SoH), SoC and the loading and unloading rate. The results show that the prediction model together with FCM obtains an average prediction error of 8.69%. In Ref. [37] a prediction model of EVs range based on the combination of identification and prediction of driving cycles was proposed. The logic of this study consists of first using the clustering model, FCM, together with the kernel principal component feature parameters method to identify the driving cycle. Subsequently, the data obtained from the previous process are related to energy consumption through a fuzzy rule. Similarly, to improve the prediction accuracy of driving range, in this study, the Markov algorithm and back propagation neural network are used to predict future driving conditions. The results indicate that the application of this model has better convergence and accuracy than other prediction methods.

9.3.4 Speed

Another important EV characteristic is speed, where, in addition to being coveted by EV users, this is crucial for models of battery SoC estimation and EV trip estimation [39]. However, speed prediction is a complicated task since it varies in a stochastic way [40].

Several studies rely on ANN to reach EV speed prediction, such as Refs. [39, 41]. In Ref. [39], two long short-term memory (LSTM) models are presented: one univariate and one multivariate, to predict the speed of an EV. The dataset used to build and refine the two models was collected from three real-world urban routes using the VSimRTI simulator. In the case of the univariate LSTM, it only considers the respective EV speed and timestamps. In addition to the features considered in the univariate model, the multivariate one also considers road ID, maximum road speed, gradient, road type, and traffic conditions. The results of both models were compared, and it is shown that the multivariate LSTM outperforms the univariate model for short- and long-term predictions. In the study presented in Ref. [41] a two-level model is proposed that uses ANN as the first model to predict the average traffic speed of segments of the target road using historical data. Subsequently, the data obtained from the first model are used as inputs in the second model, the hidden Markov model, which builds a relationship between traffic speeds and EV speed. In this way, the second model elaborates the prediction of the EV speed, where the results show that this two-level model, compared to other models, has good accuracy. Another example of application of a deep learning model is [8]. In this study a deep neural network model was proposed based on essential factors such as the historical speed of the vehicle and its corresponding acceleration, driving information, location, and driving date, to perform the prediction of the EV speed in the short term. This study considers two years of data from five EVs located in Beijing, China, to perform the tests, and the results showed that the combination of historical EV speed data and driving date, i.e. whether it is weekend or weekday, has the best prediction with a root mean square error (RMSE) of 1.5298. Similarly, the method itself is shown to have good robustness and generalization ability.

There are also studies that apply fuel cell EV energy management strategies based on speed prediction, as seen in Ref. [42]. This study, aiming to minimize equivalent hydrogen consumption and inhibit fuel cell degradation, proposes a speed prediction model predictive control strategy, where speed prediction is performed by the exponential smoothing law – Markov predictor.

This strategy was compared with three others, and it was found that this strategy has good robustness, where it can provide good results in less calculation time.

9.3.5 Electric Vehicle Battery

One of the most important aspects to consider in EVs is their respective battery, where the efficiency of the battery and the battery management system (BMS) influence EV performance. In general, BMS establishes a connection between the EV and the battery pack itself, where it acts to optimize the battery, ensuring that it is protected and functioning well. Most of the data acquired by BMS vary between the SoH, SoC, voltage, and temperature [43, 44].

There are several studies that address the prediction of EVs' SoC. For example, Ref. [45] based on a V2G system, aims at coordinating the charging and discharging process of EVs. Thus, this study presents a strategy that combines a hybrid particle swarm optimization and ANN model responsible for the SoC prediction and a novel charging and discharging control algorithm. The respective results show that the hybrid particle swarm optimization and ANN model performs an accurate prediction with RMSE of 0.9320%, and the charging and discharging control algorithm correctly performs its control. Another example where six machine learning algorithms are implemented and compared to predict SoC is Ref. [46]. In this study, the ANN, support vector machine, linear regression, Gaussian process regression, ensemble bagging, and ensemble boosting were used. All these algorithms used as inputs, the voltage, the current and the SoC of the battery. The results conclude that ANN and Gaussian process regression are the best models with an mean square error (MSE) and RMSE of (0.0004, 0.001 70) and (0.023, 0.041 18), respectively, showing that these can contribute to the creation of an efficient EV BMS.

Some works, like Refs. [47, 48] study the EV battery SoH prediction. The work presented in Ref. [48] predicts the SoH and remaining useful life of battery packs. In this study, a Gaussian process dual regression model was proposed to perform SoH prediction over the whole battery life cycle and remaining useful life prediction near the end of battery life. The results showed that both predictions are accurate and reliable with a meager computation time. Another example is Ref. [47], which presents a new LSTM neural network model that allows predicting the SoH of EV batteries. This study also considers a battery degeneration model for real scenarios, obtained by correlating different data such as SoH, air temperature, mileage, among others. The experimental results show that the model, on the one hand, allows to forecast SoH in real-time accurately, and, on the other hand, also allows to obtain the degradation path of the battery capacity. Similarly, the proposed LSTM model was compared with other models, and showed a superior performance.

Some studies also focus on multi-parameter prediction, as demonstrated for the first time in Ref. [49]. In this study, the combination of the LSTM model with a synchronous multi-parameter prediction model of the battery system was proposed, which has the purpose of accurately predicting the parameters of battery temperature, voltage and SoC. Furthermore, the presented methodology considered the impact of external factors on the battery, such as driver behavior and weather, to aid the accuracy of the mixed model. In order to test and validate the presented model, one year of data from electric taxis in Beijing was used, and the results show that the proposed model is promising, predicting with reasonable accuracy the respective parameters.

9.3.6 Charging Station Planning

Considering the increasing integration of EVs, the good positioning of CSs is essential to facilitate EV users' access. Thus, obtaining an optimal location to install a CS is crucial, and the positioning of the CS can be influenced by the high demand on the grid, the traffic density, and the range of EVs [50, 51].

In Ref. [50], a model was proposed to estimate the various optimal locations to install CS in Ankara, Turkey. The model is based on DM methods, such as clustering, where it considers the average range and the average number of EVs on the road as parameters. This study, starting from the road map of Ankara and the respective energy available on the different streets of the area, firstly determines the number of CS that could be installed. Subsequently, through the Spectral clustering model and Gaussian mixture model, the amount of CS is used to determine the optimal locations for CS installation. The results demonstrate that the proposed model is practical and reliable.

Another example is Ref. [51], which, based on data collected in Seoul, South Korea, proposes a model that selects the best locations for CS. This study uses parameters such as the number of inhabitants and workers and the number of guest facilities, which are considered the main factors influencing the optimal CS location. The strategy of this study consists of using the clustering method, latent profile analysis, and different machine learning techniques such as elastic net, random forest, support vector machine, and extreme gradient boosting to determine which are the best locations to install CS. In this way, with this strategy, it is possible to select the areas with greater need to install the CS, efficiently assigning the resources to assist the users of EVs and the integration of EVs.

There are also studies that address the CS allocation problem, considering the renewable energy resources, e.g. Ref. [52]. Considering the uncertainties related to the conventional demand and the EVs, this study presents a mixed-integer linear programming model for the CS and renewable energy resources allocation problem, in which the objective is to minimize investment and operational costs. In order to test the proposed model, this study defined scenarios based on the K-Means model in order to achieve computational tractability. The experimental results demonstrate that the installation of CS in a good location enables more effective use of renewable energy resources, reducing the costs of the expansion plan and CO_2 emissions.

9.3.7 Summary

Considering Sections 9.3.1–9.3.6, it is possible to verify the importance of the DM models in the context of the transport sector, where the rapid integration/adoption of EVs can be assisted/supported. Furthermore, through this section, it is possible to verify that large amounts of data about EVs can be collected, especially with the aid of advances in IoT technologies. However, analyzing and processing this data is a time-consuming and complicated task, especially when the data has different formats and categories.

In this context, DM models are crucial in analyzing and processing data, as they allow extracting new information, making the best use of the relationships between different types of data and information of different natures. The extracted information/data, according to the problem in the cause, can be the solution itself or can be the input to models of optimization, for example. Moreover, the application of DM models makes it possible to obtain solutions fast and efficiently, which is a vital aspect to consider when solving EV-related problems.

Thus, in general, it can be concluded that the integration of EVs can be enhanced and supported by the integration of DM models, where different grid entities such as EV users, EVs, and even power grid operators can benefit.

9.4 Conclusions

Nowadays, EVs are becoming more and more prominent, as they enable the emergence of new technologies and opportunities that can benefit different entities in the power grid. In this way, to highlight the impact of EVs, this chapter provides an overview of the current and future status of EVs integration regarding the progression of the global EV market, the economic and socio-demographic aspects, and EV policies. Furthermore, this chapter also includes a review of DM models and applications to solve, directly or indirectly, EV-related problems, namely, problems in the context of the EV charging behavior, EV user behavior, driving range, speed, EV batteries, and CS planning.

Considering this study, it is possible to verify that the integration of EVs worldwide has been increasing significantly, but we are still far from reaching the goals set in the international agreements. The integration of EVs is strongly influenced by the buyers' interest, which varies on the one hand according to personal characteristics such as education, gender, age, location, occupation, and financial power. On the other hand, it varies according to EV-related aspects such as brand, components, and price.

A possible way to improve EV integration is through policies to benefit and encourage different entities involved in this process. Thus, policymakers should pay special attention to the support they provide to EV users, where it is suggested that the subsidies provided should be in line with the financial power of EV users so that they have more opportunities. Likewise, it is suggested that companies should also be duly supported in order to facilitate the transition from ICEVs manufacturing to EVs.

Finally, with this chapter, it can also be concluded that the application of DM models allows assisting, directly or indirectly, different entities integrated into the grid, where they can obtain benefits both at the energy efficiency and economic level. In other words, the impact of the high integration of EVs can be controlled by means of the application of DM models, thus contributing to the electrification of the transport sector.

References

1 Sanguesa, J.A., Torres-Sanz, V., Garrido, P. et al. (2021). A review on electric vehicles: technologies and challenges. *Smart Cities* 4: 372–404. https://doi.org/10.3390/smartcities4010022.

2 Amamra, S.-A. and Marco, J. (2019). Vehicle-to-grid aggregator to support power grid and reduce electric vehicle charging cost. *IEEE Access* 7: 178528–178538. https://doi.org/10.1109/ACCESS.2019.2958664.

3 Khan, S.U., Mehmood, K.K., Haider, Z.M. et al. (2021). Coordination of multiple electric vehicle aggregators for peak shaving and valley filling in distribution feeders. *Energies* 14: 352. https://doi.org/10.3390/en14020352.

4 Borges, N., Soares, J., and Vale, Z. (2017). Multi-objective particle swarm optimization to solve energy scheduling with vehicle-to-grid in office buildings considering uncertainties. *IFAC-PapersOnLine* 50: 3356–3361. https://doi.org/10.1016/j.ifacol.2017.08.523.

5 Jia, Q.-S. and Long, T. (2020). A review on charging behavior of electric vehicles: data, model, and control. *Control Theory and Technology* 18: 217–230. https://doi.org/10.1007/s11768-020-0048-8.

6 Ebrahimi, M. and Rastegar, M. (2020). Data-driven charging load estimation of behind-the-meter V2G-capable EVs. *IEEE Transactions on Industry Applications* 1–1. https://doi.org/10.1109/TIA.2020.3012111.

7 Lv, Z., Qiao, L., Cai, K., and Wang, Q. (2021). Big data analysis technology for electric vehicle networks in smart cities. *IEEE Transactions on Intelligent Transportation Systems* 22: 1807–1816. https://doi.org/10.1109/TITS.2020.3008884.

8 Shahriar, S., Al-Ali, A.R., Osman, A.H. et al. (2020). Machine learning approaches for EV charging behavior: a review. *IEEE Access* 8: 168980–168993. https://doi.org/10.1109/ACCESS.2020.3023388.

9 Gryparis, E., Papadopoulos, P., Leligou, H.C., and Psomopoulos, C.S. (2020). Electricity demand and carbon emission in power generation under high penetration of electric vehicles. A European Union perspective. *Energy Reports* 6: 475–486. https://doi.org/10.1016/j.egyr.2020.09.025.

10 IEA. (2020). Global EV Outlook. https://www.iea.org/reports/global-ev-outlook-2020. All rights reserved (accessed 1 June 2022).

11 IEA. (2021). Global EV Outlook. https://www.iea.org/reports/global-ev-outlook-2021. All rights reserved (accessed 1 June 2022).

12 Liu, R., Ding, Z., Wang, Y. et al. (2021). The relationship between symbolic meanings and adoption intention of electric vehicles in China: the moderating effects of consumer self-identity and face consciousness. *Journal of Cleaner Production* 288: 125116. https://doi.org/10.1016/j.jclepro.2020.125116.

13 Yang, J. and Chen, F. (2021). How are social-psychological factors related to consumer preferences for plug-in electric vehicles? Case studies from two cities in China. *Renewable and Sustainable Energy Reviews* 149: 111325. https://doi.org/10.1016/J.RSER.2021.111325.

14 Moeletsi, M.E. (2021). Socio-economic barriers to adoption of electric vehicles in South Africa: case study of the Gauteng Province. *World Electric Vehicle Journal* 12: 167. https://doi.org/10.3390/WEVJ12040167.

15 Jena, R. (2020). An empirical case study on Indian consumers' sentiment towards electric vehicles: a big data analytics approach. *Industrial Marketing Management* 90: 605–616. https://doi.org/10.1016/j.indmarman.2019.12.012.

16 Lee, J.H., Hardman, S.J., and Tal, G. (2019). Who is buying electric vehicles in California? Characterising early adopter heterogeneity and forecasting market diffusion. *Energy Research & Social Science* 55: 218–226. https://doi.org/10.1016/j.erss.2019.05.011.

17 Nickkar, A., Shin, H.-S., and Farkas, A. (2019). Analysis of ownership and travel behavior of women who drive electric vehicles: the case of Maryland, preprint, arXiv:1911.03943.

18 Vassileva, I. and Campillo, J. (2017). Adoption barriers for electric vehicles: experiences from early adopters in Sweden. *Energy* 120: 632–641. https://doi.org/10.1016/J.ENERGY.2016.11.119.

19 Esteves, J., Alonso-Martínez, D., and de Haro, G. (2021). Profiling Spanish prospective buyers of electric vehicles based on demographics. *Sustainability* 13: 9223. https://doi.org/10.3390/SU13169223.

20 Ball, C.S., Vögele, S., Grajewski, M., and Kuckshinrichs, W. (2021). E-mobility from a multi-actor point of view: uncertainties and their impacts. *Technological Forecasting and Social Change* 170: 120925. https://doi.org/10.1016/J.TECHFORE.2021.120925.

21 Wu, Z., Shao, Q., Su, Y., and Zhang, D. (2021). A socio-technical transition path for new energy vehicles in China: a multi-level perspective. *Technological Forecasting and Social Change* 172: 121007. https://doi.org/10.1016/J.TECHFORE.2021.121007.

22 Cui, L., Wang, Y., Chen, W. et al. (2021). Predicting determinants of consumers' purchase motivation for electric vehicles: an application of Maslow's hierarchy of needs model. *Energy Policy* 151: 112167. https://doi.org/10.1016/J.ENPOL.2021.112167.

23 Hasan, S. (2021). Assessment of electric vehicle repurchase intention: a survey-based study on the Norwegian EV market. *Transportation Research Interdisciplinary Perspectives* 11: 100439. https://doi.org/10.1016/J.TRIP.2021.100439.

24 Krishnan, V.V. and Koshy, B.I. (2021). Evaluating the factors influencing purchase intention of electric vehicles in households owning conventional vehicles. *Case Studies on Transport Policy* 9: 1122–1129. https://doi.org/10.1016/J.CSTP.2021.05.013.

25 Huang, Y. and Qian, L. (2021). Consumer adoption of electric vehicles in alternative business models. *Energy Policy* 155: 112338. https://doi.org/10.1016/J.ENPOL.2021.112338.

26 Li, X., Sun, D., Li, W. et al. (2020). Data-driven mobility on demand fleets charging demand modeling. In: *2020 IEEE/IAS Industrial and Commercial Power System Asia (I&CPS Asia)*, 76–81. IEEE https://doi.org/10.1109/ICPSAsia48933.2020.9208406.

27 Lu, Z., Peiyi, W., Ping, C. et al. (2019). Customer segmentation algorithm based on data mining for electric vehicles. In: *2019 IEEE 4th International Conference on Cloud Computing and Big Data Analysis (ICCCBDA)*, 77–83. IEEE https://doi.org/10.1109/ICCCBDA.2019.8725737.

28 Wang, Y., Cai, S., Shen, Y. et al. (2019). The analysis of electrical vehicles charging behavior based on charging big data. In: *2019 IEEE 4th International Conference on Cloud Computing and Big Data Analysis (ICCCBDA)*, 63–67. IEEE https://doi.org/10.1109/ICCCBDA.2019.8725776.

29 Chung, Y.-W., Khaki, B., Li, T. et al. (2019). Ensemble machine learning-based algorithm for electric vehicle user behavior prediction. *Applied Energy* 254: 113732. https://doi.org/10.1016/j.apenergy.2019.113732.

30 Bozorgi, A.M., Farasat, M., and Mahmoud, A. (2017). A time and energy efficient routing algorithm for electric vehicles based on historical driving data. *IEEE Transactions on Intelligent Vehicles* 2: 308–320. https://doi.org/10.1109/TIV.2017.2771233.

31 Feng, X. and Hu, J. (2020). Research on the identification and management of vehicle behaviour based on Internet of things technology. *Computer Communications* 156: 68–76. https://doi.org/10.1016/j.comcom.2020.03.035.

32 Li, X., Zhang, Q., Peng, Z. et al. (2019). A data-driven two-level clustering model for driving pattern analysis of electric vehicles and a case study. *Journal of Cleaner Production* 206: 827–837. https://doi.org/10.1016/j.jclepro.2018.09.184.

33 Jahangir, H., Tayarani, H., Ahmadian, A. et al. (2019). Charging demand of plug-in electric vehicles: forecasting travel behavior based on a novel Rough Artificial Neural Network approach. *Journal of Cleaner Production* 229: 1029–1044. https://doi.org/10.1016/J.JCLEPRO.2019.04.345.

34 Bi, J., Wang, Y., Sai, Q., and Ding, C. (2019). Estimating remaining driving range of battery electric vehicles based on real-world data: a case study of Beijing, China. *Energy* 169: 833–843. https://doi.org/10.1016/J.ENERGY.2018.12.061.

35 Varga, B., Sagoian, A., and Mariasiu, F. (2019). Prediction of electric vehicle range: a comprehensive review of current issues and challenges. *Energies* 12: 946. https://doi.org/10.3390/en12050946.

36 Zhao, L., Yao, W., Wang, Y., and Hu, J. (2020). Machine learning-based method for remaining range prediction of electric vehicles. *IEEE Access* 8: 212423–212441. https://doi.org/10.1109/ACCESS.2020.3039815.

37 Pan, C., Dai, W., Liao, C. et al. (2017). Driving range estimation for electric vehicles based on driving condition identification and forecast. *AIP Advances* 7: 105206. https://doi.org/10.1063/1.4993945.

38 Wang, D., Yang, F., Gan, L., and Li, Y. (2019). Fuzzy prediction of power lithium ion battery state of function based on the fuzzy c-means clustering algorithm. *World Electric Vehicle Journal* 10: 1. https://doi.org/10.3390/WEVJ10010001.

39 Malek, Y.N., Najib, M., Bakhouya, M., and Essaaidi, M. (2021). Multivariate deep learning approach for electric vehicle speed forecasting. *Big Data Mining and Analytics* 4: 56–64. https://doi.org/10.26599/BDMA.2020.9020027.

40 Yan, M., Li, M., He, H., and Peng, J. (2018). Deep learning for vehicle speed prediction. *Energy Procedia* 152: 618–623. https://doi.org/10.1016/J.EGYPRO.2018.09.220.

41 Jiang, B. and Fei, Y. (2017). Vehicle speed prediction by two-level data driven models in vehicular networks. *IEEE Transactions on Intelligent Transportation Systems* 18: 1793–1801. https://doi.org/10.1109/TITS.2016.2620498.

42 Quan, S., Wang, Y.X., Xiao, X. et al. (2021). Real-time energy management for fuel cell electric vehicle using speed prediction-based model predictive control considering performance degradation. *Applied Energy* 304: 117845. https://doi.org/10.1016/J.APENERGY.2021.117845.

43 Castano, S., Serrano-Jimenez, D., and Sanz, J. (2016). BMS influence on Li-ion packs characterization and modeling. In: *2016 IEEE 16th International Conference on Environment and Electrical Engineering (EEEIC)*, 1–6. IEEE https://doi.org/10.1109/EEEIC.2016.7555760.

44 Peng, S., Chen, C., Shi, H., and Yao, Z. (2017). State of charge estimation of battery energy storage systems based on adaptive unscented Kalman filter with a noise statistics estimator. *IEEE Access* 5: 13202–13212. https://doi.org/10.1109/ACCESS.2017.2725301.

45 Aljanad, A., Mohamed, A., Khatib, T. et al. (2019). A novel charging and discharging algorithm of plug-in hybrid electric vehicles considering vehicle-to-grid and photovoltaic generation. *World Electric Vehicle Journal* 10: 61. https://doi.org/10.3390/wevj10040061.

46 Chandran, V., Patil, C.K., Karthick, A. et al. (2021). State of charge estimation of lithium-ion battery for electric vehicles using machine learning algorithms. *World Electric Vehicle Journal* 12: 38. https://doi.org/10.3390/WEVJ12010038.

47 Hong, J., Wang, Z., Chen, W. et al. (2021). Online accurate state of health estimation for battery systems on real-world electric vehicles with variable driving conditions considered. *Journal of Cleaner Production* 294: 125814. https://doi.org/10.1016/J.JCLEPRO.2021.125814.

48 Hu, X., Che, Y., Lin, X., and Deng, Z. (2020). Health prognosis for electric vehicle battery packs: a data-driven approach. *IEEE/ASME Transactions on Mechatronics* 25: 2622–2632. https://doi.org/10.1109/TMECH.2020.2986364.

49 Hong, J., Wang, Z., Chen, W., and Yao, Y. (2019). Synchronous multi-parameter prediction of battery systems on electric vehicles using long short-term memory networks. *Applied Energy* 254: 113648. https://doi.org/10.1016/j.apenergy.2019.113648.

50 Catalbas, M.C., Yildirim, M., Gulten, A., and Kurum, H. (2017). Estimation of optimal locations for electric vehicle charging stations. In: *2017 IEEE International Conference on Environment and Electrical Engineering and 2017 IEEE Industrial and Commercial Power Systems Europe (EEEIC/I&CPS Europe)*, 1–4. IEEE https://doi.org/10.1109/EEEIC.2017.7977426.

51 Lee, J., An, M., Kim, Y., and Seo, J.-I. (2021). Optimal allocation for electric vehicle charging stations. *Energies* 14: 5781. https://doi.org/10.3390/en14185781.

52 de Lima, T.D., Franco, J.F., Lezama, F. et al. (2021). Joint optimal allocation of electric vehicle charging stations and renewable energy sources including CO_2 emissions. *Energy Informatics* 4 (2): 1–18. https://doi.org/10.1186/S42162-021-00157-5.

Part IV

Forecasting

10

A Multivariate Stochastic Spatiotemporal Wind Power Scenario Forecasting Model

Wenlei Bai[1]*, Duehee Lee*[2]*, and Kwang Y. Lee*[3]

[1]*Oracle Energy and Water, Oracle America Inc., Austin, TX, USA*
[2]*Electrical Engineering Department, Konkuk University, Seoul, Korea*
[3]*Electrical and Computer Engineering Department, Baylor University, Waco, TX, USA*

Acronyms

AR	Autoregression
DPCA	Dynamic principal component analysis
GDFM	Generalized dynamic factor model
ERCOT	Electric Reliability Council of Texas
FA	Factor analysis
MAE	Mean absolute error
MC	Monto Carlo simulation
OPF	Optimal power flow
PDF	Probability density function
PSD	Power spectral density
RMSE	Root mean square error
VAR	Vector autoregression

Nomenclature

$\boldsymbol{X}$	Observed wind power matrix
N	Number of wind farms
T	Number of observation periods
χ	Common component of wind power matrix
ξ	Idiosyncratic component of wind power matrix
t	Time $t = 1, \ldots, T$
$\boldsymbol{\phi}$	Dynamic factors
Q	Dimension of the dynamic factors
L	Delay operator
$\boldsymbol{A}(L)$	Factor loading
M	Number of lags in factor loading
ε	White noise
$\boldsymbol{C}(L)$	Coefficient matrix

Intelligent Data Mining and Analysis in Power and Energy Systems: Models and Applications for Smarter Efficient Power Systems, First Edition. Edited by Zita Vale, Tiago Pinto, Michael Negnevitsky, and Ganesh Kumar Venayagamoorthy.

δ	Dynamic shocks
$\boldsymbol{H}$	$Q{\times}Q$ matrix that preserves the correlation structure
$\Sigma^X(f_m)$	Fourier transform of the covariance matrix of the *residual* data $\boldsymbol{X}$ at frequency f_m
$\Sigma^\chi(f_m)$	Fourier transform of the covariance matrix of the common component χ at frequency f_m
Q	Largest Q eigenvalues of a matrix.
$\boldsymbol{\Omega}^\chi(f_m)$	the diagonal matrix whose diagonals are from the top largest Q eigenvalues of $\Sigma^X(f_m)$
$\boldsymbol{G}^\chi(f_m)$	Corresponding eigenvectors associated with $\boldsymbol{\Omega}^\chi(f_m)$
$\boldsymbol{\Omega}^\xi(f_m)$	Diagonal matrix whose diagonals are the remaining N–Q eigenvalues of $\Sigma^X(f_m)$
$\boldsymbol{G}^\xi(f_m)$	Corresponding eigenvectors associated with $\boldsymbol{\Omega}^\xi(f_m)$
$\boldsymbol{\Gamma}^\xi$	Covariance matrix of ξ
T	Transpose operator
$\boldsymbol{G}^\chi(L)$	Corresponding eigenvectors associated with the top largest Q eigenvectors of the covariance matrix of common component χ at all lags
$\boldsymbol{B}(L)$	Filter
k	Future k step to forecast
$\hat{\chi}$	Estimated values of χ
$\hat{\xi}$	Estimated values of ξ
$\hat{\boldsymbol{\phi}}$	Estimated values of $\boldsymbol{\phi}$
$\hat{\varepsilon}$	Estimated values of ε

10.1 Introduction

Wind power forecasting has two major approaches in terms of output: point forecasting and probabilistic forecasting. Point forecasting gives a single future wind power output. On the contrary, probabilistic forecasting gives a conditional distribution of future wind power output [1], so that system operators and traders can utilize a much broader set of information. The wind power distribution can be represented as quantiles, interval forecasts, probability density functions (PDFs), and scenarios.

Extensive research on probabilistic wind power forecasting has been conducted [2]. Monteiro et al. in [3] gave a comprehensive review of methodologies for probabilistic wind power forecasting, which includes physical models, statistical models, and combined models. A traditional, physical time-series model, such as the autoregressive model (AR), captures the temporal correlation in wind. This model predicts the distribution of future wind power as a linear combination of previous and current data and white noise errors [4]. A statistical space-time model was proposed in [5]. It considered the terrain, wind direction, and wind speed as the input data.

However, few of the previously mentioned works have focused on the stochastic process with multiple wind farms even though the spatial correlation among wind farms has played an important role for the reliability of the power system. For example, wind farms that have a positive spatial correlation and share the same transmission lines might congest the transmission lines, and the transmission congestion could curtail the electricity supply [6].

The uncertainty of wind power forecast can have a tremendous impact on the system operations even by small errors. Such impact can be reflected in ancillary services such as reactive power compensation, voltage regulation, and operating reserves. Therefore, independent system operators (ISO) in competitive electricity markets seek for optimization methods for managing the system under reliable and economically friendly conditions by utilizing wind power [7].

Even though deterministic operation approaches such as unit commitment (UC) or optimal power flow (OPF) are appropriate for short-term generation scheduling [8], with large wind energy penetration, stochastic/probabilistic approaches are in urgent need of development due to the intermittent nature of wind power. The representation of wind power forecasting uncertainty in the UC problem has been investigated [9]. It is shown that the stochastic approach has advantages over deterministic one in providing a rational and adaptive way to provide operating reserves. The security-constrained UC with volatile wind power generation was formulated and the wind power was represented by several possible scenarios [9]. Wind power scenarios are generated through Monte Caro simulation assuming that the forecast error is under normal distribution. However, such assumption fails to consider the correlation among wind farms. In addition, the computational cost is very high because the MC simulation does not consider the correlations among scenarios, and therefore an extensive number of scenarios needs to be generated to fully represent combinations of various wind farms.

To build accurate forecasting models, economists and forecasters have tried to analyze the relationship between explanatory and response variables. Explanatory variable is any factor that influences the response variables, and response variable (wind power output in this study) is the quantity we want to know. The first challenge of analyzing the relationship is when an unobservable explanatory variable arises, such as weather conditions (e.g. when wind speed, wind direction, and air mass cannot be measured in some cases.) in wind power system. The factor analysis overcomes this obstacle by assuming that response variables are loaded by the linear combination of unobservable factors with different weighted parameter, which is a factor loading [10]. In the factor analysis, the factors and factor loadings are estimated by analyzing the response variables. The second challenge is to design multi-dimensional explanatory and response variables by considering the correlation structure among them. The multivariate time-series model can design the explanatory and response variables through the polynomial matrix, where the correlation structure is preserved. However, as the size of dimension increases linearly, the number of parameters increases exponentially, so the advantage of the factor analysis to represent the system parsimonious is faded away. Therefore, the possible solution to both obstacles is to construct the multivariate time-series model on factors in order to reduce the number of variables and describe the dynamic relationship between the factor and factor loadings. Therefore, the generalized dynamic factor model (GDFM) becomes one of solutions [11]. The GDFM was used to generate wind power and load scenarios for transmission expansion planning [12]. The model can address the spatial and temporal correlation among wind farms in nearby areas where wind power is affected by similar weather conditions, transmission lines, and movements over time. The pre-processing step of the GDFM is to extract from the observation data the common component, which is directly affected by weather conditions, so that the GDFM ignores unrelated errors obtained in a measurement process. There are multiple types of wind generation scenarios: generating point values, scaling time series, and utilizing the power curve of wind farm [9]. Yet, as mentioned above, these scenarios did not systematically consider the correlation structure among raw data. In this work, we extend the previous GDFM to forecast the short-term probabilistic wind power and represent the distribution of wind power forecasts as multiple wind power scenarios. The role of the GDFM is to analyze these co-movements and synthesize correlated wind power forecasts and reduce computational cost. In this paper, GDFM is proposed to generate wind power forecasting scenarios based on the power output data of 23 wind farms. The forecasting 24 hours wind power scenarios are used in the stochastic dynamic AC OPF problem. The advantage of stochastic optimization is that system operators can find the optimal expected benefits by fully capturing the uncertainty information of wind power.

10.2 Generalized Dynamic Factor Model

The GDFM utilizes the factor analysis (FA) to reduce the data dimension by representing data in terms of the common latent variables, called factors, so that the computational burden of the multivariate time-series analysis can be reduced significantly. Then, the explanatory variables are modelled as linear combinations of the latent factors. Finally, the explanatory variables are represented by the product of factor loadings and dynamic factors [13]. The stochastic correlation structure is inherited in the factor loadings [14]. Furthermore, since the dynamic factors are driven by uncorrelated white noise signals, the GDFM can generate arbitrary numbers of forecasting scenarios by varying the dynamic factors. Details will be introduced in the following section. In all, the advantages of the GDFM are as follows: (i) the FA is used to overcome the "curse of dimensionality," which happens when we model the high dimensional time series through the vector autoregressive model; (ii) the co-movement between time series is used to overcome the curse of dimensionality, so that all scenarios in the GDFM share the spatial and temporal correlation and statistical characteristics that the actual wind power outputs have; and (iii) arbitrary numbers of wind power scenarios can be generated from white noise signals.

10.2.1 Derivation of the GDFM

The factor analysis model, which is known as the static factor model, has been applied widely in forecasting energy, electricity load, and economic indices. For wind power forecasting, those latent variables can be wind speed, air density, or wind direction. However, the factor analysis model does not consider the temporal correlation when modeling the latent variables (factors) [15]. On the contrary, the dynamic model, which evolved from the static factor model, can consider the spatial and temporal correlations. In this study, the spatial correlation is measured among various wind farms. Here, "dynamic" means there are time lags in the factor loadings.

It is assumed that wind power outputs consist of the three components: common component, periodic component, and idiosyncratic component. The first one follows the correlated dynamic movement, the second one follows the daily trend, and the last one is a measurement error [13]. In this section, we introduce how the wind power measurement can be decomposed into common, periodic, and idiosyncratic components through the GDFM based on [13]. Furthermore, wind power forecasting scenarios are synthesized by using the GDFM. In the pre-processing step of extracting periodic component, we normalize the wind power outputs by their capacities and standardize the normalized wind power outputs. The standardized signals have mean values of zero and standard deviation values of one. Then, the periodic components are estimated by averaging wind powers at the same hour, and the periodic components are subtracted from the standardized signals. The residual signals are used for generating GDFM. The next step is to decompose the residual signal into the common component, which is a shared dynamic movement among wind farms and assumed to be affected by weather conditions, and the idiosyncratic component, which is assumed to be a measurement error. The common component is the model that describes correlated dynamic movement among wind farms.

As mentioned previously, the first assumption in the GDFM is that the residuals (the seasonality of the observation has been removed) of observed wind power matrix $\boldsymbol{X} \in \mathfrak{R}^{N\times T}$ can be decomposed into two components, the common component $\chi \in \mathfrak{R}^{N\times T}$ and the idiosyncratic component $\xi \in \mathfrak{R}^{N\times T}$ [11]. The residual of wind power can be obtained by subtracting the seasonality from the observed wind power. The variable N is the number of wind farms and T is the number of observation periods. The common component represents the portion of wind power that is driven by the latent variables, such as air stream or wind flows, so that nearby wind farms will have similar common components. The idiosyncratic component represents the remaining portion of wind power that could be an instant and unpredictable movement or measurement error of each wind power output, so that each wind farm has a different idiosyncratic component.

The column vector $\boldsymbol{X}_t = \{x_{1t}, x_{2t}, \cdots, x_{Nt}\}$ represents the wind power of N wind farms at time t, where $t = 1, \cdots, T$. Therefore, the column vector X_t at time t can be decomposed as:

$$\boldsymbol{X}_t = \chi_t + \xi_t \tag{10.1}$$

where $\chi_t \in \Re^N$is the *common component* and $\xi_t \in \Re^N$is the *idiosyncratic component* at time t. By the definition, we also assume that χ and ξ are uncorrelated as:

$$E\left[\chi_j \xi_k^T\right] = 0 \tag{10.2}$$

where $j, k = 1, \cdots, T$.

This decomposition is performed in frequency domain, in other words, we decompose the cross power spectral density (CPSD) of the residual signals in $\boldsymbol{X}$ into the CPSD of the common components in χ_t and the CPSD of the idiosyncratic components in ξ through the dynamic principal component analysis (DPCA) [14]. Like the static principal component analysis (PCA), the DPCA is a dimension reduction tool which decomposes the covariance matrix into two matrices at each frequency. In detail, only the Q number of large eigenvalues and their corresponding eigenvectors of the CPSD of signals in $\boldsymbol{X}$ at the given frequency are converted to the CPSD of common components in χ_t, and the rest of the eigenvalues and their corresponding eigenvectors are converted to the CPSD of idiosyncratic components in ξ_t. The Q represents the dimension of the dynamic factors $\phi \in \Re^{Q\times T}$, which are unobserved factors, such as wind speed or air mass, and they are loaded/linearly combined dynamically through the dynamic factor loading with different time lags. Therefore, the dynamic factor represents the dynamic movement of the common component. It should be noted that by selecting only Q ($Q < N$) eigenvectors, we can reduce the dimension of the GDFM. After the CPSD of $\boldsymbol{X}$ is decomposed into the CPSDs of common and idiosyncratic components, by applying the inverse Fourier transform to the CPSDs of χ_t and ξ_t, we can generate $\chi_t$and ξ_t, respectively.

To summarize, the second assumption is that the χ_t is assumed to be driven by the process of Q ($Q < N$) latent variables $\phi \in \Re^{Q\times T}$, which is called the dynamic factor. The common component is loaded into each wind power through the factor loading $\boldsymbol{A}(L) \in \Re^{N\times Q\times M}$, which is an $N\times Q$ matrix polynomial of M lags. Therefore, the $\chi_t \in \Re^N$ can be decomposed into the multiplication of the dynamic factor $\boldsymbol{\Phi} \in \Re^{Q\times T}$and factor loading $\boldsymbol{A}(L)$ as:

$$\chi_t = \boldsymbol{A}(L)\phi_t = \boldsymbol{A}_0\phi_t + \cdots + \boldsymbol{A}_M\phi_{t-M} \tag{10.3}$$

where the L is the delay operator, and $\boldsymbol{A}_m \in \Re^{N\times Q}$. Furthermore, $\boldsymbol{A}(L)$ is a notation for polynomial matrix, and it is defined as:

$$\boldsymbol{A}(L) = \boldsymbol{A}_0 L^0 + \cdots + \boldsymbol{A}_M L^M \tag{10.4}$$

where L^i is the delay operator. The multiplication between L^i and ϕ_t means that the ϕ_t is delayed by i, i.e. ϕ_{t-i}. Since $Q < N$, the χ_t is rank deficient.

The third assumption is that the dynamic factors ϕ are in a time-series structure; thus, vector autoregressive (VAR) model can be used to model ϕ:

$$\phi_t = C_1\phi_{t-1} + \cdots + C_R\phi_{t-R} + \varepsilon_t \tag{10.5}$$

which can also be rearranged into:

$$\varepsilon_t = \boldsymbol{C}(L) = C_0\phi_t - C_1\phi_{t-1} - \cdots - C_R\phi_{t-R} \tag{10.6}$$

where ε_t is the column vector of uncorrelated white noise $\varepsilon \in \Re^{Q\times T}$, C_0 is identity, R is the order of the model, and $C(L) \in \Re^{Q\times Q\times (R+1)}$ is a coefficient matrix. Another way to look at Eq. (10.6) is that the ϕ is driven by noise ε; however, the noise term has the same spatial correlation as in the actual residual data, while the goal here is to formulate a series of uncorrelated noise to drive ϕ so that many scenarios can be synthesized. Guided by this idea, we extract the correlation structure from ε with the Cholesky decomposition [6]:

$$\varepsilon_t = \boldsymbol{H}\boldsymbol{\delta}_t \tag{10.7}$$

where $\boldsymbol{\delta} \in \Re^{Q\times T}$ is called the dynamic shocks, and it is a series of spatially and temporally uncorrelated white Gaussian noise with zero mean and variance one. In addition, the $\boldsymbol{H}$ is the $Q\times Q$ matrix that preserves the correlation structure.

Finally, by combing Eqs. (10.3), (10.5), (10.6), and (10.7), the dynamic factor is represented as:

$$\chi_t = \boldsymbol{A}(L)[\boldsymbol{C}(L)]^{-1}\boldsymbol{H}\boldsymbol{\delta}_t \tag{10.8}$$

which leads to the comprehensive form of the GDFM. Then the question of how to estimate the polynomial matrices $\boldsymbol{A}(L)$ and $\boldsymbol{C}(L)$ arises naturally, and the procedure is discussed below.

10.2.2 Estimation of the GDFM

The estimation of the GDFM starts from estimating the $\Sigma^X(f_m) \in \Re^{N\times N\times F}$, which is the Fourier transform of the covariance matrix of the *residual* data $\boldsymbol{X}$. As mentioned in 10.2.1, the technique for the decomposition of $\boldsymbol{X}$ is DPCA.

PCA is defined as an orthogonal linear transformation that transforms the data to a new coordinate system such that the greatest variance by some scalar projection of the data comes to lie on the first coordinate (called the first principal component), the second greatest variance on the second coordinate, and so on. It is commonly used for dimensionality reduction by projecting each data point onto only the first few principal components to obtain lower-dimensional data while preserving as much of the data's variation as possible. The first principal component can equivalently be defined as a direction that maximizes the variance of the projected data. The i-th principal component can be taken as a direction orthogonal to the first principal components that maximizes the variance of the projected data. It can be shown that the principal components are eigenvectors of the data's covariance matrix. Thus, the principal components are often computed by eigen-decomposition of the data covariance matrix or singular value decomposition of the data matrix. PCA is the simplest of the true eigenvector-based multivariate analyses and is closely related to factor analysis. DPCA is to process each data set associated with time. Figure 10.1 illustrates the basic PCA.

PCA of a multivariate Gaussian distribution centered at (2,4) with a standard deviation of 3 in roughly the (0.866, 0.5) direction and of 1 in the orthogonal direction. The vectors shown are the eigenvectors of the covariance matrix scaled by the square root of the corresponding eigenvalue, and shifted so their tails are at the mean. Therefore, those the vector on which the projection is 3 is the first principal component, and its orthogonal vector is the second principal component.

DPCA for GDFM is explained as following. At each frequency f_m, we can split $\Sigma^X(f_m)$ into $\Sigma^\chi(f_m) \in \Re^{N\times N\times F}$, which is the Fourier transform of the covariance matrix of the common component χ, and $\Sigma^\xi(f_m) \in \Re^{N\times N\times F}$, which is the Fourier transform of the covariance matrix of the idiosyncratic component ξ. Furthermore, the F represents the number of frequencies.

This splitting process can be described through the eigenvalue decomposition as:

$$\sum^{X}(f_m) = \boldsymbol{G}^\chi(f_m)\boldsymbol{\Omega}^\chi(f_m)\boldsymbol{G}^\chi(f_m)^T + \boldsymbol{G}^\xi(f_m)\boldsymbol{\Omega}^\xi(f_m)\boldsymbol{G}^\xi(f_m)^T \tag{10.9}$$

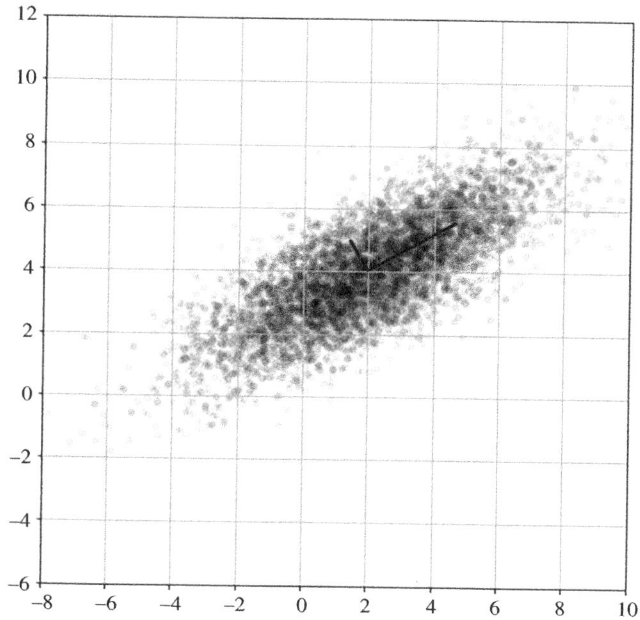

Figure 10.1 Change of basis in PCA.

where the $\boldsymbol{\Omega}^{\chi}(f_m) \in \Re^{Q\times Q\times F}$ is the diagonal matrix whose diagonals are from the top largest Q eigenvalues of $\Sigma^{X}(f_m) \in \Re^{N\times N\times F}$, and the $\boldsymbol{G}^{\chi}(f_m) \in \Re^{N\times Q\times F}$ is the corresponding eigenvectors associated with $\boldsymbol{\Omega}^{\chi}(f_m)$. The $\boldsymbol{\Omega}^{\xi}(f_m) \in \Re^{(N-Q)\times(N-Q)\times F}$ is the diagonal matrix whose diagonals are the remaining N–Q eigenvalues of $\Sigma^{X}(f_m) \in \Re^{N\times N\times F}$, and the $\boldsymbol{G}^{\xi}(f_m) \in \Re^{N\times(N-Q)\times F}$ is the corresponding eigenvectors associated with $\boldsymbol{\Omega}^{\xi}(f_m)$. Therefore, in (9) the eigenvalues and eigenvectors of the $\Sigma^{X}(f_m) \in \Re^{N\times N\times F}$ is spliced into the largest Q eigenvalues and their corresponding eigenvectors and the remaining eigenvalues and their corresponding eigenvectors. Then, by calculating the inverse Fourier transform of each frequency response, the χ and ξ in (10.1) are estimated. Since the only data we can observe is the wind power output data, it is reasonable to assume that dynamic factors can be obtained through filter $\boldsymbol{B}(L) \in \Re^{Q\times N\times M}$ $\boldsymbol{C}(L) \in \Re^{Q\times Q\times(R+1)}$ as

$$\phi_t = \boldsymbol{B}(L)\boldsymbol{X}_t \tag{10.10}$$

To forecast wind power scenarios, we should estimate the $\boldsymbol{A}(L)$ and $\boldsymbol{B}(L)$. Rearranging (10.1), (10.3) and (10.10), the idiosyncratic component ξ_t can be written as

$$\xi_t = \boldsymbol{X}_t - \boldsymbol{A}(L)\boldsymbol{B}(L)\boldsymbol{X}_t \tag{10.11}$$

The goal is to find proper $\boldsymbol{A}(L)$, $\boldsymbol{B}(L)$ to minimize the sum of the variance of ξ_t for all t. In other words, we want to minimize the diagonal terms of covariance matrix of ξ,

$$\boldsymbol{\Gamma}^{\xi} = E[(\boldsymbol{X} - \boldsymbol{A}(L)\boldsymbol{B}(L)\boldsymbol{X})(\boldsymbol{X} - \boldsymbol{A}(L)\boldsymbol{B}(L)\boldsymbol{X})^{T}] \tag{10.12}$$

where $\boldsymbol{\Gamma}^{\xi}$ is the covariance matrix of ξ, and T is the transpose operator. The problem becomes to minimize the trace of $\boldsymbol{\Gamma}^{\xi}$. The spirit of estimating the ξ is in the same line with splitting eigenvalues in (10.9). Therefore, the solutions for this problem are obtained by the DPCA and Courant-Fisher Theorem [16]: $\boldsymbol{A}(L)$ equals to $\boldsymbol{G}^{\chi}(L) \in \Re^{N\times Q\times F}$, which is the corresponding eigenvectors associated

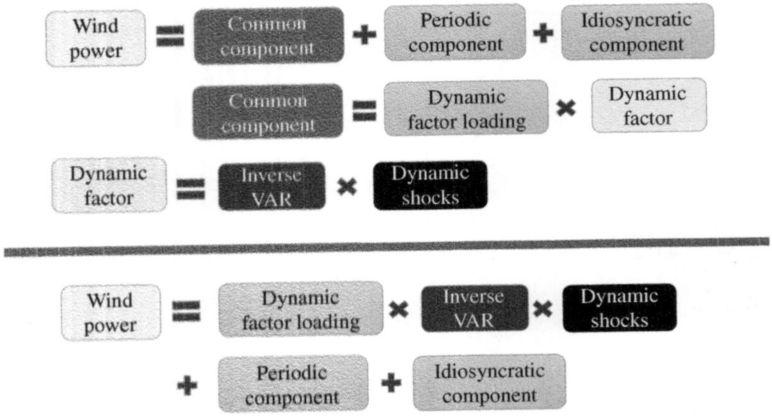

Figure 10.2 The structure of GDFM.

with the top largest Q eigenvectors of the covariance matrix of common component χ at all lags, and $\boldsymbol{B}(L)$ equals to $(\boldsymbol{G}^{\chi}(L))^T \in \Re^{Q\times N\times F}$. By Eq. (10.3) and (10.10), the χ_t can be estimated by

$$\chi_t = \boldsymbol{G}^{\chi}(L)(\boldsymbol{G}^{\chi}(L))^T\boldsymbol{X}_t \tag{10.13}$$

Again, according to (10), since $\boldsymbol{B}(L)$ equals $(\boldsymbol{G}^{\chi}(L))^T$, the ϕ_t is estimated by

$$\phi_t = (\boldsymbol{G}^{\chi}(L))^T\boldsymbol{X}_t \tag{10.14}$$

The next step is to synthesize χ_t as a function of $\boldsymbol{\delta}_t$, in other words, $\boldsymbol{C}(L)$ needs to be estimated. For given VAR process as shown in (10.5), the coefficients can be estimated by Yule-Walker equation. Such estimation process is detailed in reference [17]. In all, the structure of GDFM is summarized in Figure 10.2.

Until now, we define the GDFM and introduce the way to estimate the parameters in the GDFM from the observation data. In the next subsection, we explain the way to forecast the future wind power by using the GDFM.

10.2.3 Forecast of the GDFM

The overall process to decompose the wind power into various components in the GDFM is described in Figure 10.2. It is worth to mention that the common component of wind power consists of deterministic components (factor loadings and inverse VAR process) and stochastic component (random dynamic shocks). Such structure implies that wind power prediction scenarios are driven by random dynamic shocks $\boldsymbol{\delta}$.

As mentioned above, our first contribution is to forecast future wind power scenarios of multiple wind farms through the GDFM, which was originally developed to generate current or future wind power scenarios. The key idea is to have zero future error as the best estimate for the dynamic shocks. For the GDFM, the $\chi^t_{t_k}$ can be estimated as:

$$\chi^t_{t+k} = \boldsymbol{A}(L)[\boldsymbol{C}(L)]^{-1}\boldsymbol{H}\boldsymbol{\delta}^t_{t+k} \tag{10.15}$$

where k is the future steps past the end of the observed series. The superscript t is to be read as "given data up to time t". Therefore, to forecast over the next 24 hours, the columns in the dynamic shocks, $\delta_{t+1}, \cdots, \delta_{t+24}$ are set to zero. The forecast result is provided in Figure 10.3. We also compare the performance of the GDFM to the performance of the persistent forecasting model and the performance of the autoregressive (AR) model with order two [18]. In the persistent model, the last observed value is considered as a forecasted value.

Furthermore, the formulation in (10.15) shows that multiple future wind power scenarios can be generated simultaneously by using the common characteristics of correlated scenarios so that we can reduce the computational cost and time to forecast future wind power scenarios. This clearly proves our second contribution. We also compare the performance of the GDFM to performances of other forecasting models, such as the persistent model and AR model. As shown in the Figure 10.3, there are 30 forecasted future scenarios plotted in dark gray lines, the forecasts from the AR-2 are plotted in a gray dot line, and the forecasts from the persistent model are plotted in a magenta rhombus line. Furthermore, the historical wind power is plotted in a black line prior to the hour 2160, and the actual future wind power is plotted in a black square line. The forecast can capture the trend of actual wind power, and the generated scenarios are used by stochastic programming to make decisions under the high penetration level of intermittent wind power. The root mean square errors (RMSEs) of forecasting models with respect to the forecasting horizons are plotted

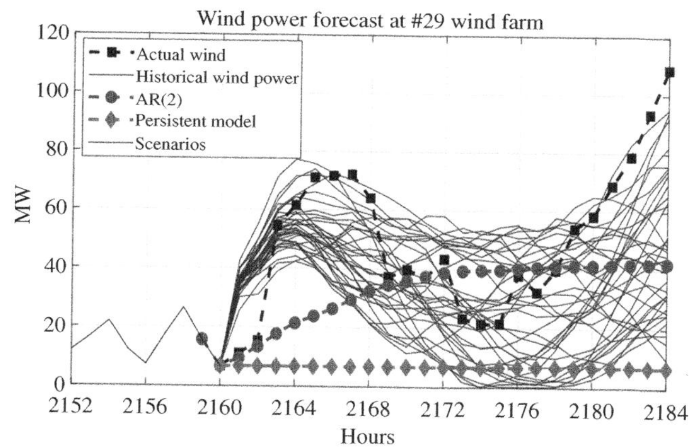

Figure 10.3 Wind power forecast at #29 wind farm.

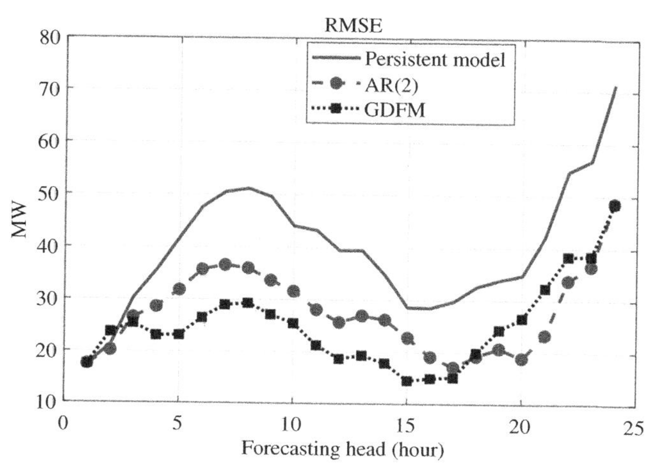

Figure 10.4 The RMSE performance of forecasting models.

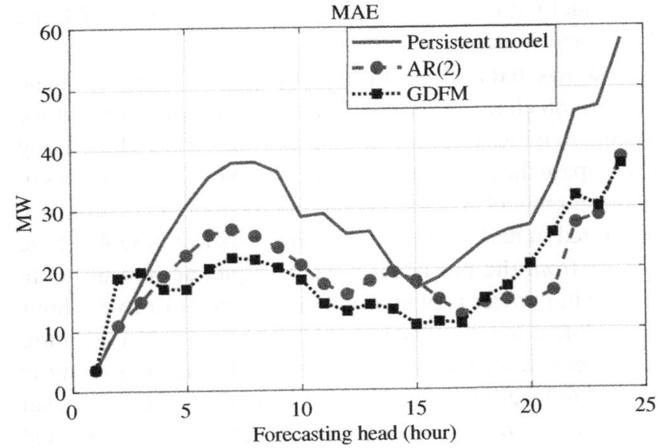

Figure 10.5 The MAE performance of forecasting models.

Table 10.1 Average performance of forecasting models.

Methods	Persistent model	AR-2	GDFM
RMSE [MW]	39.9734	27 746	25.045
MAE [MW]	28.5546	1932	18.05

in Figure 10.4, and the mean absolute errors (MAEs) of the forecasting models with respect to the forecasting horizons are plotted in Figure 10.5. In Figures 10.4 and 10.5, the forecasts of the persistent model are plotted in a dark gray line, the forecasts of the AR-2 are plotted in a gray circle line, and the forecasts of the GDFM are plotted in a black square line. When the performance of the GDFM is measured, the dynamic shock is assumed to be zero so that we can have a single scenario.

In Figures 10.4 and 10.5, we can observe that the performance of the GDFM is higher than those of other forecasting models. Furthermore, when the performance is measured by the RMSE, the performance difference between the GDFM and other forecasting models is significant. It means that the GDFM can forecast wind power outputs that are not extremely far from actual wind power outputs. For some forecasting horizons, the performance of the GDFM is lower than other forecasting models, but for most of forecasting horizons, the GDFM performs better than other forecasting models. Besides, it should be emphasized that the GDFM can keep the stochastic correlation structure among wind power outputs, but other forecasting models cannot.

We also calculate the average RMSE and MAE in Table 10.1. It is clearly shown that the GDFM has the best performance among three forecasting models.

In this subsection, we forecast the future wind power scenarios by having zero dynamic shocks. In the next subsection, we verify whether the forecasted scenarios are spatially and temporally correlated. The forecasting accuracy of the GDFM scenarios will be tested in the next section.

10.2.4 Verification of the GDFM

The procedures of synthesizing scenarios can be summarized as following: (i) $\widehat{\chi}$, $\widehat{\xi}$, and $\widehat{\phi}$ are estimated from data set $\boldsymbol{X}$ (the 'hat' denotes for the estimated value); (ii) since dynamic factors ϕ can

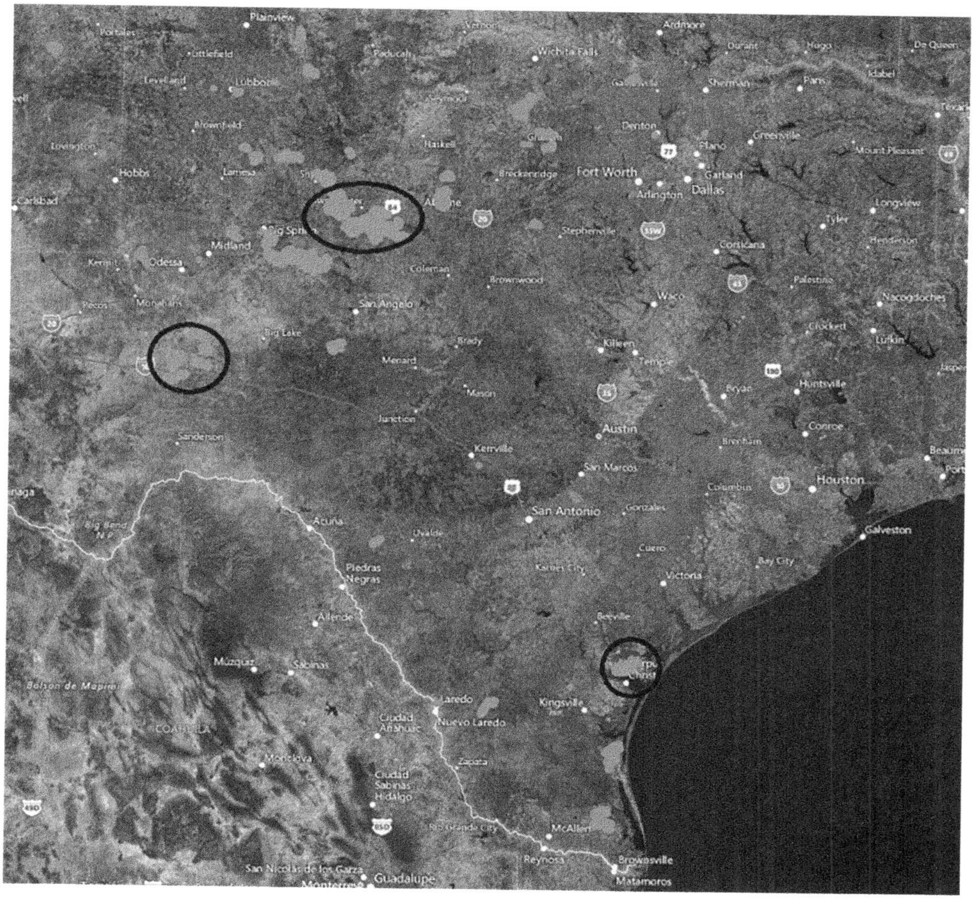

Figure 10.6 Wind farms in Texas. Source: Bai et al. [19] from MDPI.

be modeled as a VAR process, the noise term in VAR model can be estimated as $\hat{\varepsilon}$; (iii) Scenarios of $\hat{\chi}$ are synthesized by giving dynamic shocks δ, which are uncorrelated noises into the model as denoted by (10.7); (iv) The $\hat{X}$ is finally synthesized by adding $\hat{\chi}$ with $\hat{\xi}$, and the estimation of the original data can be obtained by adding seasonality and reversing the normalization process.

Figure 10.6 gives the wind farms in Texas, which are marked in gray. There were $N = 23$ wind farm observation data obtained from the Electric Reliability Council of Texas (ERCOT) being chosen. They are grouped in three regions, Mountain King (left), Sweetwater (top), and Corpus Christi (right), as circled in Figure 10.6. The hourly data is from January 1st to March 31st in 2013 for a total of $T = 24 \times 90 = 2160$ hours. The forecasting period is 24 hours in the next day in the historical data, which is April 1st, 2013.

The original data is pre-processed as follows: (1) for each wind farm, the average wind power output across the total of measured period is subtracted from the raw wind power output; (2) the residuals are divided by the standard deviation of the wind power output of each single wind farm; and (3) the diurnal pattern of the wind power output of each single wind farm is extracted. The Steps 1 and 2 are the standardizing process and the Step 3 is the process of removing seasonality. The periodic component at each hour is estimated by averaging wind power outputs measured at

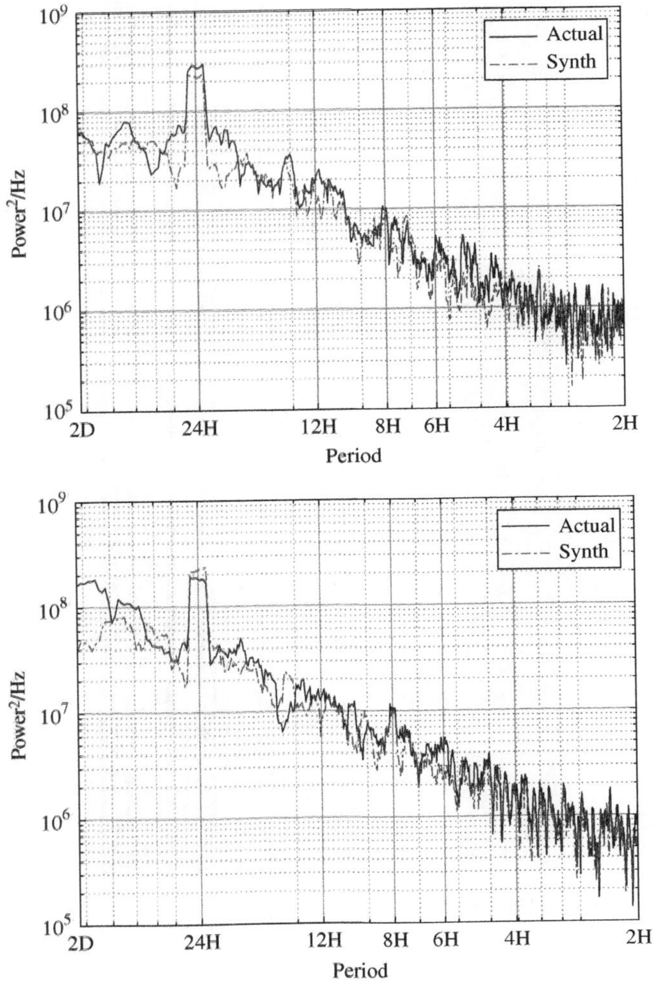

Figure 10.7 The PSD of actual wind power and synthesized wind power at Papalote Creek1 (top) and Papalote Creek 2 (bottom).

the same hour. The process to generate the periodic component is as follows. First, the wind power outputs $\boldsymbol{X} \in \Re^{23\times2160}$ are reshaped by 24 hours and 90 days as $\boldsymbol{Y} \in \Re^{23\times24\times90}$.Then the mean of each hour over 90 days is stacked as $\boldsymbol{Z} \in \Re^{23\times24}$, and the daily pattern is estimated by averaging the reshaped wind power at every hour. Finally, the daily pattern is subtracted from the $\boldsymbol{X}$. It should be noted that after synthesizing preliminary normalized scenarios using residuals, the reverse process of pre-processes is applied to the preliminary normalized scenarios to the obtain final scenarios. The model is also verified in Figure 10.7 with the power spectral density (PSD) plot of the scenario in the frequency domain. The PSD indicates the contributions of different frequency components. Note that the x-axis is changed to "period," the reciprocal of "frequency" for easier interpretation. It is found that the frequency response of the wind power data from the Papalote Creek 1 and Papalote Creek 2 has a very strong component corresponding to "24-hour period". Similarly, the synthesized PSD corresponds with the actual wind data. Figure 10.8 shows the comparison between actual wind power and its common component at the Papalote Creek 1 and Creek 2 (Corpus Christi

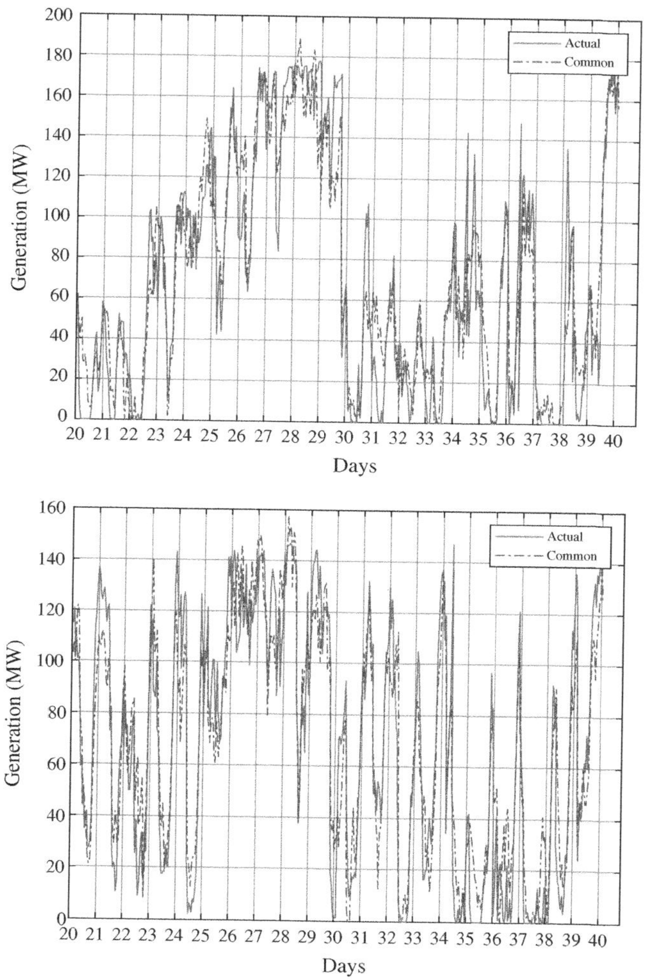

Figure 10.8 Actual wind power and its common component at Papalote Creek 1(top) and Papalote Creek 2 (bottom).

region). It is observed that the common component in a gray line can catch the trend of the actual wind power, i.e. it is similar with the actual measurements in terms of ramp, maximum, minimum, and overall shape.

As described earlier, one of the advantages is that the GDFM can capture the co-movement of synthesized scenarios, i.e. the synthesized scenarios of different wind farms have similar correlations as the actual wind power outputs do.

Figures 10.9 and 10.10 have demonstrated this with the correlation coefficients of 0.97526 and 0.96059, respectively, for the actual wind power and the synthesized scenarios of SweetWater 4A and SweetWater 4B wind farms, which are two very close wind farms.

Similarly, Figures 10.11 and 10.12 show the correlations between Papalote Creek 1 (Corpus Christi region) and Lorraine 1 (SweetWater region) for both actual and generated scenarios. The farther the wind farms are located, the less correlated the wind power outputs are. Therefore,

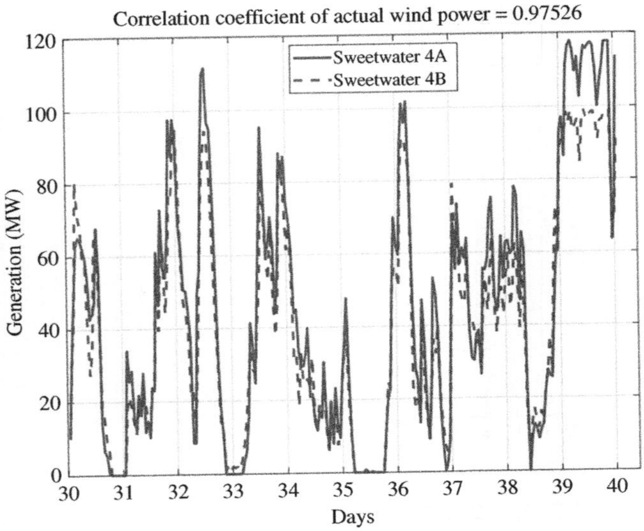

Figure 10.9 Actual wind power scenarios from the Sweetwater 4A and Sweetwater 4B.

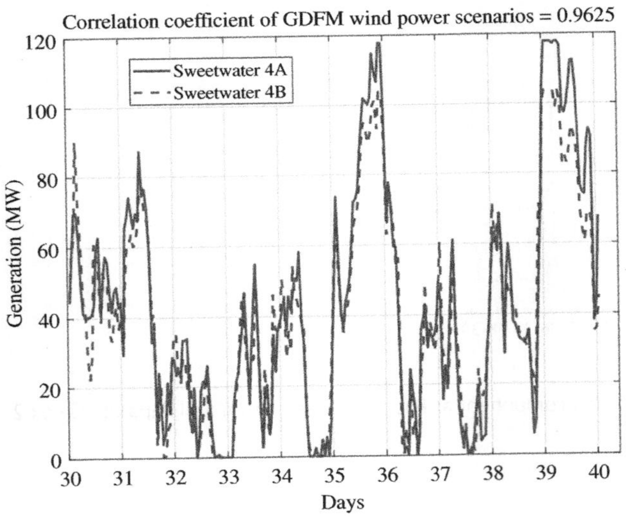

Figure 10.10 Forecasted wind power scenarios from the Sweetwater 4A and Sweetwater 4B.

since the distance between the Papalote Creek 1 and Lorraine 1 is around 400 mi, the correlation coefficient between these two wind farms is low.

In this subsection, we see that the GDFM scenarios are spatially and temporally correlated. In the next section, we formulate the stochastic and dynamic AC OPF problem where the GDFM scenarios will be used.

10.2.5 Application of the GDFM

In this section, the stochastic and dynamic AC OPF problem is formulated. The objective of the AC OPF problem is to minimize fuel costs by determining the optimal reactive and active power outputs, while satisfying voltage and apparent power constraints. Furthermore, the dynamic OPF

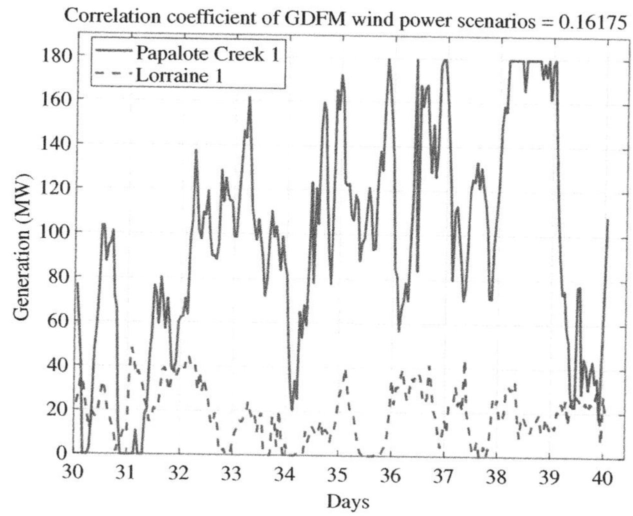

Figure 10.11 Actual wind power scenarios from uncorrelated wind farms, Papalote Creek 1 and Lorraine 1.

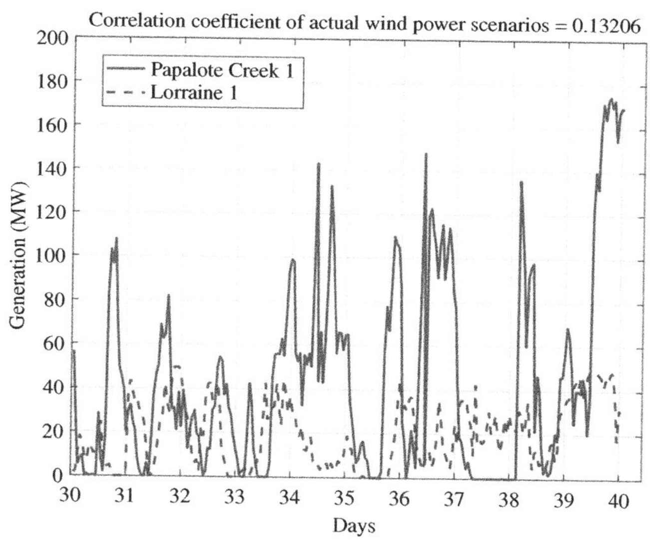

Figure 10.12 Forecasted wind power scenarios from uncorrelated wind farms, Papalote Creek 1 and Lorraine 1.

problem solves this AC OPF over a period. Moreover, the stochastic OPF problem solves the AC OPF while considering the uncertainty stemming from the stochastic wind power output. In the stochastic OPF problem, the uncertainties are captured by scenarios, and thus the mathematical formulation is

$$\text{Min} f(x, u) \tag{10.16}$$

$$\text{s.t. } g(x, u) = 0 \tag{10.17}$$

$$h(x, u) \leq 0 \tag{10.18}$$

where vector u represents decision/control/independent variables over 24 hours, which includes generator real power P_G except the power at slack bus, generator bus voltage V_G, transformer tap T, and shunt compensator Q_C at selected buses. The vector x is state/dependent variables, which include the real power P_{G1} at a slack bus, voltages V_L at a load bus, the reactive power Q_G at a generator bus, and loadings S_L of transmission lines.

The objective function is the sum of total fuel costs and real power losses. The fuel cost is

$$F_1 = \sum_{t=1}^{24} \sum_{i=1}^{N_G} (a_i P_{Git}^2 + b_i P_{Git} + c_i) \tag{10.19}$$

where a_i, b_i, c_i, and P_{Git} denote for the quadratic fuel cost coefficients and real power at the i-th unit at time t, and N_G is the number of generators. The real power loss is due to the power flowing through transmission lines. The objective function for power loss can be mathematically formulated as follows:

$$F_2 = \sum_{t=1}^{24} \sum_{k=1}^{N_l} \left\{ \frac{r_k}{r_k^2 + x_k^2} \left[V_{it}^2 + V_{jt}^2 - 2V_{it}V_{jt}cos(\delta_{it} - \delta_{jt}) \right] \right\} \tag{10.20}$$

where the total number of transmission lines is N_l, the resistance and reactance of the transmission line k that links bus i and j are r_k and x_k, respectively; V_{it}, V_{jt}, δ_{it}, and δ_{jt} are the voltages and angles at bus i and j at time t, respectively. In this work, F1 and F2 are the two cost functions considered for case studies.

The equality constraints g from (10.17) are the AC power flow balance equations at each bus representing that the power flowing into that specific bus is equal to the power flowing out, which are defined as:

$$\begin{aligned} P_{it} &= V_{it} \sum_{j=1}^{N} V_{jt} Y_{ij} \cos(\delta_{it} - \delta_{jt} - \theta_{ij}) \\ Q_{it} &= V_{it} \sum_{j=1}^{N} V_{jt} Y_{ij} \sin(\delta_{it} - \delta_{jt} - \theta_{ij}) \quad \forall t, \forall i, \forall j \end{aligned} \tag{10.21}$$

where N is the total number of buses, P_{it} and Q_{it} are the injected real and reactive power at bus i at time t, Y_{ij} and θ_{ij} are the elements in Y admittance matrix. Inequality constraints h in (10.18) include generator limits, ramp rate constraints of generators, tap position of transformers, shunt capacitor constraints, security constraints, load bus voltage and transmission line flows [20].

The essence of tackling a stochastic problem is to make a decision that can perform well in average regarding all scenarios caused by the stochastic process. Therefore, the stochastic dynamic AC OPF is formulated to minimize the expected value of the objective function over all scenarios as:

$$FF = \sum_{s=1}^{S} \text{prob}^s \sum_{t=1}^{24} F_t^s \tag{10.22}$$

where F_t is cost function from (10.19) or (10.20) at time t, probs is the probability of scenario s which quantifies the likelihood of that scenario, and S is the total number of scenarios. Since we assume that every scenario is independent with others, the probability for each scenario is $1/S$. Thus, the stochastic dynamic AC OPF can be summarized to minimize (10.22) while complying with the equality and inequality constraints. In this section, we have conducted several case studies to evaluate the effectiveness of the proposed method. The proposed problem is a highly non-linear and non-convex problem where traditional mathematical programming is hard, if not

Table 10.2 Comparison of computing time and cost in case 1.

Methods	Time (mins)	Cost ($)
GDFM	30.3	3,120,322
Random	120.3	3,254,221

impossible, to solve. We proposed a heuristic method, modified artificial bee colony (MABC) to tackle the problem (Bai and Lee, 2017), which is not scope of the book chapter. The stochastic processes are implemented in the modified IEEE 118-bus test system. In the modified system, there are 23 wind farms located at three regions, which are circled in gray in Figure.10.11, based on the approximate wind farm geographic locations. The computer used for simulation has 3.4 GHz Intel core i7 Processor and 32 GB RAM. The AC power flow is calculated by the MATPOWER package [21].

Case 10.1 Quadratic Fuel Cost Minimization

Case 10.1 is a standard OPF problem with the cost function in quadratic form. We can find the data for the IEEE 118-bus test system with control variable limits and fuel cost coefficients in [22]. There is a total of 130 control variables, which consist of 53 real power outputs at PV buses, voltage magnitudes of all 54 generator buses, nine transformer taps, and 14 shunt capacitors. The objective in this case is to minimize the total generator fuel cost in (10.19). Transformer taps and shunt capacitors are discrete variables, which increases the complexity to the problem. Table 10.2 summarizes the comparison results. The load profile was for 21 December 2014 obtained from ERCOT and is scaled to fit the size of this problem. Note that 15 scenarios are generated by the GDFM for each farm, and that they are correlated with each other. However, if random scenarios are used to represent the uncertainty, it would have been 15^{23} scenarios without certain scenario reduction techniques, which is not feasible to solve. In this study, we chose 100 uncorrelated random scenarios. Because GDFM can provide correlated scenarios, only a limited number of scenarios needed to be generated so that the computational cost for stochastic optimization is significantly reduced compared with random scenarios. Table 10.2 also compares the performance of the GDFM with the random scenarios in computational time (mins) and total fuel cost ($), where it is found that by using the GDFM, the computational cost and generation cost can be reduced compared with using random scenarios.

Case 10.2 Loss Minimization

In this case, the objective is to minimize the total real power loss defined in (10.20). The control and state variables are identical to the previous case. From the Table 10.3, a similar conclusion is drawn that the OPF based on the GDFM scenarios has a lower generation cost and power loss than the randomly generated scenarios since the forecasted GDFM scenarios can represent the spatially and temporally correlated actual wind power.

10.3 Conclusion

This work shows that the complex stochastic dynamic AC OPF under high wind power penetration can be solved based on the wind power forecasting scenarios of the GDFM. The GDFM is

Table 10.3 Comparison of computing time and loss in case 2.

Methods	Time (min)	Loss (MW)
GDFM	42.5	1128.4
Random	110.3	1178.5

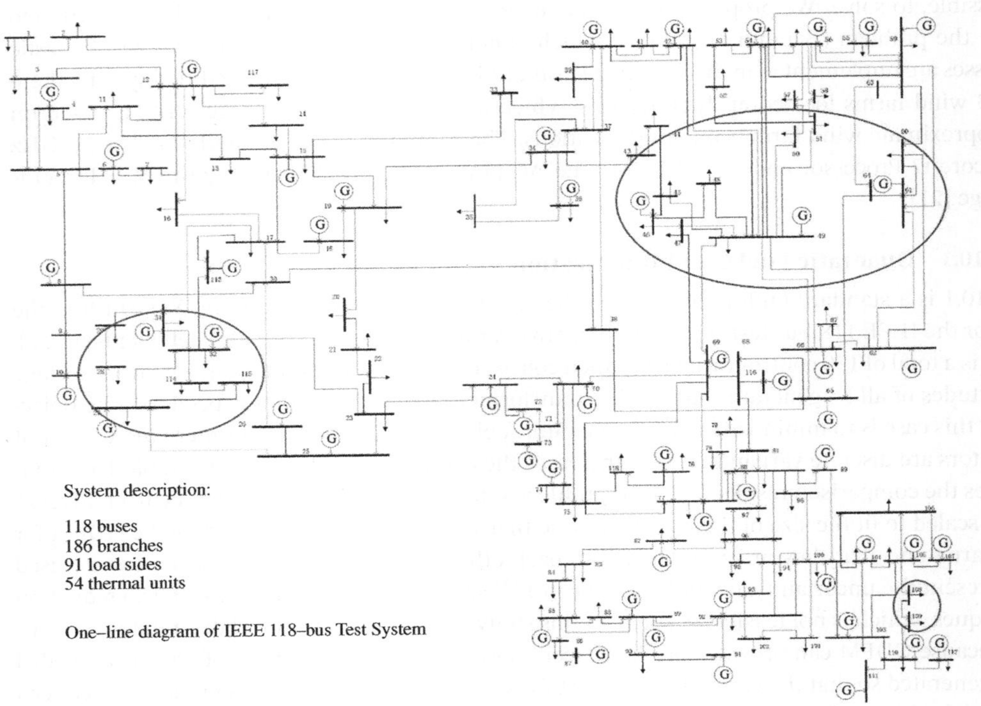

Figure 10.13 Modified IEEE 118-bus system.

verified in both time and frequency domains. The forecast accuracy has outrated the basic persistent model and AR-2 in RMSE and MAE metrics. Since the GDFM scenarios are spatially and temporally correlated, the benefits of using scenarios by GDFM are as follows: (i) only a limited number of scenarios need to be generated to represent the uncertainty so that the computational cost for stochastic optimization can be greatly reduced; and (ii) correlated scenarios can represent a more realistic forecasting where all wind farms in a close area are spatially correlated under the same weather pattern. The GDFM is tested on the IEEE 118-bus system integrated with 23 wind farms (Figure 10.13). It is shown that with the GDFM scenarios, computational cost can be significantly reduced compared with using random scenarios because in order to preserve the underlining correlations among wind farms, large number of random scenarios must be generated if using random scenarios. The computational cost, for the first case, has reduced up to 2.97 times compared with random scenarios case, the second case has reduced up to 1.6 times. Future work can be focused on extending GDFM to electrical load data, since load data has strong spatiotemporal correlation as well especially in distribution level side data.

References

1 A. Botterud, Z. Zhou, and J. Wang. (2011). Use of Wind Power Forecasting in Operational Decisions; Technical Report ANL/DIS-11-8; Argonne National Laboratory: Chicago, IL, USA.

2 Sun, M., Feng, C., and Zhang, J. (2020). Multi-distribution ensemble probabilistic wind power forecasting. *Renewable Energy* 148: 135–149.

3 Monteiro, C., Bessa, R., and Miranda, V. (2009). Wind Power Forecasting: State-of-the-Art; Technical Report ANL/DIS-10-1; Argonne National Laboratory: Chicago, IL, USA.

4 Latimier, R., Bouedec, E., and Monbet, V. (2020). Markov switching autoregressive modelling of wind power forecast errors. *Electric Power Systems Research* 189: 106641–106647.

5 Xie, L., Gu, Y., Zhu, X., and Genton, M.G. (2014). Short-Term Spatio-temporal wind power forecast in robust look-ahead power system dispatch. *IEEE Trans. Smart Grid* 5: 511–520.

6 Morales, J., Mínguez, R., and Conejo, A. (2010). A methodology to generate statistically dependent wind speed scenarios. *Applied Energy* 87: 843–855.

7 Khattam, W., Hegazy, Y., and Mansour, R. (2003). Uncertain power flow analysis of electrical distributed generation systems. In: *IEEE Conf., Power Engineering Society, General Meeting*, 1141–1144.

8 Lyu, J.K., Heo, J.H., Park, J.K., and Kang, Y.C. (2013). Probabilistic approach to optimizing active and reactive power flow in wind farms considering wake effects. *Energies* 6 (11): 5717–5737.

9 Wang, J., Botterud, A., Bessa, R. et al. (2011). Wind power forecasting uncertainty and unit commitment. *Applied Energy* 88: 4014–4023.

10 Burns, A.F. and Mitchell, W.C. (1946). *Measuring Business Cycles*. National Bureau of Economic Research.

11 Forni, M., Hallin, M., Lippi, M., and Reichlin, L. (2005). The generalized dynamic factor model. *Journal of the American Statistical Association* 100: 830–840.

12 Lee, D. and Baldick, R. (2017). Load and wind power scenario generation through the generalized dynamic factor model. *IEEE Trans. Power Syst.* 32: 400–410.

13 Lee, D., Lee, J., and Baldick, R. (2014). Wind power scenario generation for stochastic wind power generation and transmission expansion planning. In *Proceedings of the 2014 IEEE PES General Meeting, Conference Exposition*, National Harbor, MD, USA, 27–31 July 2014.

14 Forni, M. and Gambetti, L. (2010). The dynamic effects of monetary policy: a structural factor model approach. *Journal of Monetary Economics* 57: 203–216.

15 Nieuwenhuyze, C. V. (2006). A Generalised Dynamic Factor Model for the Belgian Economy—Useful Business Cycle Indicators and GDP Growth Forecasts; Technical Report 80; National Bank of Belgium Working Paper; National Bank of Belgium: Brussels, Belgium.

16 Brillinger, D.R. (1981). *Time Series: Data Analysis and Theory*. Philadelphia, PA, USA: SIAM.

17 Reinsel, G.C. (2012). *Elements of Multivariate Time Series Analysis*. Berlin/Heidelberg, Germany: Springer and Business Media.

18 Brockwell, P.J. and Davis, R.A. (2016). *Introduction to Time Series and Forecasting*, 3e. Cham: Springer.

19 Bai, W., Lee, D., and Lee, K.Y. (2017). Stochastic dynamic AC optimal power flow based on a multivariate short-term wind power scenario forecasting model. *Energies* 10: 2138–2157.

20 Bai, W. and Lee, K.Y. (2016). Modified optimal power flow on storage devices and wind power integrated system. In *Proceedings of the 2016 IEEE Power and Energy Society General Meeting (PESGM)*, Boston, MA, USA, 17–21, pp. 1–5.

21 Zimmerman, R.D., Murillo-Sanchez, C.E., and Thomas, R.J. (2011). MATPOWER: steady-state operations, planning, and analysis tools for power systems research and education. *IEEE Transactions on Power Systems* 26: 12–19.

22 Lee, K.Y., Park, Y.M., and Ortiz, J.L. (1985). A united approach to optimal real and reactive power dispatch. *IEEE Transactions on Power Systems* PAS-104: 1147–1153.

11

Spatiotemporal Solar Irradiance and Temperature Data Predictive Estimation

Chirath Pathiravasam[1,2] *and Ganesh Kumar Venayagamoorthy*[1,3]

[1] *Real Time Power and Intelligent Systems Laboratory, Holcombe Department of Electrical and Computer Engineering, Clemson University, Clemson, SC, USA*
[2] *Department of Electrical Engineering, University of Moratuwa,, Katubedda, Sri Lanka*
[3] *School of Engineering, University of KwaZulu-Natal, Durban, South Africa*

Acronyms

DER	Distributed energy resources
ED	Economic dispatch
HVAC	Heating, ventilation, and air conditioning
IDW	Inverse distance weighted
MAE	Mean absolute error
MLP	Multi-layer perceptron
NWP	Numerical weather prediction
PV	Photovoltaic
PWS	Physical weather station
RMSE	Root of the mean of the square error
SIT	Solar irradiance and temperature
VWS	Virtual weather station

11.1 Introduction

Power system operation poses a unique application for multiple time horizon weather predictions ranging from very long to very short term to reduce uncertainty. Weather parameters such as rainfall, wind speed, and solar irradiation affect the power system functions such as annual generator maintenance schedule and power generation planning.

The power system is increasingly becoming susceptible to short-term weather variations due to the increased renewable energy penetration, namely solar and wind power generation. The cost of power generation to meet urgent power requirements is very high as the energy is derived from emergency power generation units such as diesel generators. Therefore, predictability of short-term weather variations can make system operators prepared for forthcoming energy deficiencies and surpluses. It can reduce the high-cost emergency power need and avoid a penalty for tie-line power deviations.

Intelligent Data Mining and Analysis in Power and Energy Systems: Models and Applications for Smarter Efficient Power Systems, First Edition. Edited by Zita Vale, Tiago Pinto, Michael Negnevitsky, and Ganesh Kumar Venayagamoorthy.

A study done by the National Renewable Energy Laboratory on the impacts of variability and uncertainty in solar photovoltaic generation [1] showcases the advantage of performing energy dispatch in five-minute intervals utilizing shorter time horizon forecasts in place of conventional one-hour dispatch interval. It reduces the regulation reserve requirement and operational costs. Power generation follows the power demand to maintain system reliability, and demand-side management can control required power generation. The ambient temperature is the most correlated weather parameter to the power demand. A higher fraction of the power system demand comes from heating, ventilation, and air conditioning (HVAC) loads. In Kampelis et al. [2], a case study on the temperature control of a building is discussed, and the influence of ambient temperature on thermal comfort and the HVAC power demand is illustrated. Generally, the temperature variations are less compared to solar irradiance.

Artificial intelligence applications in the power system are extensively studied [3]. Economic dispatch (ED) is one of the popularly studied applications. With the emergence of retail markets, faster economic dispatch will be required to reduce the risk of distributed energy resources (DERs) [4, 5]. Very short-term weather parameter predictions will be essential in ED to reduce the uncertainty in dispatches. In Wang et al. [6, 7], deep convolutional neural network-based solar power and wind power predictions are discussed. The prediction horizons of the studies are from 15-minute to two hours. Weather prediction studies found in literature can be broadly divided into model-based weather predictions such as Numerical Weather Prediction (NWP) and statistical methodologies such as neural networks. In Tiwari et al. [8], short-term solar irradiance prediction is performed using NWPs. The accuracy of NWPs in complex terrains gets reduced as the resolution of the data points becomes insufficient, and NWPs need comparatively high computational resources [9]. In Maqsood and Abraham [10], weather forecasting analysis of different connectionist and statistical models is done. The authors have implemented multilayer perceptron (MLP), Elman recurrent neural network and radial basis function network to analyze the efficiencies of wind speed and temperature forecast. NWP-based forecasts evaluate the physical properties of the terrain in a fine resolution to infer weather predictions. However, a typical statistical weather prediction does not use sufficient spatial information to improve predictions. A Spatiotemporal prediction network created with non-linear approximators such as neural networks can improve the accuracy of predictions. It can communicate spatial data among the stations in the network to increase spatial awareness.

In Jayawardene and Venayagamoorthy [11], utility-scale solar irradiance prediction of utility-scale solar photovoltaic (PV) plants is discussed. Herein, an echo state network is used to predict solar irradiance for very short-term predictions. In Pathiravasam and Venayagamoorthy [12, 13], the authors describe several computational approaches entailing neural networks for wind speed prediction using spatiotemporal wind speed and wind direction. The authors have analyzed wind speed predictions of 15-minute, 30-minute, and one hour ahead for possible applications in the economic dispatch of power systems with wind plants. Wind speed predictions are made using MLP and recurrent neural networks as computational units of the prediction network.

Locations of weather measurements influence the accuracy of spatiotemporal predictions. In Pathiravasam et al. [14], the particle swarm optimization algorithm-based input selection technique is used to choose the optimal inputs for the prediction network. However, irregular weather station sites cause prediction networks to have suboptimal options available to choose from. Therefore, interpolation-based data synthesis is implemented to provide additional virtual weather stations (VWSs) to choose from for the predictor inputs. It can be considered as providing more relevant data to the networks with the fusion of available data. In Yonekura et al. [15], weather predictions were made for a location by linearly interpolating the available predictions closer to the new locations. In such an estimation, spatial weather variations are assumed to be linear. The accuracy of the estimation relies on the distances between the estimated locations and their parent locations. Therefore, a central value is more suitable for use as the estimated point. In this study, the Fermat point [16] of a triangle formed by three weather stations is used as the estimated point to increase the accuracy.

The contributions of this chapter are:

1. A method to improve the accuracy of spatial weather data estimation by increasing the density of total number of weather stations consisting physical and VWSs has been developed.
2. A framework for a step-ahead prediction of estimated weather data has been developed.

The rest of the chapter is as follows: in Section 11.2 weather estimation algorithms are explained. In Section 11.3, the distributed weather forecasting framework is presented, results are presented and discussed in Section 11.4, and the chapter is summarized in Section 11.5.

11.2 Virtual Weather Stations

Weather measurements construct a continuous field of data, and spatial interpolation methods can be used to determine weather of any location with known data in the proximity. Spatial interpolators include deterministic approaches such as inverse distance weighted (IDW), spline and global polynomial interpolation, and geostatistical approaches such as Kriging.

In this study, IDW is used to estimate solar irradiance and temperature (SIT) using a set of physical weather stations (PWSs). IDW follows Tobler's 1st law of geography: "Everything is related to everything else, but near things are more related than distant things" [17]. It utilizes the distance from an estimated location (virtual) to physical locations to discriminate the weightings of the measurements at physical locations as follows:

$$\tilde{y}_r(k) = \sum_i w_{ir} \times y_i(k) \tag{11.1}$$

$$w_{ir} = \frac{(d_{ir})^{-p}}{\sum_i (d_{ir})^{-p}} \tag{11.2}$$

where $\tilde{y}_r$ is estimated value at rth location, w_{ir} is weight for the value at ith location, d_{ir} is distance between locations i and r is power to the inverted distances. A p value higher than 1 would yield an estimation with more prominence to closer measurements. d_{ir} is a distance measurement and the L2 norm (Euclidean norm) is a commonly used method for this measurement.

In a network of geographically distributed weather stations, locations are determined by the installation's practical aspects, such as cost, altitude, and clearance requirements. However, sporadically distributed weather stations might not provide dynamic system information. On the other hand, data estimation methods such as IDW blindly utilize distance values between locations without any directional information. Therefore, a secondary set of proxy weather stations (VWSs) can be used to improve the distributed nature and the availability of weather measurements spatially.

In this study, artificial generation of weather data is proposed as a solution to improve the quality of the input data of the networks discussed in Section 11.2. Therefore, the positioning of VWSs is critical for the predictions. A central point of a triangle formed by three PWSs would be a good choice as it can provide evenly distributed contributions to estimating VWS estimations. Therefore,

VWSs will be placed on the Fermat point [16] of triangles formed by PWSs. The Fermat point is defined as the point with the minimum distance to all the vertices. A particle swarm optimization algorithm is used to determine the Fermat point. The data generation for the new VWS is done based on the distance of the new location from nodal vertices.

The Fermat point coordinates can then be used to estimate with triangular IDW. It should be noted that a VWS can be used as a node to produce the next VWS. Therefore, weather data estimation at the VWSs should be done iteratively. An array of physical and virtual weather data can be estimated in a matrix multiplication based on the weights of the adjacent weather stations as follows:

$$\overline{Y} = [\mathrm{Y}; \widetilde{\mathrm{Y}}] = W \times Y \tag{11.3}$$

$$W = [W_1; W_2; \dots ; W_n] \tag{11.4}$$

$$W_i = \sum_{k=1}^{3} \omega_{i,n_{ik}} \times W_{nik}, \quad i > n_{ik} \in N_i \tag{11.5}$$

$$\omega_{i,j} = \begin{cases} 0 \quad , & \text{if } j \notin N_i \\ \frac{(d_{ij})^{-p}}{\sum_r (d_{ir})^{-p}}, & \text{if } j, r \in N_i \end{cases} \tag{11.6}$$

where N_i is the set of neighbors contributing to the estimation of ith weather station (virtual). $\omega_{i,j}$ is the weight use to find the contribution from j to i. $\overline{Y}$ is comprised of weather measurements and estimations for PWSs (Y) and VWSs ($\widetilde{Y}$).

VWSs are generated using combinations of PWSs and VWSs. Therefore, it is expected to showcase different accuracy levels depending on the combinations used. Figure 11.1 shows how VWSs are created as combinations of PWSs and VWSs: L0 stations are physical ones, while L1, L2, L3, and L4 are virtually generated ones. Categorization is based on the closeness of the estimation to a PWS. In Figure 11.1, combinations of two stations are shown though three weather stations are

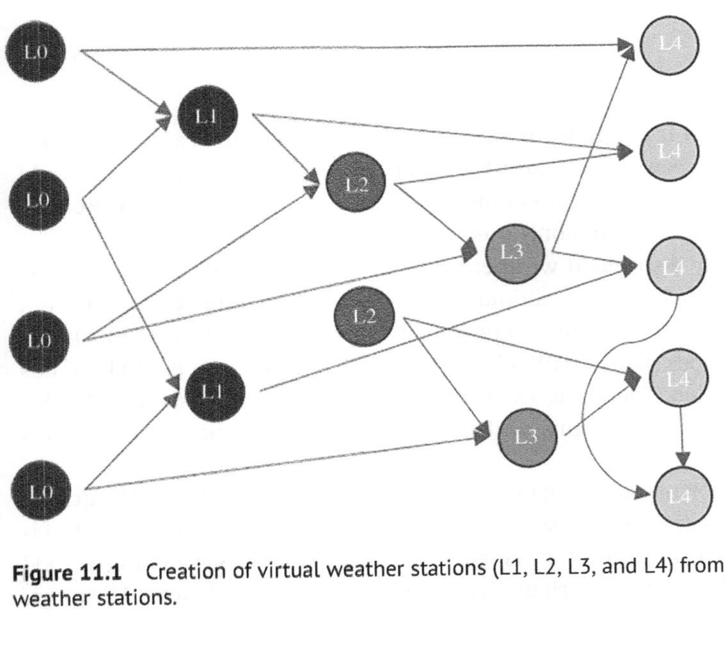

Figure 11.1 Creation of virtual weather stations (L1, L2, L3, and L4) from virtual and/ or physical (L0) weather stations.

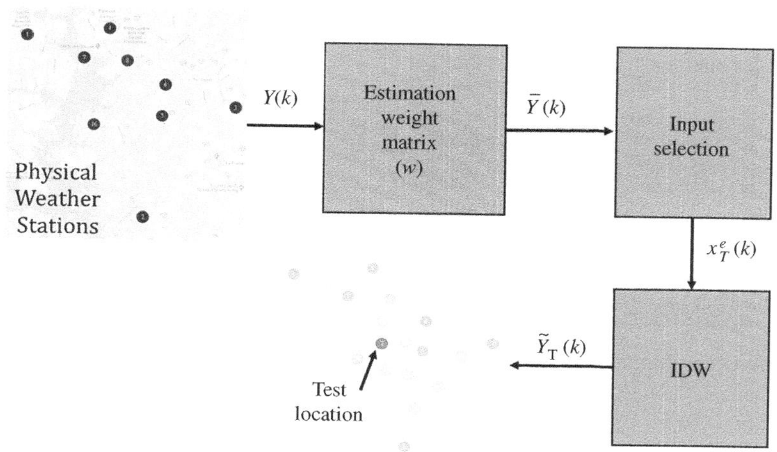

Figure 11.2 Weather estimation process.

used in this study. L1 is generated from two physical stations (L0), and thus, it is expected to have a more accurate estimation quality for its location than the estimation quality of L2, L3, and L4 virtual stations.

Physical and VWSs can be used to estimate weather data for many locations within the region using an IDW algorithm. However, the neighbor combination used in the IDW algorithm affects the accuracy. Two strategies in selecting neighbors are selecting a fixed number of closest neighbors or neighbors within a fixed radius. The best input combination can be optimized for maximum accuracy for each location.

The overall data estimation process is shown in Figure 11.2. $\widetilde{y_T}$ and x^e_T are the estimated weather data array and inputs to the IDW estimation at test location T respectively. x^e_T can be selected based on the proximity of physical and VWSs to the test location.

11.3 Distributed Weather Forecasting

Weather is a natural phenomenon with a very high spatiotemporal correlation. Embedding many spatiotemporal measurements in forecasting models can improve predictions and be computationally expensive. Intuitively, distributed and computationally light models of spatially distributed weather data can efficiently perform the task. In this chapter, spatiotemporal weather predictions are performed using neural networks.

11.3.1 Spatiotemporal Prediction Network

The prediction network consists of connected predictors, where each predictor represents a weather predictor of a location. In this chapter, weather predictions are made using predictors comprising MLP neural networks. MLP is a feedforward neural network with input, hidden, and output layers. The input layer consists of weather data measurements and estimations of selected locations, and the output is the prediction for predictor's location. Equations for the test location's prediction are as follows:

$$\hat{y}_T(k + \Delta k) = f_T\left(x^p_T(k)\right) \tag{11.7}$$

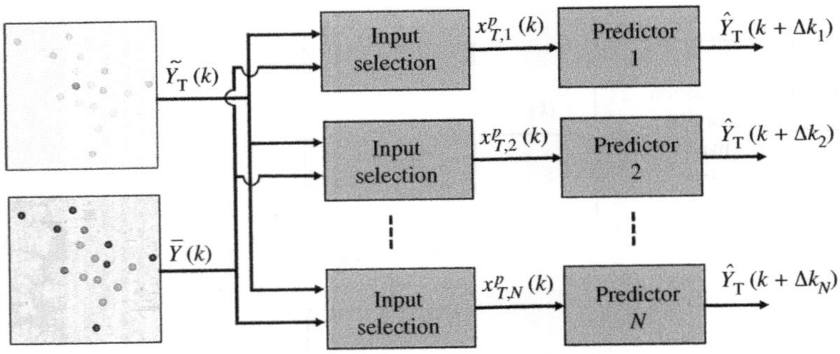

Figure 11.3 Weather data prediction framework.

$$x^p_T(k) = [v_T(k), v_T(k-\Delta k), \ldots, v_T(k - m \times \Delta k)] \tag{11.8}$$

$$v_T(k) = [y_T(k), y_{T,1}(k), y_{T,2}(k), \ldots, y_{T,q}(k)] \tag{11.9}$$

where $\hat{y}_T(k+\Delta k)$ and $y_{T,i}(k)$ are Δk ahead prediction and value of ith neighbor at time k of test location respectively. x^p_T is the input vector comprising of measurements/estimations from neighboring weather stations used in the prediction. m and q are the number of past measurements/estimations and the number of spatial neighbors used as inputs respectively. f_T is the MLP neural network which transforms spatiotemporal input vector into the prediction. Figure 11.3 shows the prediction framework. Multiple predictors (N) can be implemented to predict weather in multiple time horizons as shown in the figure.

11.3.2 Computational Units

The computational units in the prediction network are heterogeneous in construction due to the difference in inputs. The inputs of the computational units consist of spatiotemporal weather measurements/estimations of selected locations. Spatial dynamics of weather is learnt through these weather inputs.

11.4 Results and Discussion

A study was done with weather data of eight locations downloaded from Solcast website [18]. Weather data belonging to seven locations are used to generate VWSs. Weather data of the remaining location are used to serve as the testing node for estimation and predictions.

Accuracy statistics used for this study are mean absolute error (MAE) and root of the mean of the square error (RMSE). Equations for MAE and RMSE are as follows:

$$MAE = \frac{1}{N_k}\sum_k |\ \hat{y}_T(k+\Delta k) - y_T(k)\ | \tag{11.10}$$

$$RMSE = \sqrt{\frac{1}{N_k}\sum_k (\hat{y}_T(k+\Delta k) - y_T(k))^2} \tag{11.11}$$

where N_k is the number of samples.

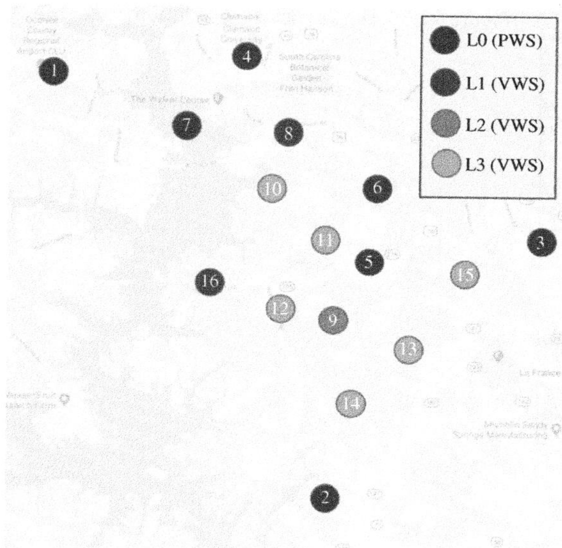

(a)

WS	Parent nodes	WS type
1	—	L0
2	—	L0
3	—	L0
4	—	L0
5	—	L0
6	—	L0
7	—	L0
8	(4, 6, 7)	L1
9	(2, 6, 8)	L2
10	(4, 7, 9)	L3
11	(6, 9, 10)	L3
12	(2, 7, 11)	L3
13	(2, 3, 9)	L3
14	(2, 9, 13)	L3
15	(3, 5, 13)	L3
16	(1, 2, 7)	L1

(b)

Figure 11.4 (a) Locations of physical and virtual weather stations, (b) parent nodes and the weather station level of the VWSs.

Results are compared against the persistent model, which uses the latest measurement/estimation as the prediction. It is a widely used benchmark statistic for short-term predictions. Therefore, the equation for $\hat{y}_T(k + \Delta k)$ is as follows:

$$\hat{y}_T(k + \Delta k) = y_T(k) \tag{11.12}$$

11.4.1 Weather Data Estimation

Figure 11.4 depicts the spatial distribution of PWSs and VWSs. There are nine locational data generated using the VWS algorithm in addition to seven PWSs. Figure 11.5 shows the weight matrix, which can be used to estimate VWS data with physical weather data. The first seven rows of the matrix constitute an identity matrix that calculate physical weather measurements without using any additional inputs.

Location T is the testing node. Weather data for location T is estimated using three cases as follows:

Case 1 — No estimation: Data value at the closest neighbor is the best guess in the absence of estimation.

Case 2 — Closest triangle: three points with physical weather data are selected to estimate with the IDW algorithm.

Case 3 — PWS and VWS data: pre-estimated VWS data was used in addition to physical data to estimate with the IDW algorithm.

There are different dynamics in weather around the year. Therefore, the accuracy of weather estimation cannot be determined solely by data from one time-period of the year. A study was done for three weeks at different times of the year to compare results. Three weeks are the first weeks of January, May, and September, respectively. Tables 11.1 and 11.2 show the MAE and the RMSE of solar irradiance and temperature estimations.

		PWS						
		1	2	3	4	5	6	7
PWS	1	1.00	0.00	0.00	0.00	0.00	0.00	0.00
	2	0.00	1.00	0.00	0.00	0.00	0.00	0.00
	3	0.00	0.00	1.00	0.00	0.00	0.00	0.00
	4	0.00	0.00	0.00	1.00	0.00	0.00	0.00
	5	0.00	0.00	0.00	0.00	1.00	0.00	0.00
	6	0.00	0.00	0.00	0.00	0.00	1.00	0.00
	7	0.00	0.00	0.00	0.00	0.00	0.00	1.00
VWS	8	0.00	0.00	0.00	0.38	0.00	0.31	0.31
	9	0.00	0.30	0.00	0.11	0.00	0.50	0.09
	10	0.00	0.09	0.00	0.34	0.00	0.15	0.42
	11	0.00	0.13	0.00	0.15	0.00	0.55	0.17
	12	0.00	0.29	0.00	0.09	0.00	0.31	0.31
	13	0.00	0.40	0.25	0.06	0.00	0.25	0.05
	14	0.00	0.56	0.09	0.06	0.00	0.25	0.05
	15	0.00	0.14	0.41	0.02	0.33	0.09	0.02
	16	0.27	0.27	0.00	0.00	0.00	0.00	0.46

Figure 11.5 Data estimation weight matrix.

Table 11.1 Solar irradiance estimation results in W/m^2.

Estimation Case	Week 1MAE	Week 2MAE	Week 3MAE	Week 1RMSE	Week 2RMSE	Week 3RMSE
1	6.82	24.79	8.98	13.32	43.74	31.71
2	7.49	29.70	9.75	14.78	52.34	39.17
3	6.02	20.46	6.70	11.86	37.28	25.85

Table 11.2 Temperature estimation results in °C.

Estimation Case	Week 1MAE	Week 2MAE	Week 3MAE	Week 1RMSE	Week 2RMSE	Week 3RMSE
1	0.188	0.189	0.145	0.197	0.201	0.183
2	0.207	0.232	0.145	0.216	0.249	0.183
3	0.146	0.144	0.069	0.153	0.154	0.088

The results show that estimations with the help of VWS (Case 3) have always stood out as the best estimations. It suggests that estimation with VWS is the most reliable and resilient estimation.

11.4.2 Weather Data Prediction

PWSs and VWSs are used to create a spatiotemporal prediction network to predict solar irradiance and temperature. Figure 11.6 depicts the connectivity among weather stations. Arrowheads show the directions of predictor-to-predictor communication.

Weather predictions for location T can be made using (say) locations 2, 5, and 7 (without estimation). However, the locations are not regularly positioned and might not provide inputs with a high correlation for the prediction models. With data estimation, different location combinations

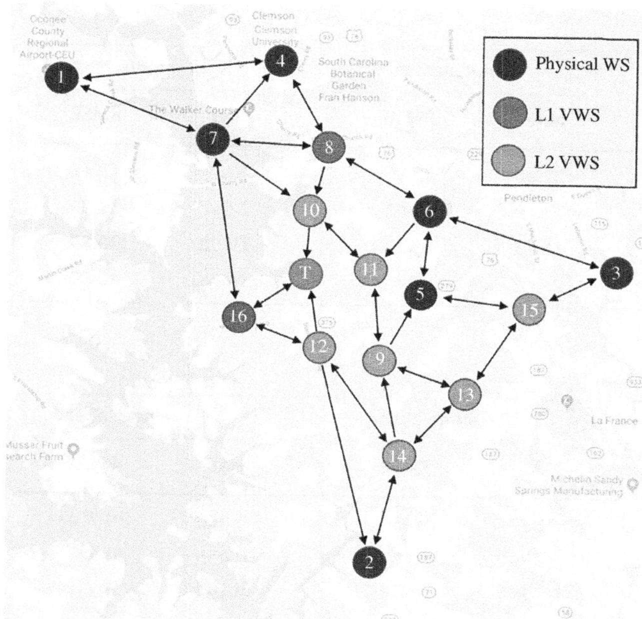

Figure 11.6 Connectivity of locations for weather predictions in the prediction network.

Figure 11.7 Input selection for prediction of the testing node in Cases 1, 2, and 3.

can be tried to find the best spatial inputs to the network. In this study following locations are considered spatial inputs (Figure 11.7):

Case 1	Locations: $[2_{L0}, 6_{L0}, 7_{L0}]$
Case 2	Locations: $[9_{L2}, 10_{L3}, 16_{L1}]$
Case 3	Locations: $[10_{L3}, 12_{L3}, 16_{L1}]$

Tables 11.3 and 11.4 show the results of 5, 10, and 15-minute ahead predictions of solar irradiance and temperature, respectively. It can be seen from the results that Case 3 predictions

Table 11.3 Solar irradiance prediction results in W/m^2.

	5-minute ahead		10-minute ahead		15-minute ahead	
	MAE	RMSE	MAE	RMSE	MAE	RMSE
Case 1	18.31	21.15	22.90	29.86	46.74	53.11
Case 2	25.36	32.58	24.79	30.62	27.17	37.52
Case 3	**9.95**	**14.12**	**16.21**	**28.13**	**19.02**	**30.85**
Persistent	13.53	16.85	25.65	31.18	36.74	42.87

Table 11.4 Temperature prediction results in °C.

	5-minute ahead		10-minute ahead		15-minute ahead	
	MAE	RMSE	MAE	RMSE	MAE	RMSE
Case 1	**0.042**	**0.054**	0.337	0.429	0.626	0.764
Case 2	0.175	0.220	0.370	0.457	0.455	0.625
Case 3	0.128	0.161	0.241	0.303	**0.178**	**0.237**
Persistent	0.102	0.129	**0.207**	**0.257**	0.315	0.387

produce better results for solar irradiance predictions. Case 3 predictors have inputs from one L1 and two L3 weather stations. It should be noted that VWSs are constructed with a fusion of multiple locations, whereas PWS produces measurements for one location (though the accuracy is high for the location). Therefore, Case 3 predictions have more spatial information as inputs than Cases 1 and 2.

Temperature does not vary significantly within the prediction horizon, making predictions less useful than solar irradiance. Measurement/estimation errors also become significant, resulting in inconclusive prediction accuracy. Figures 11.8 and 11.9 illustrate the predictions of solar irradiance and temperature for a day using Case 3 neighbors, respectively.

Solar irradiance and temperature measurements can be categorized into rising, peak, and falling regions for further analysis. Both measurements rise in the morning, reach a peak during the mid of the day, and fall in the afternoon/evening in general. Table 11.5 summarizes the results for three categories.

11.5 Summary

In this chapter, a framework for spatiotemporal solar and temperature predictive estimation has been presented. A data estimation algorithm is presented, which increases the granularity of data by strategically creating VWSs. The network of weather stations is evenly distributed in the terrain. VWSs are created using selected physical/virtual weather stations (WS), and different levels are defined for different combinations.

The IDW algorithm is used to estimate weather for any location within the network of WSs.

Distributed weather predictions are made using the physical and virtual weather data using a spatiotemporal prediction network, and results are presented for estimation and predictions of solar

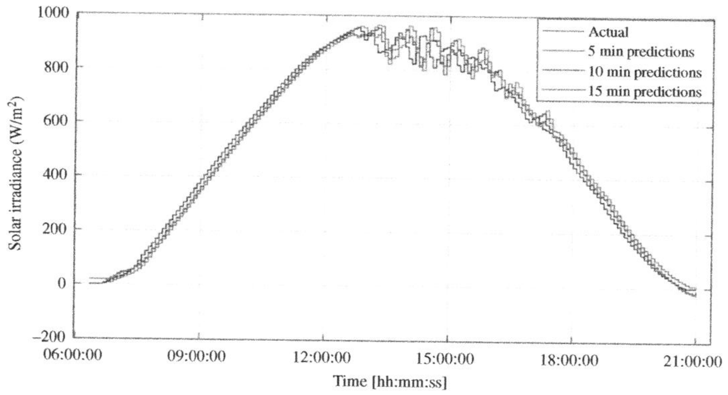

Figure 11.8 5-minute, 10-minute, and 15-minute ahead solar irradiance predictions using spatiotemporal networks.

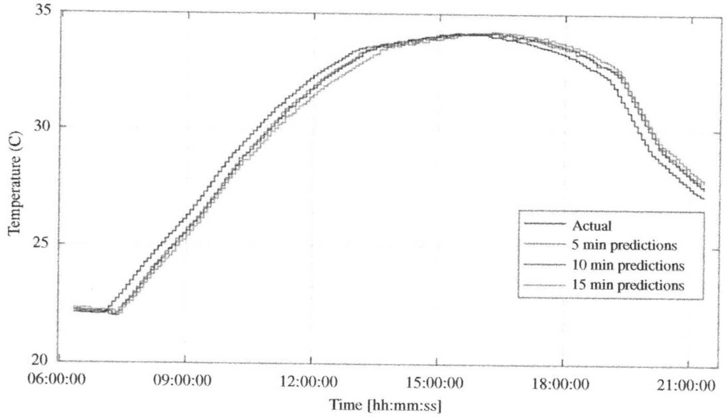

Figure 11.9 5-minute, 10-minute, and 15-minute ahead temperature predictions using spatiotemporal network.

Table 11.5 MAE and RMSE of three regions for Case 3 predictions.

		5-minute ahead		10-minute ahead		15-minute ahead	
		MAE	RMSE	MAE	RMSE	MAE	RMSE
Solar Irradiance	Rising	6.248	7.166	13.87	16.30	30.38	34.72
	Peak	13.23	18.90	27.72	38.32	34.52	42.87
	Falling	3.366	6.415	5.345	10.39	13.25	22.87
Temperature	Rising	0.158	0.178	1.080	1.241	0.268	0.293
	Peak	0.044	0.057	0.212	0.276	0.079	0.102
	Falling	0.052	0.060	0.647	0.776	0.068	0.088

irradiance and temperature. VWS-based data estimations are more accurate and resilient compared to others. Prediction results are shown for different input combinations. The input combination from L3 VWSs has better accuracy compared to those from only PWSs (L0). This is probably a result of spatiotemporal fusion of data used to create later stages of VWS creation. Temperature prediction results are observed to be erratic due to low variations and possible high measurement/estimation errors. Therefore, the persistent model is well suited for short-term temperature predictions.

The data predictive estimation framework can predict short-term variations of DERs with minimal PWSs. This framework is scalable and economical for distribution systems with a proliferation of DERs.

Acknowledgment

This work is supported by the National Science Foundation (NSF) of the United States, under grants IIP #1312260 and CNS #2131070, and the Duke Energy Distinguished Professorship Endowment Fund. Any opinions, findings and conclusions or recommendations expressed in this material are those of author(s) and do not necessarily reflect the views of National Science Foundation and Duke Energy.

References

1 Ela, E., Diakov, V., Ibanez, E., and Heaney, M. (2013). Impacts of variability and uncertainty in solar photovoltaic generation at multiple timescales. Technical report NREL/TP-5500-58274.

2 Kampelis, N., Ferrante, A., Kolokotsa, D. et al. (2017). Thermal comfort evaluation in HVAC demand response control. *Energy Procedia*.

3 Mishra, M., Nayak, J., Naik, B., and Abraham, A. (2020). Deep learning in electrical utility industry: a comprehensive review of a decade of research. *Engineering Applications of Artificial Intelligence*.

4 Golmohamadi, H. and Keypour, R. (2017). Retail energy management in electricity markets: Structure, challenges and economic aspects- a review. *Technology and Economics of Smart Grids and Sustainable Energy*.

5 Abdi, H., Fattahi, H., and Lumbreras, S. (2018). What metaheuristic solves the economic dispatch faster? a comparative case study. *Electrical Engineering* 100: 2825–2837.

6 Wang, H., Yi, H., Peng, J. et al. (2017). Deterministic and probabilistic forecasting of photovoltaic power based on deep convolutional neural network. *Energy Conversion and Management* 153: 409–422.

7 Wang, H.Z., Li, G.Q., Wang, G.B. et al. (2017). Deep learning based ensemble approach for probabilistic wind power forecasting. *Applied Energy* 188: 56–70.

8 Tiwari, S., Sabzehgar, R., and Rasouli, M. (2018). Short term solar irradiance forecast using numerical weather prediction (NWP) with gradient boost regression. *2018 9th IEEE International Symposium on Power Electronics for Distributed Generation Systems (PEDG)*, IEEE.

9 Goger, B., Rotach, M.W., Gohm, A. et al. (2016). Current challenges for numerical weather prediction in complex terrain: Topography representation and parameterizations. *2016 International Conference on High Performance Computing & Simulation (HPCS)*, IEEE.

10 Maqsood, I. and Abraham, A. (2007). Weather analysis using ensemble of connectionist learning paradigms. *Applied Soft Computing* 7: 995–1004.

11 Jayawardene, I. and Venayagamoorthy, G.K. (2016). Spatial predictions of solar irradiance for photovoltaic plants. *2016 IEEE 43rd Photovoltaic Specialists Conference (PVSC)*, IEEE.

12 Pathiravasam, C. and Venayagamoorthy, G.K. (2016). Spatio-temporal characteristics based wind speed predictions. *2016 IEEE International Conference on Information and Automation for Sustainability (ICIAfS)*, IEEE.

13 Pathiravasam, C. and Venayagamoorthy, G. (2018). Comparison of learning cellular computational networks with EKF and CPSO for multi-location wind speed prediction. *2018 IEEE Symposium Series on Computational Intelligence (SSCI)*, IEEE.

14 Pathiravasam, C., Arunagirinathan, P., Jayawardene, I., et al. (2020). Spatio-temporal distributed solar irradiance and temperature forecasting. *2020 International Joint Conference on Neural Networks (IJCNN)*, IEEE.

15 Yonekura, K., Hattori, H., and Suzuki, T. (2018). Short-term local weather forecast using dense weather station by deep neural network. *2018 IEEE International Conference on Big Data (Big Data)*, IEEE.

16 Xiaoming, X. and Chuan, Y. (2014). Three-dimensional localization algorithm of APIT based on fermat-point divided for wireless sensor networks. *2014 Seventh International Symposium on Computational Intelligence and Design, IEEE.*

17 Tobler, W. (1970). A computer movie simulating urban growth in the Detroit region. *Economic Geography* 46 (2): 234–240.

18 Solcast (2021). Solar irradiance data, https://solcast.com (accessed 23 September 2022).

12

Application of Decomposition-Based Hybrid Wind Power Forecasting in Isolated Power Systems with High Renewable Energy Penetration

Evgenii Semshikov, Michael Negnevitsky, James Hamilton, and Xiaolin Wang

School of Engineering, College of Science and Engineering, University of Tasmania, Hobart, Tasmania, Australia

12.1 Introduction

Traditionally, islanded and remote area communities rely on diesel generators to satisfy their electricity needs. Energy consumption in such areas is of a repetitive nature and has an acceptable level of predictability. To ensure the balance between generation and load, generating units are dispatched in accordance with a certain objective (e.g. minimization of operational costs and/or emissions constrained by system security) [1].

In recent years, isolated communities have been shifting toward renewable energy (RE) generation to decrease their dependency on diesel fuel [2]. The most abundant renewable energy sources, wind and solar, are internment and stochastic in nature. Although renewable technologies have the potential to reduce electricity costs and greenhouse emissions, they introduce uncertainties to power system operation posing challenges to system stability and reliability [3].

Challenges related to RE integration become particularly acute in IPSs with high wind and solar penetration. To facilitate reliable operation, such systems often rely on enabling technologies such as energy storage or demand side management. Capacity-oriented energy storage, such as pumped hydroelectric power [4], hydrogen storage [5], and compressed air energy storage (CAES) [6], are shown to be effective for energy shifting. Power quality can be improved with the help of batteries [7], supercapacitors [8], and superconducting magnetic energy storage (SMES) [9] devices with fast time response. Flywheels are useful for increasing the system inertia in power grids with high renewable penetration [10], whereas some authors presented the ability of batteries to emulate the inertial response [11].

When designing IPSs, decisions should be made on the size, number, and type of grid components and the strategy how to dispatch them [12]. In this chapter, we primarily focus on the later, the operational side of the problem, and assume that the system composition is a given. A number of innovative Australian IPS projects describe successful integration of enabling technologies and control architectures [13]. High RE penetration (>60% p.a.) is achieved in the King Island IPS (Australia). To accommodate high instantaneous wind energy penetration (i.e. greater than instantaneous demand), fast-response resistors (dump load) are used to spill the excess of generation converting RE into grid-connected reserves [14]. The diesel uninterruptible power supply (D-UPS) – a synchronous machine coupled to a diesel engine or a flywheel via an automated mechanical clutch – is employed during sudden drops of wind generation [15].

Predictive-dispatch algorithms based on the short-term load and RE forecasting are used to coordinate the operation of IPSs and increase the effectiveness of integrating enabling technology

Intelligent Data Mining and Analysis in Power and Energy Systems: Models and Applications for Smarter Efficient Power Systems, First Edition. Edited by Zita Vale, Tiago Pinto, Michael Negnevitsky, and Ganesh Kumar Venayagamoorthy.

(e.g. dump loads or uninterruptible power supply [UPS]) [16, 17]. Persistence models (PMs) have proven to be quite effective when dealing with very short-term predictions and are used in many practical IPSs [18]. For wind-based generation, persistence models are based on the notion that the present power provided by the source is highly correlated with its value in the immediate future [19]. Such forecasting systems, however, may present uncertainties to system operation during under- or over-forecasting of wind power outputs.

The nature of wind power forecasting is similar to pattern recognition and can be learned by artificial neural networks (ANNs). Significant research is dedicated to designing ANNs for wind power forecasting [20]. In this chapter, we present a three-step methodology to develop a renewable energy forecasting system, integrate it with an existing dispatch strategy, and evaluate its performance within high RE penetration IPSs. In the first step, a forecasting system is developed using historical wind power measurements. For illustrative purposes, we adopt a state-of-the-art forecasting technique based on the hybrid variational mode decomposition long short-term memory (VMD-LSTM) ANN [21]. In the second step, the developed forecasting system is integrated into the dispatch algorithm of an IPS. Using one year of real-world system operation (the King Island IPS, Australia), the developed forecast-based dispatch algorithm is compared to conventional system control strategies. In the final step, the chapter investigates whether the accuracy of the presented forecasting system is sufficient (i.e. further improvements may not produce any noticeable benefit) or redundant (i.e. less accurate forecasting techniques achieve similar results). The developed methodology is generic enough and can be applied to a number of systems.

This chapter focuses on the problems related to uncertainties caused by inaccurate forecasts that may lead to subsequent financial losses. While discussion and illustrations may center around the importance of forecasting in isolated power systems (e.g. islands, remote, and rural areas), similar trends can be seen in large-scale interconnected power systems. The chapter is structured in the following way. We first look at how to design the wind power forecasting system, where in Section 12.2, the pre-processing techniques is presented. In Section 12.3, the principle of the forecasting system is discussed. Section 12.4 outlines the energy management control strategy with an integrated forecasting system. Case studies, shown in Section 12.5, are based on the real data from King Island (Australia). Section 12.6 concludes the chapter.

12.2 Decomposition Techniques

Wind power time-series forecasting is based on the ability of artificial intelligence to learn patterns and trends from the historical data. In order to improve the performance of such systems, hybrid models are used. Hybrid models take advantage of a range of methods aimed at maximizing the ability of forecasting systems to recognize hidden signal features and consequently learn from them. Examples of hybrid models are presented in Ref. [22]. Among the hybrid models, decomposition-based approaches have been reported to be quite effective. A comprehensive review of decomposition-based methods is presented in Ref. [20].

Typical wind power time series is stochastic and volatile in nature and has a highly nonstationary spectral content. By decomposing the original signal into multiple subseries, a set of less irregular signals can be generated. Each constituent signal characterizes the original time series within a particular frequency bandwidth.

Such subseries are not burdened with the randomness of lower or higher harmonics and, therefore, are less chaotic and more predictable. A generic decomposition-based forecasting system is depicted in Figure 12.1.

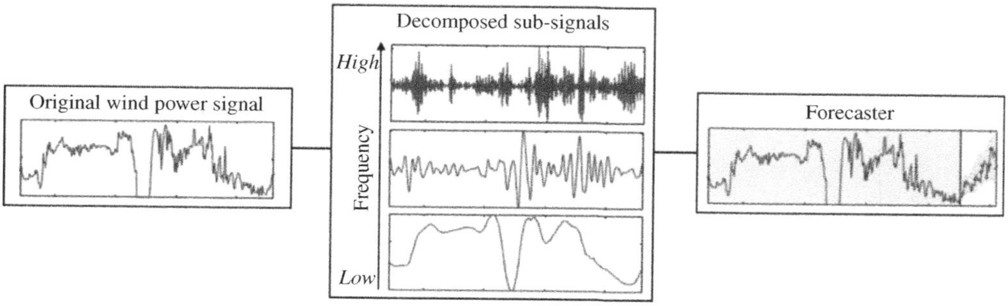

Figure 12.1 Generic decomposition-based forecasting system.

12.2.1 Variational Mode Decomposition

Variational mode decomposition (VMD) is a modern adaptive and fully intrinsic method that relies on Wiener filtering, Hilbert transform and frequency mixing, and heterodyne demodulation [23]. The VMD methods allow to obtain K narrowband intrinsic mode functions (IMFs) u_k from an original signal $x(t)$.

$$x(t) = \sum_{k=1}^{K} u_k(t) \tag{12.1}$$

Each IMF is modulated around a particular amplitude and concentrated around a central frequency f_k. The constituent signal u_k has a slowly varying envelope A_k and instantaneous frequency $\phi'_k(t)$, and can be written in the following form:

$$u_k(t) = A_k(t)\cos(\phi_k(t)) \tag{12.2}$$

In order to find the IMF and its central frequency, the optimization problem with the following augmented Lagrangian expression is solved:

$$L(u_k(t), f_k, \lambda(t)) = \alpha \sum_{k=1}^{K} \left\| \frac{\mathrm{d}}{\mathrm{d}t}\left[\left(\delta(t) + \frac{j}{2\pi}\right) * u_k(t)\right] e^{-i2\pi f_k t} \right\|_2^2 + \left\| x(t) - \sum_{k=1}^{K} u_k(t) \right\|_2^2 + \left\langle \lambda(t), x(t) - \sum_{k=1}^{K} u_k(t) \right\rangle \tag{12.3}$$

where the inner product $\langle p(t), q(t)\rangle = \int_{-\infty}^{\infty} p(t)q(t)\mathrm{d}t$ and notation $\|p(t)\|_2^2 = \langle p(t), p(t)\rangle$ is the squared 2-norm.

The first term of Eq. (12.3) performs regulation and can be described by three steps:

1. Obtain a positive spectrum and analytical signal for each IMF using the Hilbert transform.
2. Demodulate the analytic signal to the baseband.
3. Find the solution of a squared 2-norm of the signal derivative obtained in step 2.

The second and third terms of Eq. (12.1) ensure the reconstruction accuracy, where α is the quadratic penalty coefficient. More detailed description of the VMD decomposition method is presented in Ref. [23].

12.2.2 Decomposition of Wind Power Time Series

Wind-powered generators are considered to be non-dispatchable sources of electricity. Their outputs have both fluctuating and random components. The former forces power system operators to

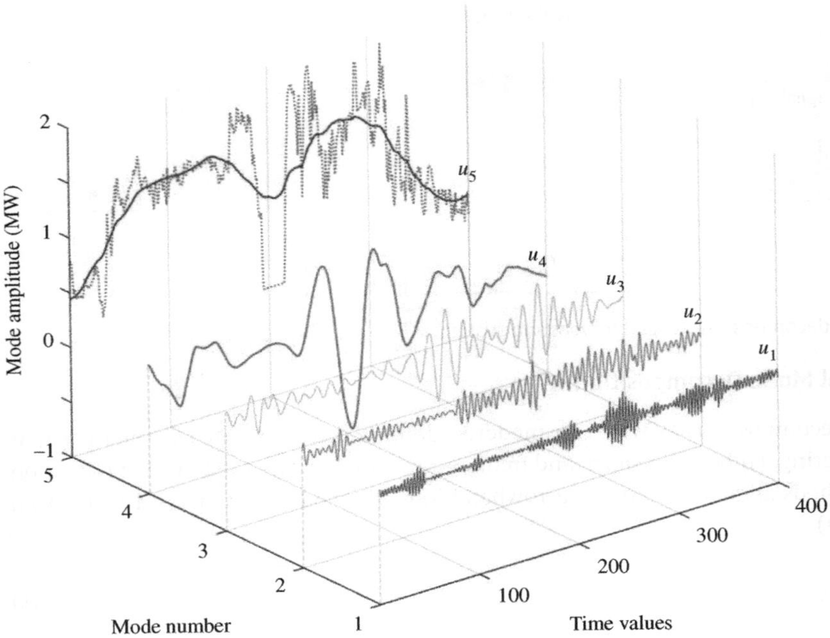

Figure 12.2 Five-mode wind power decomposition (excessive).

schedule controllable generating units in advance preventing undersupply or overgeneration scenarios. The later requires allocation of additional spinning reserves so that any random change in wind is accompanied by an immediate response from regulation service providers [24].

Wind power time-series data obtained from an isolated power system (King Island, Australia) are used in this chapter for illustrative purposes. The data originally sampled with 3-second resolution was upscaled to represent 10-minute variations. The signal – 400 samples – corresponding to almost three days of system operation is presented in Figure 12.2 (dotted line). Using the VMD method, the signal is decomposed into five layers. Mode u_1 approximates the trend of the original signal. Mode u_2 represents wind power fluctuations in a midrange frequency band. Finally, modes u_3, u_4, and u_5 reflect random high-frequency low-amplitude changes.

Increasing the number of decomposition modes may lead to a better understanding of a signal, unveiling detailed signal characteristics in various frequency bands. Although it is beneficial for some applications (filter designing, signal depth analysis, etc.), the excessive number of decomposition layers may be redundant or counterproductive when dealing with wind power forecasting [21]. Decomposed wind power modes with high center frequencies (e.g. mode u_4 or u_5 in Figure 12.2) do not necessarily capture the nature of wind power variations. In power systems, high-frequency random fluctuations can be caused by electronic switching devices, system faults, or other sources of high harmonics. Even if such stochastic disturbances can be forecasted, they do not reflect the physical changes in wind power outputs and represent pure white noise of the signal. In addition, high-frequency components produced by excessive decomposition (see u_5 in Figure 12.2) have many zero or near zero values playing little or no role compared to other modes. Secondly, when the original wind power signal is decomposed in needless number of modes, the lower-most frequency component (u_1 in Figure 12.2) might become too smooth, missing significant information required to represent real signal trends.

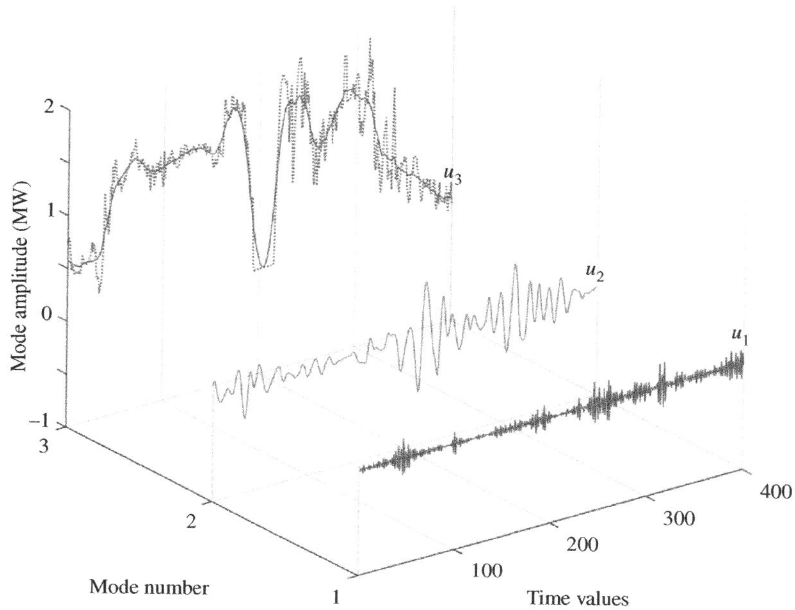

Figure 12.3 Three-mode wind power decomposition (according to physical signal properties).

When trying to determine the number of decomposition modes, one should analyze the physical property of the studied signal. In accordance with wind power characteristics, three-layer decomposition can be utilized (as illustrated in Figure 12.3) [21]. The first layer constituent mode u_1 is the low-frequency signal that shows the trend of the wind power production.

The second layer u_2 in Figure 12.3 reflects the fluctuation characteristic of the wind source. Such fluctuations happened over a shorter period of time but might still have a high amplitude. The third layer u_3 is the random component responsible for high-frequency stochastic changes in wind speed defined as a "burr of wind power." As a result, a three-mode decomposition breaks the original wind power time series into the long-term trend (significant for a day-ahead forecasting), short-term fluctuation (hour-ahead forecasting), and ultra-short randomness (5–15 minutes forecasting). These three components can then be used to train ANNs. In Section 12.3, the long short-term memory (LSTM) ANN is introduced.

12.3 Decomposition-Based Neural Network Forecasting

The problem of wind power forecasting is similar to the problem of pattern recognition, where the task is to study a vast set of patterns and try to identify one during the forecasting process. The wind power output has a time-dependent characteristic, meaning that the current data sample has a certain correlation to the future samples. ANNs are capable of learning and recreating the patterns. LSTM is a type of recursive ANN that is designed specifically for time-series. LSTM can memorize the pattern over a long period while learning new features in the sample. This section presents the principle of LSTM training and forecasting processes and the way decomposition methods (e.g. VMD) can improve them.

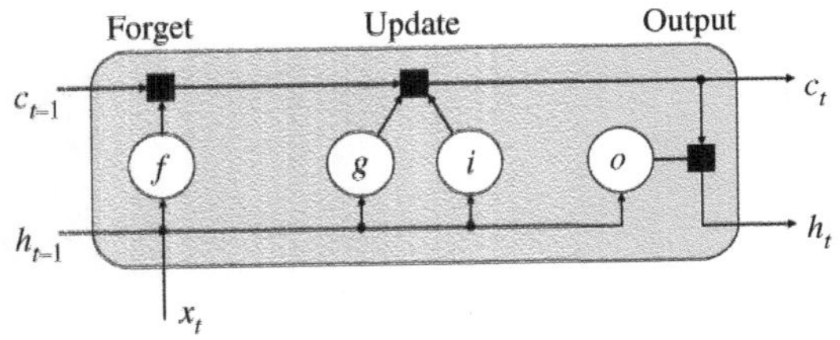

Figure 12.4 Long short-term memory block.

12.3.1 Theory Behind LSTM

LSTM ANN is an extension of a recurrent neural network (RNN). It has been designed to overcome the memory limitations of RNNs, which are unable to store the sequential information for more than a few steps back [25]. To cope with the long-term dependencies, LSTMs are equipped with the so-called "gated" cells that enhance the network selective ability to preserve or ignore information. This is done by assigning the weights to the information passing through different gates during the training phase. There are three LSTM gates, namely, "forget," "input," and "output" gates (Figure 12.4). The former gate holds the ability to decide whether a part of existing information can be removed. The input gate determines what part of the new information needs to be added to the memory. The output gate defines the extent to which the existing part of the memory contributes to the output.

To give a formal definition of LSTM, assume the vector $x = (x_1, x_2, x_3, \ldots, x_T)$ to be a time series of length T. The following equations define the three gates and the memory cells at each time step t:

The input gate:

$$i_t = \sigma_g(W_i x_t + R_i h_{t-1} + b_i) \tag{12.4}$$

The forget gate:

$$f_t = \sigma_g(W_f x_t + R_f h_{t-1} + b_f) \tag{12.5}$$

The cell candidate:

$$g_t = \sigma_c(W_g x_t + R_g h_{t-1} + b_g) \tag{12.6}$$

The output gate:

$$o_t = \sigma_g(W_o x_t + R_o h_{t-1} + b_o) \tag{12.7}$$

The cell memory state:

$$c_t = f_t \odot c_{t-1} + i_t \odot g_t \tag{12.8}$$

The hidden state:

$$h_t = o_t \odot \sigma_c(c_t) \tag{12.9}$$

where σ_c and σ_g denote the state (sigmoid) and gate (hyperbolic tangent) activation functions; W, R, and b are input weights, the recurrent weights, and the bias of each component, respectively; operator $\odot$ is the element-wise multiplication of vectors.

12.3.2 VMD-LSTM for Wind Power Forecasting

Wind power forecasting plays an important role in many areas of power system operation [26]. Predicting several steps into the future is of a particular importance when making decision on how

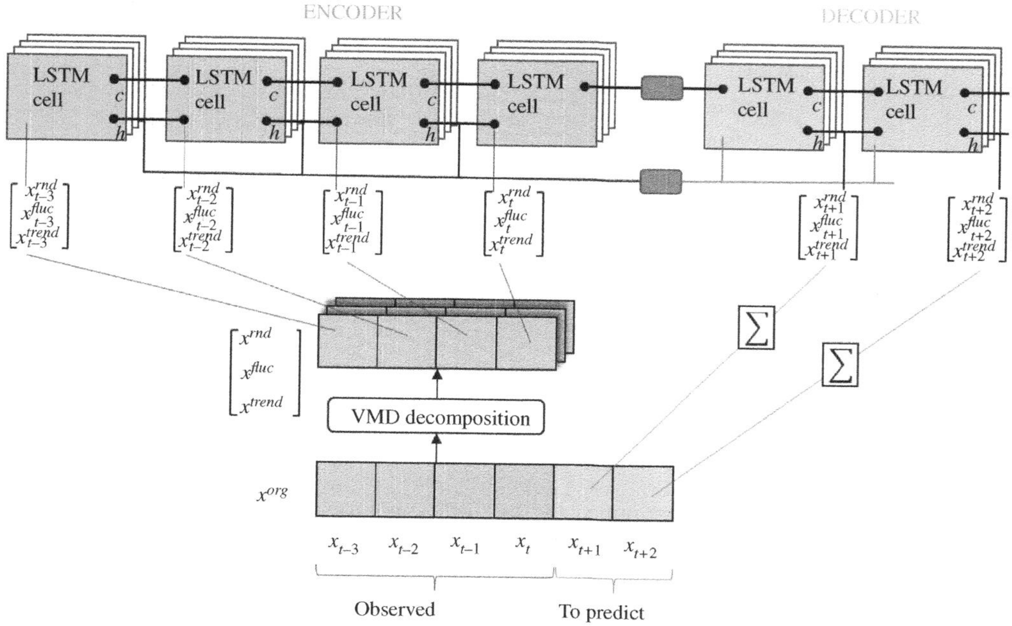

Figure 12.5 Long short-term memory block.

to dispatch a portfolio of generators. Researchers usually attempt to extend the multistep forecast by looking further into the future, adding new points in the end of the continues time series (e.g. 15-, 30-, 60-minute ahead forecast complemented by a 2-hour prediction point) [21].

Under the real-time operation apart from the distant future, it is important to have a clear picture of the immediate variations. LSTM neural networks have an ability to extract hidden patterns of a signal with a temporal correlation. Therefore, they can be used to predict short-term variations along with the overall trend of a sequence [27]. Schematic representation of the VMD-based LSTM wind power forecasting is presented in Figure 12.5. The scheme illustrates the "many-to-many" architecture where in order to predict several points into the future (provided by decoder), the sequence of input points is used (accepted by encoder). For illustrative purposes in Figure 12.5, four input time steps are presented, each containing three features (i.e. long-term trend, short-term fluctuation, and ultra-short randomness) provided by VMD decomposition. The LSTM cells are connected by solid lines (passing c and h states) representing the memory of the forecast model. Thus, the model retains the previous knowledge (through c_{t-1} and h_{t-1}) while learning the information from the current wind power data (x_t). Each output point x_{t+1} and x_{t+2} (two in this illustration) is comprised by adding three predicted components. It should be noted that the long-term component, the signal trend, would play a higher role when forecasting the distinct future, while random and fluctuating components have a bigger effect on predicting immediate variations.

12.4 Forecast-Based Dispatch in Isolated Power Systems

Forecasting is an important and cost-effective tool for integrating RES into power systems. To coordinate different components of an IPS, the three-level control hierarchy can be utilized. The three regulation stages, namely, primary, secondary, and tertiary control, follow the inertial

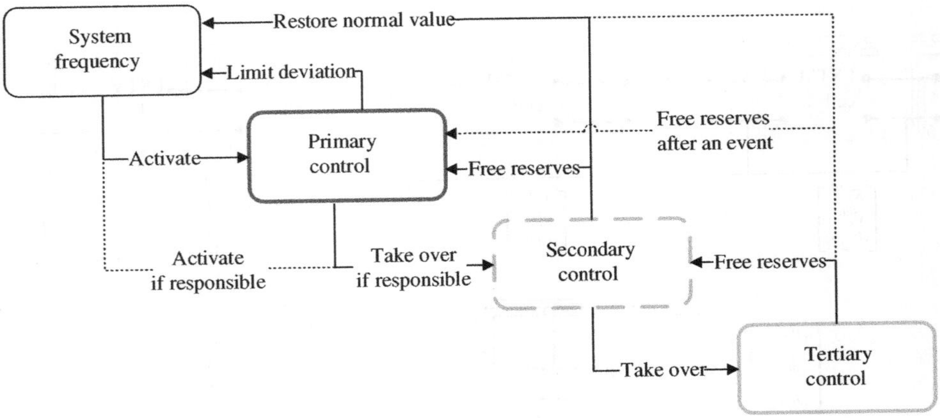

Figure 12.6 Control hierarchy. Source: Semshchikov et al. [29]/with permission of IEEE.

response [28]. Inertial response is supplied by the units synchronized to the system frequency. Primary and secondary responses are carried out by the units on the regulation and load following duties. Tertiary response implements dispatching actions enabling economic and reliable operation. Control actions are executed in a sequence of steps, each having its distinct characteristics and purposes, as shown in Figure 12.6.

12.4.1 Control Strategy

Control strategies are implemented within the tertiary control scheme. Under economic dispatch RE is represented as a non-dispatchable energy resource [29]. Therefore, by tracking minimum operational cost, the common practice is to utilize as much RE as possible and to adjust the power output of scheduled generation to accommodate for system fluctuations. Economic dispatch runs an optimization algorithm tracing minimum operational costs in the system. For the case of IPSs, three distinct dispatchable elements are considered: UPS, diesel generator, and dump load. The first one is engaged only as a last resort (i.e. back-up under contingency situation) and, therefore, is implemented within the primary scheme only. The operational costs of the other two may be evaluated according to the following cost functions [30, 31]:

$$F_g = a_g + b_g P_G + c_g P_G^2 \tag{12.10}$$

$$F_{DL} = r_u P_{DL} \tag{12.11}$$

where P_G and P_{DL} represent power levels for diesel generator and the dump load, respectively; a_g, b_g, and c_g are constants of a fossil fuel quadratic cost function; r_u is the electricity rate applied to the dump load.

The dump load cost function (12.11) represents the costs of RE not produced as a result of RE spill. However, there are three possible scenarios of the dump load operation (Figure 12.7) [29]:

Case 1: Diesel generators operate within their operational limits; frequency is controlled by diesel generators. In that case, the operation of the dump load is not economical and results in increased diesel consumption (dump load introduce additional load in the system). To penalize this use of the dump load, its operational cost is linked to the diesel generator operational cost. Thus, the electricity rate is defined as the diesel generator operational cost multiplied by the positive real number – penalty coefficient (p): $r_u - pF_g$.

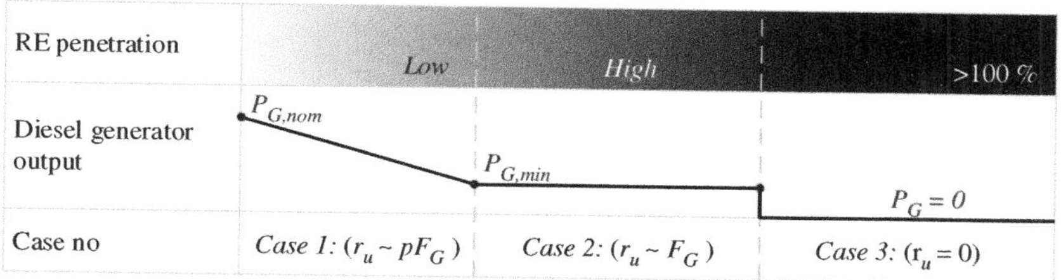

Figure 12.7 Electricity rates applied to the dump load depending on its operation. Source: Adapted from Semshchikov et al. [29].

Case 2: Diesel generator operates close to its lower limit (at about 30% of the nominal rating); frequency is controlled by the dump load. In that case, part of RE is utilized by the dump load and additional diesel fuel is used to supply loads. As a result, electricity rate applied to the dump load is governed by the diesel generator operational cost: $r_u - F_g$, such that:

$$F_{DL} = \begin{cases} r_u P_{DL}, & P_{DL} \leq P_G \\ r_u P_G, & \text{otherwise} \end{cases} \tag{12.12}$$

Case 3: Diesel generators are switched off; frequency is controlled by the dump load. In that case, the dump load is used to control the system frequency and provide the required security margin. Diesel generator operational costs are, obviously, not affected because all generators are switched off, and thus the electricity rate associated with the dump load is zero: $r_u = 0$.

The economic dispatch objective function is expressed as follows:

$$\min\,[F_g(P_G) + F_{DL}(P_{DL})] \tag{12.13}$$

subject to:

$$\begin{aligned} P_G + P_{RE} - P_L - P_{DL} &= 0 \\ y_g P_{G,min} \leq P_G &\leq y_g P_{G,max} \\ 0 \leq P_G &\leq P_{DL,max} \end{aligned} \tag{12.14}$$

where P_{RE} is the power provided by RES, P_L is the system load; $P_{G,min}$ and $P_{G,max}$ are lower and upper limits of the diesel generator output, y_g is a binary variable equal to 1 if the generator is on and 0 otherwise; $P_{DL,max}$ is an upper limit of a dynamic dump load.

For a more robust schedule, operational reserve constraints have to be added. Reserves provided by committed generators and the dump load have to guarantee the stability if the wind power and load forecasts are incorrect. For the system to operate in the zero-diesel mode (all diesel generators are disconnected) the following condition has to be satisfied:

$$P_{RE} - P_L \geq hR_{reg} \tag{12.15}$$

or

$$(y_g - 1)M \geq P_{RE} - P_L - hR_{reg} \tag{12.16}$$

where M is a big real number, R_{reg} is regulation reserve requirement used to account for uncertainties created by forecasting; h is a real number used to prevent excessive switching of diesel units and provide some security margin (e.g. $h = 1.5$ when going from "on" to "off" state and $h = 1$ when going from "off" to "on" state).

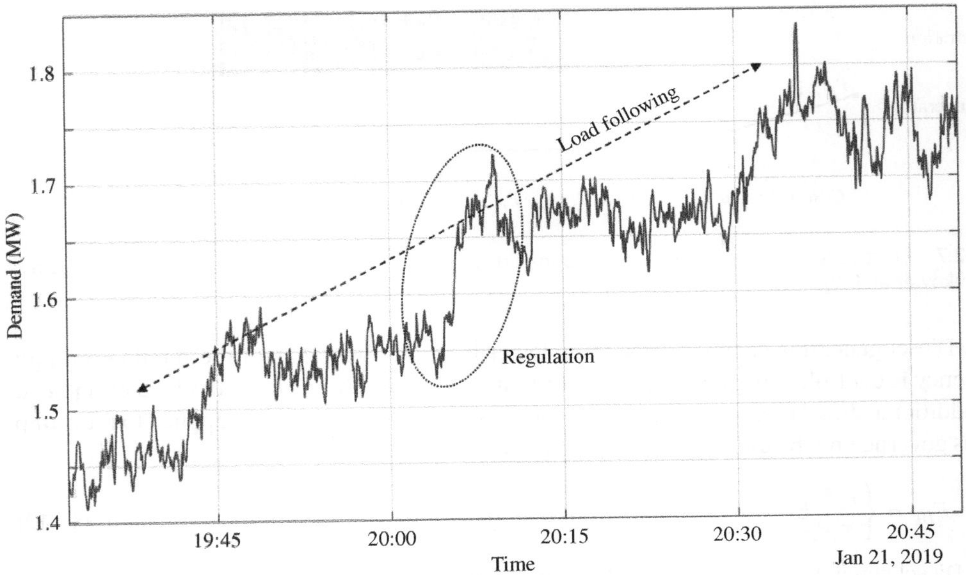

Figure 12.8 Regulation reserve requirements (King Island power system).

12.4.2 Regulation and Load Following Reserves

The approach for determining regulating and load following reserve requirements is based on a statistical analysis, which allows dynamic allocation depending on the system condition [32]. In the considered case, there are three main parts that constitute reserve requirements – load variability, wind variability, and forecast uncertainty.

The requirement for load following and regulation reserves due to variations in the aggregated electric demand is demonstrated in Figure 12.8 (this example features the King Island power system, Australia) [13]. These variations can be roughly separated into two components. The first component is linked with slower trends that are easy to predict and are generally followed by the control system. Load following is achieved by adjusting outputs of all units using economic dispatch and load forecasting (over periods from several minutes to several hours). The second component represents fast and random variations and is a concern of generating units on regulation duties (Figure 12.8 considers variations within 10-minute intervals).

The variability of load by itself from one 10-minute interval to the next is shown in Figure 12.9. The n-sigma method is selected to determine regulation reserve requirements [33]. Typically, n is selected to cover for most of the occurrences, where $n = 3$ would cover 99% of the variations.

With high renewable energy penetration, variabilities in the system increase dramatically. This can be seen in Figure 12.9, where the 10-minutes variability of a net load (wind power plus demand) is presented.

Variability of wind power depends greatly on the wind conditions. For example, a sudden change in wind speed can be followed by a prolonged episode of high wind variability [34]. Moreover, more variability is observed when a wind power plant operates in the midrange of their capacity. At low production levels, the variance is smaller because the power output is small. At high production levels, the variance is also smaller because many wind turbines are operating above

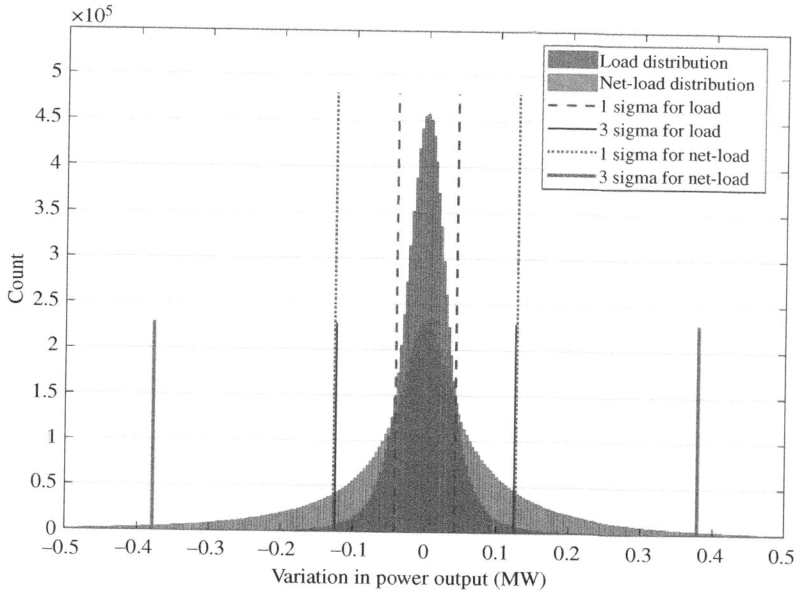

Figure 12.9 Distribution of load and net-load variability.

rated wind speed and, as a result, wind speed fluctuations are not translated into changes of wind power output.

Using actual data, it can be seen that the standard deviations within 10-minute intervals plotted against hourly production levels can be fitted into a quadratic function (Figure 12.10). Such sub-hourly variability drives dynamic reserve requirements.

In order to dispatch the future generation, a very short-term forecast (10-minute ahead) can be utilized. For wind generation, which is stochastic and internment in nature, forecasting errors create greater uncertainties resulting in need to increase the requirement for regulation reserves.

Variabilities of wind power and uncertainties of a PM can be demonstrated on the example of the King Island IPS, as shown in Figure 12.11. PM forecast used here assumes that the average production of the last 10-minute interval is the forecast for the next 10-minute interval. The system should have enough reserves to accommodate variations within the scheduled interval (i.e. 10-minute in case of King Island IPS) and uncertainty caused by inaccurate forecasting. In Figure 12.11, the persistence model forecasting error is shown by the red (gray) shaded area corresponding to the value *Error_PM* and the 10-minute variability is depicted by the green (light gray) shaded area.

Using VMD-LSTM-based forecasting, scheduling decision can be made based on a complete sequence forecasted 10-minute into the future. As a result, there is no need to separately account for the 10-minute variability. This, along with the smaller value of forecasting error, results in lower operational uncertainties and reduced reserve requirements. Using the same principle as for wind variability, the maximum expected error is observed to occur during the midrange of the wind power plant production. The empirical expected error characteristic can be approximated using a quadratic function, where the input is the average hourly production and the output is the wind generation forecast error. Error characteristics for persistence model forecast and VMD-LSTM based forecasting are presented in Figure 12.12.

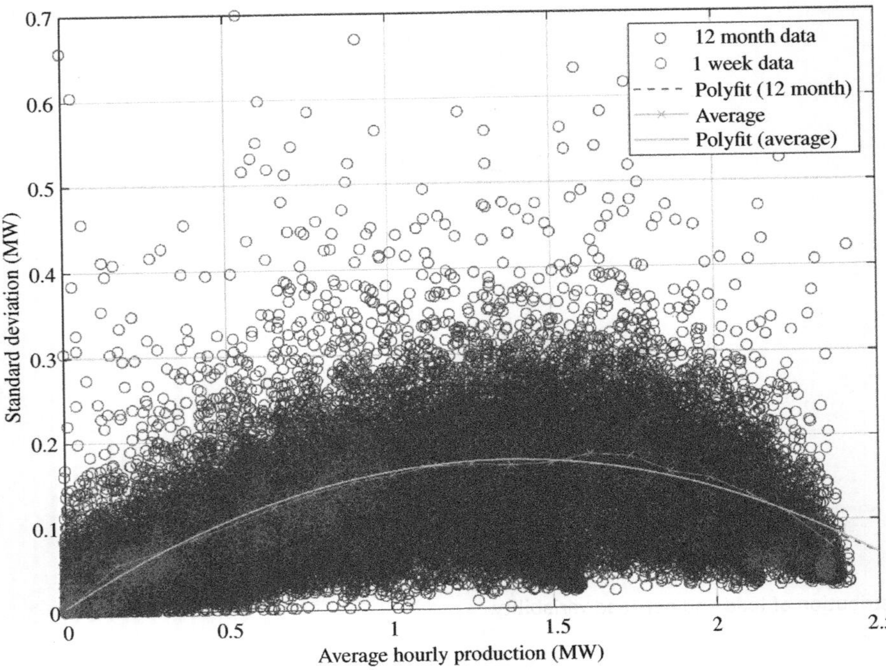

Figure 12.10 Wind variability as a function of hourly production (King Island IPS).

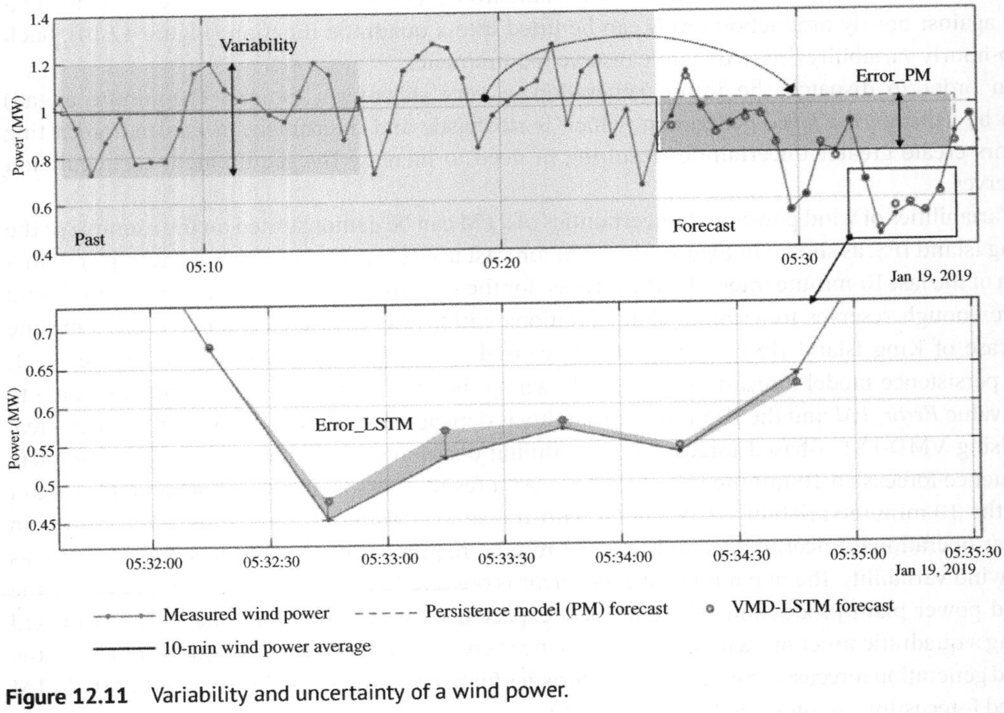

Figure 12.11 Variability and uncertainty of a wind power.

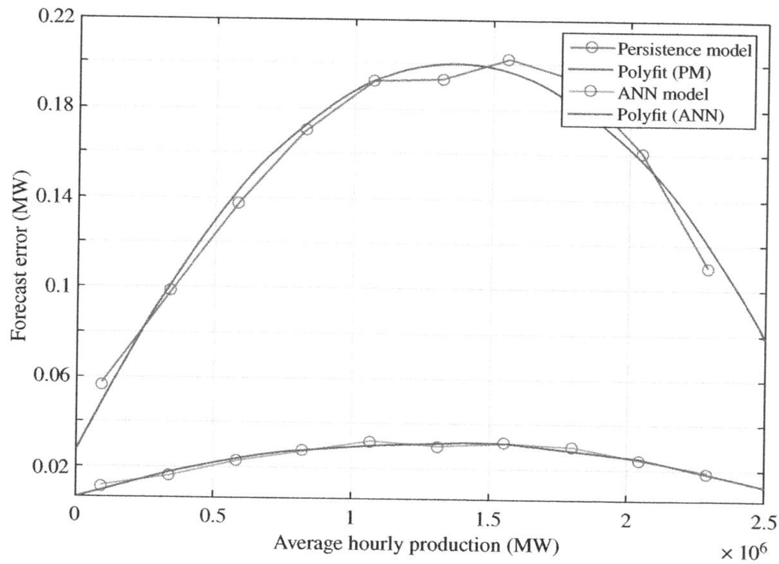

Figure 12.12 Wind power forecast error as a function of average hourly production.

12.5 Case Studies

Forecasting can affect multiple power system operational procedures such as scheduling, unit commitment, power dispatch, reserve management, and real-time balancing. In this section, we demonstrate the impact of increasing very-short-term forecasting accuracy under real-time balancing of IPSs with a high share of RE generation. During the real-time operation, the IPS is dispatched according to the algorithm presented in Section 12.5. Two approaches are compared, the forecast-based dispatch complemented by the persistence model and the dispatch algorithm based on VMD-LSTM forecasting system. The comparative studies are performed on the basis of the King Island power system (described in Section 12.5.1) considering one year of system operation.

12.5.1 King Island Isolated Power System

The system consists of diesel generators, wind turbines, a dump load, a solar array, a battery energy storage system (BESS), and UPS (Figure 12.13).

The proposed IPS configuration was modeled in MATLAB and validated against actual measurements obtained over a 12-month period. To coordinate the behavior of diesel generators and dump load, the dispatch strategy described in Section 12.5 was used. Its algorithm along with validation studies were presented in Ref. [29]. According to Semshchikov et al. [29], the battery operation was shown to be very expensive when used for energy arbitrage. In this chapter, we evaluate the economic benefit of increasing very short-term forecasting by itself; therefore, the battery (that has high effect on system costs) was removed from studies. As mentioned before, the UPS system is considered within primary control scheme and is not included within economic dispatch. The cost of diesel generation is estimated using the quadratic cost function with coefficients fitted to match experimental data [29, 35].

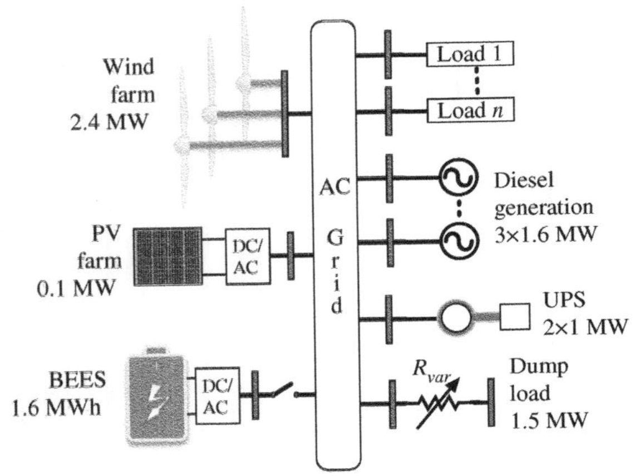

Figure 12.13 The King Island IPS.

In King Island IPS, RE penetration reaches 100%, when the system runs in a ZDO mode (all diesel generators are switched off). To facilitate ZDO, the IPS energy management system can be complemented by very short-term forecasting. Uncertainties caused by forecasting error require additional reserves (see Figure 12.12 to see the example of uncertainties expected from ANN and persistence-based forecasts). In this chapter, operational reserves are allocated every 10 minutes, whereby the feasibility of ZDO is determined. Generally, the allocation of running reserves and cold, or standby reserves can be done on a 10-minute basis [36]. In IPSs, 10 minutes is the typical time required for diesel generators to warm up or cool down.

The ZDO mode of operation is possible when a system has a robust technique of dealing with excessive RE. For example, the King Island power system is equipped with a set of dynamic resistors (Dump Loads) that convert the excess of RE into fast-response grid-connected reserves. The dump load dissipates this energy to provide the balance between generation and load acting as a regulator of system frequency under high RE conditions. For reliable system operation, ZDO can be achieved only when enough power reserved by the dump loads and forecasted to stay above certain threshold in the distinct future. It should be noted that the long-term trend is assumed to be consistent given the nature of the wind speed variations.

In this section, three IPS case studies are presented. Case Study I presents a benchmark scenario of running an IPS using no forecasting system. Case Study II investigates persistence model forecasting. Case Study III considers VMD-LSTM-based forecasting system integrated within system control strategy as a means to facilitate zero-diesel operation.

12.5.2 Case Study I (Control Strategy with No RE Forecast)

Case Study I presents a scenario of a real-time system operation without short-term RE forecasting. Practical IPSs with high RE penetration are usually equipped with a range of enabling technologies to allow for ZDO. Case Study I considers the most pessimistic operation of the system when the only enabling technology, the dump load, is engaged solely to deal with the generation excess. Since no RE forecast is available, for reliability reasons, dynamic resisters cannot be used for reserve provision. As a result, at least one diesel generator must always stay connected to the grid in order to respond to unforeseen changes in wind or solar generation. To illustrate the control strategy, one day operation of the island IPS is considered (Figure 12.14).

According to Figure 12.14, with increasing the wind power generation, the diesel generator lowers its output. The engine low-load limit (30% of the rated capacity) is reached at around 10 am.

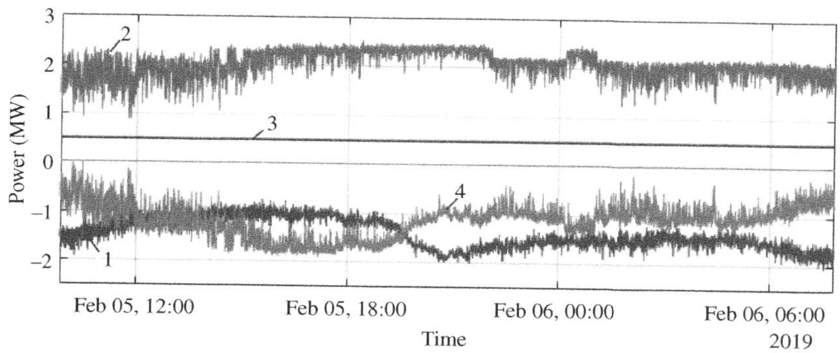

Figure 12.14 One day of an IPS operation under no forecasting (ZDO mode is not supported). 1 – load, 2 – wind power, 3 – diesel power, 4 – curtailed generation.

For security reasons, the engine cannot be disconnected. As a result, the excess of electrical power is dissipated by the dump load. Although this strategy ensures reliable operation of the IPS, it results in high spills of renewable energy.

12.5.3 Case Study II (Control Strategy Involving Persistence Model RE Forecast)

Under this scenario, a 10-minute-ahead persistence forecasting is used to enable ZDO when instantaneous RE penetration is higher than 100%. Persistence forecasting provides a relatively accurate information of the average RES power output in the next dispatch interval. Any deviations from the average value or the uncertainty presented by forecasting error should be corrected by the unit on regulation duties.

Real-time system balancing under predictive control strategy is illustrated in Figure 12.15a. When regulating reserves stored within the dump load exceed reserve requirement, the system shifts into the ZDO mode (Figure 12.15a, February 05 from 15:00 to 20:00). During ZDO, dynamic resistors provide both primary and secondary responses controlling system frequency. Reserve requirements are determined in such a way that when the forecasted wind power output is faulty or net-load fluctuations are presented, the dump load adjusts its power consumption ensuring the balance between generation and demand. Based on the PM forecasting, higher RE penetration is achieved.

12.5.4 Case Study III (Control Strategy Involving VMD-LSTM-Based RE Forecast)

Regulation reserve requirements can be reduced when the information of the future wind power outputs is complete and presented by a time series of multiple points within the next dispatch interval. In this case, the reserve requirement is determined mainly by the forecasting error, and there is no need to keep excessive capacities reserved to cope with wind power fluctuations. As the result, the ZDO mode can be extended. An example of the system operation under VMD-LSTM-based forecasting is presented in Figure 12.15b.

The regulation reserve requirement has been reduced, resulting in an earlier transaction to the ZDO (around 14:00, February 05) and an additional interval of 100% RE generation around midnight. Given smaller forecasting error and sufficient information about the future state, the system can accept more renewable energy compared to the previous cases.

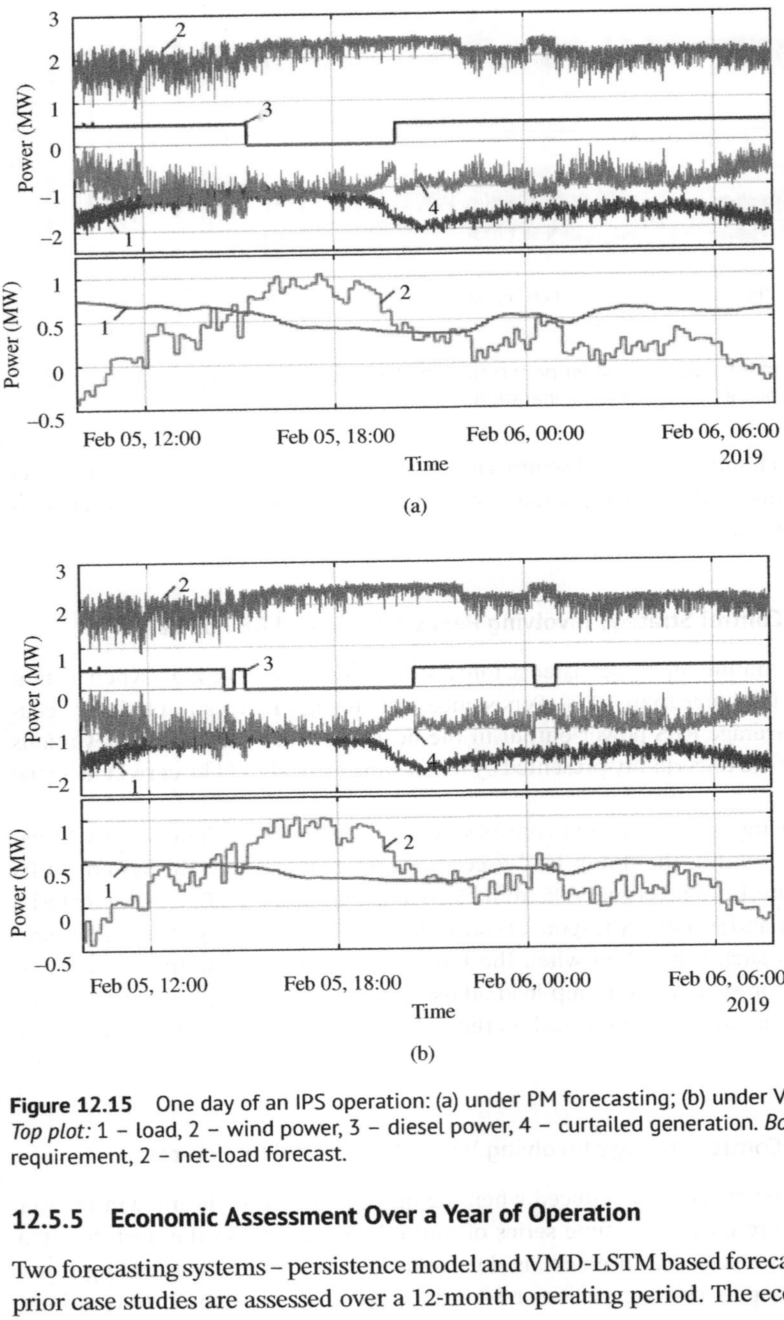

Figure 12.15 One day of an IPS operation: (a) under PM forecasting; (b) under VMD-LSTM forecasting. *Top plot:* 1 – load, 2 – wind power, 3 – diesel power, 4 – curtailed generation. *Bottom plot:* 1 – reserve requirement, 2 – net-load forecast.

12.5.5 Economic Assessment Over a Year of Operation

Two forecasting systems – persistence model and VMD-LSTM based forecasting – considered in the prior case studies are assessed over a 12-month operating period. The economic analysis is based on the actual load and resource data.

Results are presented in Table 12.1. The results are compared to the baseline case when forecasting is not utilized and to the case of using perfect forecasting system. It can be observed that improving forecasting accuracy results in a significant decrease in reserve requirements (column 4) reducing generation provided by diesel engine (see column 2). In column 3, the energy spilled by resistive load is presented. Last three columns details RE penetration and diesel costs.

Table 12.1 The 12-month comparison of control strategies.

Considered forecasting system	Energy diesel (MWh/yr)	Energy dump-load (MWh/yr)	Energy reserve requirements (MWh/yr)	RE penetration (%)	Diesel cost ($)	MWh cost ($/MWh p.a.)
No forecast	5912.3	2368.9		41.7	579 591	126.5
Persistence model	5788.6	2245.3	3760.2	42.9	568 293	124.1
VMD-LSTM model	5599.9	2056.5	1082.7	44.8	551 057	120.3
Perfect forecast	5338.0	1794.7	0.0	47.4	527 140	115.1

12.6 Conclusions and Discussions

This chapter presents a framework for integrating a forecasting system within an isolated power system. As an example, a real isolated power system, King Island (Australia), with an average load of 1.5 MW is considered. The hybrid VMD-LSTM forecasting system is applied to the IPS energy management system. By decomposing the original wind power measurements into three components (long-term trend, fluctuation component, and signal randomness), the ability of forecasting system to extract features and recognize hidden patterns is improved. Besides the higher forecasting accuracy, the decomposition-based ANN is shown to produce more comprehensive forecasting information used within dispatching procedure (VMD-LSTM system produce multiple forecasted points for the next dispatch interval as opposed to a single averaged point obtained by the persistence model). As a result, reduction in regulating reserve requirements is observed.

The performance of the VMD-LSTM-based forecasting technique is compared with the forecasting based on the persistence model. It was demonstrated that even though a crude forecasting system based on a persistence model compared to no forecasting cannot significantly increase RE penetration, its application allows to reduce the diesel consumption (the diesel cost is reduced from US$126.5 to US$124.1/MWh). Improved forecasting accuracy of VMD-LSTM-based system results in a further decrease in operational costs (US$120.3/MWh) where higher renewable penetration is observed (44.8% compared to 41.7% in the no-forecast case). Further improvement in forecasting accuracy results in reduced operational costs, where under perfect forecasting the expected diesel consumption cost is US$115.1/MWh and RE penetration is 47.4%.

References

1 Kariniotakis, G. (2017). *Renewable Energy Forecasting: From Models to Applications*. Woodhead Publishing.

2 *A path to prosperity: Renewable energy for islands*. 22nd session of the Conference of the Parties (COP 22) to the United Nations, Technical report, 7–18, 3edn, November 2016.

3 Alam, M.S., Al-Ismail, F.S., Salem, A., and Abido, M.A. (2020). High-level penetration of renewable energy sources into grid utility: challenges and solutions. *IEEE Access* 8: 190277–190299.

4 Deane, J.P., Gallachóir, B.P.Ó., and McKeogh, E.J. (2010). Techno-economic review of existing and new pumped hydro energy storage plant. *Renewable and Sustainable Energy Reviews* 14 (4): 1293–1302.

5 Gutiérrez-Martín, F., Confente, D., and Guerra, I. (2010). Management of variable electricity loads in wind – hydrogen systems: the case of a Spanish wind farm. *International Journal of Hydrogen Energy* 35 (14): 7329–7336.

6 Sundararagavan, S. and Baker, E. (2012). Evaluating energy storage technologies for wind power integration. *Solar Energy* 86 (9): 2707–2717.

7 Tang, X., Hu, X., Li, N. et al. (2016). A novel frequency and voltage control method for islanded microgrid based on multienergy storages. *IEEE Transactions on Smart Grid* 7 (1): 410–419.

8 Delille, G., Francois, B., and Malarange, G. (2012). Dynamic frequency control support by energy storage to reduce the impact of wind and solar generation on isolated power system's inertia. *IEEE Transactions on Sustainable Energy* 3 (4): 931–939.

9 Molina, M.G. and Mercado, P.E. (2011). Power flow stabilization and control of microgrid with wind generation by superconducting magnetic energy storage. *IEEE Transactions on Power Electronics* 26 (3): 910–922.

10 Nikolic, D. and Negnevitsky, M. (2018). Practical solution for the low inertia problem in high renewable penetration isolated power systems. In: *Proc. 2018 IEEE Power & Energy Society General Meeting (PESGM)*, pp. 1–5.

11 Alipoor, J., Miura, Y., and Ise, T. (2015). Power system stabilization using virtual synchronous generator with alternating moment of inertia. *IEEE Journal of Emerging and Selected Topics in Power Electronics* 3 (2): 451–458.

12 Mazzola, S., Vergara, C., Astolfi, M. et al. (2017). Assessing the value of forecast-based dispatch in the operation of off-grid rural microgrids. *Renewable Energy* 108: 116–125.

13 Hamilton, J., Negnevitsky, M., Wang, X., and Lyden, S. (2019). High penetration renewable generation within Australian isolated and remote power systems. *Energy* 168: 684–692.

14 Doolla, S. and Bhatti, T. (2006). Load frequency control of an isolated small-hydro power plant with reduced dump load. *IEEE Transactions on Power Systems* 21 (4): 1912–1919.

15 Amiryar, M.E. and Pullen, K.R. (2017). A review of flywheel energy storage system technologies and their applications. *Applied Sciences* 7 (3): 286.

16 Abdulla, K., Hoog, J.d., Muenzel, V. et al. (2018). Optimal operation of Energy storage systems considering forecasts and battery degradation. *IEEE Transactions on Smart Grid* 9 (3): 2086–2096.

17 Zheng, Y., Zhao, J., Song, Y. et al. (2018). Optimal operation of battery Energy storage system considering distribution system uncertainty. *IEEE Transactions on Sustainable Energy* 9 (3): 1051–1060.

18 Dutta, S., Li, Y., Venkataraman, A. et al. (2017). Load and renewable energy forecasting for a microgrid using persistence technique. *Energy Procedia* 143: 617–622.

19 Wu, Y. and Hong, J. (2007). A literature review of wind forecasting technology in the world. In: *Proc. 2007 IEEE Lausanne Power Tech*, pp. 504–509.

20 Qian, Z., Pei, Y., Zareipour, H., and Chen, N. (2019). A review and discussion of decomposition-based hybrid models for wind energy forecasting applications. *Applied Energy* 235: 939–953.

21 Han, L., Zhang, R., Wang, X. et al. (2019). Multi-step wind power forecast based on VMD-LSTM. *IET Renewable Power Generation* 13 (10): 1690–1700.

22 Ghofrani, M. and Suherli, A. (2018). *Time series and renewable energy forecasting*. IntechOpen, vol. 10.

23 Dragomiretskiy, K. and Zosso, D. (2014). Variational mode decomposition. *IEEE Transactions on Signal Processing* 62 (3): 531–544.

24 Ela, E., Milligan, M., and Kirby, B. (2011). *Operating Reserves and Variable Generation*. Golden, CO: National Renewable Energy Lab. (NREL).

25 Gers, F.A., Schmidhuber, J., and Cummins, F. (2000). Learning to forget: continual prediction with LSTM. *Neural Computation* 12 (10): 2451–2471.

26 Botterud, A., Zhou, Z., Wang, J. et al. (2012). Demand dispatch and probabilistic wind power forecasting in unit commitment and economic dispatch: a case study of Illinois. *IEEE Transactions on Sustainable Energy* 4 (1): 250–261.

27 Liu, H., Mi, X., and Li, Y. (2018). Smart multi-step deep learning model for wind speed forecasting based on variational mode decomposition, singular spectrum analysis, LSTM network and ELM. *Energy Conversion and Management* 159: 54–64.

28 Wu, F.F., Moslehi, K., and Bose, A. (2005). Power system control centers: past, present, and future. *Proceedings of the IEEE* 93 (11): 1890–1908.

29 Semshchikov, E., Negnevitsky, M., Hamilton, J.M., and Wang, X. (2020). Cost-efficient strategy for high renewable energy penetration in isolated power systems. *IEEE Transactions on Power Systems* 35 (5): 3719–3728.

30 Sharma, R.K., Mudaliyar, S., and Mishra, S. (2018). Power management and economic load dispatch based control of hybrid PV-battery-diesel standalone AC system. In: *Proc. 2018 IEEMA Engineer Infinite Conference (eTechNxT)*, pp. 1–6.

31 Azevedo, R.d., Cintuglu, M.H., Ma, T., and Mohammed, O.A. (2017). Multiagent-based optimal microgrid control using fully distributed diffusion strategy. *IEEE Transactions on Smart Grid* 8 (4): 1997–2008.

32 Corbus, D., King, J., Mousseau, T. et al. (2010). *Eastern Wind Integration and Transmission Study*, CP-550-46505, vol. 13, 1–8. NREL http://www.nrel.gov/docs/fy09osti/46505.pdf.

33 Holttinen, H., Milligan, M., Ela, E. et al. (2012). Methodologies to determine operating reserves due to increased wind power. *IEEE Transactions on Sustainable Energy* 3 (4): 713–723.

34 Potter, C.W., Grimit, E., and Nijssen, B. (2009). Potential benefits of a dedicated probabilistic rapid ramp event forecast tool. In: *Proc. 2009 IEEE/PES Power Systems Conference and Exposition*, pp. 1–5.

35 Hamilton, J., Negnevitsky, M., Wang, X. et al. (2017). Utilization and optimization of diesel generation for maximum renewable Energy integration. In: *Smart Energy Grid Design for Island Countries*, pp. 21–70. Springer.

36 Energy, G. (2010). *Western Wind and Solar Integration Study*. Golden, CO: National Renewable Energy Lab. (NREL).

Part V

Data Analysis

13

Harmonic Dynamic Response Study of Overhead Transmission Lines

Dharmbir Prasad[1], Rudra P. Singh[1], Irfan Khan[2], and Sushri Mukherjee[3]

[1]*Asansol Engineering College, Department of Electrical Engineering, Maulana Abul Kalam Azad University of Technology, West Bengal, India*
[2]*Department of Design and Development, Supreme & Co. Pvt. Ltd., Kolkata, West Bengal, India*
[3]*Department of Electrical Engineering, Indian Institute of Technology Delhi, Hauz Khas, New Delhi, India*

Acronyms

DoF	Degree of freedom
EBM	Energy balance method
FDM	Finite difference method
FEA	Finite elemental analysis
FEM	Finite element method
NEPC	New England Power Company
NIIPT	High Voltage Direct Current Power Transmission Research Institute, Saint-Petersburg, Russia

Nomenclature

A, B	Amplitude of sub-conductor – A, $-B$
c	Conductor self-damping coefficient
c_s	Damping coefficient of clamp
D	Conductors' diameter
E	Conductor modulus of elasticity
EI	Conductor flexural rigidity
f	Frequency of vibration
$f_n(x, t)$	Force function of the nth spacer damper
g	Acceleration due to gravity (m/s^2)
I	Moment of inertia of the conductor cross-section about the neutral axis
k	Stiffness of the system
k_s	Vertical stiffness of clamp
$[K]$	Stiffness matrix
L	Length of the conductor
m	Conductor mass per unit length

Intelligent Data Mining and Analysis in Power and Energy Systems: Models and Applications for Smarter Efficient Power Systems, First Edition. Edited by Zita Vale, Tiago Pinto, Michael Negnevitsky, and Ganesh Kumar Venayagamoorthy.

M	Shear force
M_s	Mass of spacer damper
$[M]$	Mass matrix of the system
$p(x, t)$	Aeolian excitation
P_C	Power dissipated by conductor due to self-damping
$P(X)$	Spatial component of the steady-state wind force dimensionless
Q	Momentary force
t	time
T	Tension appear in a span
S	Centroid of M_s
x	Distance along the conductor
x_n	Position of nth spacer damper
y	Transverse displacement of the conductor which is the function of t and x
y_A, y_B, y_s	Displacement of sub-conductor A
$y_0(x)$	Distribution function of conductor displacement amplitude
β	Phase difference between sub-conductor – B and subconductor – A.
α	Phase difference between spacer and sub-conductor – A
ω	Angular frequency of vibration

13.1 Introduction to Methodology

In the present scenario, there is a continuous surge in power demand for everyday energy requirements. Overhead transmission line structures are widely applicable for delivering electricity across the country. However, design of steel-based transmission structures under the wind load variation acts as a crucial role in power system reliability [1, 2]. Randomly appearance of dynamic loads is increasing with altitudes from ground level. This further complicates structural modeling under variable wind loading conditions. An installation layout for the proposed spacer damper has been presented in Figure 13.1.

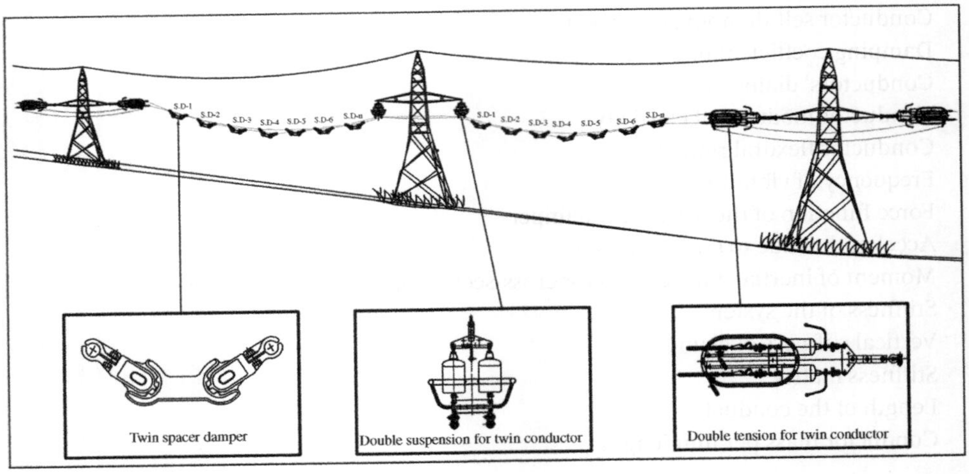

Figure 13.1 Installation layout for twin spacer damper.

13.1.1 General

In order to reduce the impact of vibration disturbances to an ignorant dimension, the T&D Committee of the IEEE power group presented a study on line vibration measurement problems in 1965 [3]. Transmission lines vibrations can be briefed by a comparative review of basic conductor motions as in Table 13.1.

13.1.2 Selection Aspects of Dampers

In order to damp transmission lines conductors' vibrations and maintain a safer distance between them, a spacer damper should be properly designed and developed. Thus, the standards (refer to Table 13.2) for various designed and developed criteria are listed below as in [3].

The major accessory associated with the current study is given in Table 13.3. In order to ascertain the suitability of a spacer damper to control the conductors' oscillatory behavior of a particular line and to protect it, the following quantitative information is required:

- data on the wind-induced power from field tests
- conductor dissipating (i.e. self-damping) power

Table 13.1 Comparison of basic conductor motions.

Particulars	Aeolian vibration	Conductor galloping	Wake induce vibration
Affected overhead line types	All	All	All
Approximate frequency range (Hz)	3–150	0.08–3	0.15–10
Approximate range of Vibration amplitude (peak to peak)	0.01 to 1 × *d*	5 to 3000 × *d*	Rigid body mode: 0.50 to 20 × *d*
Weather condition favoring conductor motion			
Wind character	Steady	Steady	Steady
Velocity (m/s)	1–7	7–8	4–18
Conductor surface	Bare/uniformly iced	Asymmetrically iced	Bare/dry
Damage			
Approximate time required for severe damage to develop	3 mo to 20 yr	1–48 hours	4–18 hours
Direct cause of damage	Metal fatigue due to cyclic loading	High dynamic loads	Conductor clashing accelerated wear in hardware
Line components most affected by damage	Conductor and shield wire strands	Conductor all hardware, insulators, structures	Suspension hardware spacers, dampers, conductor strands
Design conditions affecting conductor motion	Line tension, conductor self-damping, use of dampers and armor rods	The ratio of vertical natural frequencies to torsional natural frequencies, sag ratio, and support conditions	Sub-conductor-separation, the tilt of bundle, sub-conductor arrangement, sub-span staggering

Table 13.2 Standards for various design aspects.

Sl.	Design aspect	Standard
1	Materials, protective coatings requirements for manufacturing, and various test (laboratory and field) to check effectiveness and stability of the dampers	IEC 60826:2017
2	Design criteria citing. Requirements and tests for fittings of overhead transmission lines	IEC 61284:1997
3	Spacers and spacer dampers for twin horizontal bundle conductors	IS 10162-1982

Table 13.3 Specifications of major accessories.

Sl.	Accessory	Properties	Specifications
1	Conductor	Number of conductors	2
		Conductors' spacing	450 ± 5 mm (Twin)
		Conductors' type	ACSR Mallard
		Conductors' OD	28.95 mm
		Standard weight per unit length	1.8392 kg/m
		Conductors' UTS	171.25 kN
		Number and diameter of strands	Al: 30 × 4.135 mm; St: 19 × 2.482 mm
		Modulus of elasticity	78 500 N/mm^2
		Coefficient of linear expansion/degree	18.0×10^{-6}
2	Twin spacer damper	Materials and count of bolt, nut, washer	HDG Steel, 2 set
		Materials and counting for clamp body, keeper and arm	Al alloy, Keeper: 2, Body: 1, Arm: 2
		Materials for damping cushion	Polyethylene polymers, 2
		Weight of assembly	2.2 kg (approximately)

- performance evaluation of the spacer damper
- placement of dampers across the line
- conductor fatigue damage criterion for the safe operation
- cumulative fatigue damage possibilities

13.1.3 Literature Review

In 1932, Monroe and Templin have conducted field tests to pursue vibration and tension analysis [4]. They have suggested that suspended conductors are free to vibrate and by its curtailment in tension, and by a change in cross-section of the conductors will restrict vibration. Then, Speight [5] has introduced the torsion damper to restrict conductors' vibration in the transmission lines. In literature [6], authors have made fatigue failure test of conductor and optimal positioning of torsional dumbbell dampers along with armor reinforcing rods installation. In its continued study,

Wagner et al. [7] have applied Stockbridge vibration damper considering two DoF for transmission line damping study. They have studied dynamic stress too for understanding transmission line reliability. Immediately, in the next year, Hearnshaw [8] has adopted a spacer damper for damping performance study. They have ascertained damping by a field test. Literature [9] illustrates an application of flutter theory for wake-induced vibration review in the overhead conductors. Thereafter, in [10], the author has performed wind tunnel tests in a laboratory to evaluate dampers' effectiveness under wake-induced vibration. In one of the literature, authors of [11] have conducted a power spectrum study to evaluate conductors' vibration impact on the transmission line considering the terrain and climatic factors. However, Dhotarad et al. [12] have applied FEM and FDM techniques for vibration study of overhead lines. The IEEE standard (namely IEEE Std563-1 978) is formulated to ascertain a self-damping feature, permissible limit of amplitude variation, the appearance of wind power, wind dissipation by conductor, and supporting structure [3].

Sathikh and Chari [13] have evaluated dampers' performance by two unique methods, viz direct absolute power and relative wave power methods. In [14], the author has assumed clamps' rotary motion and thus calculated bending strains while applying damper to the overhead conductor. Authors from research works of [15] have shown various aspects of vibration like fluid dynamics, vortex-induced vibration. He has pursued single conductor behaviors as Aeolian response and uncertainty of wind energy input. Literature [16] has presented an iteration-based procedure to evaluate vibration amplitude under changing tension conditions. He has also compared span impact for both field and laboratory tests. In this regard [17], the Gaussian process has been adopted to understand the randomness of aerodynamics effects. They have tried to find swift damping of oscillations considering torsional motion and swing of the conductor. And in the same year, in the literature [18], a study on spacer dampers' influence on Aeolian has been conducted while generalized stiffness calculation was also done.

Then, in literature NEPC organization has undertaken dampers' design performance evaluation for HVDC transmission line [19]. Foata and Noiseux [20] have investigated the effects of twin bundle asymmetries to correlate ACSR conductors' damping behavior. Then, a group of researchers has adopted an analytical method to highlight the dynamic response of the conductor under excitation of turbulence wind conditions [21]. Corriga et al. [22] have proposed Ervik's concept to calculate dampers' parameters of the damper to avail damping feature under Aeolian vibration damping. Literature [23] presents wind energy impact on damping of the Drake conductor in laboratory experiment works by the wind tunnel test. They have found maximum tension as 18% for Drake conductor, which found to be self damped by Cardinal ACSR conductor. Again in the same year, Sunkle and Christian [24] have studied ADSS cables' Aeolian vibration tension analysis and corresponding damping features. In the succeeding year, Richardson [25] has studied the energy gap in between wind entering and lost energy associated with conductor triple bundle damping using Stockbridge vibration damper. Then, a laboratory works, two methods, e.g. resonance vibration across span and force resonance using shaker, have been illustrated. Also, a comparison between forwarding and background traveling waves has been analyzed [26]. In the literature [27], authors have illustrated a conductor as a cylinder with a cross-section of the circular structure. They have performed wake-induced vibration based on the tunnel test data. In [28], authors have suggested Stockbridge vibration damper for damping of conductors' vibration. They have presented the whole scenario by differential equation describing it as a motion of the tensioned beam.

Thus, anti-vibration hardware has been installed after towers inspection for reliable delivery of power [29]. In continuation with it, literature [30] highlights a study on the endurance of dampers for conductor oscillations along with their dynamic response. Hagedorn et al. [31] have the applied single perturbation method and energy balance method to find critical bending strains, which further helps in transmission line design optimization. An effective protection way for 345 kV transmission line has been introduced in [32]. Authors have studied conductors' failure possibilities due to the galloping phenomenon. An experience of a 500 kV DC transmission line for post-failure and re-installation over 400 KM line has been presented in [33]. In literature [34], a numerical model of quad spacer damper dynamics and corresponding relevant applicable tests have been conducted to evaluate vibration impact on the bundle conductors and suspension arrangements of transmission infrastructure. In one another literature, the transmission line vibration reliability of cold climatic regions has been conducted. Authors have analyzed vibration due to icing, hoarfrost, and snowfall atmospheric conditions [35]. Then in reference [36], the vortex-induced vibration has been considered as a two-dimensional problem to co-relate with conductors harmonic oscillations.

The FDM method has been applied to evaluate Aeolian-based vibration excitation and self-damping behavior of two bundle conductors along with spacer damper [37]. Again, conductors' non-linear dynamic behavior has been studied at a different height to understand initial equilibrium conditions under skew wind circumstances [38]. Guerard et al. [39] have proposed a self-damping monitoring device (namely Ampacimon) across the field for conductors across the power line nearby the Dead Water Fell test site. In the study of [40], a group of researchers has proposed a non-linearity concept based on Hydro-Québec damper to find the dynamic response of the conductor. Dulhunty [41] has suggested the application suitability of SVD for various types of conductors, e.g. ACSR, AAC, and AAAC conductor. However, a rational design-based formulation has been suggested to overcome a gap between theory and practice of overhead lines [42]. In [43], authors suggested Stockbridge vibration damper for OPGW cable application. They have also tested fatigue failure and endurance limit check across the transmission line span. Foti and Martinelli [44] have tested an enhanced model for self-damping of the stranded conductors in a unified way under the Aeolian vibrations scenario. They have adopted a mono-modal steady-state computation approach for vibrations.

In continuation with that, Mukherjee et al. (2021) have proposed a Hexa spacer damper to improve the damping reliability of 765 kV transmission lines [45]. Thereafter, researchers have checked spacer dampers' selection criteria in the South African region while considering force-displacement and frequency as their objective [46]. In paper [47], a new anti-galloping concept-oriented restraining and dissipation device has been introduced to overcome fatigue failure of the UHV overhead line. Peng et al. [48] have suggested the FRSD technique for parameter-dependent linear dynamic study of multi-structure. In this literature, a piezoelectric-based energy harvester to communicate resonance data of smart transmission lines has been proposed [49]. A quad spacer damper has been proposed in [50], to improve the transmission line reliability of the 765 kV transmission line. A study on structural flexibility and its damping feature has been studied in [51]. It also highlights the damping application of fraction dampers along with other dampers. Li et al. [52] have studied effectiveness of spacer tuned mass damper system for controlling the vibration.

13.2 Problem Formulation

The small-amplitude vibrations, which occur in the 5–60 Hz frequency range, with amplitude of 0.5–2 times the conductor diameter, arise manly on a vertical plane of the conductor and may cause fatigue in the long run to damage in the conductors. The most widely used method to offset the effect of Aeolian vibrations is to install dampers near the span extremities, which dissipate the force exerted by the wind to the conductor [20]. The design aspect and corresponding mathematical formulation have been discussed in the subsequent segments.

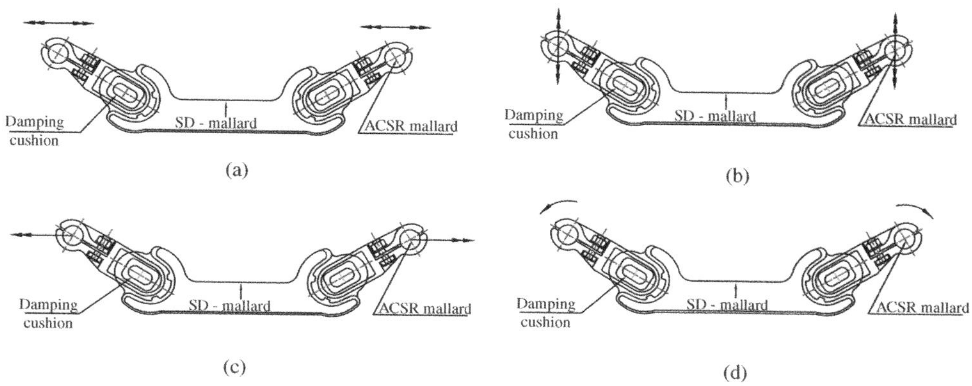

Figure 13.2 Representation of twin spacer dampers' relative movement.

13.2.1 Design Aspect

This literature mainly covers dampers' optimal positioning behind the generation of a practical acceptance curve for evaluation of the suitability of a twin spacer damper to be used in an overhead line (as shown in Figure 13.2). Corresponding to its different directional movements, the coordinates representation is portrayed in Figure 13.3.

13.2.2 Mathematical Modeling

For the considered twin spacer damper' relative movement as elaborated above, the mathematical formulation is discussed in this segment. During Aeolian vibrations, conductor self-damping acts as a major source of energy dissipation and it may be formulated as in (13.1) [37]:

$$P_C = \frac{E_C}{l} = Kf^{\beta}\left(\frac{2y_0}{D}\right)^{\alpha} \tag{13.1}$$

For a given Aeolian excitation frequency, the conductor behavior could be formulated as in (13.2) [39].

$$p(x,t) = p(x)e^{i\omega t} = P(X)e^{i\pi t} \tag{13.2}$$

Further, the wind-induced motions perceptibility of a conductor is presented by (13.3) [51].

$$\text{Perceptibility} = \frac{1}{\sqrt{\text{mass} \times \text{stiffness} \times \text{dampness}}} \tag{13.3}$$

Generally, a transmission line is treated as a beam with less flexural stiffness and its length is usually larger (Figure 13.3). Thus, the transverse motion of a single conductor can be given as in (13.4) [37],

$$\text{EI}\frac{\partial^4 y}{\partial x^4} + m\frac{\partial^2 y}{\partial t^2} + c\frac{\partial y}{\partial t} - \text{TEI}\frac{\partial^2 y}{\partial x^2} = p(x,t) + \sum^{n} \delta(x - x_n) f_n(x,t) \tag{13.4}$$

$$y(x,t) = y_0(x)e^{i\omega t} \tag{13.5}$$

Now, substituting (13.5) in (13.4), the governing equation for the transverse or vertical motion of a single conductor may be reformulated as in (13.6) [37],

$$S_\text{p}\frac{d^4 Y}{dX^4} - 4\pi^2 I_\text{p} Y + i.2\pi C_\text{p} Y - T_\text{p}\frac{d^2 Y}{dX^2} = \frac{1}{\gamma}\left(P(X) + \sum^{n} \delta(X - X_\text{n}) f_\text{n}(X)\right) \tag{13.6}$$

where, $Y = \frac{y_0}{D}, X = \frac{x}{L}, \tau = Ft, = mg, S_\text{p} = \frac{EID}{\gamma L^4}, I_\text{p} = \frac{DF^2}{g}, T_\text{p} = \frac{DT}{\gamma L^2}, C_\text{p} = \frac{cDF}{\gamma}$.

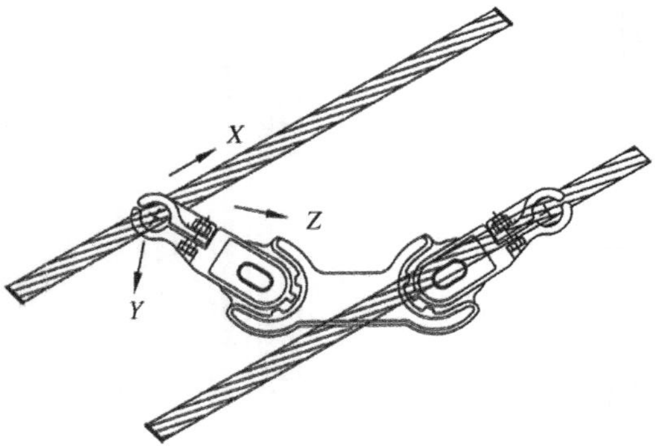

Figure 13.3 Coordinates representation of twin spacer dampers' movements.

These displacements can be expressed in the following form when the conductor vibrates steadily (13.7) [37],

$$y_A = Ae^{i\omega t}, y_B = Be^{i\omega t}e^{i\beta}, y_s = Se^{i\omega t}e^{i\alpha} \tag{13.7}$$

The motion equation of the system most easily formulated by directly expressing the equilibrium of all forces acting on the spacer mass using d'Alembert's principle as given by (13.8) [37],

$$M_S\ddot{y}_s + C_s(\dot{y}_S - \dot{y}_A) + C_s(\dot{y}_S - \dot{y}_B) + k_s(y_S - y_A) + k_s(y_S - y_B) = 0 \tag{13.8}$$

Then the interaction force between spacer and sub-conductors can be written in the form (13.9) [37],

$$\left.\begin{aligned} f_A &= C_S(\dot{y}_S - \dot{y}_A) + k_S(y_S - y_A) = y_A(iC_S\omega\Delta - iC_S\omega + k_S\Delta - k_S) + y_B(iC_S\omega\Delta + k_S\Delta) \\ f_B &= C_S(\dot{y}_S - \dot{y}_B) + k_S(y_S - y_B) = y_B(iC_S\omega\Delta - iC_S\omega + k_S\Delta - k_S) + y_A(iC_S\omega\Delta + k_S\Delta) \end{aligned}\right\} \tag{13.9}$$

Calculating the interaction force in the position of every spacer, substituting the force into (13.9), and then the governing equation of two-bundled conductor with spacers can be obtained. The FEA module of CATiA provides two options for damping:

Modal damping: Modal Damping is a fraction of the critical damping. The critical damping is computed as follows (13.10) [3]:

$$C_r = 2\sqrt{mk} \tag{13.10}$$

Rayleigh damping: The Rayleigh damping is applied in this study to signify the transmission line damping. This damping is evaluated as indicated below in (13.11) [3]:

$$[C] = \alpha[M] + \beta[K] \tag{13.11}$$

13.3 Numerical Analysis

The harmonic dynamic response is a case of cyclic load analysis in CATiA. This is based on the principle that any continuous load will produce a sustained cyclic response from the structural system. Harmonic response analysis provides leverage to predict the sustained dynamic behavior of the structure of interest. This method is utilized to find the steady-state response of a linear structure to loads that vary sinusoidal (harmonically) with respect to time. The goal is to determine the structure's response at various frequencies and plot a response quantity (such as displacement

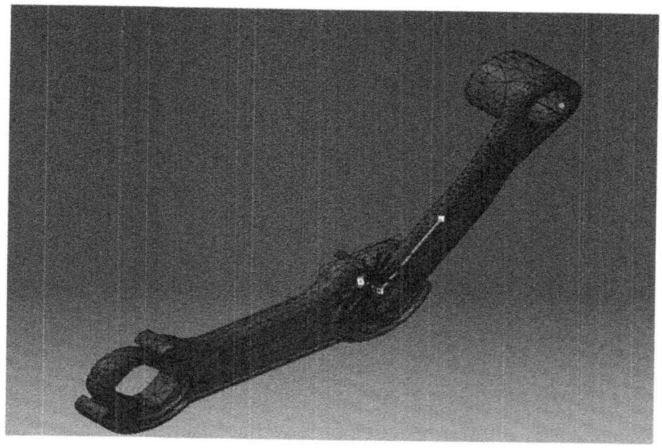

Figure 13.4 Node location vicinity in the meshed model.

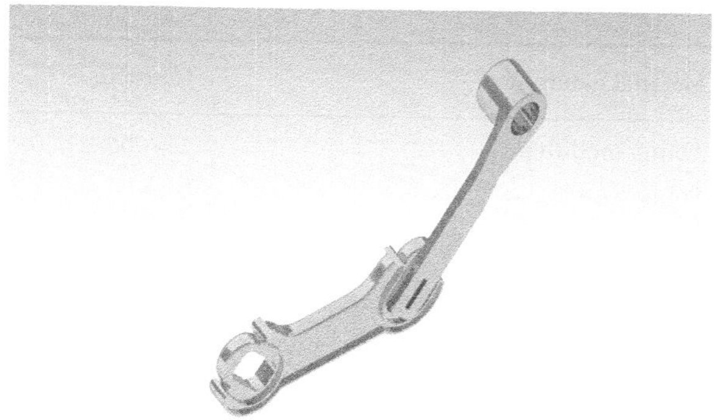

Figure 13.5 Model rendering of the twin-spacer damper.

and velocity) versus frequency graph at different nodes. "Peak" responses are then identified on the graph and strains developed due to those are evaluated further. The above-described study is conducted for a frequency range of 1–100 Hz (standard Aeolian vibration test frequency) and evaluation is done for different phase lags (due to virtue of placement on the test span of 250 m). The computer model can keep track of displacement, velocity and acceleration of all nodes in the model of the spacer damper, but our interest lies on only a particular node adjoining the clamping arrangement of the damper to the conductor. This is because its displacement is a direct indication of conductor displacement due to the cyclic loads. The node is identified by a number and is portrayed in Figures 13.4 and 13.5.

13.3.1 Simulation Inputs

13.3.1.1 Model Description

Finite elemental analysis requires a three-dimensional model of the product being tested and this model is meshed accordingly. The mesh sizes are adjusted to meet sensitivity and accuracy requirements of the model under varying load, keeping computational effort in mind. The model is represented below in Figure 13.6. For computational simplicity, the effect of vibration is only analyzed on one conductor in a selected arm, assumed other arm to be rigidly clamped during analysis. The properties of the elastomeric bush for the considered model, significant to the FEA simulations are presented in Table 13.4.

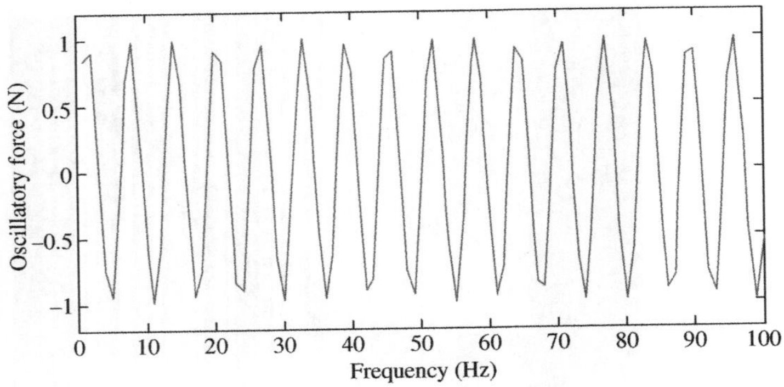

Figure 13.6 Characteristics of load modulation (frequency range: 0–100 Hz).

Table 13.4 Elastomeric bushing properties.

Material nature	**Isotropic**
Young modulus	2e+006Pa
Poisson ratio	0.49
Density	910 kg/m^3
Thermal expansion	1.62×10^{-4} K^{-1}

13.3.1.2 Load Excitation

The load that is being applied to the clamps in the static case is taken as 500 N and is modulated through the frequency range of 0–100 Hz using a frequency modulation approach; this presents the varying component of the altering load whereas the static load is merely the amplitude. The modulation is done sinusoidal by feeding .xls input to the simulation model as given in Table 13.5.

13.3.1.3 Span Wise Phase Lag

The phase lag is calculated for every twin spacer damper applied in the 250 m spans and a standard wavelength can be selected for the Aeolian vibration for calculating the phase lag from path lag (refer to Tables 13.6 and 13.7). However, every Aeolian vibration has its own wavelength and it will virtually be impossible to calculate for every wave and its corresponding phase lags on the dampers, so it takes a generalized approach by calculating lag on each spacer damper with reference to the first one as follows:

No. of spacer dampers implemented in the span = n (assuming equally spaced).
Total span length = l
Position of ith damper from the start of span = x,
Phase lag for ith damper = $\left(\frac{x}{l}\right)$ 360°.

For test case of 250 m: The damper arrangement is in the following order (refer to placement chart).

Table 13.5 Frequency modulation set given as input to simulation (frequency range: 0–100 Hz, type: sinusoidal).

Frequency (Hz)	Oscillatory force (N)	Frequency (Hz)	Oscillatory force (N)
1	0.841470985	31	−0.40403765
2	0.909297427	32	0.551426681
3	0.141120008	33	0.99991186
4	−0.7568025	34	0.529082686
5	−0.95892428	35	−0.42818267
6	−0.2794155	36	−0.99177885
7	0.656986599	37	−0.64353813
8	0.989358247	38	0.296368579
9	0.412118485	39	0.963795386
10	−0.54402111	40	0.74511316
11	−0.99999021	41	−0.15862267
12	−0.53657292	42	−0.91652155
13	0.420167037	43	−0.83177474
14	0.990607356	44	0.017701925
15	0.65028784	45	0.850903525
16	−0.28790332	46	0.901788348
17	−0.96139749	47	0.123573123
18	−0.75098725	48	−0.76825466
19	0.14987721	49	−0.95375265
20	0.912945251	50	−0.26237485
21	0.836655639	51	0.670229176
22	−0.00885131	52	0.986627592
23	−0.8462204	53	0.39592515
24	−0.90557836	54	−0.55878905
25	−0.13235175	55	−0.99975517
26	0.76255845	57	0.436164755
27	0.956375928	58	0.992872648
28	0.270905788	59	0.636738007
29	−0.66363388	60	−0.30481062
30	−0.98803162	61	−0.96611777

(continued)

Table 13.5 (Continued)

Frequency (Hz)	Oscillatory force (N)	Frequency (Hz)	Oscillatory force (N)
62	−0.7391807	90	0.893996664
63	0.1673557	91	0.105987512
64	0.920026038	92	−0.77946607
65	0.826828679	93	−0.94828214
66	−0.02655115	94	−0.24525199
67	−0.85551998	95	0.683261715
68	−0.89792768	96	0.983587745
85	−0.17607562	97	0.379607739
86	−0.92345845	98	−0.57338187
87	−0.82181784	99	−0.99920683
88	0.035398303	100	−0.50636564
89	0.860069406		

Table 13.6 Calculated Rayleigh damping and modal frequencies of the model.

Sl.	Frequency (Hz)	Sl.	Frequency (Hz)
1	85.1751	6	840.134
2	225.459	7	876.071
3	269.185	8	895.737
4	536.445	9	925.876
5	787.862	10	926.637

Table 13.7 Phase lag for dampers placed in test span 250 m.

Distance of dampers (m)	$\left(\frac{x}{l}\right)$ (Placement to span)	Phase lag (°)
26.5 (Reference damper)	Reference damper	0
42	0.168	60.48
91	0.364	131.04
147	0.588	211.68
197	0.788	283.68
223.5	0.894	321.84

13.3.2 Analysis Findings

The outputs for each damper are represented by the following graphs, Graph 1 is for "displacement vs. frequency," while Graph 2 is for "velocity vs. frequency" and Graph 3 is for the "acceleration vs. frequency" of the node as mentioned earlier (this node displacement represents the conductors' displacement at that location) (refer to Figures 13.7–13.12). The maximum displacement from the frequency domain graph is taken for each spacer damper is then tabulated and converted to micro-strains (for Mallard conductor) to check for compliance with the limit of 150 micro-strains (industry-accepted limit) as shown in Table 13.8. The graphs for corresponding phase lags (as mentioned in Table 13.8) are given below.

Here, the graphical outcome includes phase response of acceleration, velocity, and displacement with various DoF in *x*, *y*, and *z* directions at different nodes. The vibration of conductor is represented the amplitude of displacement. The maximum displacement yielded in this case is highlighted in Table 13.8, which is 5.82E−08, when taken as the net deformation of the

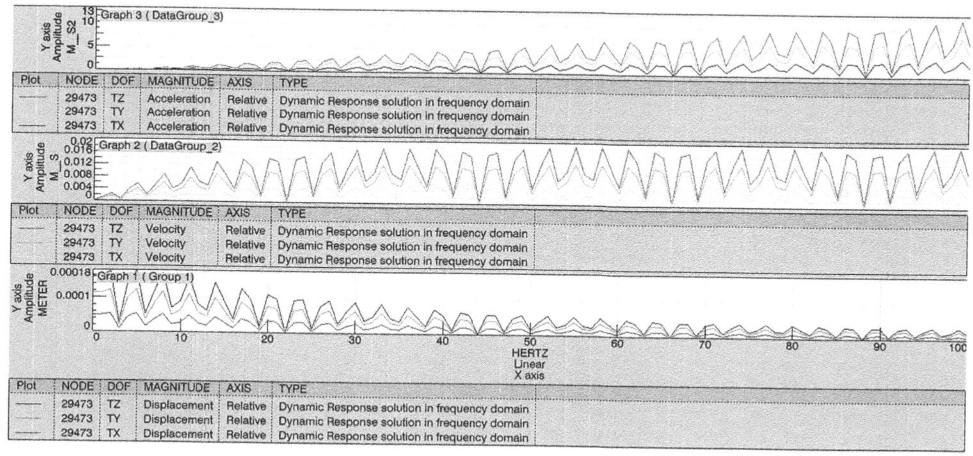

Figure 13.7 Damping characteristics (phase lag: 0°).

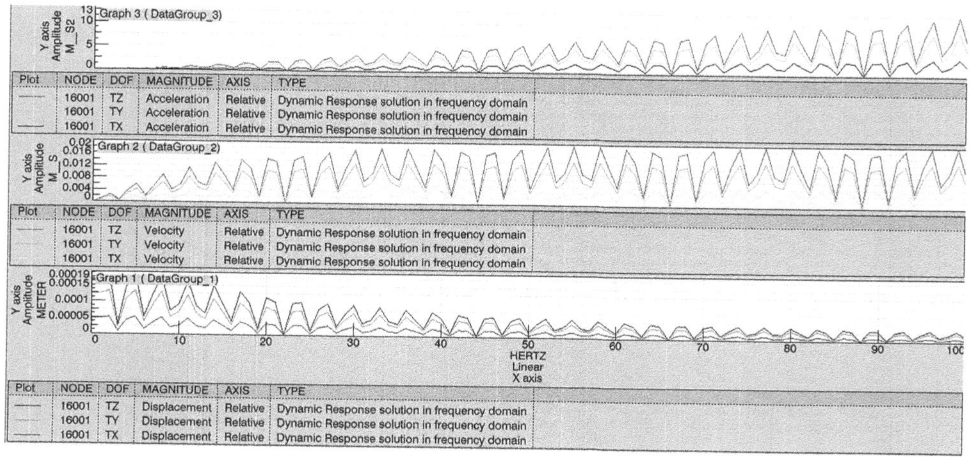

Figure 13.8 Damping characteristics (phase lag: 60.48°).

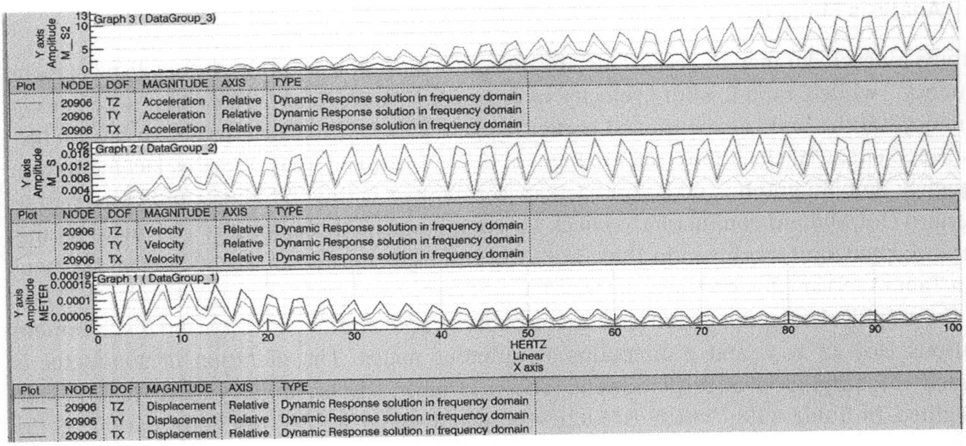

Figure 13.9 Damping characteristics (phase lag: 131.04°).

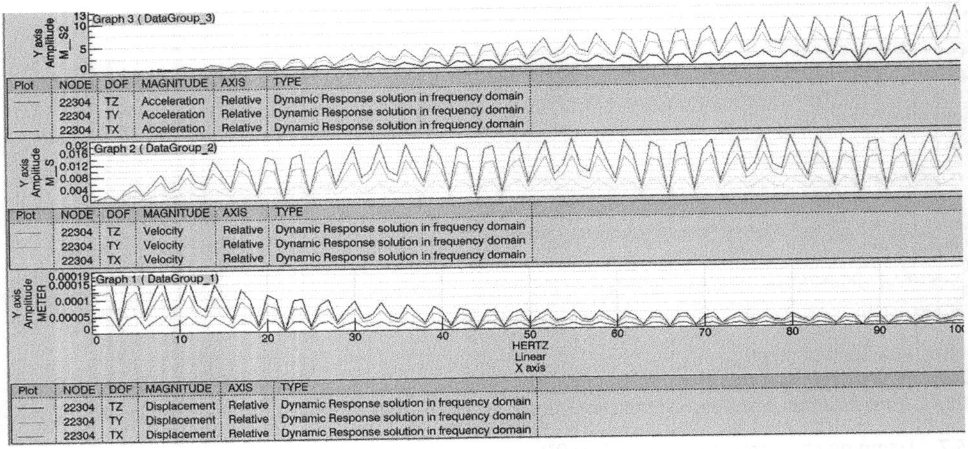

Figure 13.10 Damping characteristics (phase lag: 211.68°).

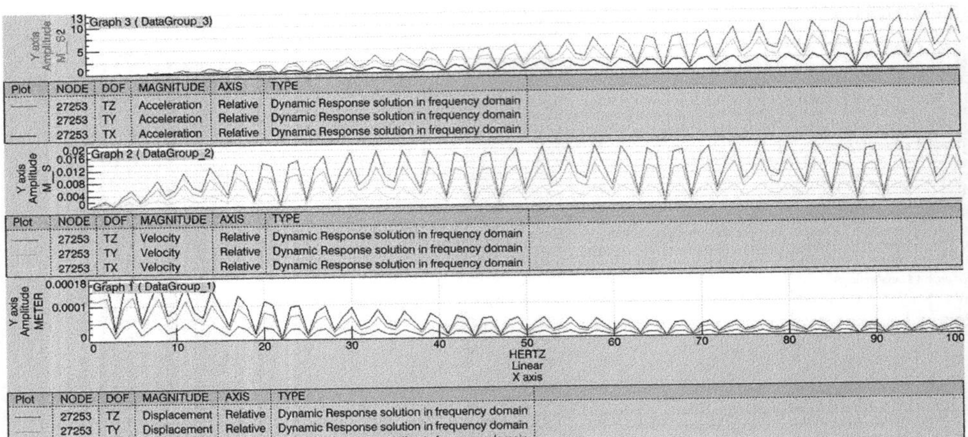

Figure 13.11 Damping characteristics (phase lag: 283.68°).

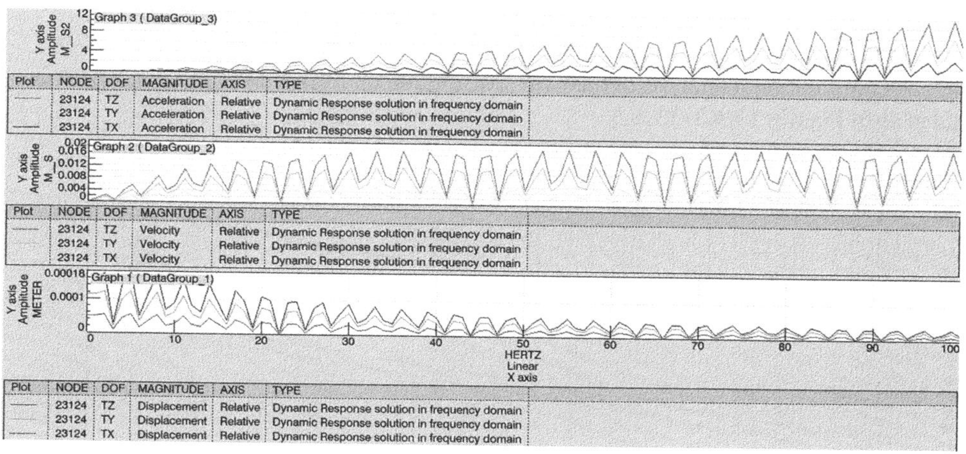

Figure 13.12 Damping characteristics (phase lag: 321.84°).

Table 13.8 Summary of displacements.

Phase lag (°)	T_x (max) (m)	T_y (max) (m)	T_z(max) (m)	Net displacement (m)
0	0.00005	0.00012	0.00018	4.93E−08
60.48	0.00005	0.00014	0.00019	5.82E−08
131.04	0.00005	0.00013	0.000195	5.74E−08
211.68	0.00005	0.00011	0.00017	4.35E−08
283.68	0.000025	0.00012	0.00018	4.74E−08
321.84	0.00002	0.00012	0.00018	4.72E−08

conductor's cross-section, yields 20.9 micro-strains ($\mu\varepsilon$). For the conductor, type refers to ACSR Mallard conductor.

13.4 Conclusion

The interruption of electricity due to the failure of transmission lines can trigger social and economic calamities. A suspended conductor of overhead power lines is lightweight and has lightly damped structural elements, which are prone to vortex-induced vibrations under the action of low or moderately strong winds. These vibrations, also called "Aeolian vibrations" in the community of transmission lines engineers, are characterized by low-amplitude, high-frequency flexural oscillations. Aeolian vibrations can induce significant bending stresses on the conductors and their support structures, leading to a reduction of life expectancy of the power line. The abatement of the severity of Aeolian vibrations is hence a crucial issue both in the design of new electrical transmission lines and in the retrofitting of existing ones. The test results indicate a maximum strain of 20.9 $\mu\varepsilon$ was induced for a test span of 250 m, which is well below the compliance limit of 150 micro-strains in the case of Aeolian vibrations in the frequency domain of 0–100 Hz (refer to force excitation). These findings assure applicability of twin spacer damper to even longer span if properly placed as suggested by placement chart given in appendix section.

13.A Appendix

After dynamic computational analysis of the twin spacer damper, its numbers and placement can be concluded as in Figures 13.A.1–13.A.6.

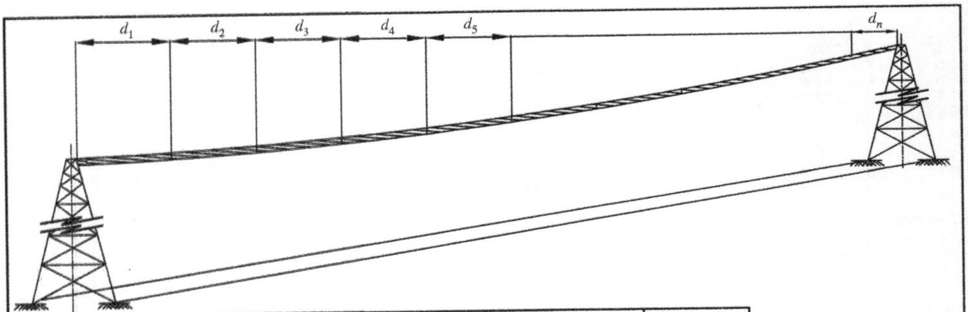

Sl.	Span length (m)	Range (m)	Spacing (m)					No. of spacer damper
			d_1	d_2	d_3	d_4	d_n	
1	150–240	150	24	46	56	NR	24	3
		155	24	49	58	NR	24	
		160	25	50	60	NR	25	
		165	25	52	63	NR	25	
		170	26	54	64	NR	26	
		175	26	56	67	NR	26	
		180	19.5	45	51	45	19.5	4
		185	20.5	46	52	46	20.5	
		190	21.5	47	53	47	21.5	
		195	22.5	48	54	48	22.5	
		200	23.5	49	55	49	23.5	
		205	24.5	50	56	50	24.5	
		210	25.5	51	57	51	25.5	
		215	26.5	52	58	52	26.5	
		220	27.5	53	59	53	27.5	
		225	28.5	54	60	54	28.5	
		230	29.5	55	61	55	29.5	
		235	30.5	56	62	56	30.5	
		240	31.5	57	63	57	31.5	

Figure 13.A.1 Placement layout for span length: 150–240 m.

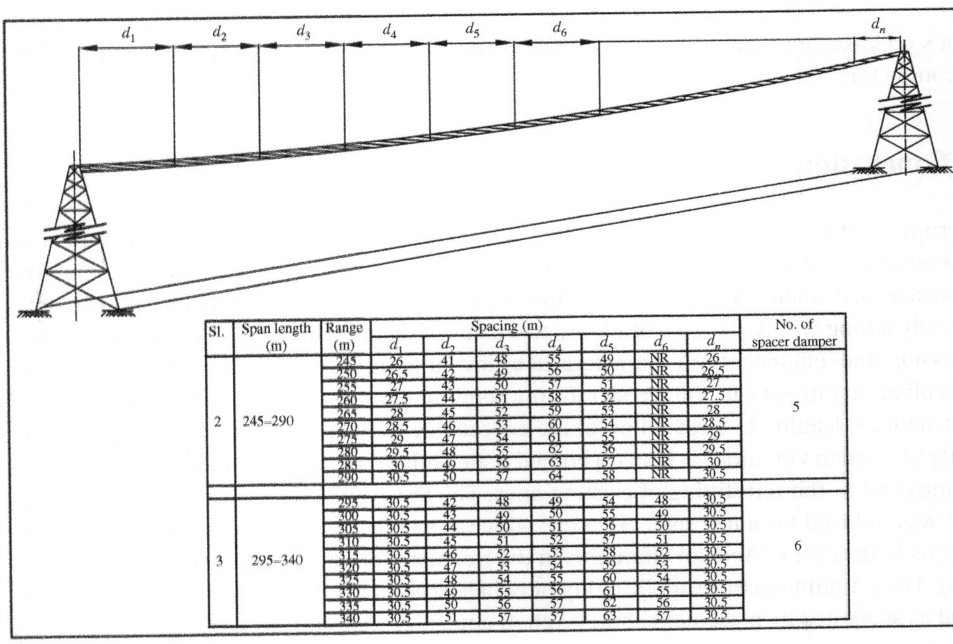

Sl.	Span length (m)	Range (m)	Spacing (m)							No. of spacer damper
			d_1	d_2	d_3	d_4	d_5	d_6	d_n	
2	245–290	245	26	41	48	55	49	NR	26	5
		250	26.5	42	49	56	50	NR	26.5	
		255	27	43	50	57	51	NR	27	
		260	27.5	44	51	58	52	NR	27.5	
		265	28	45	52	59	53	NR	28	
		270	28.5	46	53	60	54	NR	28.5	
		275	29	47	54	61	55	NR	29	
		280	29.5	48	55	62	56	NR	29.5	
		285	30	49	56	63	57	NR	30	
		290	30.5	50	57	64	58	NR	30.5	
3	295–340	295	30.5	42	48	49	54	48	30.5	6
		300	30.5	43	49	50	55	49	30.5	
		305	30.5	44	50	51	56	50	30.5	
		310	30.5	45	51	52	57	51	30.5	
		315	30.5	46	52	53	58	52	30.5	
		320	30.5	47	53	54	59	53	30.5	
		325	30.5	48	54	55	60	54	30.5	
		330	30.5	49	55	56	61	55	30.5	
		335	30.5	50	56	57	62	56	30.5	
		340	30.5	51	57	57	63	57	30.5	

Figure 13.A.2 Placement layout for span length: 245–340 m.

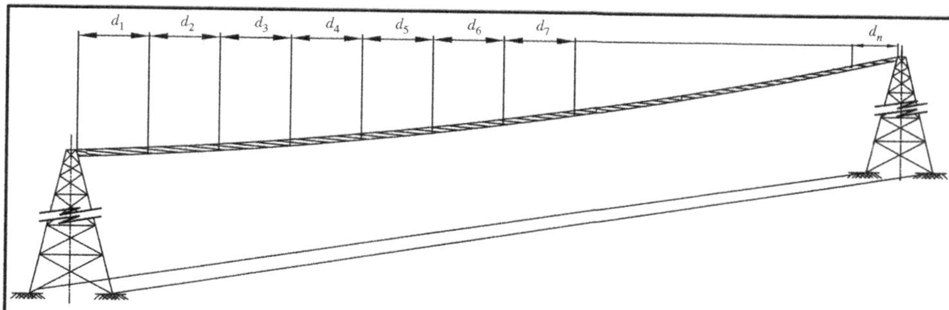

Sl.	Span length (m)	Range (m)	Spacing (m)								No. of spacer damper
			d_1	d_2	d_3	d_4	d_5	d_6	d_7	d_n	
3	345–395	345	30.5	41	46	61	45	40	51	30.5	7
		350	30.5	42	47	61	46	41	52	30.5	
		355	30.5	43	48	61	47	42	53	30.5	
		360	30.5	44	49	61	48	43	54	30.5	
		365	30.5	45	50	61	49	44	55	30.5	
		370	30.5	46	51	61	50	45	56	30.5	
		375	30.5	47	52	62	51	47	55	30.5	
		380	30.5	48	53	63	52	47	56	30.5	
		385	30.5	49	54	64	53	48	56	30.5	
		390	30.5	50	55	65	54	49	56	30.5	
		395	30.5	50	56	66	55	50	57	30.5	

Figure 13.A.3 Placement layout for span length: 345–395 m.

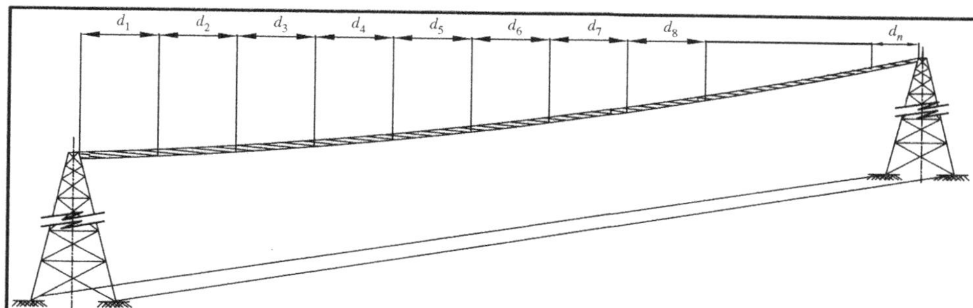

Sl.	Span length (m)	Range (m)	Spacing (m)									No. of spacer damper
			d_1	d_2	d_3	d_4	d_5	d_6	d_7	d_8	d_n	
4	400–450	400	29	42	49	58	60	51	45	37	29	8
		405	30	51	44	49	55	50	44	52	30	
		410	30	51	45	50	56	51	44	53	30	
		415	30	52	46	51	57	52	44	53	30	
		420	30	52	46	52	59	53	46	53	30	
		425	30	52	47	53	59	54	47	53	30	
		430	30	53	48	54	60	53	47	55	30	
		435	30	54	49	55	61	54	47	55	30	
		440	30	54	49	56	62	55	49	55	30	
		445	30	55	50	56	62	55	50	57	30	
		450	30	56	51	57	63	56	51	56	30	

Figure 13.A.4 Placement layout for span length: 400–450 m.

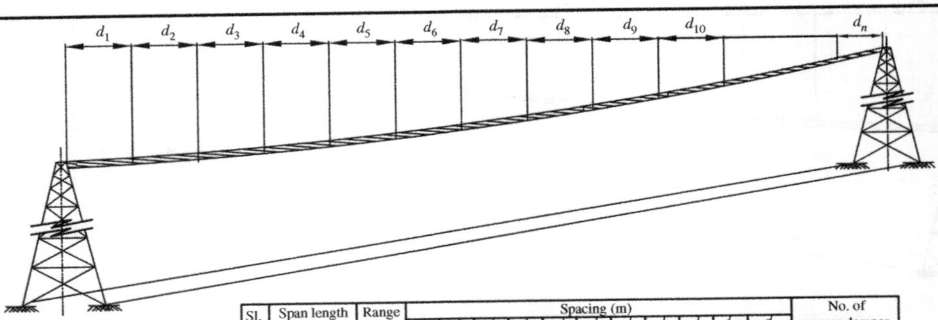

Sl.	Span length (m)	Range (m)	Spacing (m) d_1	d_2	d_3	d_4	d_5	d_6	d_7	d_8	d_9	d_{10}	d_n	No. of spacer damper
5	455–500	455	30	50	40	45	51	58	50	44	57	NR	30	9
		460	30	50	41	46	51	59	51	45	57	NR	30	
		465	30	50	42	47	52	60	51	46	57	NR	30	
		470	30	50	43	48	53	60	52	47	57	NR	30	
		475	30	50	44	49	54	60	53	47	58	NR	30	
		480	30	50	45	50	55	61	53	48	58	NR	30	
		485	30.5	50	45	50	56	62	54	49	58	NR	30.5	
		490	30.5	51	45	50	56	63	55	50	59	NR	30.5	
		495	30.5	51	46	51	57	64	56	50	59	NR	30.5	
		500	30.5	51	46	52	58	65	57	51	59	NR	30.5	
6	505–550	505	30	51	41	46	51	57	50	44	55	50	30	10
		510	30	51	42	47	52	58	50	45	55	50	30	
		515	30	51	43	48	53	59	51	45	55	50	30	
		520	30	52	44	49	53	60	51	46	55	50	30	
		525	30	52	44	49	54	61	52	47	56	50	30	
		530	30	52	44	49	55	62	53	48	56	51	30.5	
		535	30	52	45	50	56	63	54	48	56	51	30.5	
		540	30	53	45	50	56	63	55	49	57	52	30.5	
		545	30.5	52	46	51	57	64	56	50	57	51	30.5	
		550	30.5	52	46	52	58	65	57	51	57	51	30.5	

Figure 13.A.5 Placement layout for span length: 455–500 m.

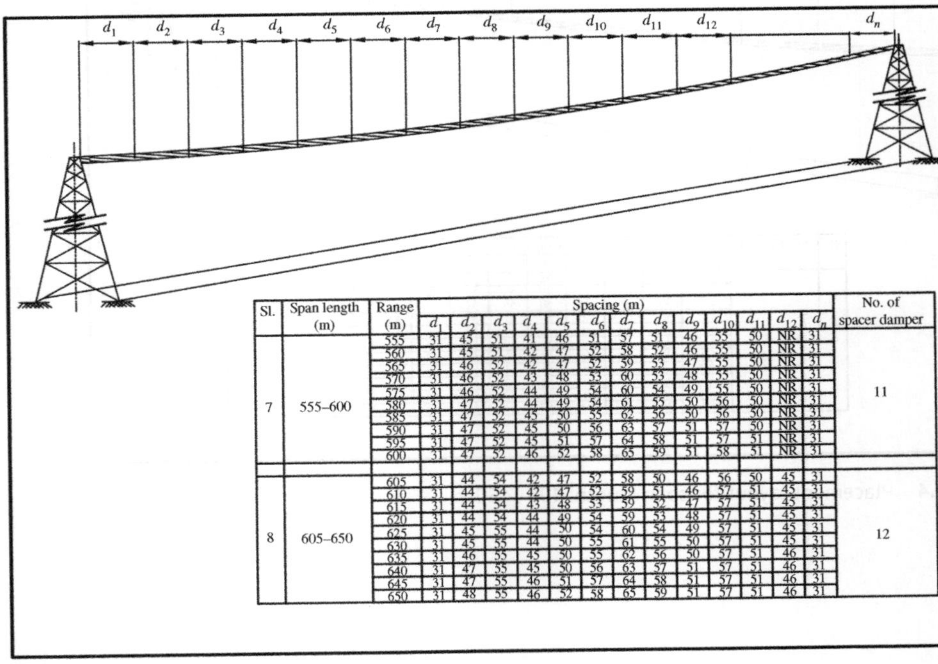

Sl.	Span length (m)	Range (m)	Spacing (m) d_1	d_2	d_3	d_4	d_5	d_6	d_7	d_8	d_9	d_{10}	d_{11}	d_{12}	d_n	No. of spacer damper
7	555–600	555	31	45	51	41	46	51	57	51	46	55	50	NR	31	11
		560	31	45	51	42	47	52	58	52	46	55	50	NR	31	
		565	31	46	52	42	47	52	59	53	47	55	50	NR	31	
		570	31	46	52	43	48	53	60	53	48	55	50	NR	31	
		575	31	46	52	44	49	54	60	54	49	55	50	NR	31	
		580	31	47	52	44	49	54	61	55	50	56	50	NR	31	
		585	31	47	52	45	50	55	62	56	50	56	50	NR	31	
		590	31	47	52	45	50	56	63	57	51	57	50	NR	31	
		595	31	47	52	45	51	57	64	58	51	57	51	NR	31	
		600	31	47	52	46	52	58	65	59	51	58	51	NR	31	
8	605–650	605	31	44	54	42	47	52	58	50	46	56	50	45	31	12
		610	31	44	54	42	47	52	59	51	46	57	51	45	31	
		615	31	44	54	43	48	53	59	52	47	57	51	45	31	
		620	31	44	54	44	49	54	59	53	48	57	51	45	31	
		625	31	45	55	44	50	54	60	54	49	57	51	45	31	
		630	31	45	55	44	50	55	61	55	50	57	51	45	31	
		635	31	46	55	45	50	55	62	56	50	57	51	46	31	
		640	31	47	55	45	50	56	63	57	51	57	51	46	31	
		645	31	47	55	46	51	57	64	58	51	57	51	46	31	
		650	31	48	55	46	52	58	65	59	51	57	51	46	31	

Figure 13.A.6 Placement layout for span length: 555–650 m.

References

1 Mukherjee, S., Prasad, D., Khan, M.I. et al. (2019). Hexa spacer damper for vibration energy decaying of 765 kV transmission line. In: *2019 Innovations in Power and Advanced Computing Technologies (i-PACT)*, vol. 1, 1–6. IEEE. doi: 10.1109/i-PACT44901.2019.8960087.

2 Mukherjee, S., Chattaraj, S., Khan, M.I. et al. (2019). Transmission line restoration using ERS structure. *Advances in Science, Technology and Engineering Systems Journal* 4 (5): 394–400. https://doi.org/10.25046/aj040551.

3 Chattaraj, S., Mukherjee, S., Khan, M.I. et al. (2019). Reliability improvement of 400 kV transmission line using Quad Spacer Damper and Quad Rigid Spacer. In: *2019 Impact of Changing Energy Mix in the Power Sector (ICIEMPS-2019)*, The Institution of Engineers (India), Electrical Engineering Division, West Bengal State Centre, 1–5. ISBN: 978-81-942561-2-0.

4 Monroe, R.A. and Templin, R.L. (1932). Conductor vibration on transmission lines-II. *Electrical Engineering* 51 (8): 562–568. https://doi.org/10.1109/EE.1932.6429934.

5 Speight, J.W. (1941). Conductor vibration-theory of torsional dampers. *Transactions of the American Institute of Electrical Engineers* 60 (10): 907–911. https://doi.org/10.1109/T-AIEE.1941.5058198.

6 Sproule, J.E. and Edwards, A.T. (1959). Progress toward optimum damping of transmission conductors. *Transactions of the American Institute of Electrical Engineers. Part III: Power Apparatus and Systems* 78 (3): 844–847. https://doi.org/10.1109/AIEEPAS.1959.4500448.

7 Wagner, H., Ramamurti, V., Sastry, R.V.R., and Hartmann, K. (1973). Dynamics of Stockbridge dampers. *Journal of Sound and Vibration* 30 (2): 207–220. https://doi.org/10.1016/S0022-460X(73)80114-2.

8 Hearnshaw, D. (1974). Spacer damper performance-a function of in-span positioning. *IEEE Transactions on Power Apparatus and Systems* PAS-93 (5): 1298–1309. https://doi.org/10.1109/TPAS.1974.293854.

9 Price, S.J. (1975). Wake induced flutter of power transmission conductors. *Journal of Sound and Vibration* 38 (1): 125–147. https://doi.org/10.1016/S0022-460X(75)80023-X.

10 Bouche, R.R. (1977). Damper vibration attenuation for overhead lines. *Electric Power Systems Research* 1 (1): 59–65. https://doi.org/10.1109/61.400836.

11 Davall, P.W., Gupta, M.M., and Ukrainetz, P.R. (1978). Mathematical analysis of transmission line vibration data. *Electric Power Systems Research* 1 (4): 269–282. https://doi.org/10.1016/0378-7796(78)90013-5.

12 Dhotarad, M.S., Ganesan, N., and Rao, B.V.A. (1978). Transmission line vibrations. *Journal of Sound and Vibration* 60 (2): 217–237. https://doi.org/10.1016/S0022-460X(78)80031-5.

13 Sathikh, S. and Chari, R.T. (1981). Acceptance curve – a practical means for EHV transmission line damper performance evaluation. *Electric Power Systems Research* 4 (1): 67–78. https://doi.org/10.1016/0378-7796(81)90038-9.

14 Hagedorn, P. (1982). On the computation of damped wind-excited vibrations of overhead transmission lines. *Journal of Sound and Vibration* 83 (2): 253–271. https://doi.org/10.1016/S0022-460X(82)80090-4.

15 Tsui, Y.T. (1982). Recent advances in engineering science as applied to aeolian vibration: an alternative approach. *Electric Power Systems Research* 5 (1): 73–85. https://doi.org/10.1016/0378-7796(82)90010-4.

16 Roughan, J.C. (1983). Estimation of conductor vibration amplitudes caused by aeolian vibration. *Journal of Wind Engineering and Industrial Aerodynamics* 14 (1–3): 279–288. https://doi.org/10.1016/0167-6105(83)90030-2.

17 Riaz, H., Biswas, S.K., and Ahmed, N.U. (1986). Stochastic modelling and stabilization of galloping transmission lines. *Electric Power Systems Research* 10 (2): 137–143. https://doi.org/10.1016/0378-7796(86)90006-4.

18 Jamaleddine, A. and Rousselet, J. (1986). Generalized stiffnesses of spacer-dampers for two-conductor bundles. *Electric Power Systems Research* 10 (2): 125–135. https://doi.org/10.1016/0378-7796(86)90005-2.

19 Richardson, A.S. (1990). Vibration of bundled and single conductors: a comparative case study. *Electric Power Systems Research* 18 (1): 19–30. https://doi.org/10.1016/0378-7796(90)90042-2.

20 Foata, M. and Noiseux, D. (1991). Computer analysis of the effects of asymmetries of a two-conductor bundle upon its aeolian vibration. *IEEE Transactions on Power Delivery* 6 (3): 1132–1138. https://doi.org/10.1109/61.85858.

21 Diana, G., Cheli, F., Fossati, F., and Manenti, A. (1993). Aeolian vibrations of overhead transmission lines: computation in turbulence conditions. *Journal of Wind Engineering and Industrial Aerodynamics* 46–47: 639–648. https://doi.org/10.1016/0167-6105(93)90332-I.

22 Corriga, G., Sanna, S., and Usai, G. (1994). Parametric optimization of wind-excited vibration dampers for overhead transmission lines. *Applied Mathematical Modelling* 18 (10): 577–581. https://doi.org/10.1016/0307-904X(94)90143-0.

23 Richardson, A.S. (1995). Vibration damping required for overhead lines. *IEEE Transactions on Power Delivery* 10 (2): 934–940. https://doi.org/10.1109/61.400836.

24 Sunkle, D.C. and Christian, D.F. (1995). The effects of tension on vibration of all-dielectric self-supporting fiber optic cable. In: *Proceedings of ESMO'95–1995 IEEE 7th International Conference on Transmission and Distribution Construction, Operation and Live-Line Maintenance*, 197–203. IEEE https://doi.org/10.1109/TDCLLM.1995.485057.

25 Richardson, A.S. (1996). Performance requirements for vibration dampers. *Electric Power Systems Research* 36 (1): 21–28. https://doi.org/10.1016/0378-7796(95)01008-4.

26 Schmidt, J.T., Biedenbach, G., and Krispin, H.J. (1997). Laboratory measurement of the power dissipation characteristics of aeolian vibration dampers. *IEEE Transactions on Power Delivery* 12 (4): 1614–1621. https://doi.org/10.1109/61.634181.

27 Cigada, A., Diana, G., Falco, M. et al. (1997). Vortex shedding and wake-induced vibrations in single and bundle cables. *Journal of Wind Engineering and Industrial Aerodynamics* 72: 253–263. https://doi.org/10.1016/S0167-6105(97)00247-X.

28 Wang, H.Q., Miao, J.C., Luo, J.H. et al. (1997). The free vibration of long-span transmission line conductors with dampers. *Journal of Sound and Vibration* 208 (4): 501–516. https://doi.org/10.1006/jsvi.1997.1196.

29 Kilroe, N. (2000). Aerial method to mitigate vibration on transmission towers. In: *2000 IEEE ESMO-2000 IEEE 9th International Conference on Transmission and Distribution Construction, Operation and Live-Line Maintenance Proceedings. ESMO 2000 Proceedings. Global ESMO 2000. The Pow*, 187–119. IEEE https://doi.org/10.1109/TDCLLM.2000.882820.

30 Lara-López, A. and Colı′ n-Venegas, J. (2001). Endurance of dampers for electric conductors. *International Journal of Fatigue* 23 (1): 21–28. https://doi.org/10.1016/S0142-1123(00)00072-4.

31 Hagedorn, P., Mitra, N., and Hadulla, T. (2002). Vortex-excited vibrations in bundled conductors: a mathematical model. *Journal of Fluids and Structures* 16 (7): 843–854. https://doi.org/10.1006/jfls.2002.0451.

32 Sier, R.J. and Riggio, J.C. (2003). Analysis of factors contributing to conductor failures on a 345 kV line. In: *2003 IEEE PES Transmission and Distribution Conference and Exposition (IEEE Cat.No. 03CH37495)*, vol. 3, 904–908. IEEE. doi: 10.1109/TDC.2003.1335057.

33 Smith, G.W. and Mailey, S. (2003). The Henday Radisson Dorsey 500 kV DC transmission line spacer damper rehabilitation project. In: *2003 IEEE 10th International Conference on Transmission and Distribution Construction, Operation and Live-Line Maintenance, 2003 IEEE ESMO*, 67–73. https://doi.org/10.1109/TDCLLM.2003.1196470.

34 Diana, G., Bocciolone, M., Cheli, F. et al. (2005). Large wind-induced vibrations on conductor bundles: laboratory scale measurements to reproduce the dynamic behavior of the spans and the suspension sets. *IEEE Transactions on Power Delivery* 20 (2): 1617–1624. https://doi.org/10.1109/TPWRD.2004.838473.

35 Baliberdin, L.L., Kozlova, M.A., and Shershnev, Y.A. (2005). Model group of controlled installation for melting ice on transmission line conductors. In: *2005 IEEE Russia Power Tech*, 1–5. IEEE https://doi.org/10.1109/PTC.2005.4524759.

36 Meynen, S., Verma, H., Hagedorn, P., and Schäfer, M. (2005). On the numerical simulation of vortex-induced vibrations of oscillating conductors. *Journal of Fluids and Structures* 21 (1): 41–48. https://doi.org/10.1016/j.jfluidstructs.2005.05.019.

37 Kong, D., Li, L., and Long, X. (2009). Numerical analysis of two-bundled conductor aeolian vibrations with spacer dampers. In: *2009 International Conference on Electronic Computer Technology*, 561–565. IEEE https://doi.org/10.1109/ICECT.2009.56.

38 Impollonia, N., Ricciardi, G., and Saitta, F. (2011). Vibrations of inclined cables under skew wind. *International Journal of Non-Linear Mechanics* 46 (7): 907–918. https://doi.org/10.1016/j.ijnonlinmec.2011.03.006.

39 Guerard, S., Godard, B., and Lilien, J.L. (2011). Aeolian vibrations on power-line conductors, evaluation of actual self damping. *IEEE Transactions on Power Delivery* 26 (4): 2118–2122. https://doi.org/10.1109/TPWRD.2011.2151211.

40 Langlois, S. and Legeron, F. (2013). Prediction of aeolian vibration on transmission-line conductors using a nonlinear time history model-part I: damper model. *IEEE Transactions on Power Delivery* 29 (3): 1168–1175. https://doi.org/10.1109/TPWRD.2013.2291363.

41 Dulhunty, P. (2013). Vibration dampers on AAC and AAAC conductors. In: *22nd International Conference and Exhibition on Electricity Distribution (CIRED* 2013*)*. https://doi.org/10.1049/cp.2013.0744.

42 Lu, M.L. and Chan, J.K. (2015). Rational design equations for the aeolian vibration of overhead power lines. In: *2015 IEEE Power & Energy Society General Meeting*, 1–5. IEEE https://doi.org/10.1109/PESGM.2015.7286498.

43 Gizaw, M., Davidson, I.E., Loubser, R., and Bright, G. (2016). Analyses of the vibration level of an OPGW at the catenary value of 2100m with multi-response Stockbridge dampers. In: *2016 IEEE PES Power Africa*, 107–111. IEEE https://doi.org/10.1109/PowerAfrica.2016.7556580.

44 Foti, F. and Martinelli, L. (2018). An enhanced unified model for the self-damping of stranded cables under aeolian vibrations. *Journal of Wind Engineering and Industrial Aerodynamics* 182: 72–86. https://doi.org/10.1016/j.jweia.2018.09.005.

45 Mukherjee, S., Chattaraj, S., Prasad, D. et al. (2021). Development of hexa spacer damper for 765 kV transmission lines' vibration damping. *Advances in Science, Technology and Engineering Systems Journal* 6 (1): 472–478. https://doi.org/10.25046/aj060151.

46 Kubelwa, Y.D., Swanson, A.G., Dorrell, D.G., and Papailiou, K.O. (2019). A comparative study on high-voltage spacer-damper performance and assessment: theory, experiments and analysis. *SAIEE Africa Research Journal* 110 (3): 153–166. https://doi.org/10.23919/SAIEE.2019.8732787.

47 Si, J., Rui, X., Liu, B. et al. (2019). Study on a new combined anti-galloping device for UHV overhead transmission lines. *IEEE Transactions on Power Delivery* 34 (6): 2070–2078. https://doi.org/10.1109/TPWRD.2019.2911894.

48 Peng, H., Wang, B., He, Q. et al. (2019). Multi-parametric optimizations for power dissipation characteristics of Stockbridge dampers based on probability distribution of wind speed. *Applied Mathematical Modelling* 69: 533–551. https://doi.org/10.1016/j.apm.2019.01.006.

49 Nie, X., Tan, T., Yan, Z. et al. (2020). Ultra-wideband piezoelectric energy harvester based on Stockbridge damper and its application in smart grid. *Applied Energy* 267: 114898. https://doi.org/10.1016/j.apenergy.2020.114898.

50 Agarwal, H., Mukherjee, S., Chattaraj, S. et al. (2020). Development of quad spacer damper to suppress oscillation in 765 kV transmission line. In: *2020 11th International Conference on Computing, Communication and Networking Technologies (ICCCNT)*, 1–6. IEEE https://doi.org/10.1109/ICCCNT49239.2020.9225394.

51 Roy, S. and Kundu, C.K. (2021). State of the art review of wind induced vibration and its control on transmission towers. *Structures* 29: 254–264. https://doi.org/10.1016/j.istruc.2020.11.015.

52 Li, S., Gao, W., Yurchenko, D. et al. (2022). Novel spacer-tuned-mass-damper system for controlling vibrations of hangers. *Mechanical Systems and Signal Processing* 167: 108537. https://doi.org/10.1016/j.ymssp.2021.108537.

14

Evaluation of Shortest Path to Optimize Distribution Network Cost and Power Losses in Hilly Areas: A Case Study

Subho Upadhyay[1], Rajeev K. Chauhan[1], and Mahendra P. Sharma[2]

[1]*Department of Electrical Engineering, Faculty of Engineering, Dayalbagh Educational Institute, Agra, Uttar Pradesh, India*
[2]*Department of Hydro and Renewable Energy, Indian Institute of Technology, Roorkee, Uttarakhand, India*

Acronyms

ACSR	aluminum conductor steel-reinforced
ArcGIS	geographic information system
BBO	biogeography based optimization
CCS	cycle charging strategy
C_{ct}	distribution network implementation cost
CDM	clean development mechanism
COE_{DN}	cost of energy of hybrid energy system including implemented distribution network cost
COE_T	total cost of energy of hybrid energy system
CRF	capital recovery factor
C_s	CO_2 emission
DA	Dijkstra algorithm
DEM	Digital Elevation Map
DG	distributed generation
EIR	energy index ratio
GA	Genetic Algorithm
HES	hybrid energy system
INR	Indian rupees
IWD	intelligent water drops
km	kilometer
kW	kilowatt
LFS	load following strategy
L_{SD}	sum of all the section lengths
NPV	Net Present Value
NR	Newton Raphson method
PSS	peak shaving strategy
SPV	solar photovoltaic
T_D	total distance of distribution network
TNPC	total net present cost
W_{ED}	weight of total emission from system containing only diesel generators (kg/yr)

Intelligent Data Mining and Analysis in Power and Energy Systems: Models and Applications for Smarter Efficient Power Systems, First Edition. Edited by Zita Vale, Tiago Pinto, Michael Negnevitsky, and Ganesh Kumar Venayagamoorthy.

W_{EDS} weight of total emission from system containing size of different diesel generating systems according to the selected dispatch strategy (kg/yr)

Y_k conductor cost per kilometer

14.1 Introduction

Grid interconnection and standalone energy generating systems are the only possible ways to connect and fulfil the electricity demand of distantly located remote areas. Standalone systems are preferred, where the cost of energy from the local generating system becomes less than the cost of drawing grid interconnections [1]. Suchitra et al. [2] proposed an optimally designed multi-source standalone hybrid generating system comprising of photovoltaic panels, wind generating system, batteries, and diesel generator. It is designed to minimize the emissions and cost, expressed in terms of Net Present Value (NPV) of the system while maximizing Energy Index Ratio (EIR). The system is further integrated with distribution network to reduce the total losses and voltage deviation of the distribution system. Moradi and Abedini [3] developed a combined method based on Genetic Algorithm (GA) and Intelligent Water Drops (IWD) to find the location and capacity of distributed generation (DG). The work emphasizes on reducing the power loss, improving the voltage regulation, and increasing the voltage stability. Kansal et al. [4] introduced a hybrid analytical method and heuristic search for the optimal placement of the DGs in power distribution network to reduce the power loss.

Implementation of a standalone hybrid energy system in such an area involves estimation of the available demand and resources in its vicinity [5, 6]. Scheduling of the resources to fulfil the demand is the second step. It is followed by determining optimal size of generators so as to minimize the cost of energy and maximize the energy index ratio [7]. After finding the size of the generators, these are required to be placed at optimal locations. As the load/generating centers are located in an uneven hilly surface, minimum distance among them changes as compared to places located in plains and hence the determination of shortest path between the load centers or generating stations is essential. For this purpose, biogeography-based optimization algorithm (BBO) is selected to determine the nodes through which the shortest path between the load and generating centers [7, 8] will pass. The results obtained using the BBO are compared with Dijkstra Algorithm (DA), which is a classical technique for finding shortest path and requires large search storage [9–11]. The shortest path is where the distribution system will be laid down. Newton Raphson method (NR) for Load flow analysis on the distribution system is undertaken to find out the minimum power loss. The selection of NR method is due to its accuracy and convergence compared to other load flow analysis methods [12, 13]. The load flow analysis helps to select the aluminum conductor steel-reinforced cable (ACSR) conductor for distribution system. The cost of drawing ACSR conductor is added in the overall Hybrid Energy System (HES) cost of energy [7]. The CO_2 emitted from the diesel generator [14] is also considered in terms of cost of energy in different energy management strategies according to the environment accord. Finally, the sensitivity analysis of the overall system is performed.

14.2 Design of Power Distribution Network

Designing a power distribution network for rural area requires selection of conductor based on evaluated cost and power loss associated with HES. The designing process involves extracting Digital

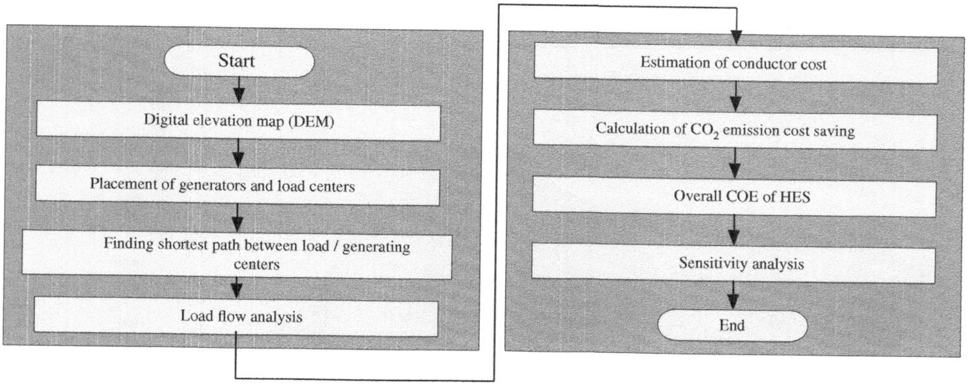

Figure 14.1 Flowchart for designing power distribution network.

Elevation Map (DEM) of the area. It is followed by placing of load and generating centers. According to the location of the load and generating centers, the shortest path between a section (i.e. two load/generating centers) is determined using BBO technique and the results are compared with Dijkstra algorithm. The total evaluated distance is applied to Newton Raphson method for load flow analysis for selecting the type of the ACSR. Selection of ACSR depends on the INR/km price and power loss derived from Newton Raphson method. The total COE (COE_T) of system is derived from implementation cost associated with distribution network, cost savings achieved in terms of CO_2 emitted from diesel generator and net present cost of HES. Finally, sensitivity analysis on the HES is evaluated. The flowchart of designing a power distribution network is shown in Figure 14.1.

14.3 Digital Elevation Map

Geographic information system (ArcGIS) is used to generate elevation of the equidistant locations obtained from MATLAB. The equidistant elevation data with 100-m cross-section are extracted and used to generate DEM. The elevation data corresponding latitude and longitude is represented by DEM as shown in Figure 14.2, along with load and generating centers.

14.4 Placement of Generators and Load Centers

The load centers, generating units, interfacing devices, and batteries are placed in different locations of study area. Figure 14.2 is represented in two-dimensional spaces as shown in Figure 14.3, to find location of the generating center. The generating center consists of solar photovoltaic (SPV), DG, interfacing devices, and battery bank. The locations of load centers are represented in DEM as shown in Figure 14.2.

Hydro potential of 15 kW is available perennially near Nailpar village. Stream flows continuously and hence, a micro hydropower system can be placed in the location. The biomass crop residue and forest foliage of 91.91 and 35.65 ton/yr, respectively, are accumulated over a year. The biomass potential can run a 30 kW biomass gasifier for 10 h/d. Biomass gasifier of 15 kW each is to be placed in Seli and Naugaon villages. Seli village is located at the center of Sirani, Tapri, Chimkholi, and Nailpar villages, hence is able to support a15 kW biomass gasifier. On the other hand, Naugaon

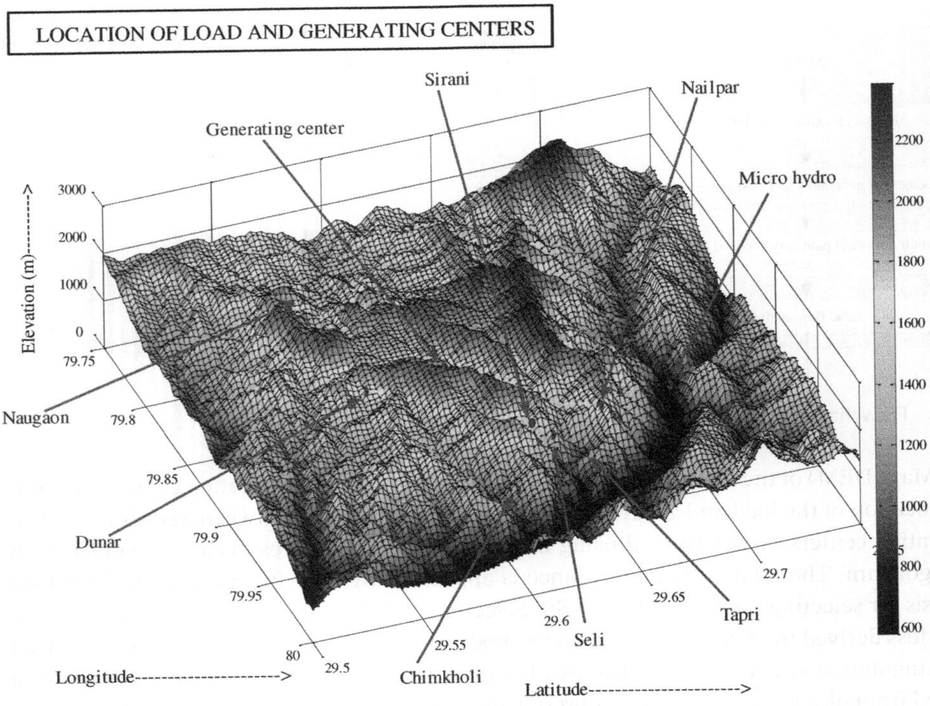

Figure 14.2 Digital elevation map of the site (12 × 10.4 km).

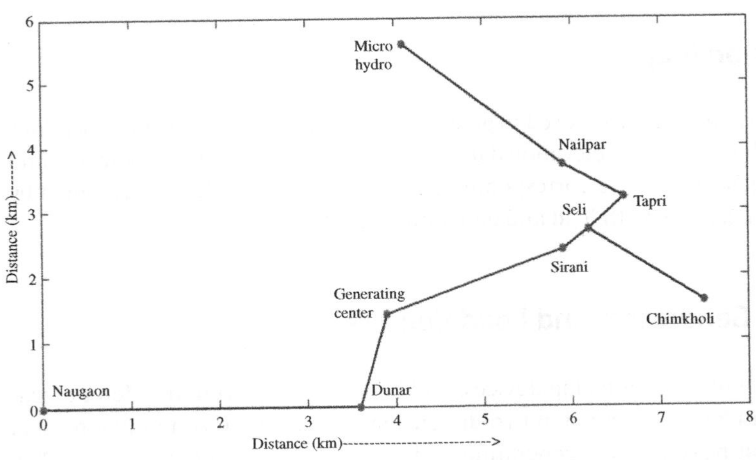

Figure 14.3 Representation of load/generating centers in a two-dimensional space.

is most distant load center of the selected area. Therefore, to maintain voltage stability another 15 kW gasifier is required to be placed. The Naugaon–Dunar cluster is the largest load center and the available biomass in the area is sufficient to run a 15 kW biomass gasifier.

Biogas of 248.82 m^3/d is available collectively in the load centers, which can run a 35 kW of biogas generating system for 9 h/d. According to the biogas availability three 10 kW generators and one 5 kW generator are placed in Dunar, Naugaon, Tapri, and Nailpar load centers, respectively.

Table 14.1 Coordinates of the selected villages and maximum load required at 22nd hour during summer.

S. No.	Load centers	Location of the load/generating centers (km)
1	Generating center	(3.9, 1.4)
2	Sirani	(5.9, 2.4)
3	Seli	(6.2, 2.7)
4	Chimkholi	(7.5, 1.6)
5	Tapri	(6.6, 3.2)
6	Nailpar	(5.9, 3.7)
7	Hydro	(4.1, 5.6)
8	Dunar	(3.6, 0)
9	Naugaon	(0, 0)

The Barycenter formula (Eqs. (14.1) and (14.2) [15]) is used to find location of generating systems such as diesel generator, SPVs, and interfacing devices such as bidirectional converters, boost converters, charge controllers, and batteries, to be placed in the DEM.

$$x = \frac{P_1 \times X_1 + P_2 \times X_2 + \dots P_n \times X_n}{P_1 + P_2 + \dots P_n} \tag{14.1}$$

$$y = \frac{P_1 \times Y_1 + P_2 \times Y_2 + \dots P_n \times Y_n}{P_1 + P_2 + \dots P_n} \tag{14.2}$$

where

$P_1, P_2, \dots P_n$ are the real peak powers at the load centers during 22nd hour of a summer day,
$X_1, X_2, \dots X_n$ are the x coordinates at these locations,
$Y_1, Y_2, \dots Y_n$ are the y coordinates at these locations.

Location of Naugaon village is considered to be located at ($x = 0$, $y = 0$) coordinate and with respect to this village, other villages are placed in x–y axis (Figure 14.3 and Table 14.1). By solving the Eqs. (14.1) and (14.2), location of generating center is $x = 3.9$, $y = 1.4$, and is marked in Figures 14.2 and 14.3.

14.5 Single Line Diagram of 9-Bus System

A single line diagram of the generating and load centers during peak load at 22nd hour of the summer season is represented in Figure 14.4. Here bus number 1 contains the generating center, bus 2 is Sirani load center, bus 3 represents Seli load and generating center, bus 4 contains Chimkholi village as load center, bus 5 is Tapri load and generating center, bus 6 is Nailpar load and generating center, bus 7 represents micro hydro site, bus 8 is Dunar load and generating center, and bus 9 is Naugaon load and generating center.

The real and reactive power available from generating center and required by the load centers at 22nd hour shown in Table 14.2. The type of conductors used for a 3-phase 4-wire distribution network is shown in Table 14.3, with their series impedance, area and weight values. According to data available the load flow analysis is developed in the Section 14.7.

Table 14.2 Real and reactive power values on the buses at peak load (22nd hour summer).

Bus number	Village name/ generation type	Voltage (pu)	Angle	Load		Generation		Generation-load	
				P_l (kW)	Q_l (kVAR)	P_g (kW)	Q_g (kVAR)	$P_{(g-l)}$ (kW)	$Q_{(g-l)}$ (kVAR)
1	SPV and diesel generators (G)	1.05	0	0	0	22.9	11.09	22.9	11.09
2	Sirani	1.05	0	−3.7	−1.79	0	0	−3.7	−1.79
3	Seli	1.05	0	−8.91	−4.32	15	7.27	6.09	2.95
4	Chimkholi	1.05	0	−4.69	−2.27	0	0	−4.69	−2.27
5	Tapri	1.05	0	−3.87	−1.87	10	4.84	6.14	2.97
6	Nailpar	1.05	0	−25.53	−12.36	5	2.42	−20.53	−9.94
7	Micro hydro generator (G)	1.05	0	0	0	15	7.27	15	7.27
8	Dunar	1.05	0	−29.46	−14.27	10	4.84	−19.46	−9.43
9	Naugaon	1.05	0	−25.93	−12.56	25	12.11	−0.93	−0.45

Table 14.3 Type of conductor and series impedance (Ω/km) [16].

Conductor (code name)	Series impedance (Ω/km)	Area (m^2)	Weight (kg/km)
WEASEL	$0.9047 + 0.308i$	36.9	128
FERRET	$0.676 + 0.298i$	49.5	172
RABBIT	$0.5404 + 0.291i$	61.7	214
MINK	$0.454 + 0.285i$	73.7	255
BEAVER	$0.365 + 0.2795i$	87.5	302
OTTER	$0.3418 + 0.2725i$	97.9	339
CAT	$0.3005 + 0.2725i$	111.3	386
HARE	$0.2722 + 0.2695i$	122.5	425
LEOPARD	$0.2177 + 0.2625i$	148.5	492

The distribution system is considered to be radial in nature and balanced.
The lower and upper voltage limits are set at 0.95 and 1.05 pu.
Load level is constant during the simulation.

14.6 Finding Shortest Path Between Load/Generating Centers

Distribution network supplies power from generating centers to load centers present in the locality. Villages and generating stations are located at different elevation due to hilly terrain and hence, shortest path is to be determined for minimizing the distribution power losses and reduce distribution network implementation cost. The shortest path between all the load/generating centers is calculated using BBO and results are compared with that obtained from DA. BBO is used to find shortest path between source and destination centers on the DEM. The working of algorithm is explained in the Section 14.6.1. Total distance between source and destination centers is the sum of the distances between intermediate nodes.

14.6.1 Objective Function

Objective function involves minimizing the total distance between source and destination centers (L_{SD}). The Minimum length depends on the sum of the lengths (l_{ij}) between all the intermediate nodes n_i and n_j,

$$L_{SD} = \sum_{a=1}^{a=m} l_{ij} \tag{14.3}$$

here n_i and n_j are the lower node adjacent upper node respectively
where a is the intermediate section.

Here, n_i node is the number of intermediate nodes in a set of latitudes or longitudes. If the number of longitudes is more than latitudes in a section, then the number of n_i becomes equal to the number of longitudes, while the position of the nodes varies along the latitudes and vice versa.

Step 1: Selecting the section comprising of source and destination nodes
Developed algorithm initiates by selecting a section containing source and destination centers. A three-dimensional representation of the area containing generating center and Sirani village as load center is presented in Figure 14.5. The distance between consecutive latitude and longitude is 100 m.

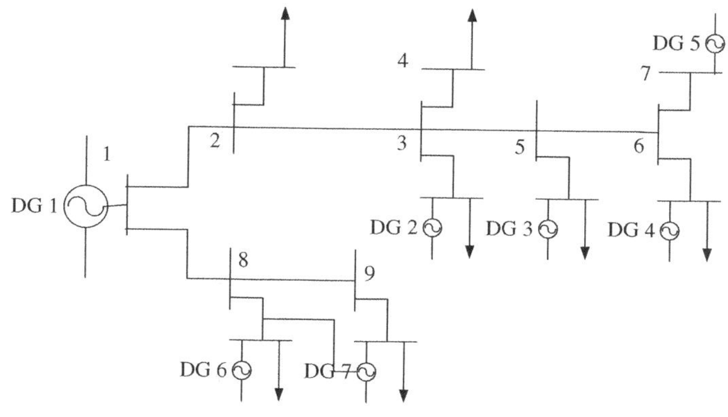

Figure 14.4 Single-line diagram of 9-bus system including generators and load centers.

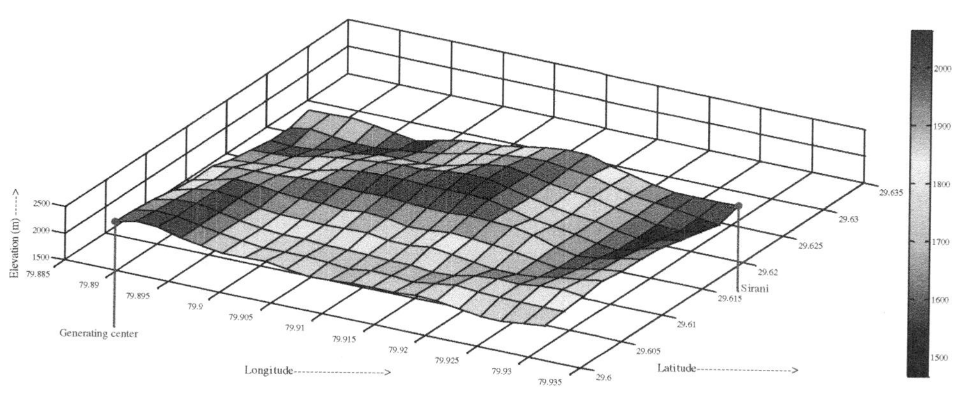

Figure 14.5 Section representation between generating center and Sirani load center.

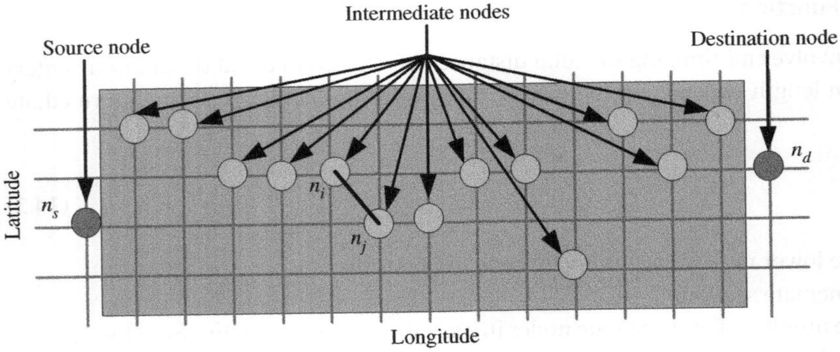

Figure 14.6 Section representation between source and destination.

Step 2: Initialize intermediate nodes

Selection of section area is followed by initializing the location of intermediate nodes along the longitude or latitude. If number of longitudes is more than the number of latitudes in the section, position of intermediate nodes is varied according to longitudes or vice versa. Number of longitudes, as shown in Figure 14.5, is more than number of latitudinal lines for the section between generating center and Sirani village. Hence, total number of intermediate nodes is equal to number of longitudes, while position of nodes varies along the latitudes. Role of BBO is to assist in selecting intermediate nodes (i.e. latitudes in this case) through which the distribution network will be constructed.

Step 3: Finding distance between intermediate nodes

If n is the node representing each village/generating center or intermediate points as shown in Figure 14.6, then $n_i{\cdot}x$, $n_i{\cdot}y$, and $n_i{\cdot}z$ are the surface coordinates of the node. Here n_s is represented as the starting node and n_d is the destination node. The projected distance from n_s to n_d is calculated as sum of distances between consecutive nodes which is given as [9]:

$$d(n_i, n_j) = \sqrt{(n_i \cdot x - n_j \cdot x)^2 + (n_i \cdot y - n_j \cdot y)^2} \tag{14.4}$$

n_i and n_j are the consecutive nodes of latitude/longitude, whose dimension is greater. The elevation difference is represented as n_i and n_j

$$\nabla(n_i, n_j) = (n_i \cdot z - n_j \cdot z) \tag{14.5}$$

Finally, the Euclidean distance between n_i and n_j in a three-dimensional space is calculated as:

$$l_{ij} = \sqrt{d(n_i, n_j)^2 + \nabla(n_i, n_j)^2} \tag{14.6}$$

l_{ij} denotes the length between nodes n_i and n_j.

Step 4: Total distance for constructing distribution network

Sum of all the lengths present in the sections provide the total distance (T_D) of constructing distribution network, while b is total number of sections.

$$T_D = \sum_{b=1}^{b=m} L_{SD} \tag{14.7}$$

where, T_D is the sum of all the section lengths (L_{SD}). The shortest path between generating center and Sirani load center is shown in Figure 14.7.

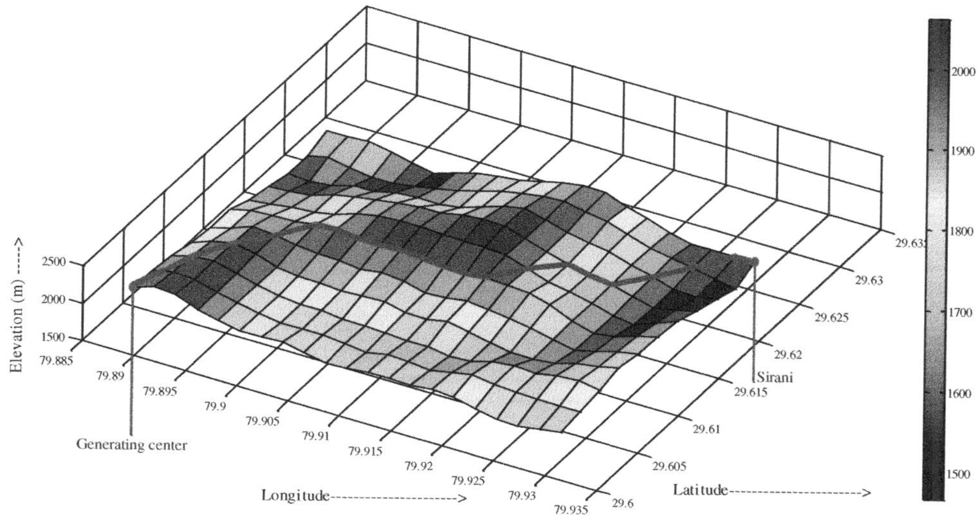

Figure 14.7 Section representation between generating center and Sirani load center with distribution network.

Table 14.4 Distances between each section using DA and BBO.

S. No.	From	To	DA (km)	BBO (km)
1	Naugaon	Dunar	4.12	4.04
2	Dunar	Generating center	1.89	1.64
3	Generating center	Sirani	3.12	2.56
4	Sirani	Seli	0.55	0.49
5	Seli	Tapri	1.18	0.93
6	Tapri	Nailpar	1.25	1.14
7	Seli	Chimkholi	2.47	2.03
8	Nailpar	Micro hydro	4.19	3.23
Total distribution network distance			18.8	16.05

14.6.2 Distribution Network Distance

The shortest distances between the villages/generating center are optimized and the results are shown in Figure 14.8 and Table 14.4. The distances of each section obtained from BBO are minimum as compared to DA algorithm as shown in Table 14.4. Naugaon to Dunar load center distance is greatest with 4.04 km, while the distance between Sirani to Seli load centers is lowest with 0.49 km. The distance between Dunar and generating center is 1.64 km, generating center to Sirani is 2.56 km, Seli to Tapri is 0.93 km, Seli to Chimkholi is 2.03 km, Tapri to Nailpar is 1.14 km, and Nailpar to micro hydro site is 3.23 km using BBO. It is the sum of all the calculated section distances. The total distribution system length remains constant in case of DA, while the overall distance value decreases with the increase in iteration count using BBO.

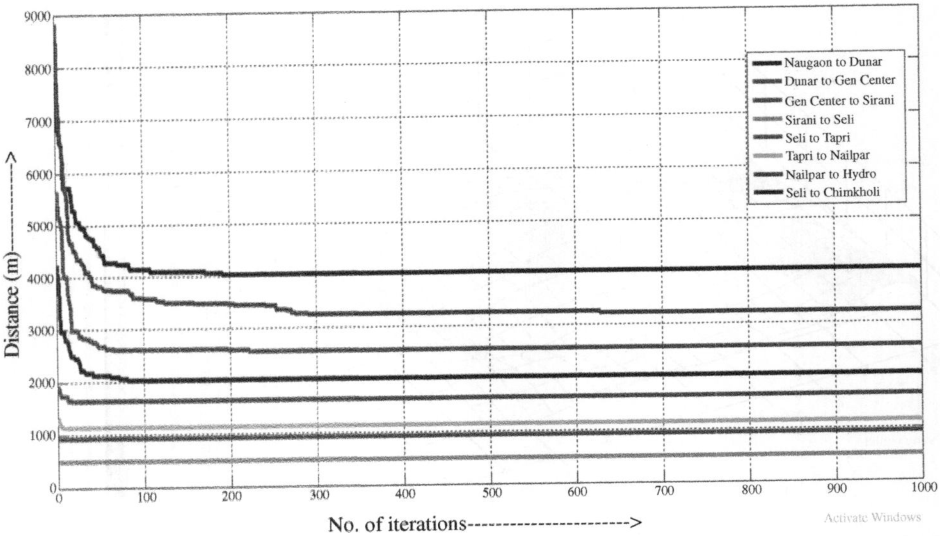

Figure 14.8 Optimized results of shortest distance between each section using BBO.

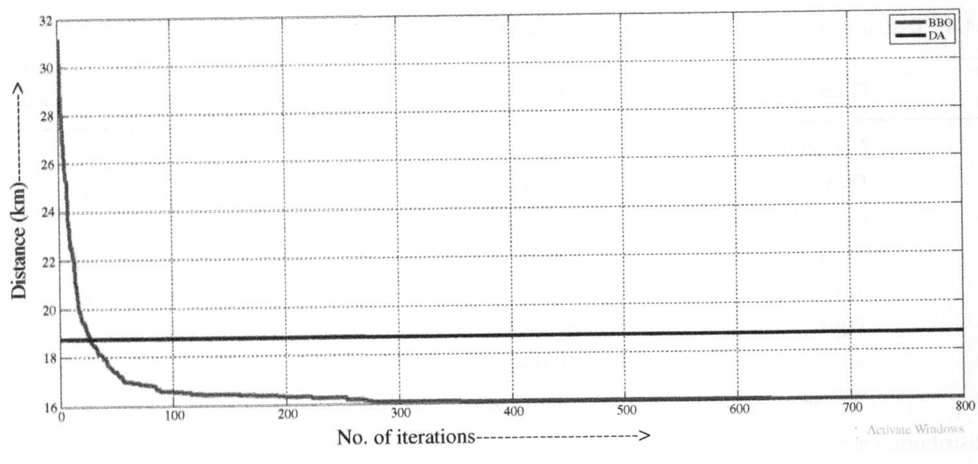

Figure 14.9 Optimized results of overall shortest distance using BBO and DA techniques.

The overall comparative distance of distribution network is 16.05 and 18.8 km in case BBO and DA, respectively, as shown in Figure 14.9.

The total distance to draw distribution network is 16.05 km in length, as obtained from BBO. A three-dimensional representation with elevation data of the area along with the shortest distance for laying electrical distribution network is shown in Figure 14.10.

14.7 Selection of Conductor Using Newton Raphson Method

The Newton Raphson method is used to check and limit violations according to assumptions considered. A 9-bus radial distribution system is considered as shown in Figure 14.4. Here I_i is the

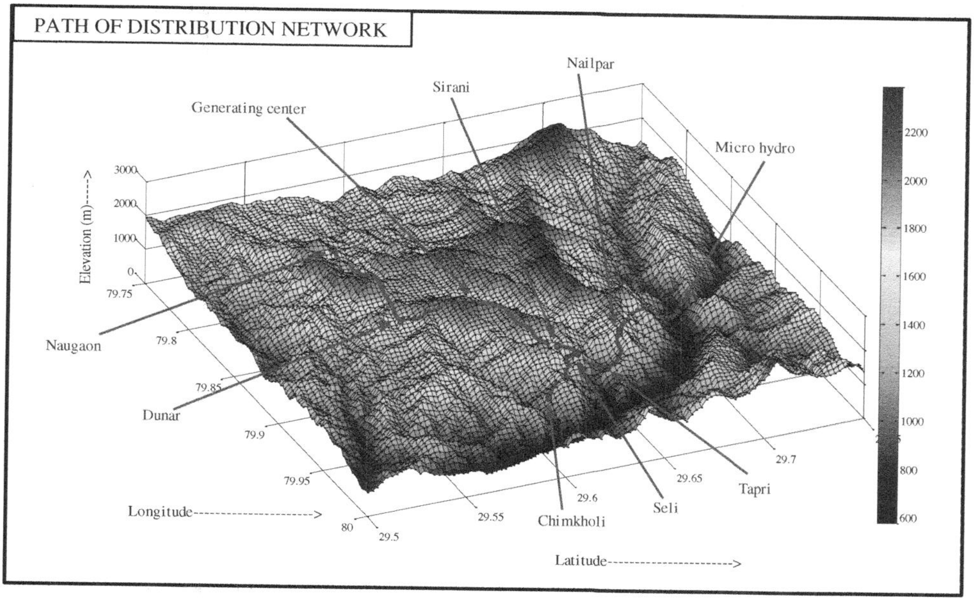

Figure 14.10 Shortest route for laying electrical distribution network. Source: Upadhyay et al. [17]/ with permission of IEEE.

current in the *i*th branch. The total power loss of system, PL can be given as [18].

$$P_L = \sum_{i=1}^{N-1} I_i^2 \times R_i = \sum_{i=1}^{N-1} \left(I_{ai}^2 + I_{ri}^2\right) \times R_i \tag{14.8}$$

Here, I_{ai} and I_{ri} are the real and imaginary components of current, respectively.

Conductor cost includes the total conductor required to be installed in the area along with the labor and other costs. Table 14.5 displays the results of total power loss and cost of laying a type of 3-phase 4-wire distributed system.

Table 14.5 ACSR conductor specification of 3-phase 4-wire system [19].

Conductor (code name)	Cost (INR/km)	Total system power loss (kW)	Total cost for implementing distribution system (INR)
WEASEL	314 000	13.91	5 039 700
FERRET	421 000	8.7	6 757 050
RABBIT	525 000	6.36	8 426 250
MINK	627 000	5.01	10 063 350
BEAVER	745 000	3.73	11 957 250
OTTER	833 000	3.41	13 369 650
CAT	947 000	2.85	15 199 350
HARE	1 042 000	2.48	16 724 100
LEOPARD	1 264 000	1.78	20 287 200

Adapted from Ref. [19].

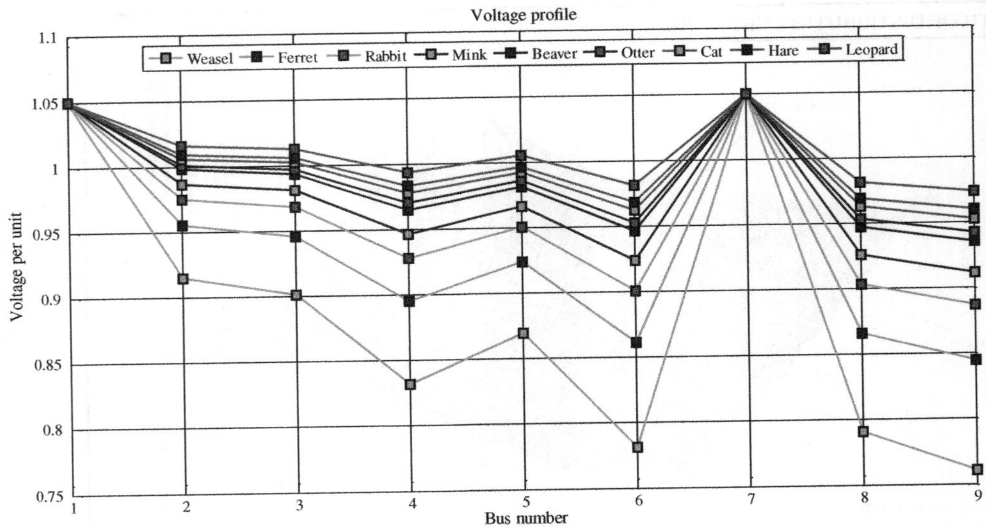

Figure 14.11 Conductors with voltage per unit values at different bus numbers.

Table 14.6 Emission of CO_2 from diesel generator for CCS, LFS, PSS, and only diesel operating system.

Dispatch strategy	Cycle charging strategy (CCS)	Load following strategy (LFS)	Peak shaving strategy (PSS)	Only diesel system
Weight (kg/yr)	14 614	6 813	21 769	422 740

The voltage profile of all the conductors is shown in Figure 14.11. The 6th, 8th, and 9th buses have minimum voltage pu value compared to other buses obtained through simulation. The selection of conductor depends on its minimum voltage of 0.9 pu at each bus.

14.7.1 Estimation of Conductor Cost

The total distance (T_D) for laying distribution network is 16.05 km as shown in Table 14.6. The maximum permissible power loss considered in the distribution system must not exceed 10% of the peak demand, while considering the voltage values of above 0.9 pu at all the buses. This forms the basis of selecting CAT conductor for power distribution network. The CAT conductor has a power loss is 2.85 kW, which is less than 10% of the peak demand of 102.1 kW, as shown in Table 14.5. The distribution network implementation cost (C_{ct}) for laying a CAT 3-phase 4-wire conductor is the product of conductor cost per km (Y_k) to the total minimum evaluated distance (T_D).

$$C_{ct} = Y_k \times T_D \tag{14.9}$$

The cost associated for laying CAT conductor is 15 199 350 INR. The COE_{DN} is the cost of energy associated with distribution network implementation cost (C_{ct}) along with net present cost of hybrid energy system. The cost of energy (COE_{DN}) thus evaluated is 7.36 INR/kWh in case of cycle

Table 14.7 Emission of CO_2 in terms of cost saved using CCS, LFS, and PSS [7].

Dispatch strategy	Cycle charging strategy (CCS)	Load following strategy (LFS)	Peak shaving strategy (PSS)
Cs (INR)	6 121 878	6 238 893	6 014 558

Source: Adapted from Upadhyay and Sharma [7].

charging strategy as shown in Table 14.6. Here total net present cost (TNPC) and capital recovery factor (CRF) values are taken from [7].

$$COE_{DN} = (C_{ct} + TNPC) \times CRF \Big/ \sum_{t=1}^{8760} E_{Gen}(t) \tag{14.10}$$

14.8 Calculation of CO_2 Emission Cost Saving

The saving of cost in terms of CO_2 emission (Cs) on account of developed HES has been calculated in terms of cost for all the three strategies. In the clean development mechanism (CDM) each ton of CO_2 is sold at US$ 11 [14]. The difference in total cost is determined by considering emission of CO_2 (ton/yr) from a standalone diesel generating system, fulfilling all the demand to the emission by diesel generator selected from these strategies (Tables 14.6 and 14.7), of the developed HES. After applying CDM, it is found that the cost saving for LFS is minimum compared to CCS and PSS as shown in Table 14.7. To determine the CO_2 emission cost saving, four set of 30 kW diesel generators can be operated, depending on the demand for an hour. To fulfil the required demand four set of diesel generators should be operated during 12th hour and from 19th to 22nd hour, while only one diesel generator can meet the demand during the morning hours as shown in Figure 14.12.

If only diesel generators are used, then total annual energy generated will be 547 590 kWh/yr. The total CO_2 emission in this case will be 422 740 kg/yr. The CO_2 emission in terms of weight for different operation strategies is represented in Table 14.6.

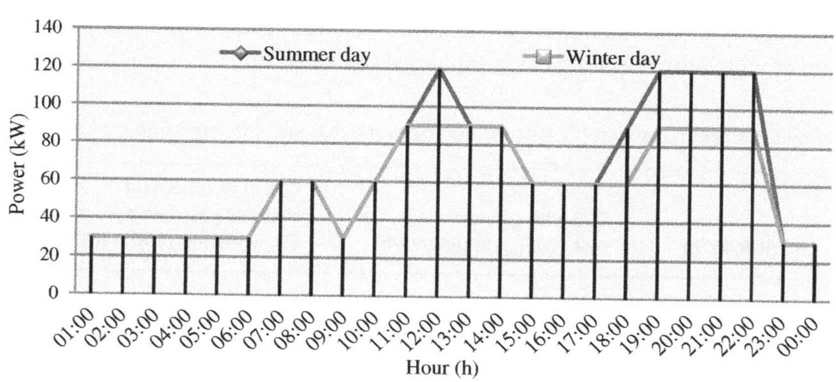

Figure 14.12 Diesel generating system operation over a summer and winter day.

The CO_2 emission cost saving is equal to the difference between emission produced by only diesel generating system and HES using dispatch strategies, with product of lifetime of the system and US$ 11 (US$ 1~60 INR) of penalty charge per ton of CO_2 emission [14].

$$C_S = \frac{(W_{ED} - W_{EDS}) \times N \times 660}{1000} \tag{14.11}$$

where

C_S the total cost saved from CO_2 emission for 25 years period (ton)

W_{ED} the weight of total emission from system containing only diesel generators (kg/yr)

W_{EDS} the weight of total emission from system containing size of different diesel generating systems according to the selected dispatch strategy (kg/yr)

N lifetime of the system (25 years)

The total cost in terms of CO_2 saved is shown in Table 14.7.

14.9 Overall Cost Estimation of Distribution System

The total cost of energy of distributed energy system is the sum of cost saved from CO_2 emission, with total net present cost of HES and distribution system implementation cost.

$$COE_T = ((C_{ct} + TNPC - C_S) \times CRF) / \sum_{t=1}^{8760} E_{Gen}(t) \tag{14.12}$$

Here, COE_T is the total cost of energy of the hybrid energy system, as given in Table 14.8. The cost of energy is minimum using CCS by applying distribution network cost and CO_2 emission cost saving. The COE_T for CCS is 7.19 INR/kWh followed by LFS with 7.27 INR/kWh and PSS with 7.5 INR/kWh.

14.10 Sensitivity Analysis

Sensitivity analysis shows the effect of HES by changing system parameters. The change in diesel fuel price, demand, availability of solar radiation and change in energy index ratio are important parameters, on account of which the effect of HES is explained in this section.

Table 14.8 Cost of energy after optimal sizing [7], distribution network cost and CO_2 emission cost saving.

Dispatch strategy	COE (INR/kWh)	COE with conductor cost (COE_{DN}) in (INR/kWh)	COE with conductor cost and cost saved from CO_2 emission (COE_T) (INR/kWh)
Cycle charging	5.64	8.23	7.19
Load following	5.76	8.32	7.27
Peak shaving	5.91	8.54	7.5

Source: Adapted from Upadhyay and Sharma [7].

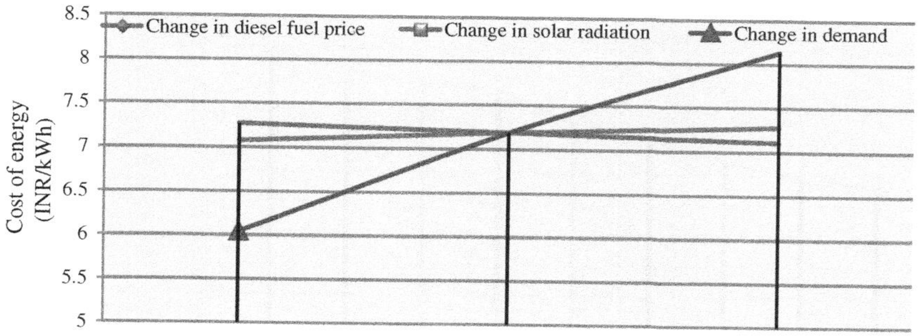

Figure 14.13 Sensitivity analysis by variation in demand (kW), diesel fuel price (INR/l), and solar radiation (kW/m^2).

14.10.1 Change in Diesel Fuel Price

In the present global scenario, the fuel prices keep on fluctuating. Hence the diesel fuel price (INR/l) is varied 10% from the present price per liter value. If the fuel price is reduced below 10% of present price, the COE of HES reduces to 7.08 INR/kWh and increases to 7.28 INR/kWh if diesel fuel price rise by 10% per liter.

14.10.2 Change in Solar Radiation

The effect of global warming has a drastic impact in the planet's weather pattern and hence the wind pattern, rainfall and solar radiation also changes. Hence the change of solar radiation impacts the output of SPV system. The decrease in solar radiation increases the pressure to select diesel generator for supplying the demand. Hence, the total COE_T of the HES increases to 7.27 INR/kWh and decreases to 7.11 INR/kWh as the solar radiation received reduced to 5% [20] and increased to 5% of the available radiation, respectively.

14.10.3 Change in Demand

The migration of the people is also becoming very evident. With the passage of time people are becoming more aware about the various opportunities available in cities. This is causing people to migrate with the reduction in the overall demand. If the demand decreases by 10% the COE_T decreases to 6.04 INR/kWh, while COE_T increases to 8.13 INR/kWh as the demand increases by 10%. Figure 14.13 shows the change in diesel fuel price, variation in solar radiation and change in demand and its impact in the overall COE_T.

14.10.4 Change in Energy Index Ratio

Energy index ratio is the reliability parameter inversely proportional to COE. If EIR is varied from 0.9 to 1, COE_T increases from 6.09 to 7.19 INR/kWh. The variation in EIR and its impact on COE is shown in Figure 14.14.

14.11 Conclusion

The article begins with placement of generators according to availability of resources and demand. DEM of the area is developed using ArcGIS to find the shortest path on the hilly area. The elevation

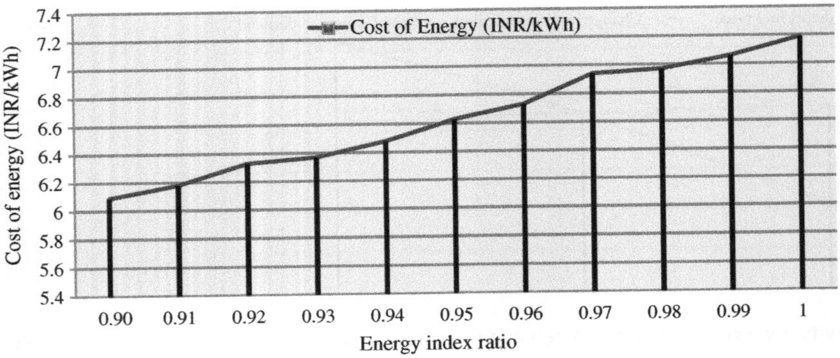

Figure 14.14 Energy index ratio vs. cost of energy.

data with 100 m cross-section is obtained and a grid of 12 × 10.4 km is generated, containing all the villages and hydro site. Shortest paths between the load/generation centers are calculated for implementing distribution network using BBO and DA. The shortest path in hilly area for laying distribution network is found minimum in case of BBO having 16.05 km whereas by utilizing DA technique it is 18.8 km.

A three-phase four-wire system consisting of a CAT conductor is selected for the distribution system using Newton Raphson method. The voltage profile of conductors above CAT cannot be selected as their voltage per unit value is below 0.95. The COE_T of the system by including the distribution network cost and clean development mechanism is 7.19 INR/kWh using cycle charging dispatch strategy.

The developed analysis provides a way to design power distribution network in a hilly area. The shortest path is needed to optimize the distance between the buses. It also provides a means to achieve conductor selection for the implementation of the distribution network. The overall scenario is evaluated in terms of cost of energy of the overall system. Finally, a sensitivity analysis is presented for understanding the variation in various parameters on the cost of energy of the system.

The model is developed on the basis of peak load of the system. Hence, fluctuations can be considered for more detailed understanding. The area is forested, hence if forest clearance is not available then the area can be marked as nonconstructed zone. In such condition, the area under the forest cover can be removed, for new evaluation of shortest path.

References

1 Gill-Wiehla, A., Miles, S., Wu, J., and Kammen, D.M. (2022). Beyond customer acquisition: a comprehensive review of community participation in mini grid projects. *Renewable and Sustainable Energy Reviews* 153: 1–14.

2 Suchitra, D., Jegatheesan, R., and Deepika, T.J. (2016). Optimal design of hybrid power generation system and its integration in the distribution network. *International Journal of Electrical Power & Energy Systems* 82 (1): 136–149.

3 Moradi, M.H. and Abedini, M. (2016). A novel method for optimal DG units capacity and location in microgrids. *Electrical Power and Energy Systems* 75 (1): 236–244.

4 Kansal, S., Kumar, V., and Tyagi, B. (2016). Hybrid approach for optimal placement of multiple DGs of multiple types in distribution networks. *Electrical Power and Energy Systems* 75 (1): 226–235.

5 Singh, R. and Bansal, R.C. (2018). Review of HRESs based on storage options system architecture and optimisation criteria and methodologies. *Renewable Power Generation IET* 12 (7): 747–760.

6 Upadhyay, S. and Sharma, M.P. (2014). A review on configurations, control and sizing methodologies of hybrid energy systems. *Renewable and Sustainable Energy Reviews* 38 (1): 47–63.

7 Upadhyay, S. and Sharma, M.P. (2016). Selection of a suitable energy management strategy for a hybrid energy system in a remote rural area of India. *Energy* 94 (1): 352–366.

8 Roy, P., Pal, S.C., Arabameri, A. et al. (2021). Climate and land use change induced future flood susceptibility assessment in a sub-tropical region of India. *Soft Computing* 25: 5925–5949.

9 Ganganath, N., Cheng, C.T., and Tse, C.K. (2015). A constraint-aware heuristic path planner for finding energy-efficient paths on uneven terrains. *IEEE Transactions on Industrial Informatics* 11 (3): 601–611.

10 Sunita and Garg, D. (2021). Dynamizing Dijkstra: a solution to dynamic shortest path problem through retroactive priority queue. *Journal of King Saud University – Computer and Information Sciences* 33 (3): 364–373.

11 Ngo, T.G., Dao, T.K., Thandapani, J. et al. (2021). Analysis urban traffic vehicle routing based on Dijkstra Algorithm Optimization. In: *Communication and Intelligent Systems* (ed. Sharma, H., Gupta, M.K., and Tomar, G.S., Lipo, W.), Lecture Notes in Networks and Systems, vol. 204. Springer, Singapore. https://doi.org/10.1007/978-981-16-1089-9_7.

12 Mohamad, H., Zakaria, Z., and Mazlan, M.Z.B. (2016). Development of GUI power system load flow analysis tool based on Newton Raphson method. In: *Proceedings of IEEE 7th International Conference on Engineering Education*, Kanazawa, Japan, 17–18 Nov. 2015, 29–34.

13 Zhang, X.S., Liu, Z., Yu, E.K., and Chen, J.C. (1998). A comparison on power flow calculation methods for distribution network. *Power System Technology* 22: 45–49.

14 UNFCCC. Potential of carbon markets. https://unfccc.int/files/cooperation_and_support/financial_mechanism/application/pdf/potential_of_carbon_matkets.pdf.

15 IET Setting Standards for Electrical Isolation. (2016). Energy efficiency of low voltage electrical installations. p. 62.

16 ACSR. (2016). Midal cables aluminium conductor steel reinforced (ACSR) specifications. https://www.midalcable.com/sites/default/files/ACSR-metric.PDF (accessed 9 June 2019).

17 Upadhyay, S., Chaudhary, A.S., Kumar, M., and Sharma, M.P. (2018). Evaluating the shortest path for laying distribution network in a hilly area. In: *8th IEEE India International Conference on Power Electronics (IICPE)*, MNIT Jaipur, India, 13–15 Dec. 2018, 1–5.

18 Viral, R. and Khatod, D.K. (2015). An analytical approach for sizing and siting of DGs in balanced radial distribution networks for loss minimization. *Electrical Power and Energy Systems* 67 (1): 191–201.

19 Administrative Staff College of India. (2010). Capital cost benchmarks for distribution business – Administrative Staff College of India report. http://www.forumofregulators.gov.in/Capital-cost-banchmarking.aspx (accessed 29 June 2019).

20 Katsigiannis, Y.A., Georgilakis, P.S., and Karapidakis, E.S. (2012). Hybrid simulated annealing-Tabu search method for optimal sizing of autonomous power systems with renewable. *IEEE Transactions on Sustainable Energy* 3 (3): 330–338.

15

Intelligent Approaches to Support Demand Response in Microgrid Planning

Rahmat Khezri[1], Amin Mahmoudi[1], and Hirohisa Aki[2]

[1]*College of Science and Engineering, Flinders University, Adelaide, SA, Australia*
[2]*Faculty of Engineering, Information and Systems, University of Tsukuba, Tsukuba, Ibaraki, Japan*

Acronyms

AC	air conditioner
AI	artificial intelligence
ANN	artificial neural network
ARM	association rule mining
BESS	battery energy storage system
CPP	critical peak pricing
DER	distributed energy resource
DG	diesel generator
DLANN	deep leaning artificial neural network
DR	demand response
DRER	distributed renewable energy resource
DSM	demand side management
ESS	energy storage system
FC	fuel cell
HPWH	heat pump water heater
ICT	information communication technology
PV	photovoltaic
RTP	real time pricing
SoC	state of charge
SVR	support vector regression
TDR	thermostatically domestic refrigerators
ToU	time of use
WT	wind turbine

15.1 Introduction

Recently, microgrids are known as inevitable parts of smart power grids by reliability improvement, resiliency enhancement, and financial sustainability of electricity supply in distribution networks.

Intelligent Data Mining and Analysis in Power and Energy Systems: Models and Applications for Smarter Efficient Power Systems, First Edition. Edited by Zita Vale, Tiago Pinto, Michael Negnevitsky, and Ganesh Kumar Venayagamoorthy.

A group of distributed energy resources (DERs) and load centers are interconnected to constitute a microgrid entity [1]. The boundary of microgrid should clearly be defined to indicate the microgrid as a single entity in the power grid which is controllable and can operate either in islanded or grid-tied mode. Microgrids have been developed in different scales from small home microgrids to large industrial or commercial microgrids [2].

Climate crisis due to global warming and climate change has led the microgrid designers to use high capacity of distributed renewable energy resources (DRERs). Variability of DRERs' output power and unpredictability of electricity consumption will enforce microgrid planners to use high capacity of energy storage system (ESS). However, high cost of energy storage technologies will boost the financial aspect of microgrids. Non-optimal small capacities of DRERs and ESSs cannot ensure the proper operation of the microgrid to supply the loads uninterruptedly. On the other side, non-optimal large capacities may increase the cost and power losses in the microgrid [3].

Demand response (DR) is the concept of changing consumption side rather than supply side [4]. Traditionally, the power balance between generation and consumption side was maintained using the changes in supply side of power systems. By the appearance of DR, however, the consumption side has the capability to change to match with the supply side. Application of DR in bulk power systems is still a serious challenge when a high power of consumers should be collected to participate in DR program. However, large industrial consumers are appropriate candidates to contribute to DR in power systems [5]. Also, in microgrids, the DR can be implemented because of the proximity of load centers to generation units and simplicity for smart grid application.

Load shifting and demand curtailment are the common manners of DR implementation in microgrids according to time-based and incentive-based programs [6]. The available DR options for load shifting are electric vehicles (EVs), air conditioners (ACs), heat pump water heaters (HPWHs), and thermostatically domestic refrigerators (TDRs). The DR programs can be significantly used for different purposes in microgrids. An important aim of DR application is to reduce the capacity of components especially ESS one. The operator can receive the data of electricity consumption, renewable generation, and electricity prices to apply the DR program most efficiently. However, due to considerable volume of received data, making the best decision for DR is challenging. To overcome this challenge, data mining and artificial intelligence (AI) techniques can be used for classification, diagnosis, segmentation, and prediction of the data [7].

The main goal of this book chapter is to explain the role of intelligent approaches to support DR in microgrids planning. For this purpose, the microgrid planning problem is first identified by indicating the optimization process, data analysis, microgrid operation, as well as generation, storage, and controllable load components. The DR strategies for microgrid applications are introduced and discussed. The artificial intelligence methods to ease DR application in microgrids are explained and their impacts on microgrid planning are described.

15.2 Microgrid Planning

Microgrid is a system with interconnections between various generation and storage components and loads. The microgrid can operate in two modes of grid-connected or islanded. Figure 15.1 a sample model of a microgrid with DERs, loads, ESSs, and control center. The components are connected to each other using electrical energy lines. The information flow is used to send/receive the data to/from the control center to loads and components. The control center is the core of microgrid to control the power flow, DR, stability, and reliability [8].

To achieve a reliable and secure operation of microgrids, the capacity of components should be correctly selected. Planning of microgrid is the problem of detecting the optimal capacity of

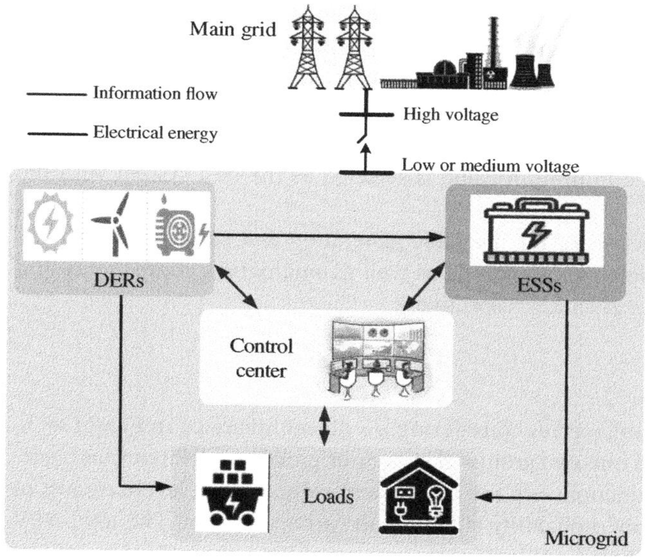

Figure 15.1 A sample model of microgrid with generation and storage components and loads.

components to achieve desired objective functions. This problem is overviewed in this section by emphasizing important characteristics which can affect the planning procedure.

15.2.1 Problem Overview

Microgrid planning is an optimization problem in essence. Figure 15.2 demonstrates a general algorithm for microgrids optimal planning with DR. The planning problem starts with data analysis.

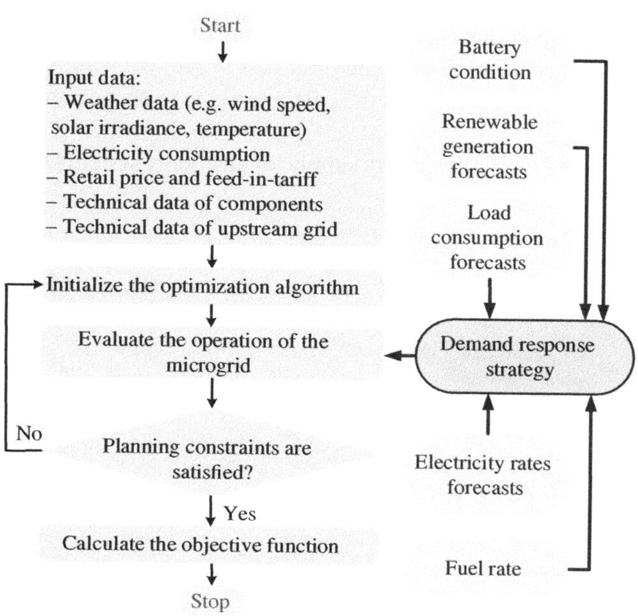

Figure 15.2 A general algorithm for microgrids optimal planning with demand response.

All the required data for solving the problem should be collected. This includes the weather data, load profile, main grid characteristics, and components data [9]. Following the data analysis, the optimization method should initialize the solving algorithm. The operation of the microgrid is evaluated. The DR strategy is implemented in the operation. Different data are needed for the DR strategy which will be discussed in Section 15.2.3. If all the planning constraints are satisfied, the objective function is calculated and the minimum value is selected as the best system with the optimal decision variables.

The decision variables are the capacities of the available generation and storage components in the microgrid. The objective functions, data analysis as well as microgrid's components and operation are discussed in Sections 15.2.2–15.2.5, respectively.

15.2.2 Objective Functions

The applicable objective functions for microgrids considering DR are summarized in Figure 15.3.

The objective functions are classified into two groups of system-oriented and DR relevant objectives. The system-oriented objective functions are the net present value (NPV), levelized cost of electricity (LCOE), loss of power supply probability (LPSP), and carbon mission. The total NPV ($\mathcal{NPV}_t$) of microgrid is expressed based on the NPV of components ($\mathcal{NPV}_c$), the NPV of electricity exchange ($\mathcal{NPV}_e$), the NPV of fuel ($\mathcal{NPV}_f$), and the NPV of DR ($\mathcal{NPV}_{dr}$).

$$\mathcal{NPV}_t = \mathcal{NPV}_c + \mathcal{NPV}_e + \mathcal{NPV}_f + \mathcal{NPV}_{dr} \tag{15.1}$$

The NPV of components is formulated based on the number of components multiplied by replacement cost (C_r^i), capital investment (C_a^i), maintenance cost (C_m^i), and salvation value (C_s^i) as follows:

$$\mathcal{NPV}_c = \mathcal{X}_i \cdot \left(\sum_{i=1}^{n} (C_a^i + C_r^i + C_m^i - C_s^i) \right) \tag{15.2}$$

where $\mathcal{X}_i$ represents the number of components, i is the type of components, and n shows the variety of available components for optimal planning.

The NPV of electricity exchange with the main grid is expressed based on the value of purchased electricity minus the sold electricity. The fuel NPV represents the total cost of fuel consumption during the project lifetime. The DR NPV is the total payment to the customers due to participation in DR program during the project lifetime.

The LCOE is a measure of the total NPV based on the electricity supply for an energy system [10]. The LCOE can be expressed as follows:

$$\mathcal{LCOE} = \frac{\mathcal{NPV}_c \cdot crf_1 + (\mathcal{NPV}_e + \mathcal{NPV}_f + \mathcal{NPV}_{dr}) \cdot crf_2}{\mathcal{E}_m} \tag{15.3}$$

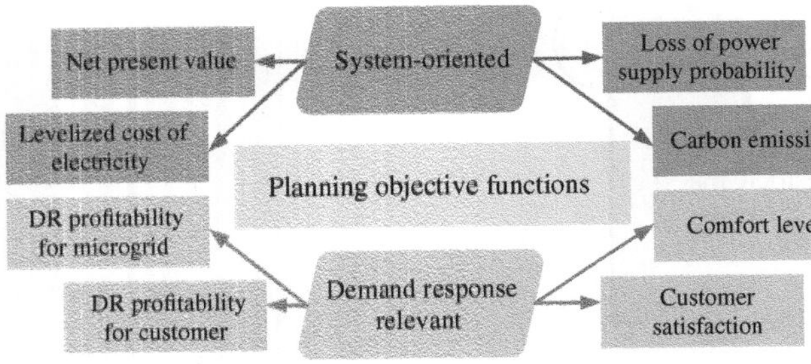

Figure 15.3 Applicable objective functions for optimal planning of microgrids considering DR.

where crf_1 and crf_2 are the capital recovery factors. $\mathcal{E}_m$ represents the total electricity supply for the microgrid loads.

The LPSP is a reliability-based objective function [11]. It is defined as the probability of failures in load supply ($\mathcal{P}_m$) by the available DERs power generation ($\mathcal{P}_c$), the power imported from the grid ($\mathcal{P}_g$), and the available ESSs' discharging power ($\mathcal{P}_b$).

$$\mathcal{LPSP} = \frac{\sum_{t=1}(\mathcal{P}_m(t) - \mathcal{P}_g(t) - \mathcal{P}_c(t) - \mathcal{P}_b(t))}{\sum_{t=1}\mathcal{P}_m(t)} \tag{15.4}$$

The carbon emission is the total greenhouse gas emissions caused by the available diesel generators in the microgrid or the electricity which is imported from the conventional generation units of main grid.

The DR relevant objective functions are customer satisfaction (CS), comfort level, and the profitability of DR for customers or microgrid. The customer satisfaction (CS) can be expressed as follows:

$$\mathcal{CS} = 1 - \frac{\sum_{t=1}(\mathcal{P}_{ad}(t) + \mathcal{P}_{sh}(t))}{\sum_{t=1}\mathcal{P}_m(t)} \tag{15.5}$$

The purpose of CS is to maximize the satisfaction of customers when DR is applied [12]. In (15.5), $\mathcal{P}_{ad}$ and $\mathcal{P}_{sh}$ represent the adjusted/curtailed and shifted power, respectively.

The comfort level of customers for DR participation can be used as an objective function by minimizing their power adjustment [13]. The customer profit from participating in DR should be maximized if the optimal planning is done in customer side. On the other hand, the profit of microgrid by proposing DR programs should be attained if the optimal planning is done by the operator for the microgrid [14].

15.2.3 Data Analysis

Data analysis is the most important piece of microgrid planning puzzle to achieve accurate solutions. Various data are required to accomplish the optimal planning. The required data are long period data for operation or constant data to initiate the optimal planning algorithm. The constant data are about the project lifetime, technical, and economic data of microgrid components, as well as interest/discount and inflation rate during the lifetime. Figure 15.4 demonstrates the long period data which should be used for optimal planning with DR. These data are discussed in the following.

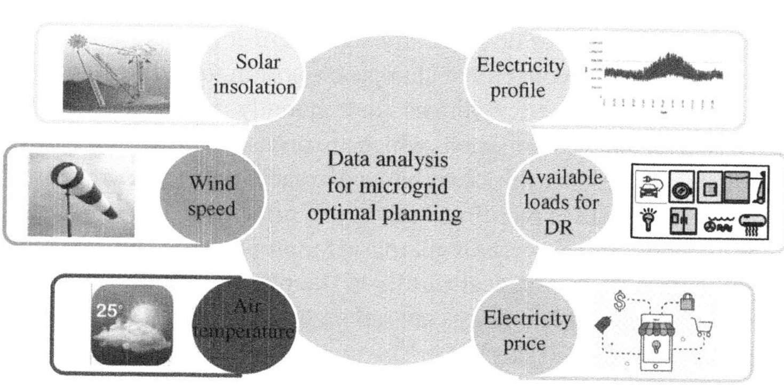

Figure 15.4 Data analysis for microgrid optimal planning with DR.

15.2.3.1 Weather Data

Weather data are required to anticipate the power generation of DRERs. Solar insolation and air temperature are the input to solar photovoltaic (PV) model to predict the solar PV generation. The wind speed data should be used in the wind turbine (WT) model. These data should be arranged based on average season generation, average monthly generation, hours, or even in shorter timeframe. To achieve accurate and practical results, shorter timeframe in longer periods is more useful. The accuracy of forecasting methods is another important item. Since the forecasting methods can attain the anticipated data based on different procedures, the most efficient method should be used for each of weather data [15].

Weather data are also needed to organize DR in the microgrid. For example, efficient solar insolation prediction can disclose the cloudy days with the minimum solar PV generation. Hence, the microgrid operator can schedule the load shifting for day-ahead operation. The same scenario may happen for WT in the microgrid. If the wind speed prediction shows low wind speed for day-ahead, then the DR can be organized by the operator for day-ahead.

15.2.3.2 Load Data

Load data contains the electricity profile and the type of controllable loads in the microgrid. The electricity profile of microgrid is usually a challenge due to hardness of prediction. For accurate microgrid planning, the electricity profile should be forecasted based on different characteristics like human behavior and weather condition. When the electricity profile is modeled, the data can be used in the operation to accomplish the planning. Like the weather data, the electricity profile can be arranged based on average season generation, average month generation, hours, or even in shorter timeframe.

The availability, type, and model of controllable loads in the microgrid are important data for DR application [16]. In the first stage, the type of controllable loads should be recognized. EVs, ACs, HPWHs, and TDRs are the mostly available controllable loads in microgrids. However, the EVs are the most flexible loads to take part in DR program. EVs are available in different levels and locations of microgrids. Due to vehicle-to-grid and vehicle-to-home technologies, EVs can be simply connected to the microgrid. Nevertheless, the EV model contains uncertainty which should be analyzed. The travel distance of EVs is required to reach accurate model of initial state-of-charge (SoC) of their battery when arrive to the destination. The arrival and departure of EVs should also be modeled. As can be seen, an efficient data analysis is required for load electricity profile and availability of loads in a microgrid.

15.2.3.3 Electricity Price

In grid-connected microgrids, the electricity prices or tariffs are significant for proper operation and, hence planning. The electricity price can be constant or changeable within time. Due to recent advancements in smart grids, the changeable electricity prices are frequently used in microgrids. The electricity prices are defined based on different programs which are discussed in Section 15.3.4. However, in recent electricity markets, the electricity price is changed in each hour or in some cases half an hour. This type of electricity price is called real-time or dynamic pricing.

The dynamic pricing can highly affect the cost analysis of microgrid for proper planning. Efficient prediction of electricity price can achieve more practical results in the planning stage. The load level of microgrid, penetration level of renewable energy generation, and policies of the market should be used for the prediction of dynamic electricity price. The prediction can be accomplished in short-term for hourly steps or in long-term for day-ahead market [17].

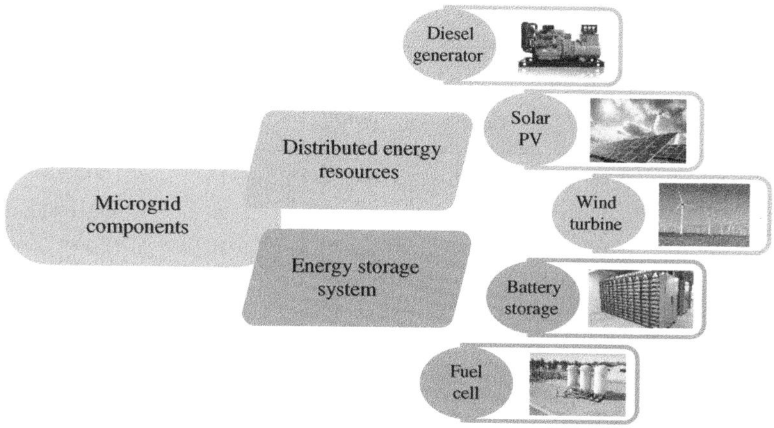

Figure 15.5 Available DER and ESS components in microgrids.

15.2.4 Microgrid Components

Microgrid components are classified into generation and storage units. Although a variety of generation and storage components have been proposed, the mostly applied and available DER and ESS components are shown in Figure 15.5. these components are discussed in the following.

15.2.4.1 Distributed Energy Resources

DERs are classified into two groups of dispatchable and non-dispatchable DERs. The dispatchable DERs are fuel-based generators. Diesel generators (DGs) are the well-known components for this group. The electricity generation depends on the fuel input to the generator. The fuel price fluctuates in the market and hence the operation cost of DG changes. Analyzing the fuel price is another part of data analysis for DR application in microgrid planning problem. On the other hand, there is not enough fuel in countries and the air pollution is a major challenge for DGs; hence, the application of DGs for microgrids has been limited in the power systems. To fill the generation capacity, the penetration level of DRERs is increased.

The non-dispatchable DERs are renewable energy resources that their electricity generation depends on the weather conditions. These resources cannot participate in control and energy scheduling of microgrids. Several forms of DRERs like PVs, WTs, tidal power, hydro power, and biogas are created for power generation. However, solar PV and WT are commonly applied in microgrids. This is due to market availability of PV and WT, technology maturity, declining cost in the market, and well-known installation procedure.

15.2.4.2 Energy Storage Systems

Energy storage systems have been broadly used in recent microgrids to compensate the variability of DRERs and to increase the reliability and profitability of microgrids. The ESSs are developed based on different technologies. The most common and market available ESSs for microgrids are battery energy storage systems (BESSs) and fuel cell (FC) storages. The BESSs are the most applicable based on their availability in different capacities for various applications. In addition, the BESSs have been developed based on different chemical compounds. Table 15.1 lists the characteristics of BESSs and FC which can be used for microgrid planning problem.

The main features of FC for using in microgrids are high energy density and a long history of integration in microgrids because of using hydrogen in the electricity generation procedure.

Table 15.1 Characteristics of BESSs and FCs for planning problem of microgrids.

Energy storage technology	Capital cost ($/kWh)	Efficiency (%)	Power rating (MW)	Discharge time	Energy density (Wh/l)	Lifetime (yr)	Lifetime (cycles)
Fuel cell	8 000–15 000	30–50	0–50	s–24 h	600	5–15	1 000
Lithium ion	700–3 000	85–95	0–100	s–h	200–400	5–20	600–1 200
Vanadium redox flow	600–1 500	80–90	0.03–30	1–10 h	20–70	5–10	12 000
Sodium sulfur	1 000–3 000	80–90	0.05–40	s–h	15–300	10–15	2 500–4 500
Lead acid	300–600	50–90	0–20	s–h	50–80	3–15	250–1 500

Source: Khezri et al. 2021 [18] and Olabi et al. [19].

The FC storage, however, has a high capital cost with a low efficiency [19]. Four types of BESSs are used in microgrids: (i) Lithium-ion, (ii) Vanadium redox flow, (iii) Sodium-sulfur, and (iv) Lead-acid. The Lead-acid battery has the lowest capital cost for investment; however, it has short lifetime with low efficiency. The Lithium-ion battery is broadly expanded in microgrids because of high efficiency with longer lifetime [18]. Type of BESS should be carefully selected in optimal planning of microgrids. The effects of DR on battery degradation are another topic which should be studied based on depth of discharge of battery.

15.2.5 Microgrid Operation

The procedure of power flow control between microgrid components, main grid, and load centers is the main purpose of microgrid operation. The microgrid operation should be carefully accomplished to realize a reliable, economic, and sustainable operation while satisfying the system constraints. Energy management system (EMS) is the core of microgrid operation. The EMS center receives data, using information flow, from available components, electricity market, and main grid to control the power flow. Smart grid technologies advancement has helped the EMS for better monitoring and data analysis. Using the recent smart grid technologies, the data can be received and sent to/from the EMS immediately [20]. Smart grid has also facilitated DR application which is discussed in Section 15.3.

15.3 Demand Response in Microgrids

DR, also referred to as the demand side management (DSM), is defined as any change in power consumption of consumers. In this section, the applications of DR in microgrids, as well as the types and characteristics of DR are discussed.

15.3.1 Overview on Demand Response Application for Microgrids

Figure 15.6 shows different applications of DR in microgrids. The DR programs can be utilized for different aims like energy scheduling, voltage control, frequency regulation, peak shaving, reliability improvement, resiliency enhancement, resource integration, and power reserve [21]. These applications can be considered as the main aim for optimal planning. For example, the DR can be

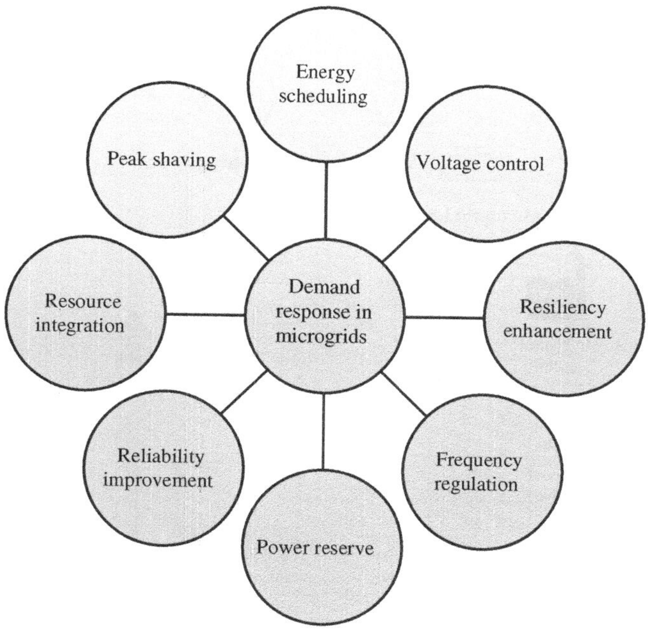

Figure 15.6 Applications of demand response in microgrids.

used for frequency regulation by changing the load level whenever the frequency changes. This can be considered in the operation and the optimal planning can be accomplished. In some cases, these aims are considered as constraints by DR application. For example, the range of frequency change is constrained when the DR is applied. Hence, the main target of DR application should be decided prior to optimal planning of components.

It is shown in Figure 15.2 that the DR program receives various data in the procedure of microgrid optimal planning. The battery condition including SoC and depth of discharge, the renewable energy generation and electricity consumption forecasts, as well as the fuel and electricity prices are the essential data for DR program. To be more accurate, Figure 15.7 shows the integration of DR in a sample microgrid. The current microgrids are becoming more complicated by integration of various components. The information of all these components should be sent to DR management system, concurrently, in order to achieve a proper DR with the minimum effect on consumers electricity supply. Due to huge data analysis in the DR management system, data mining and artificial intelligence are seriously required. When the DR management system collected all the information and analyzed the data using the intelligent methods, it decides for changing the power consumption of available controllable loads in the microgrid.

15.3.2 Demand Response: Types and Characteristics

The main target of DR application in microgrids is to make the load an active participant in power balance between supply and consumption sides. The applied DR should compete with the available supply resources for this power balance target. Generally, the DR programs allow reduction of electrical loads in response to variations of electricity price with time, or for incentive payments, or for microgrid emergency situations.

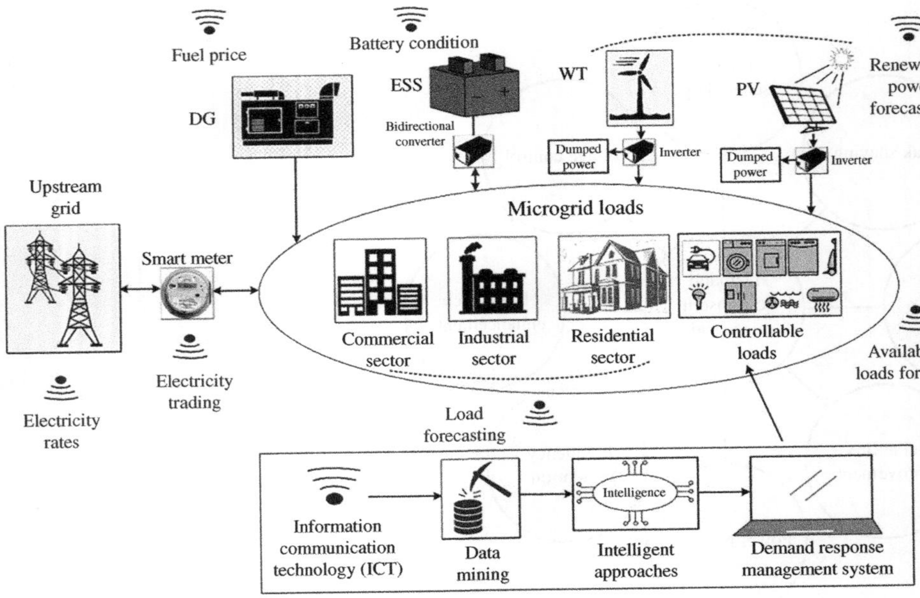

Figure 15.7 Demand response integration in a sample microgrid.

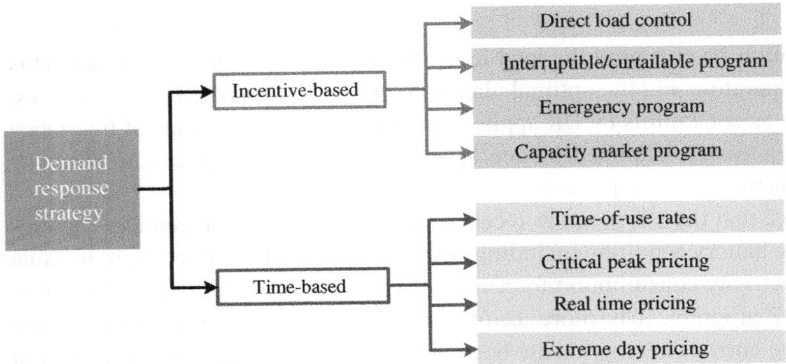

Figure 15.8 Types of demand response strategies in microgrids.

Figure 15.8 demonstrates the types of DR strategies in microgrids. There are two major types of incentive-based and time-based DR strategies which are discussed in the following Sections 15.3.3 and 15.3.4 [22].

15.3.3 Incentive DR

In this type of DR, the customers will be paid or incentivized for load shifting or load curtailment in high electricity price time or when an emergency jeopardizes the microgrid reliability [23]. The incentive-based DR strategies are classified into four groups: (i) direct load control, (ii) interruptible/curtailable program, (iii) emergency program, and (iv) capacity market program.

Direct load control program gives the ability to the operator to remotely control or shut down the controllable loads of the participant on a short notice [24]. Some typical controllable loads for direct load control are ACs and HPWHs. Hence, the direct load control program is more related to small commercial complexes, residential prosumers, and military microgrids. In interruptible/curtailable program, the consumers receive upfront rate discounts or incentive payments for DR participation. The operator asks the participants to reduce their electricity consumption to predefined ranges. If a participant does not observe the contract by exceeding the consumption from the range, then the participant will be faced penalties based on the conditions and terms in the contract [25]. In emergency program, there is a contract between utility and consumer which enables the utility to reduce the electricity consumption of the consumer in emergencies [26]. In capacity market program, the utility operator offers incentives to consumers to commit a pre-specified load reduction in the case of market contingencies. The participant receives a notification a day-ahead of events and should pay penalty if does not respond to the load reduction. The capacity market program also involves the ancillary services which allows the utility operator to reduce the electricity consumption of consumers in response to frequency regulation or power reserve.

15.3.4 Time-Based DR

The electricity price of consumer is changed within the time-based DR program. This variation of electricity price can be organized for both import and export power of consumer. Hence, the electricity price is not flat, and it fluctuates according to real time electricity cost of market. The main aim with time-based DR programs is to flatten the electricity consumption curve by offering a lower price for off-peak periods and a higher price for peak periods. The time-based DR strategies are categorized into four groups: (i) time-of-use rates, (ii) critical peak pricing, (iii) real-time pricing, and (iv) extreme day pricing [27].

The most common time-based DR is time-of-use (ToU) in which the electricity price changes two or three times during the day. Three blocks of off-peak, shoulder, and peak time are considered in ToU, where the lower rate is applied for off-peak time while the rate is penalized for peak time [28]. The electricity rate of shoulder time is in between the off-peak and peak times. In the critical peak pricing, the electricity price is increased in peak hours of critical days with high electricity price in the market. The consumers are notified in advance, generally, a day-ahead, to reduce their electricity consumption. The extreme day pricing is almost similar to critical peak pricing. However, in extreme day pricing, the electricity price is increased for the all the hours of extreme day. In the real-time pricing (RTP), the electricity rate changes in each hour or less than hour intervals. The RTP follows the electricity price of wholesale market. In RTP, the consumers are informed about the electricity price in day-ahead or hour-ahead so that they can change their electricity consumption. DR based on the RTP is known as an efficient strategy for microgrids.

15.4 Intelligent Approaches to Support Demand Response

Artificial intelligence (AI) is a broad branch of computer science which refers to preparing machines to learn from experiences to accomplish tasks where the human intelligence is typically required. Nowadays, the AI, as an interdisciplinary science with multiple methodologies, has been broadly used in different engineering fields. Using the AI technology, computers can be trained to perform specific tasks by analyzing massive data and recognizing different patterns in those data.

AI application in DR management of microgrids is becoming a vital task to achieve the optimal operation and planning. With the advent of big data for microgrids, the load centers have become not only electricity-intensive but also data-incentive. Intelligent systems using data mining techniques have been broadly used to detect patterns or trends in massive data. This section discusses the application of data mining and intelligent systems for DR application in microgrid planning.

15.4.1 Data Mining Methods in DR

Data mining technologies have been recognized as one of the most promising tools to extract valuable knowledge from massive amounts of microgrid operation data. Data mining methodologies are categorized in two groups of supervised methods and unsupervised data mining methods [29]. Figure 15.9 indicates the data mining techniques in each group.

15.4.1.1 Supervised Data Mining

Supervised data mining methods aim at learning the complicated relationship between multiple variables. In these methods, there is prior knowledge about the impacts of the existing samples on the outputs. In the DR domain, the supervised methods are generally used for prediction of renewable energy generation, electricity consumption, controllable loads availability, and electricity market prices. The supervised data mining techniques are categorized into three important groups: (i) artificial neural network (ANN), (ii) deep learning, and (iii) support vector regression (SVR).

Artificial Neural Network ANN is a main component of AI that is used to simulate the human brain functioning [30]. ANN has the capability of self-learning which enables the ANN to generate more desirable results when more input data are available. Neurons are the processing units which constitute layers in the ANN structure. These layers are input, hidden, and output layers. Number of neurons are different in each layer, and they can be in direct connection with each other.

Figure 15.10 a three-layer ANN-based predictor for microgrids. The first layer which is input layer receives the input signals (e.g. historical weather data, electricity consumption, and online monitored data). The second layer is hidden layer which weighs the inputs and then directs them using activation functions as the outputs. Output layer is the last layer that produces the predicted electricity consumption, renewable generation, electricity prices, and controllable loads availability in the microgrid.

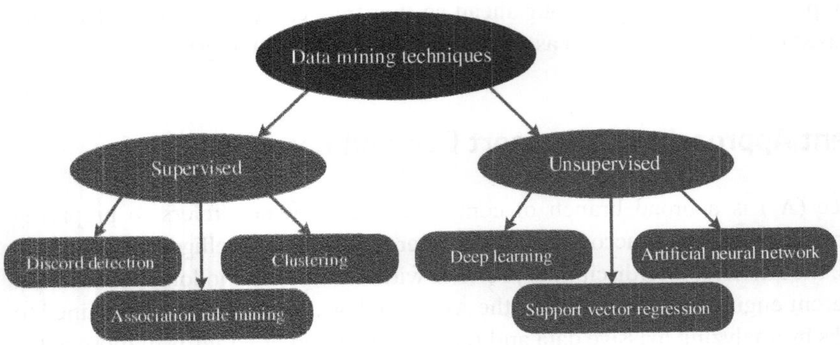

Figure 15.9 Classification of data mining techniques.

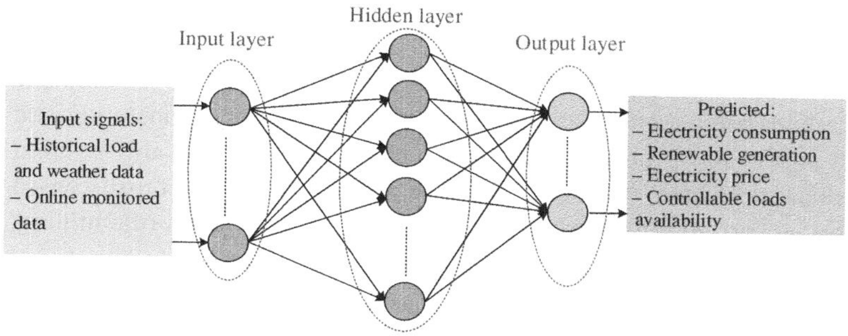

Figure 15.10 A three-layer ANN-based predictor for microgrids.

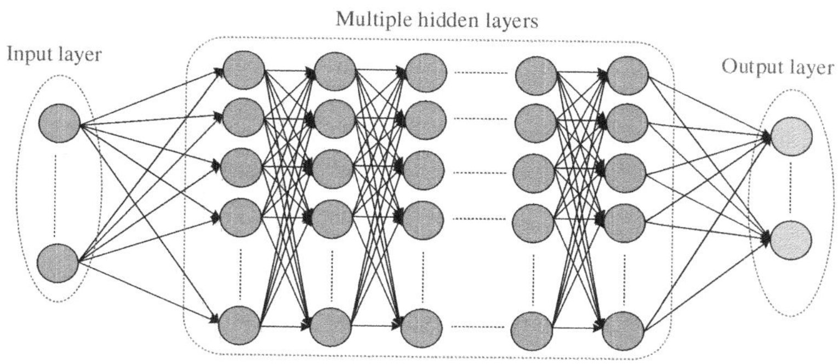

Figure 15.11 A deep learning artificial neural network structure.

Deep Learning Deep learning is a component of machine learning family based on ANN. The main aim of deep learning artificial neural network (DLANN) is to achieve a more accurate model of human brain by imitating the data processing and pattern recognition of the brain. The term "deep" for ANN refers to the fact that multiple hidden layers are used in the structure of DLANN, where the number of the hidden layers indicates the depth of DLANN [31]. In general, a simple ANN has one or two layers in the hidden layer. However, in DLANN, the number of hidden layers exceed three. Figure 15.11 demonstrates a general structure of a DLANN. Using DLANN in microgrids can achieve much accurate prediction of renewable energy generation, electricity consumption, electricity market prices, and availability of controllable loads. Therefore, the results of microgrid planning based on DLANN are more reliable.

Support Vector Regression SVR is the most popular component of support vector machine techniques. In SVR, the problem is defined by regression concept where a function should be found to approximate the data mapping of input objects to real numbers using training samples. The main feature of SVR is minimization of the generalization errors over the other mapping approaches in attaining the real numbers. In addition, SVR has shown excellent performance in nonlinear and time series predictions. The SVR methods are successfully applied for weather and load forecasting in microgrids to achieve high accuracy of predictions [32]. Like ANN and DLANN, the SVR methods can be used for DR application in microgrids by forecasting the electricity prices and availability of controllable loads.

15.4.1.2 Unsupervised Data Mining

Unsupervised data mining approaches are excellent to analyze the received data by discovering intrinsic structures, interconnectedness, and relations. In these methods, there is no available labelled outputs. These methods are used for pattern identification of microgrid operation and event detections like high-impact low-probability incidents. These methods can also be used to cope with stochastic behaviors of load, renewable, and controllable loads. The unsupervised data mining techniques are categorized into three important groups: (i) association rule mining, (ii) clustering, and (iii) discord detection.

Association Rule Mining Association rule learning (ARM) methods are rule-based techniques in machine learning for analyzing data in order to discover the relations of different variables in a massive data. The main aim of ARM is to detect and extract the rules between available variables in the problem with large amount of data like microgrid planning. The general representation of ARM can be indicated in the form of "A → B." Herein, A represents the antecedent and B indicates the consequent. In order to achieve significant patterns of data in DR application for microgrid planning, the experts should analyze all detected association rules. Several ARM methods like a priori, temporal, quantitative, and weighted ARMs have been used for microgrid operation identification [33].

Clustering Clustering is another technique of unsupervised data mining for pattern identification in microgrids. The main task of clustering methods is to making groups of data in which the objects in each group (i.e. cluster) have more similar features than the objects in other groups. Clustering methods are important techniques for analyzing statistical or historical data. Figure 15.12 demonstrates the pattern identification using data clustering methods. The objects are classified into different clusters with characteristics similarity. K-means and fuzzy c-means clustering methods are broadly used for microgrid operation and planning problems [34]. Clustering methods can also be used for pattern identification of electric vehicles to achieve a proper DR in microgrids.

Discord Detection Discord detection is used for detection of infrequently occurring sub-sequences known as discord in time-series data. Therefore, the critical task is to find the main important discord in time series data. The uncommon time sub-sequences generally express faulty operation patterns of microgrids [35].

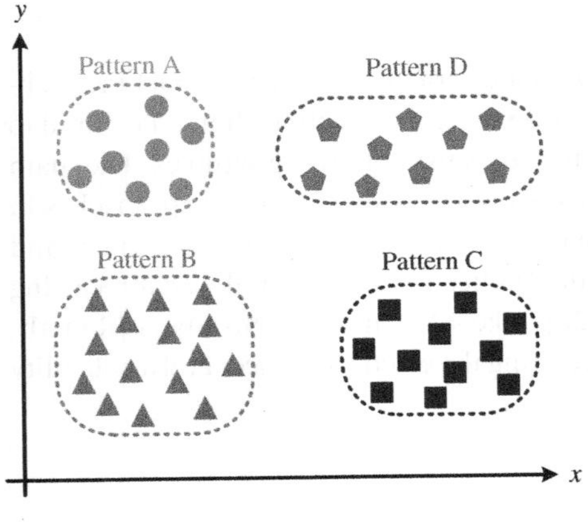

Figure 15.12 Pattern identification using data clustering methods.

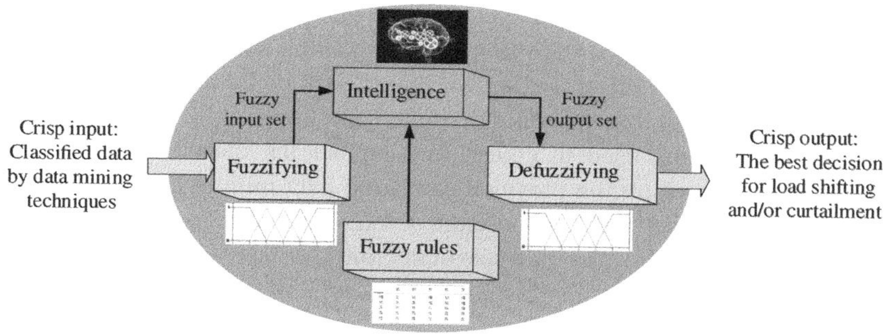

Figure 15.13 A general procedure of fuzzy logic design for DR application.

15.4.2 Fuzzy Logic-Based DR

When the received data analyzed by data mining, the DR management system decides for load shifting or load curtailment based on the available controllable loads in microgrid. By load shifting, the time of electricity consumption of shiftable loads can be rearranged. For example, the candidate loads can be shifted from off-peak time to peak load time. By load curtailment, the electricity consumption of candidate curtailable loads is interrupted. Hence, decision making of DR strategy is extremely significant. Intelligent methods like fuzzy logic can play an important role for the DR application [36].

Fuzzy logic is an approach of reasoning which resembles human reasoning. Fuzzy has the ability to provide the best decisions based on the knowledge of experts or by optimizing the fuzzy rules and membership functions. Figure 15.13 demonstrates a general procedure of fuzzy logic design for DR application. In the first step, appropriate signals should be selected as the inputs. For the DR application, the crisp inputs should contain the classified data by data mining techniques. Then, these data should be converted to the corresponding fuzzy values using membership functions in fuzzifying stage. The produced fuzzy input sets are analyzed using fuzzy rules to generate the fuzzy output sets. The fuzzy rules should be determined based on the knowledge of experts about the DR in microgrid. All the available loads should be determined and the effects of DR on comfort level of consumers should be considered to achieve a desirable rule table. Finally, the fuzzy output sets are defuzzified using membership functions to produce usable outputs for the microgrid. The output of fuzzy logic for DR application can be selected as the amount of load shedding, as well as the time and amount of load shifting.

15.4.3 Applications in Microgrid Planning

The intelligent approaches to support DR can be applied in microgrid planning problem to reduce the capacity of generation and storage components, and hence to decrease the microgrid total cost. Intelligent approach to support DR is considered in the operation of microgrid planning algorithm as shown in Figure 15.2. Recently, intelligent techniques to support DR should be applied for cost reduction, resiliency enhancement, flexibility improvement, and battery capacity reduction.

15.4.3.1 Cost Reduction

Cost reduction is the main purpose of DR application in most microgrids [37]. All cost related signals should be collected in the DR management center to train the AI techniques. These include the electricity prices in the market, the of prices of components, interest and inflation rates, and the incentive payments for the DR participators. A novel modified version of virus colony search algorithm was used to implement a price-based DR for microgrid planning based on uncertainties of renewable energy and electric vehicle [38]. It was shown that the DR could decrease the total cost of microgrid by 26% compared to the case without DR implementation.

The data mining methods should be used to collect and predict the electricity prices in hour-ahead or day-ahead market. Based on the provided data from data mining techniques, an intelligent approach for DR can, then, be used to choose the correct amount of load shedding or demand shifting in the microgrid. When the operation cost of microgrid is decreased, the planning cost reduces accordingly.

15.4.3.2 Resiliency Enhancement

Resiliency is the capability of microgrids to cope with unexpected high-impact events which cause grid outages [39]. Since these events are occurred by natural disasters, the data mining techniques can be efficiently used to predict the events. For this purpose, the historical and online weather data should be given to the data mining technique. Another event that can be happened is cyber threats in the current microgrids with massive smart grid infrastructure [40]. The data mining techniques can detect the false data and predict this malicious action. When the event is predicted, the intelligent approach for DR can be trained to make the best decision to enhance the resiliency of microgrid. This procedure can be considered in the optimal planning of microgrid to achieve a resilient system. A suitable methodology based on data clustering was developed [41]. The proposed method clustered the grid outage events using a K-medoids approach while applied DR for optimal planning of a PV-BESS microgrid. The new system resulted in a reliable microgrid for the end-user customers.

15.4.3.3 Flexibility Improvement

Flexibility is defined as the ability of microgrid to respond to changes in generation and consumption [42]. This term has been used in current power systems and microgrid with high penetration of renewable energy resources. To improve the flexibility in microgrids, it is so important to attain an accurate prediction of electricity generation of renewable resources. Then, data mining techniques should be applied to predict the cumulative power generation of DRERs in microgrids. When the prediction is prepared, the DR strategies can be applied to reconcile the generation with consumption using load shifting.

An example of intelligent approach to support DR for flexibility in optimal planning is provided in the literature [43]. A data-driven approach was developed to increase the operational flexibility for planning. The data-driven approach used the knowledge learnt from smart meter data under a risk-managed framework. It was shown that the new data-driven approach for flexibility could increase the annualized expected profit of the system.

15.4.3.4 Battery Capacity Reduction

Battery price is still not that much affordable for large capacity investment in microgrids. If the capacity of batteries increases in the planning stage, the total cost of microgrid increases as well. An efficient solution to reduce the battery capacity is to use DR in microgrid. Battery is mainly used to store the extra energy of DRERs after load supply and release that energy when there is not enough generation. An intelligent approach to support DR, by taking information from renewable generation and load consumption, can compensate the role of battery. Therefore, a concurrent application of battery and DR in the planning stage can assist to achieve lower capacity of BESS [44]. An intelligent fuzzy logic approach to support DR was proposed for planning of standalone microgrids [45]. The proposed fuzzy logic received the day-ahead forecasts of wind and solar generations as well as the SOC of battery storage. Then, based on the received data, it decided for load shifting or curtailment and the amount of load to participate in DR. The proposed technique reduced the BESS capacity and hence the cost of the microgrid. This kind of study needs more attention for microgrids planning in the future.

15.5 Conclusion

This chapter presented an overview on application of intelligent approaches to support DR for optimal planning of microgrids. The problem of microgrids planning was overviewed by indicating the optimization process, available generation and storage components, objective functions in the presence of DR, and required data analysis. Different DR techniques for microgrids were classified and discussed.

It was found that large amount of data should be conducted for microgrid planning in the presence of DR. Such a massive data needs deep and accurate analysis. Therefore, application of data mining and intelligent systems for DR integration in microgrids was discussed. The main implications and future works for the application of intelligent approaches to support DR for microgrid planning are as follows:

- It was found that, as the first stage, the current application of intelligent approaches to support DR can be focused on cost reduction in microgrid planning. For this aim, the intelligent methods can receive and analysis data of electricity market, generation cost of distributed generators, and the load consumption to minimize the microgrid cost. Deep learning methods are excellent solutions for this application of DR as a future direction.
- Resiliency of microgrids would greatly enhance by intelligent approaches for DR. This should be considered in the planning level to assign appropriate capacity for DR and ESS. In the future, the machine learning algorithms can be significantly used to predict the low-frequency high-impact incidents in microgrids to design an appropriate DR based on the prediction.
- Flexibility has become an important issue in microgrids that should be improved appropriately to avoid electricity shortage. The machine leaning techniques can be used to forecast the cumulative power demand of microgrids' loads and the cumulative power generation of renewables. When the prediction is generated, intelligent techniques like fuzzy can be used to support DR to reconcile the generation with demand. As a future task, a deep leaning ANN can be applied for demand and generation forecasting in microgrids.
- Although battery energy storages have attracted much attention for microgrids, the high cost of the battery is a barrier for further extension. Intelligent methods can support DR to efficiently reduce the capacity of battery in the planning stage of microgrids.

References

1 Lasseter, R.H. and Paigi, P. (2004). Microgrid: a conceptual solution. In: *2004 IEEE 35th Annual Power Electronics Specialists Conference (IEEE Cat. No.04CH37551)*, Aachen, Germany: IEEE, vol. 6, 4285–4290. https://doi.org/10.1109/PESC.2004.1354758.

2 Farrokhabadi, M., Canizares, C.A., Simpson-Porco, J.W. et al. (2020). Microgrid stability definitions, analysis, and examples. *IEEE Transactions on Power Systems* 35 (1): 13–29. https://doi.org/10.1109/TPWRS.2019.2925703.

3 Jalilpour, K., Khezri, R., Mahmoudi, A., and Oshnoei, A. (2019). Optimal sizing of energy storage system, Chapter 9. In: *Variability, Scalability, and Stability of Microgrids* (ed. S.M. Muyeen, S. Islam and F. Blaabjerg), 263–289. Stevenage: The Institution of Engineering and Technology (IET).

4 Su, C. and Kirschen, D. (2009). Quantifying the effect of demand response on electricity markets. *IEEE Transactions on Power Systems* 24 (3): 1199–1207. https://doi.org/10.1109/TPWRS.2009.2023259.

5 Golmohamadi, H., Keypour, R., Bak-Jensen, B. et al. (2020). Robust self-scheduling of operational processes for industrial demand response aggregators. *IEEE Transactions on Industrial Electronics* 67 (2): 1387–1395. https://doi.org/10.1109/TIE.2019.2899562.

6 Albadi, M.H. and El-Saadany, E.F. (2008). A summary of demand response in electricity markets. *Electric Power Systems Research* 78 (11): 1989–1996.

7 O'Leary, D.E. (2013). Artificial intelligence and big data. *IEEE Intelligent Systems* 28 (2): 96–99. https://doi.org/10.1109/MIS.2013.39.

8 Bihari, S.P., Sadhu, P.K., Sarita, K. et al. (2021). A comprehensive review of microgrid control mechanism and impact assessment for hybrid renewable energy integration. *IEEE Access* 9: 88942–88958. https://doi.org/10.1109/ACCESS.2021.3090266.

9 Khezri, R. and Mahmoudi, A. (2020). Review on the state-of-the-art multi-objective optimization of hybrid standalone/grid-connected energy systems. *IET Generation Transmission and Distribution* 14 (20): 4285–4300.

10 Khezri, R., Mahmoudi, A., and Haque, M.H. (2020). Optimal capacity of solar PV and battery storage for Australian grid-connected households. *IEEE Transactions on Industry Applications* 56 (5): 5319–5329. https://doi.org/10.1109/TIA.2020.2998668.

11 Combe, M., Mahmoudi, A., Haque, M.H., and Khezri, R. (2019). Optimal sizing of an AC-coupled hybrid power system considering incentive-based demand response. *IET Generation Transmission and Distribution* 13 (15): 3354–3361.

12 Chen, J., Zhang, W., Li, J. et al. (2018). Optimal sizing for grid-tied microgrids with consideration of joint optimization of planning and operation. *IEEE Transactions on Sustainable Energy* 9 (1): 237–248. https://doi.org/10.1109/TSTE.2017.2724583.

13 Anvari-Moghaddam, A., Monsef, H., and Rahimi-Kian, A. (2015). Optimal smart home energy management considering energy saving and a comfortable lifestyle. *IEEE Transactions on Smart Grid* 6 (1): 324–332. https://doi.org/10.1109/TSG.2014.2349352.

14 Khezri, R., Mahmoudi, A., and Haque, M.H. (2020). Two-stage optimal sizing of standalone hybrid electricity systems with time-of-use incentive demand response. In: *2020 IEEE Energy Conversion Congress and Exposition (ECCE)*, Detroit, MI, USA: IEEE, 2759–2765. https://doi.org/10.1109/ECCE44975.2020.9236381.

15 Aslam, S., Herodotou, H., Mohsin, S.M. et al. (2021). A survey on deep learning methods for power load and renewable energy forecasting in smart microgrids. *Renewable and Sustainable Energy Reviews* 144: 110992.

16 Honarmand, M.E., Hosseinnezhad, V., Hayes, B. et al. (2021). An overview of demand response: from its origins to the smart energy community. *IEEE Access* 9: 96851–96876. https://doi.org/10.1109/ACCESS.2021.3094090.

17 Conejo, A.J., Plazas, M.A., Espinola, R., and Molina, A.B. (2005). Day-ahead electricity price forecasting using the wavelet transform and ARIMA models. *IEEE Transactions on Power Systems* 20 (2): 1035–1042. https://doi.org/10.1109/TPWRS.2005.846054.

18 Khezri, R., Mahmoudi, A., and Aki, H. (2021). Optimal planning of solar photovoltaic and battery storage systems for grid-connected residential sector: review, challenges and new perspectives. *Renewable and Sustainable Energy Reviews* 153: 111763.

19 Olabi, A.G., Onumaegbu, C., Wilberforce, T. et al. (2021). Critical review of energy storage systems. *Energy* 214: 118987.

20 Deng, R., Yang, Z., Chow, M., and Chen, J. (2015). A survey on demand response in smart grids: mathematical models and approaches. *IEEE Transactions on Industrial Informatics* 11 (3): 570–582. https://doi.org/10.1109/TII.2015.2414719.

21 Jordehi, A.R. (2019). Optimisation of demand response in electric power systems, a review. *Renewable and Sustainable Energy Reviews* 103: 308–319.

22 Siano, P. (2014). Demand response and smart grids – a survey. *Renewable and Sustainable Energy Reviews* 30: 461–478.

23 Astriani, Y., Shafiullah, G.M., and Shahnia, F. (2021). Incentive determination of a demand response program for microgrids. *Applied Energy* 292: 116624.

24 Shabshab, S.C., Lindahl, P.A., Nowocin, J.K. et al. (2020). Demand smoothing in military microgrids through coordinated direct load control. *IEEE Transactions on Smart Grid* 11 (3): 1917–1927. https://doi.org/10.1109/TSG.2019.2945278.

25 Wang, B., Li, Y., Ming, W., and Wang, S. (2020). Deep reinforcement learning method for demand response management of interruptible load. *IEEE Transactions on Smart Grid* 11 (4): 3146–3155. https://doi.org/10.1109/TSG.2020.2967430.

26 Imani, M.H., Niknejad, P., and Barzegaran, M.R. (2018). The impact of customers' participation level and various incentive values on implementing emergency demand response program in microgrid operation. *International Journal of Electrical Power & Energy Systems* 96: 114–125.

27 Palensky, P. and Dietrich, D. (2011). Demand side management: demand response, intelligent energy systems, and smart loads. *IEEE Transactions on Industrial Informatics* 7 (3): 381–388. https://doi.org/10.1109/TII.2011.2158841.

28 Javeed, I., Khezri, R., Mahmoudi, A. et al. (2021). Optimal sizing of rooftop PV and battery storage for grid-connected houses considering flat and time-of-use electricity rates. *Energies* 14 (12): 3520.

29 Zhao, Y., Zhang, C., Zhang, Y. et al. (2020). A review of data mining technologies in building energy systems: load prediction, pattern identification, fault detection and diagnosis. *Energy and Built Environment* 1: 149–164.

30 Khezri, R., Oshnoei, A., Yazdani, A.M., and Mahmoudi, A. (2020). Intelligent coordinators for automatic voltage regulator and power system stabiliser in a multi-machine power system. *IET Generation Transmission and Distribution* 14 (23): 5480–5490.

31 Zhu, L., Hill, D.J., and Lu, C. (2020). Hierarchical deep learning machine for power system online transient stability prediction. *IEEE Transactions on Power Systems* 35 (3): 2399–2411. https://doi.org/10.1109/TPWRS.2019.2957377.

32 Chen, Y., Xu, P., Chu, Y. et al. (2017). Short-term electrical load forecasting using the Support Vector Regression (SVR) model to calculate the demand response baseline for office buildings. *Applied Energy* 195: 659–670.

33 Li, K., Liu, L., Wang, F. et al. (2019). Impact factors analysis on the probability characterized effects of time of use demand response tariffs using association rule mining method. *Energy Conversion and Management* 197: 111891.

34 Lin, S., Li, F., Tian, E. et al. (2019). Clustering load profiles for demand response applications. *IEEE Transactions on Smart Grid* 10 (2): 1599–1607. https://doi.org/10.1109/TSG.2017.2773573.

35 Park, J.Y., Wilson, E., Parker, A., and Nagy, Z. (2020). The good, the bad, and the ugly: data-driven load profile discord identification in a large building portfolio. *Energy and Buildings* 215: 109892.

36 Al-Awami, A.T., Amleh, N.A., and Muqbel, A.M. (2017). Optimal demand response bidding and pricing mechanism with fuzzy optimization: application for a virtual power plant. *IEEE Transactions on Industry Applications* 53 (5): 5051–5061. https://doi.org/10.1109/TIA.2017.2723338.

37 Pourmousavi, S.A., Nehrir, M.H., and Sharma, R.K. (2015). Multi-timescale power management for islanded microgrids including storage and demand response. *IEEE Transactions on Smart Grid* 6 (3): 1185–1195. https://doi.org/10.1109/TSG.2014.2387068.

38 Huang, S. and Abedinia, O. (2021). Investigation in economic analysis of microgrids based on renewable energy uncertainty and demand response in the electricity market. *Energy* 225: 120247.

39 Chen, C., Wang, J., Qiu, F., and Zhao, D. (2016). Resilient distribution system by microgrids formation after natural disasters. *IEEE Transactions on Smart Grid* 7 (2): 958–966. https://doi.org/10.1109/TSG.2015.2429653.

40 Venkataramanan, V., Srivastava, A.K., Hahn, A., and Zonouz, S. (2019). Measuring and enhancing microgrid resiliency against cyber threats. *IEEE Transactions on Industry Applications* 55 (6): 6303–6312. https://doi.org/10.1109/TIA.2019.2928495.

41 Tostado-Véliz, M., Icaza-Alvarez, D., and Jurado, F. (2021). A novel methodology for optimal sizing photovoltaic-battery systems in smart homes considering grid outages and demand response. *Renewable Energy* 170: 884–896.

42 Liang, L., Hou, Y., and Hill, D.J. (2019). Enhancing flexibility of an islanded microgrid with electric springs. *IEEE Transactions on Smart Grid* 10 (1): 899–909. https://doi.org/10.1109/TSG.2017.2754545.

43 Liang, H. and Ma, J. (2021). Data-driven resource planning for virtual power plant integrating demand response customer selection and storage. *IEEE Transactions on Industrial Informatics* https://doi.org/10.1109/TII.2021.3068402.

44 Wang, D., Ge, S., Jia, H. et al. (2014). A demand response and battery storage coordination algorithm for providing microgrid tie-line smoothing services. *IEEE Transactions on Sustainable Energy* 5 (2): 476–486. https://doi.org/10.1109/TSTE.2013.2293772.

45 Khezri, R., Mahmoudi, A., and Haque, M.H. (2021). A demand side management approach for optimal sizing of standalone renewable-battery systems. *IEEE Transactions on Sustainable Energy* 12 (4): 2184–2194. https://doi.org/10.1109/TSTE.2021.3084245.

16

Socioeconomic Analysis of Renewable Energy Interventions: Developing Affordable Small-scale Household Sustainable Technologies in Northern Uganda

Jens B. Holm-Nielsen[1], Achora P.O. Mamur[2], and Samson Masebinu[1]

[1]*Department of Energy Technology, Center for Bioenergy and Green Engineering, Aalborg University, Esbjerg, Denmark*
[2]*Faculty of Sociology, Environmental and Business Economics, University of Southern Denmark, Esbjerg, Denmark*

Acronyms

CBA	Cost–benefit analysis
CH_4	Methane
CO_2	Carbon dioxide
DALYS	Disability adjusted life years
DPISR	Driver pressure impact state response
EUA	European Union Allowance
GHG	Greenhouse gas
GTZ	German Technical Cooperation
GWP	Global warming potential
IRR	Internal rate of return
NGO	Nongovernmental organization
NPV	Net present value
PV	Photovoltaic
SE4All	Sustainable energy for all
SAREI	Socioeconomic analysis of renewable energy interventions

16.1 Introduction

Modern and sustainable renewable energy is essential for achieving public, national, and international goals: the Paris Agreement (UNFCCC, 2015), 17 UN Sustainable Development Goals [1, 2]. The goals, especially goals 3, 5, 7, and 13, as the "Ensuring healthy lives and promoting well-being for all ages"; "Gender equality"; "Energy access for all" and "Climate action," respectively [2–4] are critical for a modern society. Globally, about 2.7 billion people (40% of the world population) lack access to clean energy, and they depend on traditional cooking methods based on the use of open fire systems and solid biomass [5–7]. Such traditional approaches are less efficient, emitting a vast range of health-damaging pollutants, such as fine smoke particles and carbon monoxide, which increase indoor air pollution [8, 9].

The World Health Organization (WHO, 2018) [10] reported that annual global estimated premature deaths of 4.3 million are attributed to exposure to indoor air pollution [11]. Among the

Intelligent Data Mining and Analysis in Power and Energy Systems: Models and Applications for Smarter Efficient Power Systems, First Edition. Edited by Zita Vale, Tiago Pinto, Michael Negnevitsky, and Ganesh Kumar Venayagamoorthy.

reasons attributed to such premature deaths is the hazardous smoke particles from charcoal and firewood-based cooking systems typical of developing and underdeveloping countries. Notably, 600 000 inhabitants are exposed to such cooking systems in Africa, with the majority being women and children [9, 12]. Collective information confirms that women and children in sub-Saharan Africa collect substantially firewood, spending between 1 and 2 hours/day [13, 14], a time that could be used in engaging in productive activities to improve their quality of life. K.V. Ramani once stated, “The extent of physical hardship imposed on poor women in acquiring and using energy for the most basic survival needs is enslavement that denies them vital opportunities to escape their state of self-deprivation” [15]. The energy source that women spend their time collecting is mostly biomass.

Biomass accounts for 49% of the total energy consumed in Africa and 80% in sub-Saharan Africa [16]. Overreliance on solid biomass in developing/underdeveloping countries contributes to environmental degradation and pressure on forest resources [17]. In Africa, the exploitation of biomass has raised concerns about the growing demand for this source of energy, which is expected to increase by 27% by 2030 due to a high population growth rate. Several reports have highlighted the need for different sustainable and affordable renewable energy systems to mitigate the effects of excessive biomass exploitation [8, 9].

Africa’s energy consumption is shallow compared to the world average. It accounts for only 4% of the global energy demand despite being home to 13% of the global population [13, 18, 19]. This is due to low industrial activities and a high number of people without access to energy. The UN Secretary-General (2011) launched Sustainable Energy for All (SE4ALL) program to encourage governments and stakeholders in every country especially developing nations to take the responsibility for ensuring access to modern and sustainable energy services by 2030 [3, 20]. These goals are built as per 17 UN sustainable goals and UN Millennium Development Goals (MDGs, 2015)[1] [1].

The United Nations Secretary-General Ban Ki-Moon stated, “Energy is the golden thread that connects economic growth, increased social equity, and an environment that allows the planet to thrive” [20]. In this perspective, the Government of Uganda (2015) embraced and developed an Action Agenda SE4ALL guidelines with the UN goals for sustainable energy for all [20, 21]. The goals outline milestones through 2030 for building three main energy pillars: (i) Energy access for all, (ii) Energy efficiency, and (iii) Renewable energy. The Government of Uganda, together with other partners, is working toward quantitative objectives [22].

In Uganda, 91.5% of total domestic and primary energy comes from biomass, through the forms of charcoal, firewood, and crop residues, because of ease of accessibility and affordability [2]. These fuel sources are available almost everywhere and burned directly; they are cheaper than other fuels and collected at no monetary cost [23]. The energy usage category in Uganda is clearly described in Figure 16.1.

Considering that biomass is the source of energy for domestic activities in Uganda, the negative social, economic, health, and environmental effects continue to be exacerbated as a massive area of forest between 90 000 and 200 000 ha is being destroyed yearly [24, 25]. Not only time is wasted collecting biomass with the associated danger, but about 20 852 per year premature deaths have been attributed to household air pollution in Uganda [26]. To reduce the burden of sourcing for

1 MDGs: United Nations Millennium declaration signed (September 2000) to commit world leaders to combat poverty and hunger, achieve universal primary education, promote gender equality, and empower women, reduce child mortality, improve maternal health, combat HIV/AIDS, malaria, and other diseases, ensure environmental sustainability and to develop a global partnership for development (UN, 2018).

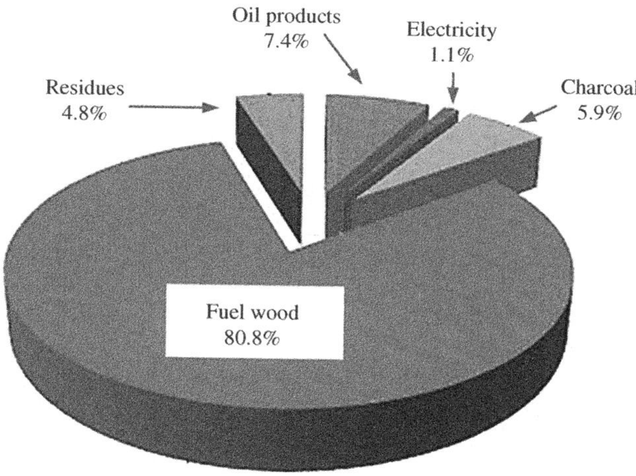

Figure 16.1 Energy use by category in Uganda. Source: Okello et al. [22]/with permission of Elsevier.

biomass and the associated indoor air pollution, preserve human life, and develop the nation, there is a need to test alternative options considering the local context. This study, therefore, aims:

1. To identify small-scale alternative renewable energy technologies for households
2. To identify the socioeconomic and environmental benefits of these alternative technologies
3. To identify hindrances and solutions toward the adoption of these technologies

To create awareness and develop cheap and efficient alternative technologies that enable households and communities to access affordable and reliable energy, this study, therefore, explores the potential technologies, socioeconomic growth, environmental importance, and level of awareness of the resident of the four districts in Northern Uganda.

16.2 Renewable Energy Technologies

A literature review was conducted to provide a broader background of alternative small-scale renewable energy technologies. To offer a more in-depth understanding, and to gain insights and information related to these technologies, the emphasis was made on family-size biogas, bio-briquettes, bio-pellets, bio-oils, solar photovoltaic (PV) cookers, and small-scale/households PV-solar power generation plants (solar home systems). These identified small-scale renewable energy technologies are in the market elsewhere, introduced and successfully implemented. However, in Uganda, some of these technologies have not been widely adopted, while some are yet to be introduced or are at the pilot stage.

16.2.1 Bio-oil

Bio-oil is a biodegradable and environmentally friendly free-flowing combustible liquid fuel extracted from biomass [27]. It provides clean and sustainable energy services to developed/developing/underdeveloping countries [28]. Bio-oil is also derived from different feedstocks such as soya bean, canola, palm oil, rapeseed, jatropha [29], waste cooking oil, animal fats, and algae [30, 31]. Jatropha, for example, is extracted mechanically through the traditional method

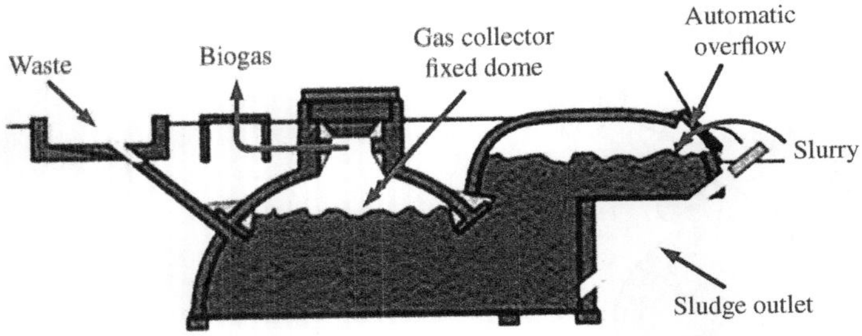

Figure 16.2 Fixed dome digester: CAMARTEC Model. Source [32].

using wooden or stone sekku (mortar and pestle), batch oil presses (cold press method), and expellers [32, 33]. These technologies have been successfully implemented in other African countries [34–37] to contribute to rural electrification [33]. However, in Uganda, it is not a developed technology for community usage. Oil crops (e.g. jatropha) can be planted on a small scale – farmers come in small groups and set up a small bio-oil extraction plant to produce their own oil and use for cooking and lighting [38].

16.2.2 Bio-pellets

Biomass briquetting is the conversion of loose biomass material into a high-density product by subjecting the material to high pressure, with or without a binder. The briquetting process involves material collection, drying, commutation, and densification, using various types of presses [22, 38]. The briquettes can be produced from any form of biomass such as industrial wastes and co-products, agricultural residues, and forest residues [39]. Briquette offers many benefits [40–43], such as for cooking in gasifier stoves, which are more efficient than traditional stoves [38]. Briquetting technology still has a fragile foothold in Uganda because of technical constraints and factors of machine cost, lack of technology sharing and knowledge, low market demand, less public awareness, availability of raw materials, and mode of production [32, 44].

16.2.3 Biogas

Biogas was first introduced in Uganda in the late 1950s [45]. Since then, many initiatives by various stakeholders for promoting the technology [46] through pilot demonstration, capacity building, and provision of subsidies to households from the government have been implemented. Despite all efforts toward the dissemination of this technology, its adoption is low [47]. Several factors limit the dissemination, such as technical skills, socioeconomic and organizational factors, among others [48–51]. Biogas system ranges from 8 to 16 m^3 for household systems [52, 53] with animal manure being the most common substrate for feeding household digesters. Two herds of cattle or eight pigs are required to produce sufficient biogas for small households [54]. A simple burner can be used for the combustion of biogas, which is conveyed from the digester to the house through pipes [38, 55]. Figure 16.2 elaborates the detail of the design of a fixed dome digester.

16.2.4 Solar Cookers

Solar cooking in Uganda was introduced in 1994 by Solar Cookers Association (SCA) with support from several organizations such as the World Wildlife Fund Switzerland. Since then, SCA has

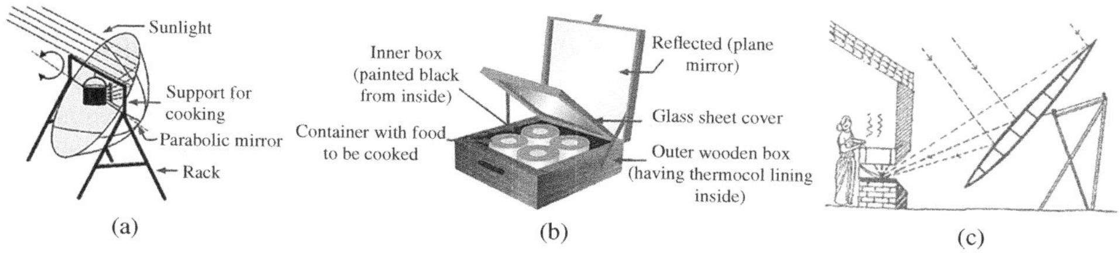

Figure 16.3 Solar cooker technologies. (a) Focusing type cooker (b) box solar cooker (c) indirect solar cooker. Source: Regattieri et al. [56]/With permission of Elsevier.

distributed over 50 000 solar cookers in Uganda, South Sudan, Eastern Congo, and Burundi. Solar cookers use the energy of solar radiation to cook food [56]. The cookers are made locally and of different sizes and capacities. Using solar cookers, in areas with the right solar radiation intensity, leads to reduction in deforestation [57, 58]. Figure 16.3 shows different types of solar cookers. Due to the dependence of this technology on solar irradiation, cooking is preferably done during the daytime, or this technology is paired with biogas to ensure evening cooking.

16.2.5 Solar PV

Solar PV technology has steadily grown since 2004 with a handful of solar companies, NGOs, and governments supporting solar projects across the country [59], mainly installed at schools, hospitals, health centers, administrative offices, and businesses [60]. Solar PV systems have been deployed in different ways ranging from minimal applications to Solar Pico Systems (SPSs) used for lighting, which should replace kerosene lamps, to power radio, TV, or phone charging to mid-scale solutions such as Solar Home Systems (SHSs) [61–63]. The scaling of solar PV has also been implemented by integrating different energy sources to create a more efficient household energy mix using solar PV and modern bioenergy to produce multiple outputs. These solutions are affordable, sustainable, reliable, and emission-free sources of energy to households in Northern Uganda with locally available resources.

16.3 Methodology

The proposed method implemented in this study incorporated a conceptual framework and field study, which comprised data collected from four districts of Northern Uganda to investigate and find out the need for small-scale renewable energy technologies. The conceptual frameworks were based on the Driver Pressure Impact State Response framework (DPISR), to investigate the impact of households' energy consumptions on the forest resources of Northern Uganda. For household renewable energy intervention, the Cost–Benefit Analysis was conducted based on the German Technical Cooperation Cost–Benefit Analysis (GTZ-CBA) for households renewable energy intervention in less developing countries.

16.3.1 Driver Pressure Impact State Response Framework

This framework was considered from a local perspective of Northern Uganda. The framework is used to find out the cause–effect relationship between the household's energy needs and the local forest resources. DPISR is a tool used to model the cause–effect relationships between environment and human system process and to develop management response [64–70]. By definition, drivers are social, economic, and environmental development that arises from human needs to satisfy their primary needs (food, energy, water, shelter) and secondary requirements (mobility, entertainment, culture) [71]. These drivers exert pressure and cause certain environmental changes, because of production and consumption, such as excessive use of resources and emission into the environment. These pressures result in a state of environmental changes, impacting human welfare (social, economic, or environmental), resulting in a social response. The response presents a set of societal and policymaker prioritization, with the aim to reduce the undesired effects as much as possible [71]. The response may be fed to drivers, pressures, state, or effects [70]. Energy need for cooking is the primary driver that leads to a household demanding firewood and charcoal for cooking. This demand significantly affects the forest resources leading to deforestation, loss of biodiversity, and harm human welfare such as the following:

1. Soil erosion and loss of soil nutrients leading to poor crop yields.
2. Sources of livelihood are destroyed more especially for the rural population, who depend on fruits and traditional medicines from the forest resources.
3. Negative effects on water sources that cause streams and wells to dry up.
4. Building materials for rural populations are destroyed [72].
5. Deforestation leading to global warming, climatic changes that affect negatively on human welfare [73].

Table 16.1 elaborates the relationship between the impact of different alternative sources of energy from local forest resources and human welfare of Northern Uganda.

16.3.2 Cost–Benefit Analysis

According to the literature [74], the WHO conducted the first CBA interventions on household cooking systems for 11 developing countries for middle-income families. The assessment was based on "household perspective" and local circumstances analysis of improved cookstove systems. The study yielded results suggesting a range of cooking system interventions [75]. Results found to be attractive include health benefits, cost reduction, reduction in CO_2 and CH_4 emissions, preservation of forest and environmental benefits [74–76]. For this study, household's energy intervention was considered from the local perspective (Northern Uganda). The CBA evaluation is based on an estimation of cost and co-benefits in terms of firewood saving, health benefits, environmental conservation at the local level and greenhouse gas emissions based on the framework developed by GTZ (the German Technical Cooperation) [77] for energy-saving stoves in Uganda. The framework was used to identify, attach values, and compare the cost and benefits of alternative modern technologies, namely, family-size biogas and solar cookers. One thing that distinguishes this study from the one carried out by GTZ is that the energy intervention is to replace the use of charcoal and firewood by introducing clean alternative sources of fuels. The choice for only two alternative technologies is linked directly to private individuals/households, and other alternatives such as liquid biofuel, briquettes, and bio-pellets require a village-scale factory associated with a high investment cost. Table 16.2 illustrates the elements considered for the CBA to solve the problem of overdependence on charcoal and firewood identified through the DPISR framework.

16.4 Application of the Method

The study was carried out in Northern Uganda, four districts of Gulu, Amuru, Nwoya, and Omoro, as shown in Figure 16.4. About 78% of the population relies on agriculture for livelihood, and most

Table 16.1 Alternative source of energy in Northern Uganda.

Driver	Pressure	State	Impact	Response
Firewood	Forest resources	Loss in bio-diversity. A decrease in forest cover. Indoor air quality. Negative health impact.	Soil erosion, and loss of soil nutrients. Low crop yields. Loss of sources of livelihood. Environmental degradation. Climate change. Health implications.	Promotion of agroforestry, and tree farming. Creation of awareness, and promotion of other clean alternative sources of fuel.
Charcoal	Forest resources	Loss of bio-diversity A decrease in forest cover	Soil erosion, and loss of soil nutrients Low crop yields Loss of sources of livelihood Environmental degradation Climate change Health implications	High taxes on charcoal businesses Promotion of agroforestry, and tree farming Creation of awareness, and promotion of other clean alternative sources of fuel. Introduction of efficient charcoal burning technology.
Biogas	–	–	A clean source of energy. Improve crop yield using bio-slurry. Environmentally friendly.	Subsidies to households. Good credit and favorable financing mechanisms. Donor funded programs toward the dissemination of biogas.
Biofuel	Land Forest resources	Low food production. High use of agro-chemicals. Deforestation	Food security. Environmental degradation. Loss of diversity. Loss of sources of livelihood. Pollution because of agrochemicals. Health implications.	An incentive for food production. Set up small-scale bio-refinery plants. Energy crop cultivation on wastelands.
Bio-briquettes / pellets	—	—	—	Step up a village-scale factory.
Solar cookers	—	—	—	Donor funded projects. Good credit and favorable financing mechanisms.

a) Source of information from authors' research and field survey work.

of them live in rural areas [78, 79]. More than a third of the population lives below US$ 1.9 a day [80], and about 91% of the population relies on biomass for cooking and paraffin for light using traditional and less fuel-efficient technologies. However, already, concerns have been raised on how biomass resources are being extracted in this area [38].

Real-time data for this study sourced from the four selected districts, as stated above, was conducted in March 2018 and from a sample population of 124 households. Questionnaires prepared

Table 16.2 Elements considered in the CBA.

Criteria	Sub-criteria	Description
Direct cost	Capital	One time investment associated with fuel transition. This includes the cost of a new stove, security deposits, required installation, and training costs.
	Operation	Cost of preparation, procurement, and storage of fuel
	Maintenance	Cost of repairs or servicing of the cooking system
Indirect cost		
	Ecological integrity	Incorporate fuel pollution effects of traditional energy technology
	Global warming potential (GWP)	GWP includes pollutants generated throughout the life cycle of fuel related to waste disposal.
	Waste management	Cost related to waste management
Total benefit	Health	Both acute and chronic health effects are caused by exposure to emissions (indoor pollutants).
	Time-saving collecting firewood for cooking	Time and opportunity cost as times saved could earn money or engage in other productive activities.
	Fatigue/frustration is reduced	Labor is required for fuel collection, procurement, preparation, and use.
Benefits associated with fuel type	Sustainability	The renewability of each fuel type will depend on the reserves and consumption rates.
	Environmental	GHG emission reduction Economic values of local forest areas
	Supply and accessibility	Depending on resource availability and the distribution network
	Alternative uses	Use of the fuel for purposes other than cooking and lighting, for example, space heating purposes.
	Employment opportunities	Jobs were created for the supply, distribution network, sales, and repairs of these technologies.

a) Source of information from author's research and field survey work.

and containing both semi-structured and open-ended ones were used to collect the data. Participants were selected at random based on two social classes (urban and rural population) and their willingness to take part in the exercise. Two social classes were classified based on the following:

1. An infrastructure that is semi-permanent or permanent building for urban households and grass-thatched huts/houses for rural households.
2. The primary source of energy used for cooking and lighting. Charcoal and electricity are the primary sources of energy for urban households, whereas firewood and paraffin are used in rural households.

Data analysis was conducted using statistical package for social science (SPSS) software with descriptive statistics, to estimate the needs for and examine the factors for adoption of modern biomass technologies, solar cooking, and small-scale PV solar power plants in households of Northern Uganda. CBA was carried out to determine the benefits and costs of the adoption of these small-scale renewable energy technologies near Northern Uganda.

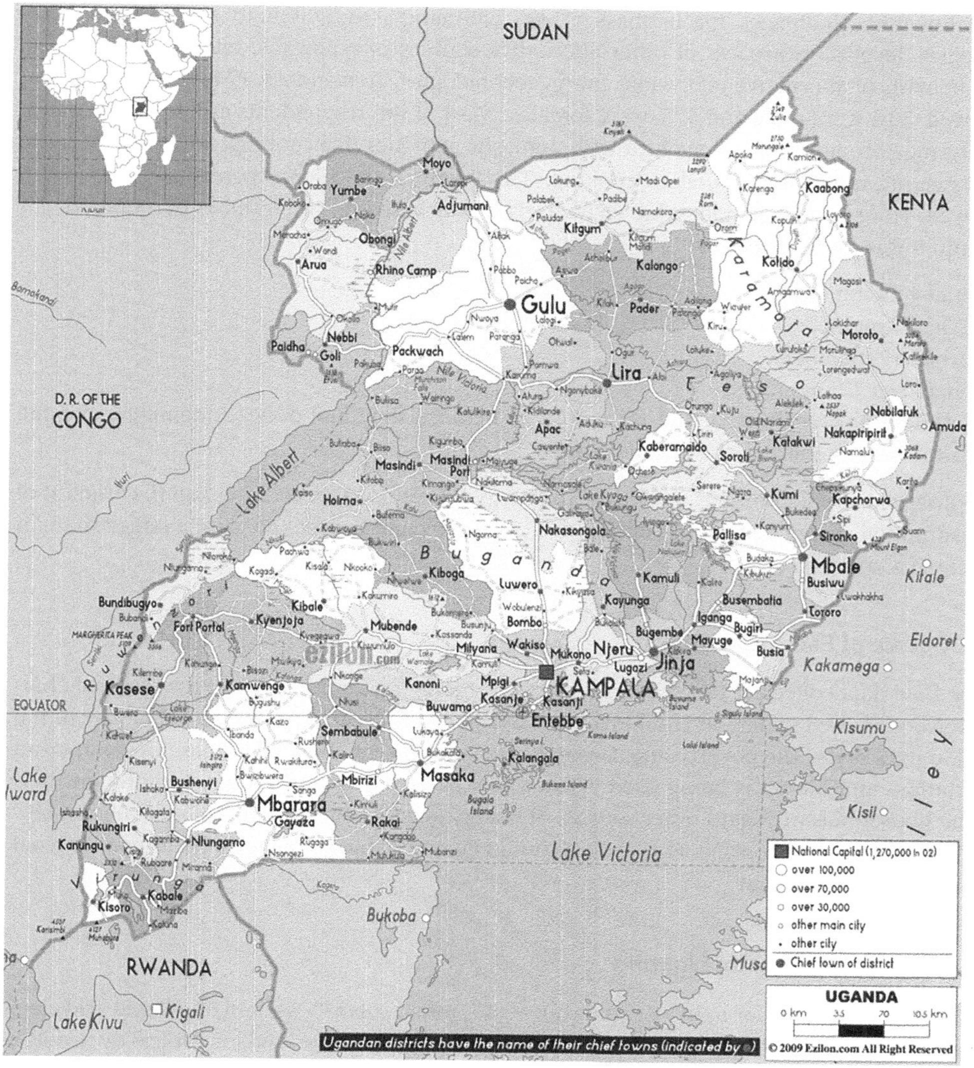

Figure 16.4 Map for capital income and resources Uganda.

16.5 Case Study Results for Product Development

From both the field data collected from the four districts of Northern Uganda and the results from the CBA framework for the household's renewable energy interventions, a solution for a product development framework is proposed in this section.

16.5.1 Field Study

Responses from the 124 sampled population/households from the four districts in Northern Uganda are used in this section. Among the questions asked and considered, where the demographic characteristics, households' sizes, main sources of energy for cooking and lighting, monthly income, challenges for biomass usage, willingness to switch to alternative energy technologies, level of awareness of other alternative sources of renewable energy technologies, and the benefits of alternative renewable energy technologies. To be noted, 124 respondents were interviewed, 33% were male and 67% were females. Most of the respondents are women because culturally, men are not directly involved with domestic activities like cooking, and they shy away from answering questions related to cooking. This justifies the bigger percentage of women interviewed. Household's demographic characteristic was based on the number of members per family, and it was found out to be:

1. 39% had 4 members,
2. 36.3% had 7 members,
3. 15.3% had 12 members,
4. 7.4% had over 12 members,
5. 2% withheld household size, and this could be because they had over 12 members, and the provision was not there in the questionnaire.

The household size determines the choice of energy source used for cooking and the amount of fuel consumption. Large households choose the cheapest source of fuel because a larger share of their income is spent on food.

16.5.2 Source of Energy for Cooking

From the field data collected, the primary sources of energy for cooking were firewood and charcoal. Firewood is used widely in rural areas for availability and affordability as a majority of the population lives below the poverty line. Urban households prefer charcoal for ease of storage and transportation, and bio-oil is being used because few households have grown jatropha on their farm. The households disregard other sources of energy given in the options (Bio-briquettes, biogas, electricity, and solar cooker). However, it should be noted that some households use charcoal and firewood interchangeably.

16.5.3 Source of Energy for Lighting

Primary sources of energy for light were paraffin 60.4%, micro solar PV system 32.3% and grid electricity 3.2%. Paraffin was the primary source of energy for light because it is economical by people's living standards. Micro-solar PV systems are made available and affordable to households in both urban and rural settings through many government incentives programs toward rural electrification. The technology is favored by the constant amount of solar irradiation throughout the year in Northern Uganda (eight hours a day). About 3.2% of the households identified grid electricity as their major source of energy for lighting. These households are all within the urban area. However, grid electricity has been connected to most of the urban and rural areas, yet it is the least source of energy identified. The low fraction of usage can be attributed to its irregularity, and high cost of monthly bills, therefore, most household's disregards grid electricity as their major source of lighting.

16.5.4 Household Income Level

The income level estimate was based only on monthly income per household which showed that 52% of the respondents earn between US$ 10–50 per month, 34% US$ 100–160 and 13% earn over

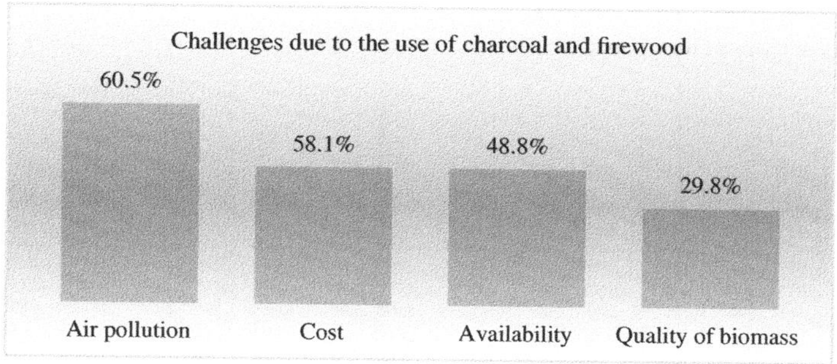

Figure 16.5 Statistics of charcoal and firewood in terms of the selected function. *Source of information from authors research and field survey work.

US$ 160 in a month. The reason for high income with 13% of the respondents is that they are engaged in trading activities or are formally employed, mostly within the urban areas. The low level of income signifies the fact that a majority of the targeted groups are not engaged in commercial activities but in subsistence farming, which is seasonal. This has made it very difficult for them to estimate their monthly income and to quantify their harvest, as they produce crops only for their consumption.

16.5.5 Challenges for Firewood and Charcoal Use

The major challenges associated with charcoal and firewood use were considered from the user's perspective. The result included indoor air pollution, poor quality of biomass, cost and availability, deforestation, scarcity and high prices of charcoal and firewood. About 91% of the population in Northern Uganda depend on charcoal and firewood with a majority using three stones open fire systems. Therefore, the result justifies the corresponding numbers of households and their percentages. About 60.5% of households identified air pollution as the major challenge. About 58.1% of households identified the cost attached to biomass energy as their challenge, with charcoal and firewood being highly-priced. Such high price makes them unaffordable to the population who live below the international poverty line. About 48.4% of households identified the availability of biomass energy as their challenge. About 29.8% of households identified the quality of biomass as their challenge. Finally, 2.2% of households gave other reasons as their challenge for using charcoal and firewood. All these challenges identified by households are for the lack of alternative sources of clean energy or the high prices of viable alternatives that are in the market. It is important to note that, in Figure 16.5, the total number of households and their percentages are not presented because of overlapping numbers of challenges identified. This implies that a household could be confronted with more than one challenge.

16.5.6 The Rank of Adoption Toward Sustainable Renewable Energy Technologies

The field data ranked participants' preferences, benefits, and factors that influenced their decision to switch to alternative energy technologies. These ranks varied across households ranging from energy security, money-saving, environmental responsibility, time-saving to health benefits. Figure 16.6 showed that the residents of the four districts in Northern Uganda will adopt sustainable renewable energy technologies. However, some factors need to be addressed: the cost of renewable energy technologies, level of income, and knowledge about these technologies. Therefore, there is a need to increase loan facilities with a low-interest rate and flexible long term repayment plan, promote donor-funded projects toward renewable energy projects, government-funded

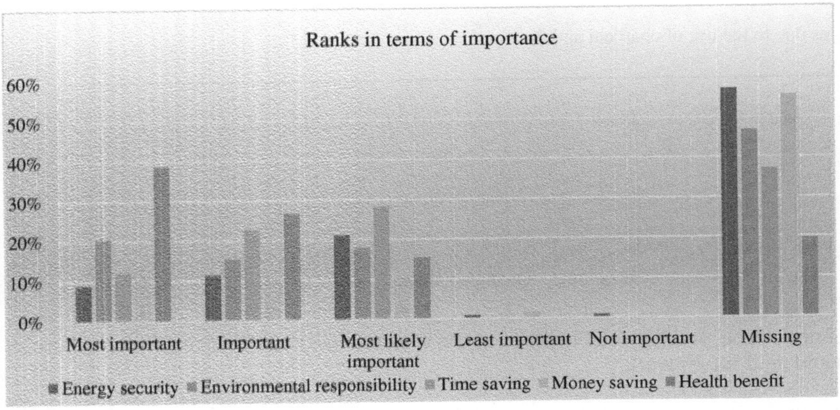

Figure 16.6 Statistics of energy security, time-saving, health benefit, environmental responsibility, and money-saving. *Source of information from authors research and field survey work.

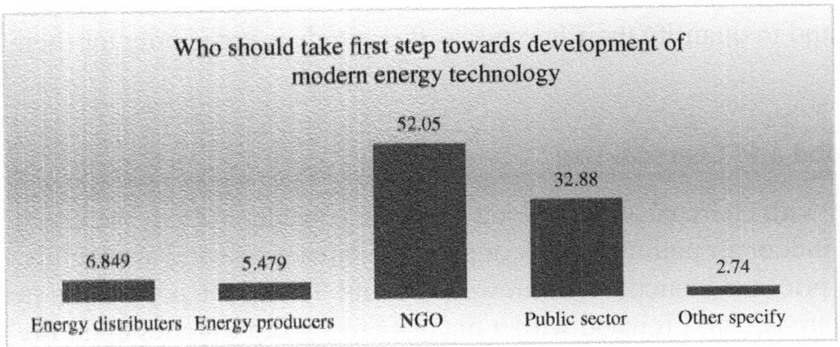

Figure 16.7 Statics of modern energy technology and development prospects. *Source of information from authors research and field survey work.

projects/interventions, increase subsidy led initiatives, and improve information dissemination to increase the adoption level of alternative technologies.

16.5.7 Household Opinions for Modern Energy Technologies

The target groups had different opinions on who should take the first step in providing an alternative energy solution. The responses varied across households based on their experience with the different agencies in the past. About 6.85% of the households preferred energy distributor to take up the first step toward the production of alternative sources of energy, 33.8% of the households preferred the public sector to take up the lead, 52% of the households preferred NGOs, 5.7% of the households preferred the ongoing energy producer. Finally, 2.1% of households made no specification. The reason for households in Northern Uganda to strongly prefer the services delivered by the nongovernmental institutions/organization includes the minimal level of bureaucracy and a high level of transparency in the NGOs implemented projects. Therefore, this is a prompting perspective for the households to prefer NGOs also, prior experience with NGOs is the free distribution of items. So, they believe if they take the first lead, then they will get cheap or freely donated technologies. Figure 16.7 shows the statistics of how the target group responded in expressing their opinion on who should facilitate modern energy technology development.

Table 16.3 Modern bioenergy and small-scale renewable energy for households in Northern Uganda.

Renewable energy for households in Northern Uganda	Number of households	Percentage (%)
Bio-pellets	7	5.6%
Bio-briquettes	14	11.3%
Liquid bio-fuel/bio-oil	24	19.4%
Biogas	17	13.7%
Solar cooker	15	12.1%
Small-scale PV solar	101	81.5%
Others/ none of the above	9	7.3%

a) Source of information from authors research and survey work.

16.5.8 Level of Awareness of the Population

Table 16.3 illustrates the total number of households and their technology preference. Some households identified more than one source of alternative energy technology; hence, the combined percentage exceeds 100%. The result shows, 5.6% of households identified bio-pellets as the most sustainable, 11.3% households identified bio-briquettes, 19.4% households identified liquid bio-fuel, 13.7% households identified biogas energy, 12.1% identified solar cookers, 81.5% identified small-scale PV solar, and 7.3% said none. Other sources of fresh energies like bio-pellets, bio-briquettes, and biogas are not highly used, because of limited knowledge of these technologies. These technologies might seem less advanced and easily affordable, therefore, there is a need for increased information dissemination and awareness about the benefit of these fairly unknown technologies. However, bio-oil would be quickly adopted for the availability of land, since people are already growing jatropha on a small scale in marginal lands.

16.5.9 Medium of Information

Households can receive information about alternative sources of renewable energy technologies through TV/radio media and scaled advertisements in local languages. Figure 16.8 shows the field-collected data for the medium of information. About 31.8% of households identified door-to-door salesperson as a medium of information, 54.55% highlighted that they receive information through advertisement on television or radio, 4.5% of households identified telephone as means of information, 9.09% indicated friends' referral, while 4.55% households chose others but did not specify. The results confirm that most households in Northern Uganda receive vital information through television or radio. The least households receive information through the telephones and friends' referrals. From field study, it is recommended that further information and awareness should be delivered to households through workshops/seminars at the village/local arrangement and public gathering as well as through TV/radios to ensure all households are informed about the alternative renewable energy technologies.

16.5.10 Promotion to Purchase Alternative Renewable Energy Technologies

In Figure 16.9, most of the respondents chose donation as the primary option that enables them to use modern and clean energy technologies. Discounts, free maintenance, long term payment,

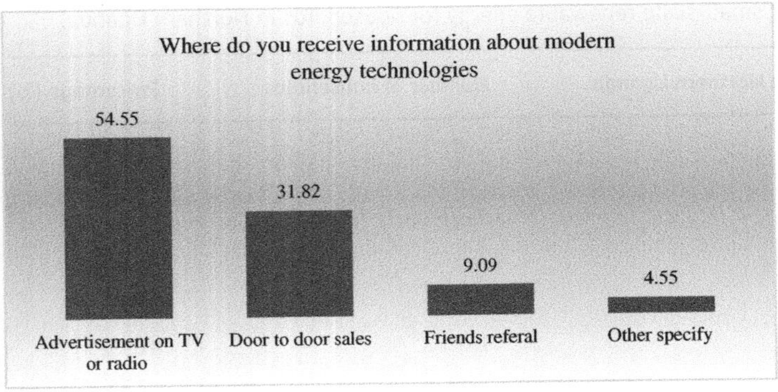

Figure 16.8 Source media for awareness in modern energy technologies. * Source of information from authors research and field survey work.

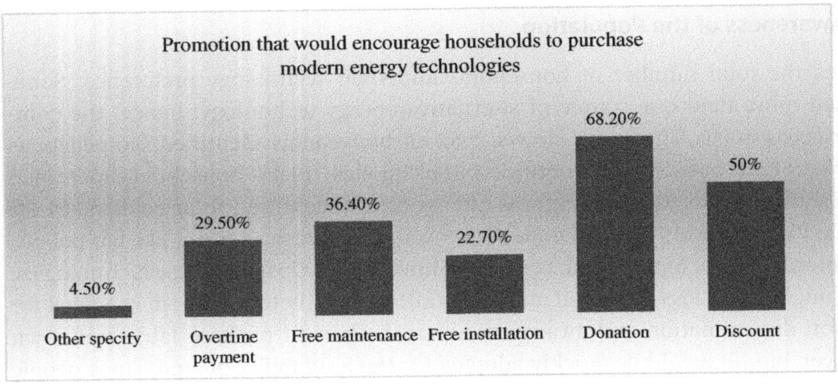

Figure 16.9 Statistics for modern energy technologies and their promotion. * Source of information from authors research and field survey work.

free installation, and other forms of promotion are factors that ease the transition to clean energy technologies. Based on respondents answers, it signifies people favored donation since the economic conditions of the local area limit free cash for other activities. This is because a high segment of society could not afford to buy the cheapest technology in the market. Therefore, the price of energy technology products should be well-thought-out considering affordability for low-income households.

16.5.11 Sources of Fund for Investment in Northern Uganda Toward Renewable Energy Technologies for Households

About 34.4% of the people interviewed will use their savings to buy alternative sources of energy. However, a high percentage opted for aided financing options such as 21.1% donation, 16.7% cost-sharing, 16.7% donor-funded, 10% credit from micro-finance, and 1% chose other sources.

The level of awareness of alternative energy sources was limited because a majority do not know they exist. Also, a majority of the population prefer alternative clean energy technologies to be donated to them; others prefer government lead investment, loans, and cost-sharing because they do not have enough money to invest. From the above responses, it is apparent that the people of Northern Uganda will adopt these alternative clean and sustainable energy technologies, but the economic condition of people is a limiting factor. This study recommends options to support the Northern Uganda public by providing loans with a low-interest rate with a long repayment term, donor-funded projects toward clean energy, government-funded projects/interventions, and subsidy-led programs.

16.6 Cost–Benefit Analysis (CBA)

General assumptions used to determine the CBA include avoided fuel cost and reduced firewood collection time. The following observations were made systematically for the analysis:

i) Fuelwood savings were valued with the average market price of firewood when purchased, and at a shadow price when collected.
ii) The shadow price of firewood collected was based on the average amount of firewood collected per hour, 5 kg [77].
iii) The market price per kg of fuelwood was 0.080 EUR. The shadow price of 1 kg of fuelwood was 0.09 EUR.
iv) The market price for charcoal was 0.15 EUR/kg.
v) The assumption was that 50% of firewood was bought and 50% collected.
vi) The mean household size is 5.5 in the region [78, 81].
vii) Collected firewood was valued at shadow price based on the mean monthly income of households in the study area, while purchased firewood was based on the market price in 2016 of the study areas.
viii) The amount of charcoal was converted by a factor of 9 (9 kg firewood = 1 kg of charcoal) [82].
ix) A shadow wage was assigned to 50% of the saved time. It was assumed that even if the household member could not transfer the time saved directly into money-making activities. That time could have been used to engage in productive activities like farming, doing house chores, and child-caring, improving the family wellbeing.
x) A monetary value was assigned called "shadow wage," derived from an average monthly income of a household in the study area [79], which is amount to 43.98 EUR/month based on the assumption that two household's members make that money in 30 days assuming 8 hours per day of work [83].
xi) Shadow wage was 0.09 EUR/hour.
xii) The corresponding shadow price of firewood collected was 0.09 EUR/kg.

16.6.1 Benefits to Better Health

This is based on averted disability adjusted life years (DALYS), for lack of data, DALYS was not calculated. However, reduced time saved for not falling sick and reduced burn cases from using traditional energy technologies was considered; this was based on the assumptions that:

i) Households that have switched to alternative renewable energy technologies will have reduced acute respiratory diseases, eyes diseases, and burn cases.
ii) On average, a household loses 8 hours in a year for burns of family members, 80 hours for acute respiratory diseases, and 16 hours for eyes problems [83].
iii) A monetary value (shadow wage) of 50% of the time saved for better health benefits was assigned to time saved from health benefits.

iv) The cost of medical bills avoided was set at zero. This is for the lack of data on standard monetary medical bills set in Uganda. Not everyone who gets sick receives treatment from a certified health facility. This scenario limits data availability.

16.6.2 Benefits on Greenhouse Gas Emissions Reduction

The assumptions on greenhouse gas emission avoided for the use of alternative technologies (bio-oil, bio briquettes, bio-pellets, biogas, and solar cookers) instead of charcoal burning, and firewood are presented.

i) A monetary value was assigned to the quantity of CO_2 and CH_4 gas emissions caused by the combustion of firewood and charcoal.
ii) The conversion factor of 1.5 kg of CO_2 for 1 kg of fuelwood burned, and the CH_4 emission factor of 4 g for 1 kg fuelwood burned was used for calculation.
iii) The economic value assumed based on European Union Allowance (EUA) cost of 1 ton of $CO_{2,\text{avoided}}$ according to the closing of April 2018 at 15.11 EUR [84].
iv) For CH_4 emission, no emission price exists but was assumed that 20 times higher price than CO_2 emission. Derived from 20 times higher CH_4 potency than CO_2, the corresponding value was 302.2 EUR per ton of CH_4.

16.6.3 Benefits of the District Forest Resources Preservation

The following hypotheses were made for this analysis:

i) 10% of firewood consumed for cooking comes from the felling of forest cover; this means a 10% savings preserve forest resources. The value of forest reserve is equal to the assumed afforestation cost of 0.0352 EUR/kg of firewood.
ii) This was derived from the average afforestation cost of a forest plantation of 1.2 million USh./ha (288.80 EUR) [85]. Based on an average stock density of 17 m^3/ha of forest cover and an average wood weight of 850 per solid m^3. Losses derived from unmarketable wood species or destruction caused by fire were not taken into consideration in the calculation for afforestation cost.

16.6.4 Outcomes of Cost–Benefit Analysis

Biomass is the primary source of energy in the study area, with a majority using less efficient technologies. The creation of awareness and adaptation of sustainable renewable energy technologies would create a massive impact on the population.

The many benefits of renewable energy systems include:

i) Saves time for women, which could engage in other activities and improve their quality of life.
ii) Reduced fuelwood consumption.
iii) Health benefits for the reduction in indoor air pollution; this will reduce household and government expenditures on health.
iv) Reduction of CO_2 and CH_4 emissions will limit global warming and climate change.
v) Preservation of the local forest, and natural resources.

Table 16.4 shows the benefits for firewood and charcoal saving from using alternative energy sources and technologies. The avoided firewood and charcoal were calculated as 2044 kg and 444 kg per household per year, respectively.

Table 16.4 Firewood and charcoal saving from the use of alternative sources of renewable energy technologies.

Annual fuel consumption per household and total annual savings	Using traditional stove (three stones and metal charcoal stoves)	Saving using alternative modern and clean technologies	Total fuel for households using an alternative source of energy technologies (124) households	Total fuel saving from the local perspectives (Northern Uganda's perspective) 150700 households (ton)
Firewood	2044 kg	2044 kg	253 456 kg	308 036.8
charcoal	444 kg	444 kg	55 056 kg	66 910.8

Table 16.5 Benefits for firewood saved.

	Total annual fuel saved per household (EUR)	Total annual fuel saved by sample population (EUR) 124 households	Total annual fuel saved Northern Uganda's perspective (EUR) 150 700
Economic benefits on firewood saved	93.96	11 651.0	14 159 772

Table 16.6 Benefits of reduced cooking time.

	Household perspective	The sample population (124)	Northern Uganda perspective (150 700)
Total Annual time saved (hours)	166.075	20 593.3	25 027 502.5
Economic benefit corresponding to time saved (EUR)	7.473	926.652	1 126 181.1

Table 16.7 Benefits to preserve the local forest resources.

	Per household perspective (EUR	Sample population (124) (EUR)	Northern Uganda (EUR)
Annual Benefits to preserve district forest	0.72	89.28	108 504

Table 16.5 shows the benefit for firewood and charcoal saved, which amounted to 93.96 EUR per household per year.

Table 16.6 shows the benefit of cooking time saved, which was calculated as 166.08 hours/year, resulting in a monetary saving of 7.47 EUR per year per household.

Table 16.7 shows the economic value to preserve the local forest resource. The economic value amounted to 0.72 EUR per household per year.

Table 16.8 Health benefits.

	Household perspective	The sample population (124)	Northern Uganda perspective (150 700)
Annual time saved (hours)	104	12 896	7 836 400
Economic benefit corresponding to time saved (EUR)	4.68	58.032	705 276

Table 16.9 Greenhouse gas emissions reduction.

	Per household perspective (EUR)	Sample population (124) (EUR)	Northern Uganda (EUR) (150700)
Annual emission saved	2520.35	312 523.1	379 816 745

Table 16.8 shows the health benefits of using alternative energy technologies, which were calculated based on the time saved. A household will lose around 104 productive hours in a year for burn, respiratory infection, and eye infections due to indoor air pollution. With the lost hours avoided due to switching to alternative energy technologies, an economic saving of 4.68 EUR per household per year, can be achieved.

Table 16.9 shows the level of CO_2 and CH_4 emissions reduction amounted to 2520.35 EUR saving in a year per household. Economic benefits to households are shown in Table 16.10.

Tables 16.4–16.10 summarize all the benefits that could be accrued because of the use of alternative sources of renewable energy technologies other than charcoal and firewood. The study showed a household could save 2627.18 EUR per year. The CBA was conducted based on two alternative technologies, biogas (20 years life span) and solar cookers (10 years) only and based on the assumption that: All benefits and costs were discounted to the beginning of 2018. The general assumption is:

Economic analysis was based on the hypotheses that:

vi) The values of the considered alternative technologies like price and the fuel-saving rate would remain constant.

vii) The alternative technologies will be replaced after their average life span.

viii) Discount rates of 5 and 10% per annum were applied to every technology.

ix) The substrate for biogas production was collected free.

To be noted, the survey comprised capital and installation cost, annual maintenance cost, and depreciation of the two alternative energy technologies.

From Tables 16.11 and 16.12, the results for both technologies showed positive net present values (NPVs) at 5 and 10% interest rates and internal rate of return (IRR) of 8% for biogas and 13% for solar cookers. Based on the benefit to cost ratio of 1.09 and 0.5 for biogas and solar, respectively, both technologies are better options for private investment as shown in Table 16.12. The results presented in Table 16.12 show a promising return on investment for households if they switch to alternative technologies.

Table 16.10 Summary of benefits.

Economic benefits considered (valued at shadow price and shadow wage)	Economic benefit per household (Euro)	Economic benefits in Euro (124) Households	Total economic benefit (Euro) Northern Uganda Perspective
Benefits for fuel savings	93.96	11 651.04	14 159 772
Benefit for reduced cooking time	7.473	926.652	1 126 181.10
Benefit for better health	4.68	58.032	705 276
Benefit for preserving district forest	0.72	89.28	108 504
Benefits for reduction in CO_2 and CH_4 emission	2 520.35	312 523.10	379 816 745
Total economic saving	**2 627.183**	**325 248.10**	**398 916 478.1**

Table 16.11 Cost of intervention for biogas and solar cooker.

Cost of intervention (EUR)	Biogas	Solar cooker
Capital and installation	963.1	300.14
Annual maintenance	38.35	12.86
Depreciation	38.35	—
Total cost	1039.8	313

Table 16.12 Result of NPV and IRR.

NPV	5%	10%	IRR	Benefit/cost ratio
Biogas	963.97	954.00	8%	1.09
Solar cooker	516.23	531.20	13%	0.5

16.7 Conclusion

This research study investigated the socioeconomics of alternative renewable energy technologies compared to the base case through a field survey of households in four districts Northern Uganda. The study found positive environmental, health, and economic benefits of the alternative renewable energy technologies; however, there are barriers to their adoption. The barriers limiting the adoption of these renewable energy technology alternatives are socioeconomic, technological pathway dependency, financial, and institutional. The framework to overcome the identified challenges includes the stimulation of agricultural economic activities by the Government of Uganda coupled with donor-funded projects on the identified technologies. Awareness programs on the alternative technologies should be coupled with technical training providing new skills to the inhabitants of the four districts in construction, operation, and maintenance. Lastly, a subsidy-led program and low-interest loan with a long-term repayment plan should be provided to support those interested in transitioning to the alternative technologies but are constrained financially. Appropriate policies, frameworks, conditions, and action plans to increase the level of adoption of the identified alternative renewable energy systems in the study areas of Northern Uganda are therefore needed.

References

1 UNEP (2018). UN-Sustainable Development Goals [cited 2018 28/02]. http://www.un.org/sustainabledevelopment/sustainable-development-goals.

2 Zaccai, S.L.E. (2017). *Making Clean Cooking Champions: Perceptions on Development of Private Actors in Uganda.*

3 UN (2018). Sustainable Development Goals:17 Goals to Transform our World [cited 2018 09/02]. http://www.un.org/sustainabledevelopment/energy.

4 Hansen, K., Breyer, C., and Lund, H. (2019). Status and perspectives on 100% renewable energy systems. *Energy* 175: 471–480.

5 IEA (2017). *World Energy Statistics 2017*. IEA.

6 Bildirici, M. and Özaksoy, F. (2016). Woody biomass energy consumption and economic growth in sub-Saharan Africa. *Procedia Economics and Finance* 38: 287–293.

7 Mottaleb, K.A. and Rahut, D.B. (2019). Biogas adoption and elucidating its impacts in India: implications for policy. *Biomass and Bioenergy* 123: 166–174.

8 Gebreegziabher, Z., Beyene, A.D., Bluffstone, R. et al. (2018). Fuel savings, cooking time and user satisfaction with improved biomass cookstoves: evidence from controlled cooking tests in Ethiopia. *Resource and Energy Economics* 52: 173–185.

9 Singh, R., Wang, X., Mendoza, J.C. et al. (2015). Electricity (in)accessibility to the urban poor in developing countries. *Wiley Interdisciplinary Reviews: Energy and Environment* 4 (4): 339–353.

10 WHO (2018). Household (Indoor) Air Pollution [cited 2018 07/02]. http://www.who.int/indoorair/en.

11 Karanja, A. and Gasparatos, A. (2019). Adoption and impacts of clean bioenergy cookstoves in Kenya. *Renewable and Sustainable Energy Reviews* 102: 285–306.

12 Vaccari, M., Vitali, F., and Tudor, T. (2017). Multi-criteria assessment of the appropriateness of a cooking technology: a case study of the Logone Valley. *Energy Policy* 109: 66–75.

13 Harris, T., Collinson, M., and Wittenberg, M. (2017). Aiming for a moving target: the dynamics of household electricity connections in a developing context. *World Development* 97: 14–26.

14 Charmes, J. (2006). A review of empirical evidence on time use in Africa from UN-Sponsored surveys. In: *Gender, Time use, and Poverty in sub-Saharan Africa* (ed. C.M. Blackden and Q. Wodon), 173. WorldBank.

15 Ramani, K.V. (2002). Energy as an instrument of Women's Economnic empowerment. *Energia News* 5 (1): 8–10.

16 Kofi-Opata, E. (2016). Spatial patterns and trends in energy use and consumption in Africa. *Perspectives on Global Development and Technology* 15 (4): 406–418.

17 Parajuli, I. (2018). Health and environmental impact assessment of biomass cook stove using integrated environmental performance score. *International Journal of Current Research* 10 (6): 8.

18 Olanrewaju, B.T. et al. (2019). A panel data analysis of renewable energy consumption in Africa. *Renewable Energy* 140: 668–679.

19 Präger, F. et al. (2019). Biomass sources for a sustainable energy supply in Ghana – a case study for Sunyani. *Renewable and Sustainable Energy Reviews* 107: 413–424.

20 UN-SE4ALL-Vision (2011). A Vision Statement by Ban Ki-moon Secretary-General of the United Nations.

21 MEMD-SEALL (2015). SE4ALL Action Agenda.

22 Okello, C. et al. (2013). Development of bioenergy technologies in Uganda: a review of progress. *Renewable and Sustainable Energy Reviews* 18: 55–63.

23 Kandel, P. et al. (2016). Consumption patterns of fuelwood in rural households of Dolakha District, Nepal: reflections from community Forest user groups. *Small-Scale Forestry* 15 (4): 481–495.

24 Lee, L.Y.-T. (2013). Household energy mix in Uganda. *Energy Economics* 39: 252–261.

25 Josephat, M. (2018). Deforestation in Uganda: population increase, forest loss and climate change. In: *Environmental Risk Assessment and Remediation* 2 (2): 46–50.

26 WHO (2015). *Climate and Health Country Profile - Uganda*. World Health Organisation.

27 Bridgwater, A.V. (2012). Review of fast pyrolysis of biomass and product upgrading. *Biomass and Bioenergy* 38: 68–94.

28 Chen, Y. et al. (2015). Thermochemical conversion of low-lipid microalgae for the production of liquid fuels: challenges and opportunities. *RSC Advances* 5 (24): 18673–18701.

29 Koh, M.Y. and T.I. Mohd. Ghazi, A review of biodiesel production from *Jatropha curcas L.* oil. *Renewable and Sustainable Energy Reviews*, 2011. 15(5): p. 2240–2251.

30 Yaakob, Z. et al. (2013). Overview of the production of biodiesel from waste cooking oil. *Renewable and Sustainable Energy Reviews* 18: 184–193.

31 Patel, M. and Kumar, A. (2016). Production of renewable diesel through the hydroprocessing of lignocellulosic biomass-derived bio-oil: a review. *Renewable and Sustainable Energy Reviews* 58: 1293–1307.

32 Sugathapala, A.G.T. and Chandak, S.P. (2013). *Technologies for Converting Waste Agricultural Biomass to Energy*. Reading Mater United Nations Environmental Program, Division of Technology, Industry, and Economics, International Environmental Technology Center Osaka, 1–229.

33 UNDESA (2007). Small-Scale Production and Use of Liquid Biofuels in Sub-Saharan Africa: Perspectives for Sustainable Development.

34 Contran, N. et al. (2016). Potentialities and limits of *Jatropha curcas L.* as alternative energy source to traditional energy sources in Northern Ghana. *Energy for Sustainable Development* 31: 163–169.

35 Mogaka, V. et al. (2014). Understanding the underlying mechanisms of recent *Jatropha curcas L.* adoption by smallholders in Kenya: a rural livelihood assessment in Bondo, Kibwezi, and Kwale districts. *Energy for Sustainable Development* 18: 9–15.

36 Grimsby, L.K., Aune, J.B., and Johnsen, F.H. (2012). Human energy requirements in *Jatropha* oil production for rural electrification in Tanzania. *Energy for Sustainable Development* 16 (3): 297–302.

37 Sekoai, P.T. and Yoro, K.O. (2016). Biofuel development initiatives in sub-Saharan Africa. *Opportunities and Challenges. Climate* 4 (2): 33.

38 Okello, C. et al. (2014). Appraising bioenergy alternatives in Uganda using strengths, weaknesses, opportunities and threats (SWOT)-analytical hierarchy process (AHP) and a desirability functions approach. *Energies* 7 (12): 1171–1192.

39 Wikipedia (2018). Pellet fuel. In: *Wikipedia*.

40 Shuma, R. and Madyira, D.M. (2017). Production of loose biomass briquettes from agricultural and forestry residues. *Procedia Manufacturing* 7: 98–105.

41 Dobson, S., Nyamweru, H., and Dodman, D. (2015). Local and participatory approaches to building resilience in informal settlements in Uganda. *Environment and Urbanization* 27 (2): 605–620.

42 Haykiri-Acma, H. and Yaman, S. (2010). Production of smokeless bio-briquettes from hazelnut shell. *Proceedings of the World Congress on Engineering and Computer Science 2010*, Vol. II.

43 Ferguson, H. (2012). *Briquette Businesses in Uganda*. The Potential for Briquette Enterprises to Address the Sustainability of the Uganda Biomass Fuel Market. GVEP International, 1–47.

44 Ståhl, M. (2008). *Improving Wood Fuel Pellets for Household Use*, 91. Karlstad University Studies.

45 Tumutegyereize, P. et al. (2017). Technical evaluation of uptake, use, management and future implications of household biogas digesters—a case of Kampala City peri-urban areas. *Computational Water, Energy, and Environmental Engineering* 06 (02): 180–191.

46 Lwiza, F. et al. (2017). Dis-adoption of household biogas technologies in Central Uganda. *Energy for Sustainable Development* 37: 124–132.

47 Walekhwa, P.N., Lars, D., and Mugisha, J. (2014). Economic viability of biogas energy production from family-sized digesters in Uganda.

48 Mulinda, C., Hu, Q., and Pan, K. (2013). Dissemination and problems of African biogas technology. *Energy and Power Engineering* 05 (08): 506–512.

49 Skovsgaard, L. and Jacobsen, H.K. (2017). Economies of scale in biogas production and the significance of flexible regulation. *Energy Policy* 101: 77–89.

50 Tucho, G. et al. (2016). Problems with biogas implementation in developing countries from the perspective of labor requirements. *Energies* 9 (9).

51 Surroop, D., Bundhoo, Z.M.A., and Raghoo, P. (2019). Waste to energy through biogas to improve energy security and to transform Africa's energy landscape. *Current Opinion in Green and Sustainable Chemistry* 18: 79–83.

52 McKendry, P. (2002). Energy production from biomass. *Bioresource Technology* 83: 37–46.

53 Walekhwa, P.N., Mugisha, J., and Drake, L. (2009). Biogas energy from family-sized digesters in Uganda: critical factors and policy implications. *Energy Policy* 37 (7): 2754–2762.

54 Berhe, T.G., Tesfahuney, R.G., Desta, G.A., and Semungus, L. (2017). Mekonnen, biogas plant distribution for rural household sustainable energy supply in Africa. *Energy and Policy Research* 4 (1): 10–20.

55 Njenga, M. et al. (2019). Innovative biomass cooking approaches for sub-Saharan Africa. *African Journal of Food, Agriculture, Nutrition and Development* 19 (01): 14066–14087.

56 Regattieri, A. et al. (2016). Innovative portable solar cooker using the packaging waste of humanitarian supplies. *Renewable and Sustainable Energy Reviews* 57: 319–326.

57 MacClancy, J. (2015). Solar Cookers. In: *The SAGE Encyclopedia of Food Issues*.

58 MacClancy, J. (2015). Solar cooking. *Food, Culture & Society* 17 (2): 301–318.

59 Hansen, U.E., Pedersen, M.B., and Nygaard, I. (2014). Review of Solar PV Market Development in East Africa.

60 Enclude, B.V., Heemskerk, L. (The Netherlands), Eenhoorn, G., Namiti, B. (2014). Market Assessment of Modern Off Grid Lighting Systems in Uganda.

61 Feron, S. (2016). Sustainability of off-grid photovoltaic systems for rural electrification in developing countries. *A Review. Sustainability* 8 (12): 1326.

62 Feron, S., Heinrichs, H., and Cordero, R.R. (2016). Sustainability of rural electrification programs based on off-grid photovoltaic (PV) systems in Chile. *Energy, Sustainability and Society* 6 (1).

63 Fransson, T. (2010). Solar PV Based Rural Electrification in Rema Rural Village, p. 116.

64 Bell, S. (2012). DPSIR=a problem structuring method? An exploration from the "imagine" approach. *European Journal of Operational Research* 222 (2): 350–360.

65 Ness, B., Anderberg, S., and Olsson, L. (2010). Structuring problems in sustainability science: the multi-level DPSIR framework. *Geoforum* 41 (3): 479–488.

66 Sun, S. et al. (2016). Sustainability assessment of regional water resources under the DPSIR framework. *Journal of Hydrology* 532: 140–148.

67 Tscherning, K. et al. (2012). Does research applying the DPSIR framework support decision making? *Land Use Policy* 29 (1): 102–110.

68 Bidone, E.D. and Lacerda, L.D. (2004). The use of DPSIR framework to evaluate sustainability in coastal areas. Case study: Guanabara Bay basin, Rio de Janeiro, Brazil. *Regional Environmental Change* 4 (1): 5–16.

69 Lin, Y.-C., Huang, S.-L., and Budd, W.W. (2013). Assessing the environmental impacts of high-altitude agriculture in Taiwan: a driver-pressure-state-impact-response (DPSIR) framework and spatial emergy synthesis. *Ecological Indicators* 32: 42–50.

70 O'Higgins, T. (2015). A Systems Approach to Environmental Management 2015.

71 Sekovski, I., Newton, A., and Dennison, W.C. (2012). Megacities in the coastal zone: using a driver-pressure-state-impact-response framework to address complex environmental problems. *Estuarine, Coastal and Shelf Science* 96: 48–59.

72 Gibson, J. (2018). Forest loss and economic inequality in the Solomon Islands: using small-area estimation to link environmental change to welfare outcomes. *Ecological Economics* 148: 66–76.

73 Jorgenson, A.K. and Burns, T.J. (2007). Effects of rural and urban population dynamics and national development on deforestation in less-developed countries, 1990–2000. *Sociological Inquiry* 77 (3): 460–482.

74 Hutton, G., Rehfuess, E. Swiss Tropical Institute with, and World Health Organization (2006). Guidelines for conducting cost–benefit analysis of household energy and health interventions.

75 Foell, W., Pachauri, S., Spreng, D., and Zerriffi, H. (2011). Household cooking fuels and technologies in developing economies. *Energy Policy* 39 (12): 7487–7496.

76 Bruce, N.G., Rehfuess, E.A., and Smith, K.R. (2011). *Household Energy Solutions in Developing Countries*. In: Nriagu, J.O. (ed.) *Encyclopedia of Environmental Health*, 3: 6275.

77 Habermehl, H. and Eschborn: GTZ (2007). *Economic Evaluation of the Improved Household Cooking Stove Dissemination Programme in Uganda*.

78 UBOS (2017). *The Uganda National Household Survey 2016/17*.

79 Uganda Bureau of Statistics (2017). Statistical Abstract. Updated on November 20th, 2017 [cited 2018 2/02/2018]. http://www.ubos.org/onlinefiles/uploads/ubos/statistical_abstracts/2017_Statistical_Abstract.pdf.

80 World Bank Uganda country profile (2018). [cited 2018 28/02]. http://databank.worldbank.org/data/Views/Reports/ReportWidgetCustom.aspx?Report_Name=CountryProfile&Id=b450fd57&tbar=y&dd=y&inf=n&zm=n&country=UGA.

81 UBOS (*2018*). National Housing Statisitcs [cited 2018 24/5]. http://www.ubos.org/onlinefiles/uploads/ubos/UNHS_12_13/2012_13%20UNHS%20Final%20Report.pdf.

82 Uganda-Energy -Situation (2018). [cited 2018 25/05]. https://energypedia.info/wiki/Uganda_Energy_Situation.

83 García-Frapolli, E. et al. (2010). Beyond fuelwood savings: valuing the economic benefits of introducing improved biomass cookstoves in the Purépecha region of Mexico. *Ecological Economics* 69 (12): 2598–2605.

84 Markets Insider (2018). [cited 2018 23/05]. http://markets.businessinsider.com/commodities/historical-prices/co2-emissionsrechte/euro.

85 NFA (2018). Vital Forestry Statistics in Uganda [cited 2018 24/05]. www.nfa.org.ug/index.php/component/fsf/?view=faq&catid=5.

Part VI

Other Machine Learning Applications

17

Non-Intrusive Load Monitoring Using A Parallel Bidirectional Long Short-Term Memory Model

Victor Andrean and Kuo-Lung Lian

Department of Electrical Engineering, National Taiwan University of Science and Technology, Taipei, Taiwan

Nomenclature

X	aggregated power signals
x_i	the element of X
t	time stamp
T	time length
Y	labeled individual appliance power
k	index individual appliance
$\hat{Y}_{(k)}$	estimated values by neural network for kth data
F_s	DNN model
W	sliding window length
ϵ	residual error
Zn_i	normalized data
i	the order of the data in array
σ	standard deviation
Zs_i	standardized data
a	boundary of uniform distribution
ΔP	power difference
ΔPv	difference of the variant power
$p(t)$	instantaneous power
α	a reflection rate which is a ratio for reflecting variation between original and variant signals
$M_l^{(i)}$	one dimensional CNN
L	size of a convolutional filter
θ_k	the kth element in a convolutional filter
M	feature map
f	the activation function
n_p	number of trainable parameters
f_t	forget gate
σ_a	sigmoid activation function
$W_{\xi x}$, $W_{\xi h}$	weight matrices for the input data(x_t) and previous output value (h_{t-1})
b_ξ	a bias vector associated with a particular neural network layer inside the LSTM

Intelligent Data Mining and Analysis in Power and Energy Systems: Models and Applications for Smarter Efficient Power Systems, First Edition. Edited by Zita Vale, Tiago Pinto, Michael Negnevitsky, and Ganesh Kumar Venayagamoorthy.

$\widetilde{C}_t$	a vector of new candidate values
i_t	input gate
o_t	output gate
n_p	number of trainable parameters
I_c	number of input channel
U	number of units used in LSTM
$Var(\theta)$	variance of initial values of learnable parameters of neural networks.
n_{in}	number of units to the layer
n_{out}	number of units out of layer
$y_t^{(i)}$	actual power of the appliance i at time t
$\hat{y}_t^{(i)}$	estimated power of appliance i at time t
$\overline{y}_t$	mean of aggregated actual power at time t

17.1 Introduction

The increase in energy consumption and the emergence of various types of loads have led to complexities in managing energy on the demand side. Therefore, energy conservation, security, and reliability are highly needed not only for consumers but also for power companies. For consumers, load monitoring is useful to suppress energy consumption, detect faulty appliances, and observe appliance usage behavior. For a utility company, load monitoring can provide information to support intelligence diagnosis, load peak shaving procedure, optimization of power scheduling. Non-intrusive load monitoring (NILM) is a load monitoring scheme that requires single sensor installed at the electrical service entry (ESE) capturing whole appliance power signals (aggregate signal). The aggregate signal is then disaggregated to obtain the power consumption of each appliance. In the U.S., most low-cost smart meters available in the market have a low sampling rate, typically 1 Hz or less [1]. With this low sampling rate, active power patterns are the common information used to estimate energy consumption. There are two methods that are frequently used to perform load disaggregation, i.e. optimization-based and learning-based approaches.

17.1.1 Optimization-Based Approach

Load identification problem can be formulated into an objective function with some constraints. Then, this model can be solved using some optimization methods such as mixed integer linear programing [2, 3] or metaheuristic methods. For instance, Egarter and Elmenreich [4] tested six different metaheuristic algorithms, i.e. evolutionary algorithm, differential evolution, particle swarm optimization, simulated annealing, cuckoo search algorithm, and firefly algorithm. The results show that the metaheuristic algorithms provide satisfying results for an appliance set where power draws are easy to separate. However, it is hard to disaggregate loads with similar power draws. Bhotto et al. [2] proposed a load disaggregation method based on integer programming. Nevertheless, this model is limited to the use of constraints that do not depend on time measurements. Suzuki et al. [5] formulates the NILM problem as an integer quadratic programming problem. The technique represents the problem, as a combination of waveforms from multiple loads. At any given period, the overall load current is represented as a superposition of the currents of the operating appliances. The overall current waveform is considered to be influenced by the waveform of each individual appliance. However, this technique requires data at a very high sampling rate. Thus, a large number of points from each cycle are necessary. The major problem of an optimization-based approach is that the accuracy greatly declines and the computational burden dramatically increases as the number of appliances increases. Moreover, the presence of unknown loads makes the model even more complex. Apart from being computationally expensive, appliances with similar or overlapping load signatures are difficult to discern using this approach.

17.1.2 Learning-Based Approach

A learning-based system is also known as a data-driven system. It requires an expert to specify some parameters such as the optimizer algorithm, number of layers and neurons, and activation functions. Such parameters are commonly called hyperparameters in machine learning field. The other parameters such as weights and filter values are called learnable parameters because they are dynamically updated during the training process.

Learning-based models are generally divided into two types namely the supervised and unsupervised. The supervised learning model is a model that requires both input and target data (labeled data) to train a model. The error values can be calculated easily by subtracting the output value yielded by the model with the target data values. On the other hand, unsupervised learning is aimed to learn principal structure or distribution of the data. Unlike supervised learning, labeled data are not necessary on unsupervised learning. In energy disaggregation, a learning-based model is expected to extract the estimated power consumption of a certain appliance (disaggregated power) from total power consumption (aggregate power). The model can be supervised learning such as the methods proposed by Biansoongnern and Plangklang [6], Zhang et al. [7], and Kelly and Knottenbelt [8] or the unsupervised learning such as that in [9].

Factorial Hidden Markov Models (FHMM) and combinatorial optimization have been developed by scholars for NILM. In FHMM, aggregate power consumption can be considered as observable states and disaggregated power consumption can be considered as hidden states. Hence, by giving a sequence of aggregate power, FHMM can infer whether a particular load is likely to be ON or OFF. In 2011, Kolter and Johnson [10] built a dataset called Reference Energy Disaggregation Data Set (REDD) and proposed FHMM to disaggregate energy of a specific appliance. However, it only achieved 47.7% accuracy. The work was then further developed by Kim et al. [11] and Zoha et al. [12], respectively in years of 2011 to 2013. Kim used variants of FHMM to develop probabilistic models of appliance behaviors. Zoha utilized real power, reactive power and power factor to recognize a specific load pattern of an appliance. However, the result indicates that the F1 score [13] significantly decrease as the number of appliances increases [14]. Another disadvantage in using FHMM model is that it can only be used for discrete state appliances [15]. In [16], the authors proposed an additive factorial approximation maximum a posterior (AFAMAP) based on iterative fuzzy C-means to adaptively obtain the number of clusters to improve the accuracy of the load disaggregation.

The uses of supervised learning are still the mainstream for the NILM studies [17]. For instance, Lin and Tsai [18] built an NILM system based on artificial neural network (ANN) using extracted features such as edge detection. Zhou et al. [19] proposed an NILM system using neural network patterns recognition (NNPR) which is a novel research direction in the field of pattern recognition. However, those methods are sensitive to the detection of false edges coming from noise or non-linear loads. Complex and inconsistent pattern as well as unintended noise makes simple ANN fail to address well the NILM problems. In 2015, Kelly and Knottenbelt [8] have applied techniques of deep learning on NILM. They proposed long short-term memory (LSTM) network to perform nonlinear regression tasks between aggregate power readings and appliance power readings at the same time stamps. They apply sequence-to-sequence approach as an input-to-output relation (IOR)[1] and use sliding windows along the aggregate readings and take a mean of the

1 Note that IOR such as sequence-to-sequence will be further explained in Section 17.2.3.

overlapping outputs. This model, however, does not work very well on dishwashers, microwaves, and washing machines which are the appliances that have multistates and rapid changing patterns. On average, the approach achieved F1 score equal to 0.38. In the same year, Mauch and Yang [15] proposed a similar architecture as the LSTM proposed by Kelly. The difference is that they divide the aggregate readings into several non-overlapping blocks to be processed individually. This technique makes the training data become less varied and it causes the model easily to be either under-fitting or over-fitting due to large window size and numerous LSTM units. Mauch and Yang did not use aggregate signal data provided by the REDD dataset to validate their method. Instead they used synthetic aggregate power signals which were obtained by summing up all the sub-metered signals. The result from validation is less satisfying except for the fridge. Moreover, the testing results were even much worse than those at the validation stage, especially for the result of the microwave. Later, this work has been improved by Krystalakos et al. [20] in 2018. They proposed a sequence-to-point approach. This model maps data from sliding windows into single points. However, it was shown that the F1 score was no more than 0.5. Kelly and Knottenbelt [8] also proposed an autoencoder (AE) for the NILM. The paper showed that this model successfully outperforms FHMM. Nevertheless, it still works poorly on device with multi-state and rapidly changing patterns such as dishwashers, microwaves, and washing machines. Zhang et al. [7] proposed a sequence-to-point IOR. They implemented a deep convolutional neural network (DCNN) as an NILM model. The idea behind this is to use information, before and after midpoint to estimate the midpoint. This method produces one single output for each window, so there is no overlapping point. Based on the results shown in the paper, this method is still not good enough to handle multi-state patterns. In 2018, Chen et al. [21] proposed a combination of gated linear unit (firstly proposed by Dauphin et al. [22]) and residual network named (GLU-Res). This approach applies sequence-to-sequence IOR. Based on our investigation, this model can perform well on multistate appliance. However, rapidly changing load still cannot be well-estimated by this model. Moreover, Kelly and Knottenbelt [8] and Mauch and Yang [15] have implemented Bidirectional Long Short-Term Memory (BLSTM) on NILM applications. The BLSTM performs quite well but it sometimes has difficulty to predict loads with similar patterns or load with random patterns.

In the time series problem, feature engineering is still highly required especially when the time series data are not stationary such as stock price [23]. Non-stationary time series data are a type of data whose mean or/and the variance change over time. One approach to deal with this kind of time series is to take the difference between the current point and the previous point. This technique is well known as differencing [24]. In the field of stock price forecasting, we need to perform even more feature engineering to help the prediction model grasp more information due to the complex underlying pattern in the data. Nabipour et al. [25] extracted 10 technical indicators, i.e. moving average (MA), weighted moving average (WMA), momentum, relative strength index (RSI), and moving average convergence/divergence (MACD) as inputs for the model. Some other papers that use technical indicators as inputs for their deep learning techniques can be found in Refs. [26–28].

In the case of NILM, some appliances like fridge have a constant pattern; hence the mean and the variance should be constant. However, most of the appliances such as a dishwasher or kitchen outlet is non-stationary since the mean and the variance would vary for a given different window.

We proposed a feature extractor module to extract statistical features (SFs) that can help the model to well identify the start and end points on rapidly changing patterns and the states on multistate loads. SFs also provide a deep learning model with stationary features. Secondly, we proposed a parallel architecture that is used to facilitate independent learning from different engineered features. A concatenation layer is added to combine information like an ensemble model. Hao and Gao [29] used multi-head or parallel CNN-LSTM branch to learn from multiple time scale features, i.e. short-, medium-, and long-term features. At the end of each branch, extracted features are concatenated and further be passed to the fully connected neural network. On the other hand, in our work, parallel structures are used for extracting feature representation from two sources, i.e. SFs and original features. The original features are passed to keep some information that might have gotten phased out during feature engineering process. Parallel branches prevent two flows of features from interrupting one to another. The idea is that the original features will not be weighted too much in the concatenation layer/ensemble layer if the information is not pertaining to the target output. The proposed method is validated with the REDD dataset. Several most recent methods such as DCNN [7], BLSTM [20], and AE [8], and GLU-Res [21] were also implemented, and compared with the proposed model.

The rest of the chapter is organized as follows. Section 17.2 describes the NILM system and the proposed data-preprocessing. Section 17.3 describes the proposed method. The validation results and discussions are described in Section 17.4. Finally, Section 17.5 provides the conclusion of the chapter.

17.2 NILM System and Data Preprocessing

Mathematically, aggregated power signals can be written as $X = (x_1, x_2, \dots, x_T)$, where t stands for time stamp and T is the time length. Targeted individual appliance power (ground truth) can be written as $Y_{(k)} = (y_{k1}, y_{k2}, \dots, y_{kT})$. Mapping function of an NILM model then can be expressed as $Y_{(k)} = F_s(X_{[t-W:t]}) + \epsilon_k$ where the estimation can be expressed by $\hat{Y}_{(k)} = F_s(X_{[t-W:t]})$, where F_s is the DNN model, W is the sliding window length, ϵ is the residual error, and k is the index individual appliance. $\epsilon_k = \hat{Y}_{(k)} - Y_{(k)}$.

Figure 17.1 describes gradual steps (shown in dashed box) before designing neural network structure. In the following subsections, data scaling, windows length selection, and IOR are explained.

17.2.1 Data Scaling

In general, there are two types of data preprocessing namely normalization and standardization. Normalization rescales the values into a range of [0,1]. The normalized data (Zn_i) are calculated by (17.1).

$$Zn_i = \frac{x_i - \min(X)}{\max(X) - \min(X)} \tag{17.1}$$

where i indicates the order of the data in the array, X is the aggregated power vector, and x_i is the element of the vector.

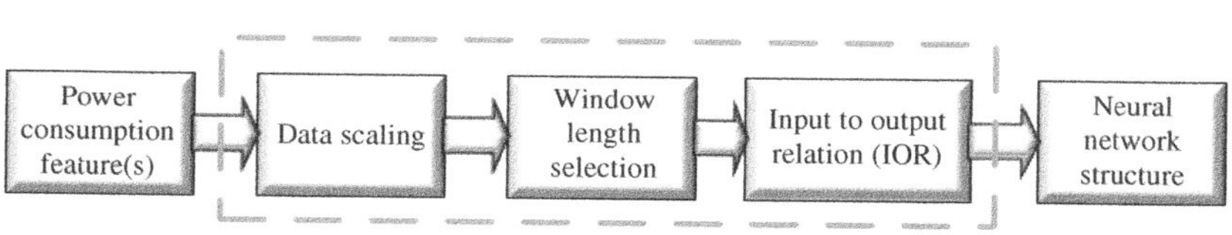

Figure 17.1 Data preprocessing.

On the other hand, standardization rescales data to have a mean equal to zero and standard deviation (σ) of 1 (unit variance). The standardized data (Zs_i) are calculated by (17.2)

$$Zs_i = \frac{x_i - \text{mean}(X)}{\sigma} \tag{17.2}$$

where $\sigma = \sqrt{\frac{\sum_{i=1}^{T}(x_i - \text{mean}(X))}{T}}$

We have re-implemented some of the most recent deep learning methods on NILM. Eventually, data preprocessing employed for this method is based on their original publication, and is listed as follows:

1. Standardized input and standardized output: DCNN employs this mode.
2. Standardized input and normalized output: BLSTM, AE, and proposed method employ this mode.
3. Normalized input and normalized output: GLU-Res employs this mode.

17.2.2 Window Length Selection

Table 17.1 shows the window length of each appliance. The longer the window, the more points get covered yet the more information will end up becoming noise due to irrelevant information that corresponds to the target. Overfitting is a condition when a model overly fits the training data so that it fails to generalize the pattern that it is supposed to learn. In other words, the model finds a way to fit to the noise pattern instead of the meaningful pattern in the data, which is something that we do not want the model to do. Hence, it is necessary to find an optimal window length to provide sufficient information to the model yet reduce the irrelevant information that may lead to overfitting.

In this work, we selected the window length based on the time taken by one complete cycle of the corresponding target appliance. Figure 17.2 shows the window length (shown in dash line) for the refrigerator and kitchen outlet. For appliances which do not exhibit repeated patterns such as lighting, any window length can be chosen.

17.2.3 Input-to-Output Relation (IOR)

There are commonly two approaches for applying a sliding window, and they are sequence-to-point and sequence-to-sequence as shown in Figure 17.3. Moreover, either sequence-to-sequence or

Table 17.1 Windows length and handpicked maximum power for each appliance.

Appliance	Window length
Refrigerator (FR)	320
Dishwasher (DW)	512
Microwave (MW)	128
Lighting (LT)	192
Kitchen outlets (KO)	256

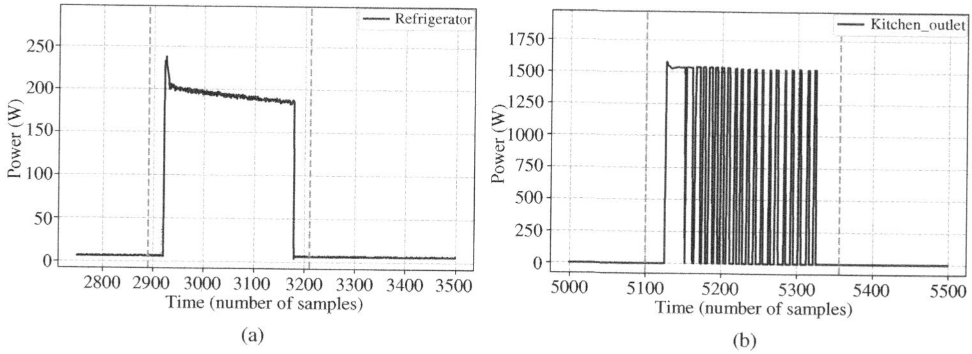

Figure 17.2 Window length for (a) refrigerator, (b) kitchen outlet.

sequence-to-point can be further varied, and there are at least four variations that can be made which are called IOR1, IOR2, IOR3, and IOR4.

17.2.3.1 IOR1

It takes sequence input from $t - W$ to t and produces single output at t as shown in Figure 17.3a, and mathematically stated in (17.3). *IOR1* is applied to the BLSTM and the proposed method.

17.2.3.2 IOR2

It takes sequence input from $t - W$ to t and produces single output at $t - W/2$ as shown in Figure 17.3b, and mathematically stated in (17.4). *IOR2* is applied to the DCNN model.

17.2.3.3 IOR3

It takes sequence input from $t - W$ to t and generates sequence output at the middle with length equal to 1/8 of length input as shown in Figure 17.3c, and mathematically stated in (17.5). *IOR3* is applied to the GLU-Res model.

17.2.3.4 IOR4

It takes sequence input from $t - W$ to t and generates sequence output with length equal to input length as shown in Figure 17.3d, and mathematically stated in (17.6). *IOR4* is applied to the AE model.

Because GLU-Res and AE produce a sequence output, there are multiple estimated values for each time step. Hence, an average for overlapping estimation points is usually taken.

$$Y_{(k)[t]} = F_s(X_{[t-W:t]}) + \epsilon_k \tag{17.3}$$

$$Y_{(k)[t-W/2]} = F_s(X_{[t-W:t]}) + \epsilon_k \tag{17.4}$$

$$Y_{(k)[t-a*W:t-b*W]} = F_s(X_{[t-W:t]}) + \epsilon_k \tag{17.5}$$

where $a{=}3.5/8$ and $b{=}4.5/8$.

$$Y_{(k)[t-W:t]} = F_s(X_{[t-W:t]}) + \epsilon_k \tag{17.6}$$

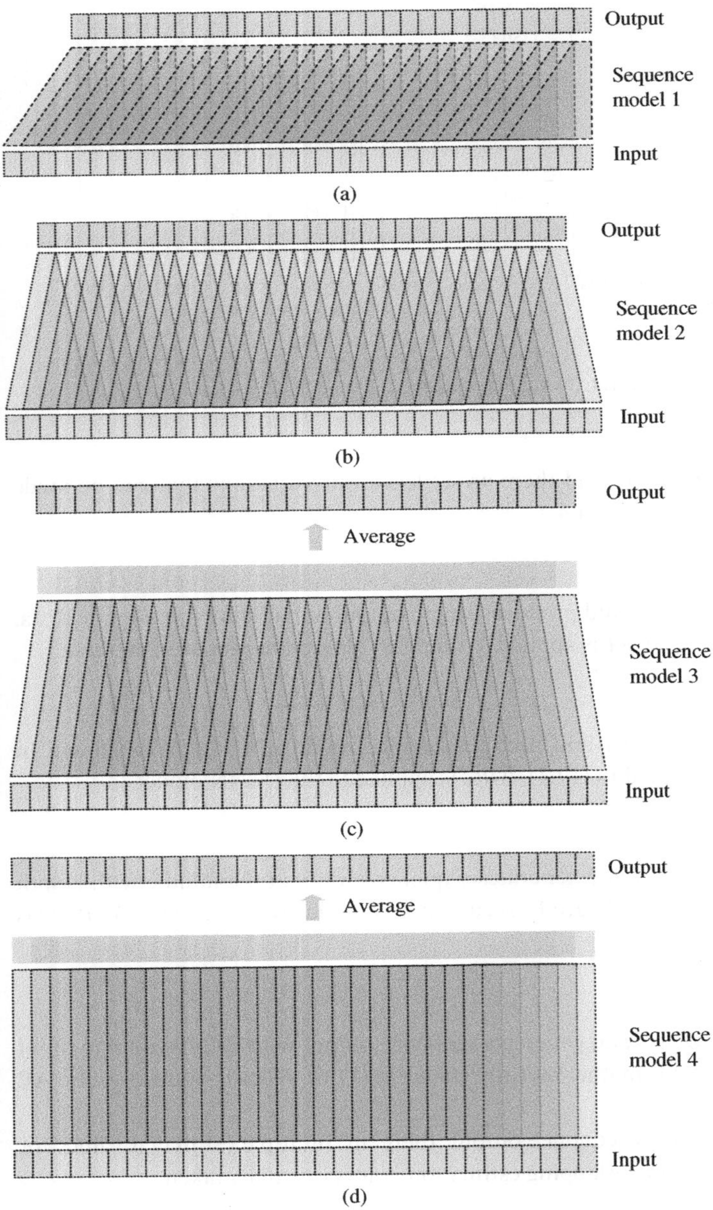

Figure 17.3 (a) IOR1, (b) IOR2, (c) IOR3, and (d) IOR4.

17.3 Proposed Method

The proposed model, named as parallel Bidirectional Long Short-Term Memory Model (PBLSTM) for energy disaggregation, is shown in Figure 17.4. The aggregate power is fed to the left branch (Block 1) of the network, while the right branch (Block 2) receives the SF(s) from the feature extractor. The individual parallel branch allows the network to individually process and extract information from both inputs. BLSTM is chosen because its superiority to handle sequence problem. The process continues to Block 3 of Figure 17.4. Concatenation layers in the end are added

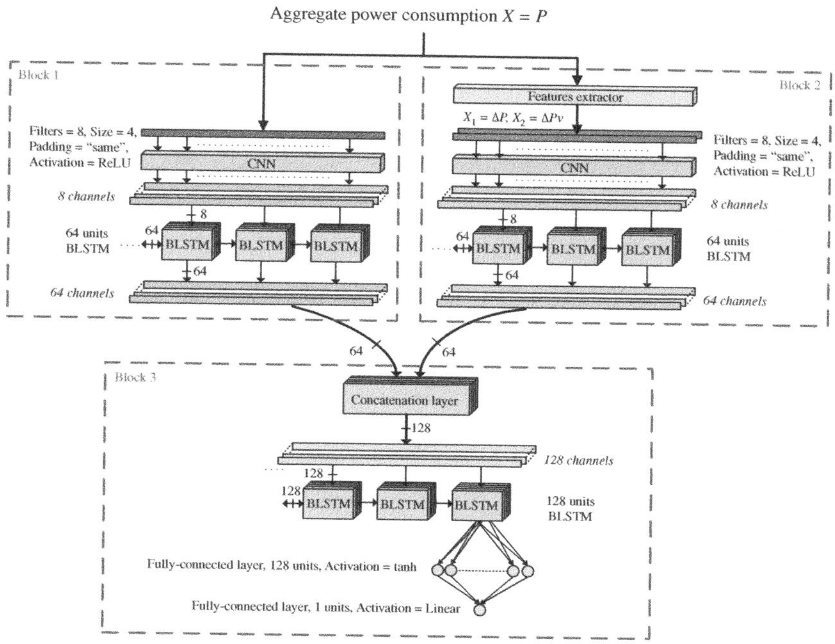

Figure 17.4 Parallel bidirectional long short-term memory model.

to ensemble the two individual models and combines the information all together. After concatenation layer, another BLSTM layer is added. Finally, a dense or also called fully connected layer is added to enhance the nonlinear fitting ability.

17.3.1 Feature Extractor

A feature extractor module is added in the model to extract several SFs from the aggregated power signal. Two SFs, which are the power difference (ΔP) and the difference of the variant power (ΔPv), are used. The difference of the variant power was proposed in [14].

ΔP can function as an event-based detectors. It extracts the state transition edges of appliances to identify the start and the end of an event. To recognize appliances that have rapidly changing patterns such as microwave or kitchen outlet, this feature will help the network to recognize the rapidly changing patterns as shown in Figure 17.5a. Mathematically, ΔP is written in (17.7).

$$\Delta P = p(t) - p(t-1) \tag{17.7}$$

On the other hand, ΔPv is to emphasize the variations in the appliance states. For appliances with multi-states, this feature can be very useful as shown in Figure 17.5b. ΔPv can be done by subtracting one variant signal from the other as shown in (17.8). α is known as a reflection rate which is a ratio for reflecting variation between original and variant signals.

$$\Delta Pv = pv_1(t) - pv_2(t) \tag{17.8}$$

$$pv_1(t) = (1-\alpha_1)pv_1(t-1) + \alpha_1 p(t)$$

$$pv_2(t) = (1-\alpha_2)pv_2(t-1) + \alpha_2 p(t)$$

where $\alpha_1 = 0.1$ and $\alpha_2 = 0.01$

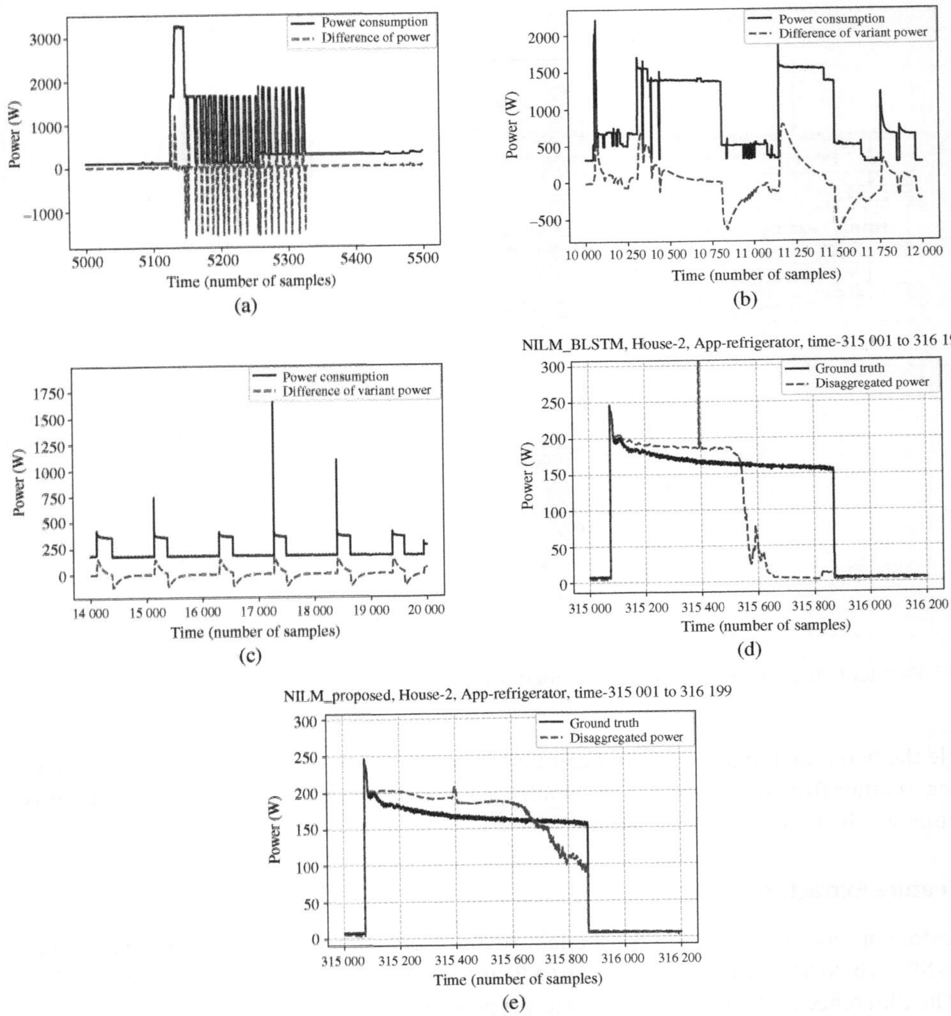

Figure 17.5 (a) Difference of power along aggregate power that consists of KO pattern, difference of variant power along aggregate power that consists of (b) DW pattern, (c) FR pattern, power estimation of (d) BLSTM, (e) PBLSTM.

In multi-state appliances, momentary noise is likely to occur and it can have a negative influence on learning. The low reflection rate can reduce the effect of momentary noise but preserve long-range variations as illustrated in Figure 17.5c. Note that when the reflection rate is too low, the original power pattern may not be reflected. Kim et al. [14] has empirically found that 0.1 or 0.01 is a proper reflection rate. The difference in variant power signal then is used as a second feature in our method.

As the window slides, the window may not always cover complete patterns of the target appliances. ΔP can help the model to recognize the current state. Figure 17.5d,e shows the difference in the estimated energy consumption using, respectively, BLSTM and proposed PBLSTM. Figure 17.5e shows that the presence of SF leads to accurate energy estimation.

Table 17.2 Statistical features.

Appliance	Feature
Refrigerator (FR)	ΔPv and ΔP
Dishwasher (DW)	ΔPv
Microwave (MW)	ΔP
Lighting (LT)	ΔPv and ΔP
Kitchen outlets (KO)	ΔPv and ΔP

Different appliances exhibit different patterns. Some features work well at certain appliances. Based on prior investigation, we found that for certain loads, a combination of the SFs can improve the model performance, while for the other loads, single SF is better. This is summarized in Table 17.2. A refrigerator mostly has on/off state and constantly changing patterns. In this case, ΔP functions as an edge detector for on and off times, whereas ΔPv reduces unintended noise and emphasizes on the variations in the corresponding states. A microwave has no long-term ON-state. However, it exhibits rapidly changing patterns. For such an appliance, we find that ΔP significantly increases the performance. Lighting means a combination of the lamps in one house. Thus, the lighting pattern contains many states. For such a complex load, we find that adding ΔP and ΔPv makes the performance improve drastically. Different from microwave, the kitchen outlet exhibits rapidly changing patterns over some random periods. For such a load, ΔP together with ΔPv can enhance the learning capability.

17.3.2 Elements of the PBLSTM

In this section, we will go through the other elements of the proposal PBLSTM, which include CNNs, BLSTMs, and dense layers.

17.3.2.1 Convolution Neural Network (CNN)

CNNs are commonly used to learn locally connected features. CNNs works by convoluting filter along feature maps. As shown in Figure 17.6, convolutional process happens in Blocks 1 and 2. CNN layer consists of eight filters. Filter size is equal to 4. Zero padding is added such that the output has the same length as the input. As shown in Figure 17.6, input consists of two channels, each convolutional filter will consist of two arrays. The arrays slide along the input features with stride equal to 1. The elements of arrays are multiplied to the elements of input features. After multiplication, the results are added up together. Two arrays result in two scalar values. Those values are added up into a single value. Finally, the value is passed to the activation function ReLU. Eight filters result in eight feature maps as the output.

In the parallel CNN, the power consumption patterns and SFs from the feature extractor module are processed separately. The purpose is to mitigate the negative influence of less related features that may potentially degrade the model's learning ability. From Figure 17.4, CNN at the left side processes the power consumption pattern, and it learns to generate locally connected features from this input. On the other hand, SFs provide useful information to identify the states and the edges

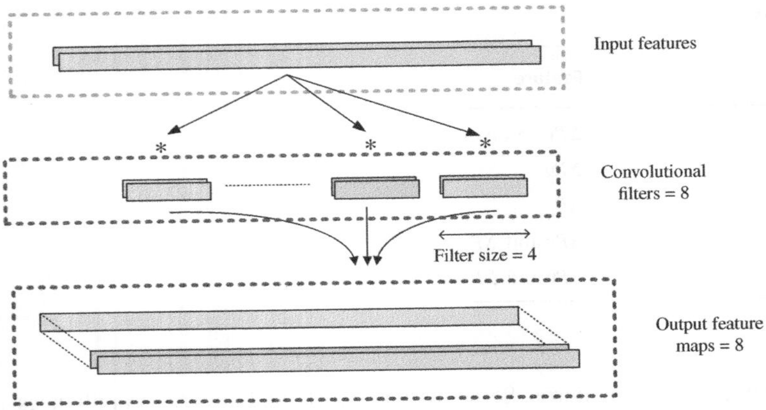

Figure 17.6 CNN structure.

of power patterns. Those features are also processed using CNN to get locally connected features. One dimensional CNN can be mathematically expressed in (17.9),

$$M_l^{(i)} = f\left(\sum_{k=1}^{L} \theta_k M_{i+k}^{l-1}\right) \tag{17.9}$$

where L is the size of a convolutional filter. θ_k is the kth element in a convolutional filter. M is the feature map, l indicates lth layer of CNNs, i indicates ith data, and f indicates the activation function. Output size after lth convolutional layer can be calculated in (17.10).

$$M_s^l = \left(\frac{M_s^{l-1} - L + 2P}{\text{Stride}}\right) \tag{17.10}$$

where P stands for total element of zero padding. *Stride* stands for the sliding step of kernel and M_s^l stands for the size of feature maps at layer l. Total trainable parameters (n_p) between two layers are counted based on (17.11).

$$n_p = (\theta_s M_s^{l-1} + 1)M_s^l \tag{17.11}$$

17.3.2.2 Bidirectional Long Short-Term Memory (BLSTM)

Standard LSTM networks process sequences in temporal order and they ignore future context. Referring to Figure 17.7a, there are three gates on LSTM. The **forget gate** (f_t) controls which information from the last cell state to be kept, and its mathematical form is stated in (17.12).

$$f_t = \sigma_a(W_{fx}x_t + W_{fh}h_{t-1} + b_f) \tag{17.12}$$

where σ_a is a sigmoid activation function to map a variable between 0 and 1. Note that $W_{\xi x}$ and $W_{\xi h}$ are the weight matrices for the input data (x_t) and previous output value (h_{t-1}). b_ξ is a bias vector associated with a particular neural network layer inside the LSTM. The subscript ξ is associated with a particular gate. For instance, W_{fx} and W_{fh} are the weight matrices for the input data and previous output value of f_t. b_f is the bias vector of f_t.

A *tanh* layer generates a vector of new candidate values ($\tilde{C}_t$), whereas **the input gate** (i_t) controls which values to be updated. The mathematical formula is shown in (17.13).

$$i_t = \sigma_a(W_{ix}x_t + W_{ih}h_{t-1} + b_i) \tag{17.13}$$

The candidates of the next state are determined by (17.14).

$$\tilde{C}_t = \tanh(W_{cx}x_t + W_{ch}h_{t-1} + b_C) \tag{17.14}$$

Finally, the current state (C_t) is calculated by (17.15).

$$C_t = f_t \otimes C_{t-1} + i_t \otimes \tilde{C}_t \tag{17.15}$$

It is based on previous state (C_{t-1}) and candidate state ($\tilde{C}_t$). As stated in (17.15), (f_t) and (i_t) serve to control the percentage information being passed. Output value (h_t) is determined by *tanh* of current cell state and controlled by **output gate** (o_t).

$$o_t = \sigma_a(W_{ox}x_t + W_{oh}h_{t-1} + b_o) \tag{17.16}$$

$$h_t = o_t \otimes \tanh(C_t) \tag{17.17}$$

Number of trainable parameters can be calculated by (17.18)

$$n_p = 4((I_c + 1)U + U^2) \tag{17.18}$$

where I_c stands for the number of input channels and U stands for number of units used in LSTM.

BLSTM networks shown in Figure 17.7b has the ability to process forward and backward context by passing sequence data from two directions using two separate hidden layers. The sequence output from forward and backward flows are concatenated all together. As shown in Figure 17.4, the BLSTM layer consisting of 64 units are placed after the CNN layer. The purpose of this layer is to process sequence information by the CNN layer. Both BLSTM layers will produces 64 feature maps. These feature maps are concatenated together in the concatenation layer and finally are fed into the second BLSTM layer, consisting of 128 units of BLSTM. In this stage, the feature maps will get processed and summarized.

In the convolution layer, the time series will be filtered and added up with the other channel to form new feature maps. Different information may interfere one another in this process. Parallel architecture allows the process to be done separately and be combined after having been summarized by LSTM layer. In addition, the proposed PBLSTM utilizes fewer BLSTM units compared to the BLSTM models.

17.3.2.3 Dense Layer

A dense or known as fully connected neural network is used to transform data from multi-dimension to one-dimension for the inference process. Fully connected neural network can

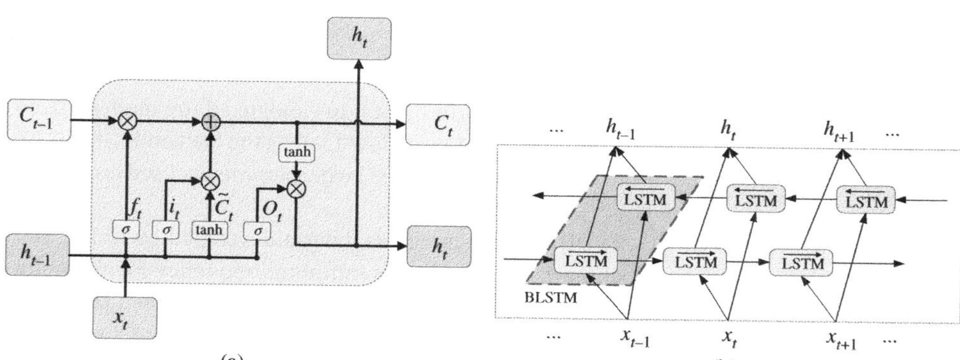

Figure 17.7 Structure diagram of (a) Long short-term memory (LSTM), (b) bidirectional LSTM.

also enhance the nonlinear fitting ability. After BLSTM layer, two dense layers are added. The first layer consists of 128 neurons and followed by *tanh* as the activation function. The second layer is the output layer. It consists of single neurons with a linear activation function. The number of trainable parameters of FCLs (n_p) can be calculated as shown in (17.19).

$$n_p = (I_s + 1)N \tag{17.19}$$

17.3.2.4 Deep Neural Network Training

For supervised learning, the target vector is subtracted from the output model to calculate loss function as shown in (17.20). Loss function applied in this work is the *mean square error*. Training process updates the weights in such a way to minimize the loss function.

$$\text{Loss} = \frac{1}{T}\sum_{l=1}^{T}(y_i - f(x_i))^2 \tag{17.20}$$

Initialization is an important factor that determine the success of training. In the past, uniform or gaussian distribution with random variance was commonly used. Later, it was found that selecting random defined variance can slow down the training. Scholars, Glorot and Bengio [30], through comprehensive research found that variance should be regulated. By defining variance in (17.21), uniform distribution (Glorot uniform) is defined as (17.22). In this work, Glorot uniform is used.

$$\text{Var}(\theta) = \frac{2}{(n_{in} + n_{out})} \tag{17.21}$$

and θ are sampled from the following uniform distribution:

$$\theta \sim U[-a, a] \tag{17.22}$$

where $a = \sqrt{\frac{6}{(n_{in}+n_{out})}}$, and n_{in} and n_{out}, respectively, are the number of units to the layer and the number of units out of layer. Mini batch gradient descent technique makes the training process more efficient. Based on investigations done previously [31, 32], the mini-batch commonly is chosen between 32 to 512. In this work, mini-batch equal to 128 is used. In the past, Adaptive Gradient Algorithm (AdaGrad) was widely used as an optimizer for neural network model. AdaGrad decays the learning rate very aggressively as the denominator grows. As a result, after a while, the frequent parameters will start receiving very small updates because of the decayed learning rate. Adaptive Moment Estimation (ADAM) proposed by Kingma and Ba [33] is simple to implement, computationally efficient, and requires little memory. Therefore, ADAM is used as the optimizer.

17.4 Validation

The multipurpose platform for this work has been built in Python to validate all the models. Deep learning models are done using Keras 2.2.2 as API with tensorflow 1.9.0 as the backend. GPU GTX NVIDIA 1070 with 8 GB of RAM was used to train the model. The Hyperparameters used are shown in Table 17.3.

Data set used in this work is REDD. REDD is an open access dataset developed by Kolter and Johnson [10]. It is known to be the benchmark for NILM. The sampling frequency for the main and appliance are 1 and 1/3 Hz, respectively. To synchronize them, we down sampled the reading of the main so that the sampling frequency is 1/3 Hz. REDD consists of data of six houses in the region of *Massachusetts*, USA. Maximum power of each individual appliance is shown in Table 17.4. These values are obtained based on REDD dataset by measuring the peak of each load.

Table 17.3 Hyperparameters.

Parameter	Specification
Input	P (Real power)
Number of epochs	100
Batch size	128
Cost function	Mean square error (MSE)
Optimizer	Adaptive moment estimation (ADAM)
Weight initialization	Glorot uniform
Preprocessing	Depend on each method

Table 17.4 Threshold and maximum power of each appliance.

Appliance	Threshold (W)	Max power (W)
Refrigerator (FR)	50	500
Dishwasher (DW)	300	1400
Microwave (MW)	200	2000
Lighting (LT)	100	300
Kitchen outlets (KO)	500	1200

Table 17.5 Household appliances description of REDD dataset.

House	Total number of appliances	Total data	Total duration (days)
1	10	406 748	14
2	8	316 840	11
3	12	376 150	13
4	11	428 076	15
5	14	77 451	2.6
6	11	192 192	6.7

Table 17.5 shows the complete description of the dataset. It has the sampling interval equal to three seconds. Table 17.6 shows a list of appliances in each house. In this chapter, houses 1 and 5 are selected for training and house 2 for testing. The total data for training and testing are 77 451 and 316 840, respectively. Five different common appliances across three different houses which are Refrigerator (FR), Dishwasher (DW), Microwave (MW), Lighting (LT), and Kitchen Outlet (KO) are selected to evaluate the performance of each model.

Five different metrics shown in (17.23)–(17.27) are included to assess the disaggregation performance of each model. Proportion of total energy correctly assigned (PTECA) is a metric benchmark formulated by Kolter and Johnson [10] when REDD dataset was first released. The equation is stated in (17.23). MAE measures mean absolute error of each data point toward the ground truth.

Table 17.6 REDD dataset.

House	Total individual appliances	Total data	Total duration (days)	Device categories
1	10	406 748	14	Oven, **refrigerator**, **dishwasher**, **kitchen outlets**, **lighting**, washer dryer, **microwave**, bathroom gfi, electric heat, stove
2	8	316 840	11	**Kitchen outlets**, **lighting**, stove, **microwave**, washer dryer, **refrigerator**, **dishwasher**, disposal
3	12	376 150	13	Outlets unknown, lighting, electronics, refrigerator, disposal, dishwasher, furnace, washer dryer, microwave, smoke alarms, bathroom gfi, kitchen outlets
4	11	428 076	15	Lighting, furnace, kitchen outlets, outlets unknown, washer dryer, stove, air conditioning, miscellaneous, smoke alarms, dishwasher, bathroom gfi
5	14	77 451	2.6	**Microwave**, **lighting**, outlets unknown, furnace, washer dryer, subpanel, electric heat, bathroom gfi, **refrigerator**, **dishwasher**, disposal, electronics, **kitchen outlets**, outdoor outlets
6	11	192 192	6.7	Kitchen outlets, washer dryer, stove, electronics, bathroom gfi, refrigerator, dishwasher, outlets unknown, electric heat, lighting, air conditioning

The performance of ON/OFF status can be assessed from F1 score. To classify the condition, a threshold is defined. The appliance is considered as ON if the appliance's power consumption at a particular time is greater than the threshold and as OFF if its power consumption is less than or equal to the threshold. The threshold value for each appliance used in this chapter is shown in Table 17.4. Precision (Positive Predictive Value) is used to detect the existence of false positive (FP), also commonly called the false alarm. The higher the precision, the less FP. Recall is to detect the existence of false negative (FN) also commonly called mis-hit. The higher the recall, the less the value of FN. Finally, F1 faithfully measures the identification accuracy of the model.

$$\mathrm{PTECA} = 1 - \frac{\sum_{t=1}^{T} |\hat{y}_t^{(i)} - y_t^{(i)}|}{2\sum_{t=1}^{T} \bar{y}_t} \tag{17.23}$$

where $y_t^{(i)}$ represents the actual power of the appliance i at time t, $\hat{y}_t^{(i)}$ represents the estimated power of appliance i at time t, $\bar{y}_t$ is the mean of aggregated actual power at time t, and $\bar{y}_t = \sum_{i=1}^{n} \overline{y_t^{(i)}}$

$$\mathrm{MAE} = \frac{1}{T}\sum_{t=1}^{T} |\hat{y}_t^{(i)} - y_t^{(i)}| \tag{17.24}$$

$$\text{precision} = \frac{TP}{TP + FP} \tag{17.25}$$

where TP represents the number of true positives and FP represents the number of false positives.

$$\text{recall} = \frac{\mathrm{TP}}{\mathrm{TP} + \mathrm{FN}} \tag{17.26}$$

where FN is the number of false negatives.

$$\mathrm{F1} = 2 \cdot \frac{\text{precision} \cdot \text{recall}}{\text{precision} + \text{recall}} \tag{17.27}$$

Table 17.7 summarizes all the disaggregation performances. The best values are shown in bold. The training was done for one network per target appliance. PTECA generally can assess the

Table 17.7 Performance comparison.

Model	App.	PTECA	MAE	Precision	Recall	F1
DCNN	FR	93.062	32.025	0.893	0.870	0.881
	DW	97.929	9.570	0.567	0.881	0.690
	MW	98.214	8.233	0.887	0.881	0.884
	LT	89.679	47.603	0.491	0.749	0.593
	KO	97.412	11.940	0.982	0.323	0.486
Average		95.259	21.874	0.764	**0.741**	0.707
GLU-Res	FR	92.445	34.874	0.913	0.847	0.878
	DW	97.522	11.447	0.448	0.999	0.618
	MW	97.829	10.008	0.495	0.759	0.599
	LT	91.014	41.444	0.447	0.706	0.548
	KO	96.781	14.850	0.850	0.192	0.313
Average		95.118	22.525	0.630	0.701	0.591
BLSTM	FR	91.858	37.569	0.942	0.757	0.839
	DW	97.945	9.485	0.504	0.601	0.548
	MW	98.178	8.400	0.688	0.826	0.750
	LT	91.734	38.123	0.415	0.717	0.525
	KO	97.032	13.692	0.814	0.629	0.709
Average		95.326	21.563	0.695	0.698	0.683
AE	FR	89.513	48.387	0.906	0.583	0.710
	DW	92.405	35.061	0.000	0.000	0.000
	MW	97.173	13.035	0.317	0.703	0.437
	LT	89.771	47.177	0.470	0.613	0.532
	KO	96.437	16.438	0.438	0.063	0.111
Average		93.060	32.020	0.426	0.392	0.358
PBLSTM	FR	93.873	28.270	0.967	0.828	0.892
	DW	98.209	8.269	0.773	0.557	0.648
	MW	98.455	7.121	0.960	0.915	0.937
	LT	93.117	31.742	0.526	0.693	0.598
	KO	98.019	9.141	0.964	0.653	0.778
Average		**96.335**	**16.909**	**0.838**	0.729	**0.771**

PTECA shown in percentage and *MAE shown in Watt*

performance of the energy estimation, whereas F1 is used to assess the performance of classifying ON/OFF status. For example, Figure 17.8 shows the energy estimation of Refrigerator for various methods. Values of PTECA of Refrigerator for DCNN, GLU-Res, BLSTM, AE, PBLSTM are 93.062%, 92.445%, 91.858%, 89.513%, 93.873%, respectively. It is evident that those with higher PTECA scores match closer to the ground truth. Another parameter to assess the performance is MAE. MAE measures the mean of the power error of a particular appliance. For example, Figure 17.9 shows the energy estimation of Kitchen outlets for various methods. The values of MAE of Kitchen outlets for DCNN, GLU-Res, BLSTM, AE, PBLSTM are 11.94, 14.850, 13.692, 16.438, 9.141, respectively. Those with lower MAE estimate energy more accurately.

GLU-Res performs poorly for appliances that exhibit rapidly changing patterns such as Kitchen outlets as shown in Figure 17.9b. This is because taking an average of overlapping points along the sliding window leads to difficulty in learning rapidly changing patterns.

F1 score measures identification accuracy of a model. The value of F1 of Refrigerator for DCNN, GLU-Res, BLSTM, AE, PBLSTM are 0.881, 0.878, 0.839, 0.710, and 0.892, respectively. Those with higher F1 identify ON/OFF load better.

PBLSTM mainly composed of BLSTM, can outperform previously proposed BLSTM model. Tables 17.8 and 17.9 have clearly indicated this fact. Note that more stars indicate better performance. For example, PTECA of a Refrigerator for BLSTM is 91.858. On the other hand, the PTECA

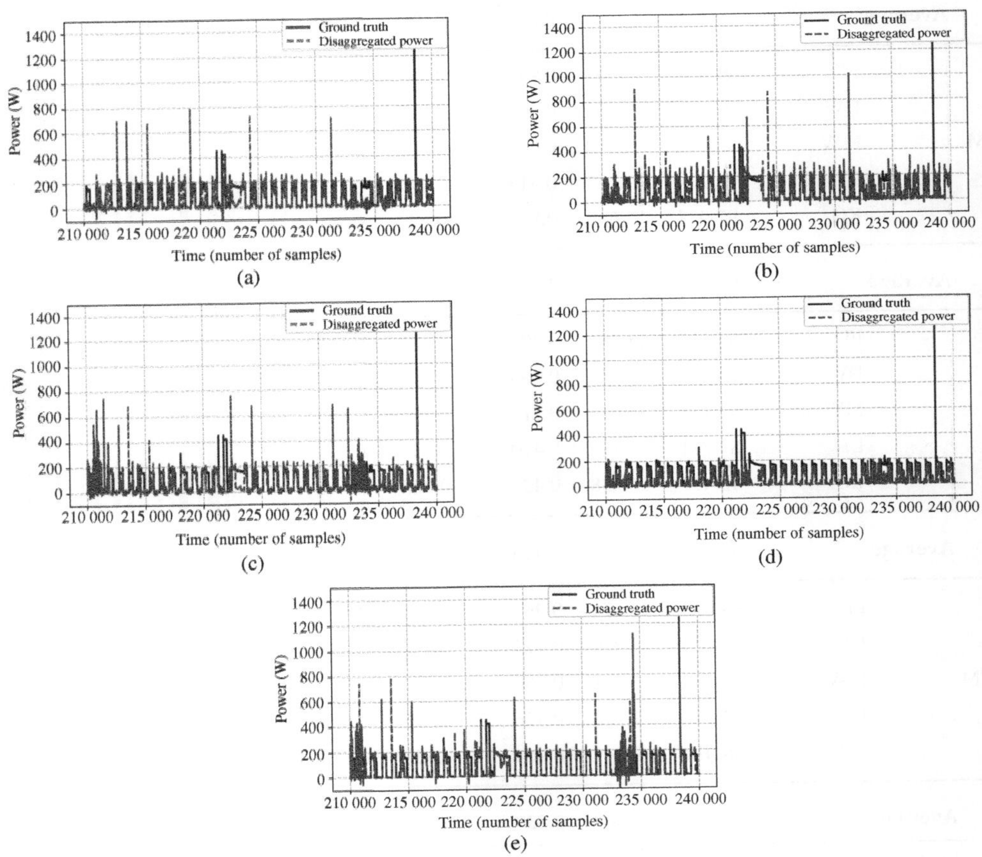

Figure 17.8 Ground truth and disaggregate power of refrigerator (FR) (a) DCNN, (b) GLU-Res, (c) BLSTM, (d) AE, and (e) PBLSTM.

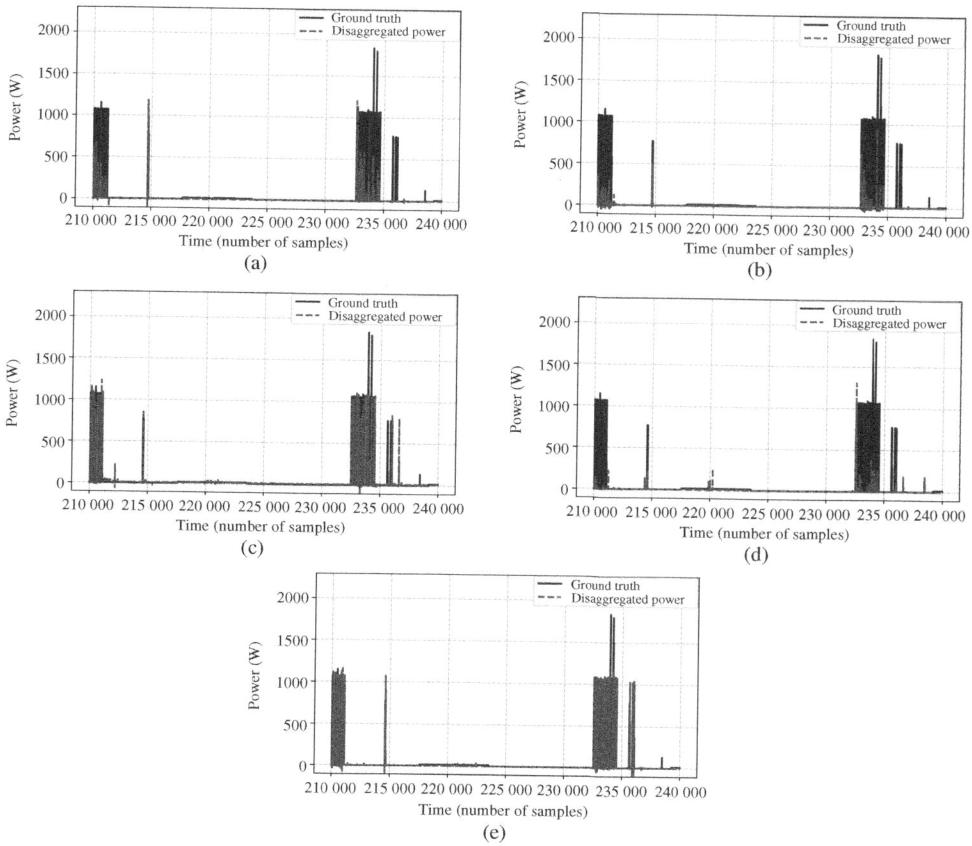

Figure 17.9 Ground truth and disaggregate power of kitchen outlet (KO) (a) DCNN, (b) GLU-Res, (c) BLSTM, (d) AE, and (e) PBLSTM.

of a Refrigerator for PBLSTM can raise the score to 93.873. Similarly, the F1 score of a Refrigerator for BLSTM is 0.750. On the other hand, PBLSTM can raise the score to 0.937.

Referring to Tables 17.9, AE performs the worst compared to the others. AE is mainly composed of three dense layers in a bottleneck-like shape [34]. The purpose is to extract target appliance patterns and eliminate those that are not the target appliance patterns. In general application, AE is used as a signal denoiser such as removing grains from an image. In this case, AE learns the most salient features of the training data. Unlike the noisy picture, in NILM, the target pattern is immersed among the other appliance patterns. In this case, AE filters out some of the target appliances. Table 17.8 shows that AE performs poorly at almost all appliances in terms of PTECA. Moreover, AE completely fails to estimate power for Dishwasher. Dishwasher pattern happens very sparsely along the training data as can be seen in Figure 17.10. This makes AE to learn the dominant features extremely difficult. Therefore, AE estimates all the power to be close to zero as shown in Figure 17.10d. Moreover, AE performs poorly at rapidly changing appliances such as Kitchen outlets as shown in Figure 17.9d. Taking an average of overlapping points along the sliding window makes the network difficult to learn such a pattern. Therefore, AE completely missed the switching event.

From the descriptions in Section 17.3, PBLSTM makes use of SFs to improve the learning capability of the model. We see that PBLSTM successfully outperform the BLSTM in PTECA and F1 score,

Table 17.8 PTECA comparison in ranking representation.

	PTECA				
	FR	DW	MW	LT	KO
DCNN	****	***	****	*	****
GLU-Res	***	**	**	***	**
BLSTM	**	****	***	****	***
AE	*	*	*	**	*
PBLSTM	*****	*****	*****	*****	*****

as shown in Tables 17.8 and 17.9, respectively. SFs help the model to recognize multi-state and rapidly changing patterns. PBLSTM estimates energy very well not only for rapidly changing pattern appliance such as Kitchen outlets as shown in Figure 17.9a but also for multi-state appliances such as Refrigerator and Dishwasher as shown in Figures 17.8a and 17.10a.

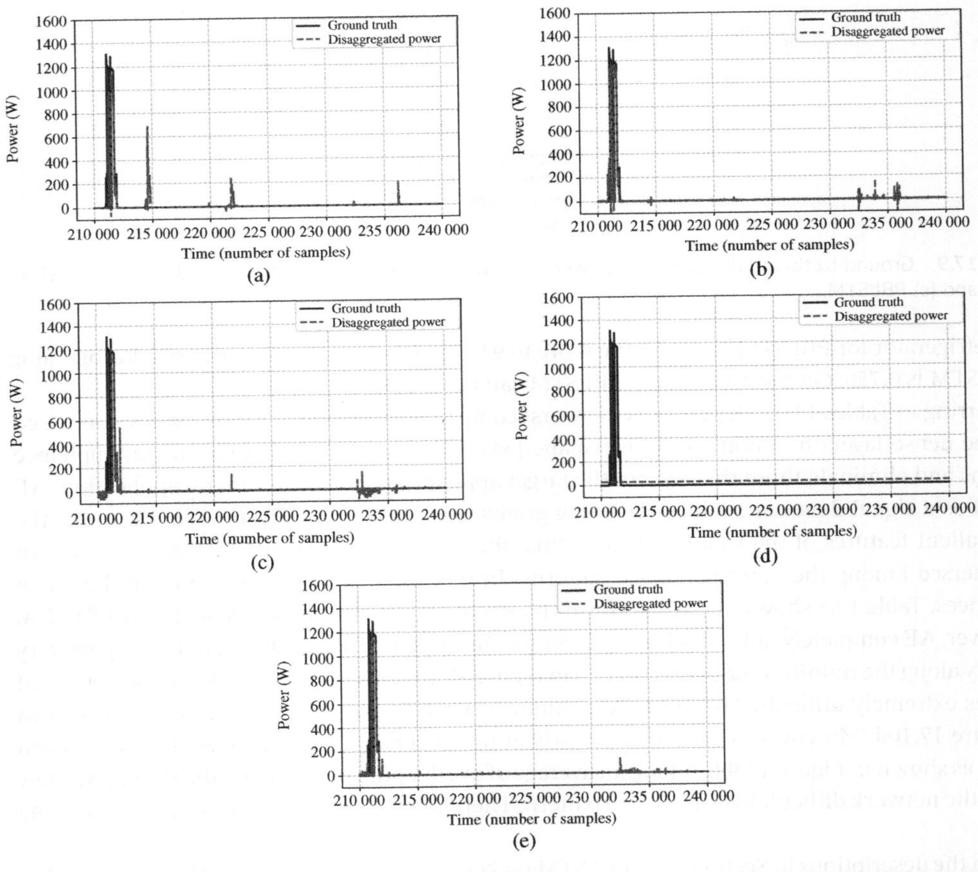

Figure 17.10 Ground truth and disaggregate power of dishwasher (DW) (a) DCNN, (b) GLU-Res, (c) BLSTM, (d) AE, and (e) PBLSTM.

Table 17.9 F1 comparison in ranking representation.

	F1				
	FR	DW	MW	LT	KO
DCNN	****	*****	****	****	***
GLU-Res	***	***	**	***	**
BLSTM	**	**	***	*	****
AE	*	*	*	**	*
PBLSTM	*****	****	*****	*****	*****

Note that for Dishwasher, the F1 score of BLSTM is 0.548 which is 0.142 lower from the highest value, achieved by DCNN. However, if PBLSTM is used, the score can be raised to 0.648, which amounts to only 0.042 lower than that of DCNN. Generally, sparse patterns may interfere with the performance of BLSTM if the pattern is sparse and every pattern exhibits different duration. Hence, the proposed PBLSTM can counteract this problem and raise the score.

Moreover, Figure 17.11 shows that the proposed PBLSTM performs the best for the appliance that has rapidly changing patterns. Comparing Figure 17.11c,e, one can see that PBLSTM

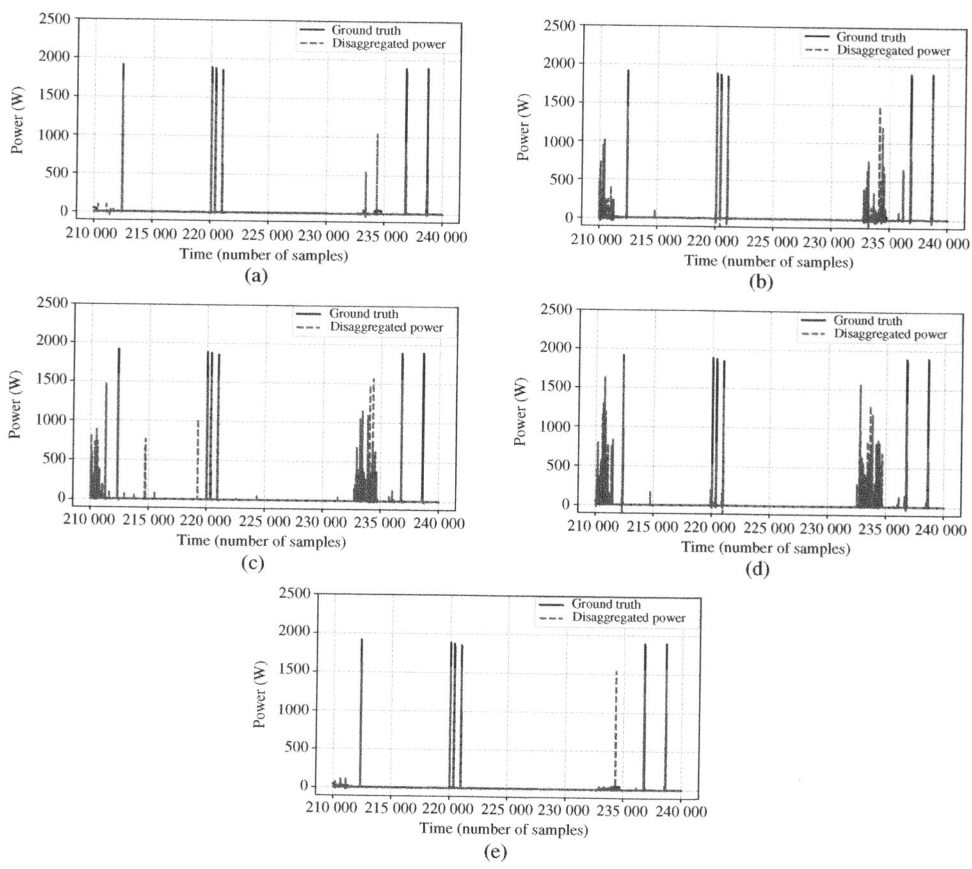

Figure 17.11 Ground truth and disaggregate power of microwave (MW) (a) DCNN, (b) GLU-Res, (c) BLSTM, (d) AE, and (e) PBLSTM.

performs much better than BLSTM. BLSTM has quite a few wrongly disaggregated power whereas PBLSMT has almost none. Finally, for appliances which do not exhibit repeated patterns such as lighting, Figure 17.12 indicates that both BLSTM and PBLSTM perform better than DCNN, GLU-Res, and AE. Close inspection on Figure 17.12c,e, one notice that PBLSTM perform slightly better than BLSTM. Hence, F1 score for PBLSTM is 0.598 whereas that for BLSTM is 0.525.

After knowing the estimation performance of individual power consumption, the estimation of the individual appliance power can be added up to be an estimation of total power. Figure 17.13a shows the ground truth of the total power consumption. Figure 17.13b–f shows the estimation of total power produced by DCNN, GLU-Res, BLSTM, AE, and PBLSTM, respectively. By comparing the estimated total power with the ground truth, it is found that the MAE of DCNN, GLU-Res, BLSTM, AE, and PBLSTM, respectively, are 76.041, 72.396, 62.735, 67.7, and 58.719 W. This result indicates that PBLSTM performs the best among the others.

Finally, average performances are shown in Figure 17.14. Figure 17.14a shows the PTECA and it is clear from the figure that the proposed method has the highest PTECA. The average of MAE for each method is shown in Figure 17.14b, and the proposed method shows the smallest amount. Figure 17.14c shows the precision, recall, and F1 score. The proposed method shows the highest

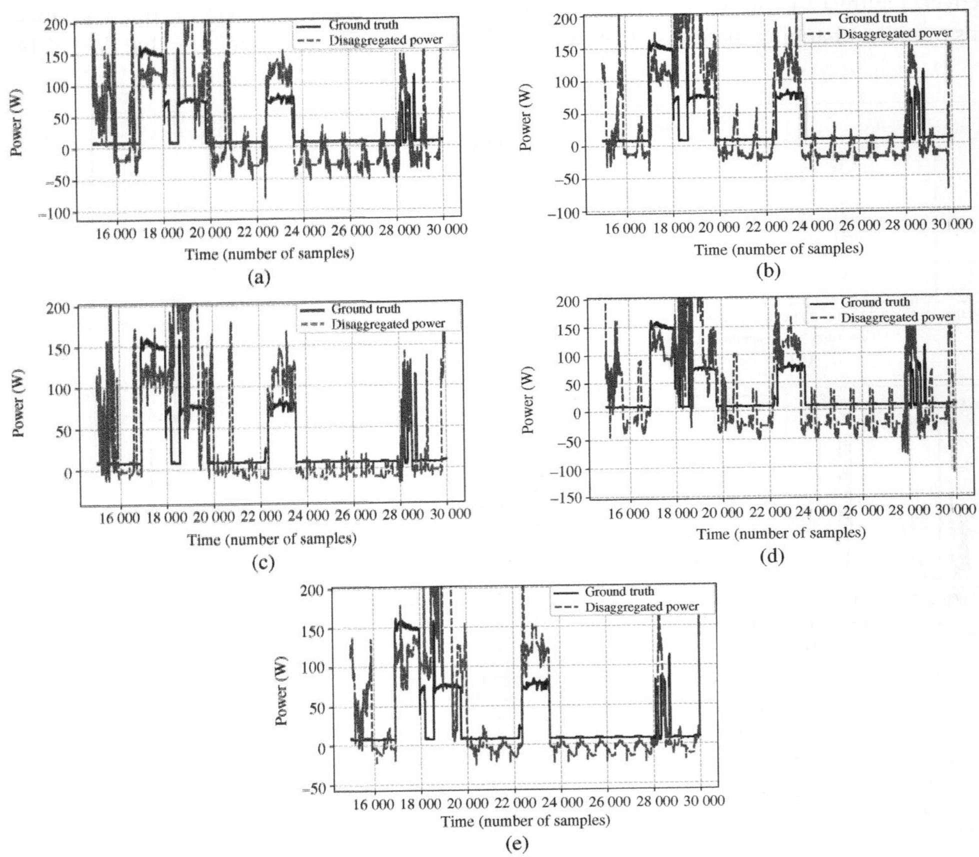

Figure 17.12 Ground truth and disaggregate power of lighting (LT) (a) DCNN, (b) GLU-Res, (c) BLSTM, (d) AE, and (e) PBLSTM.

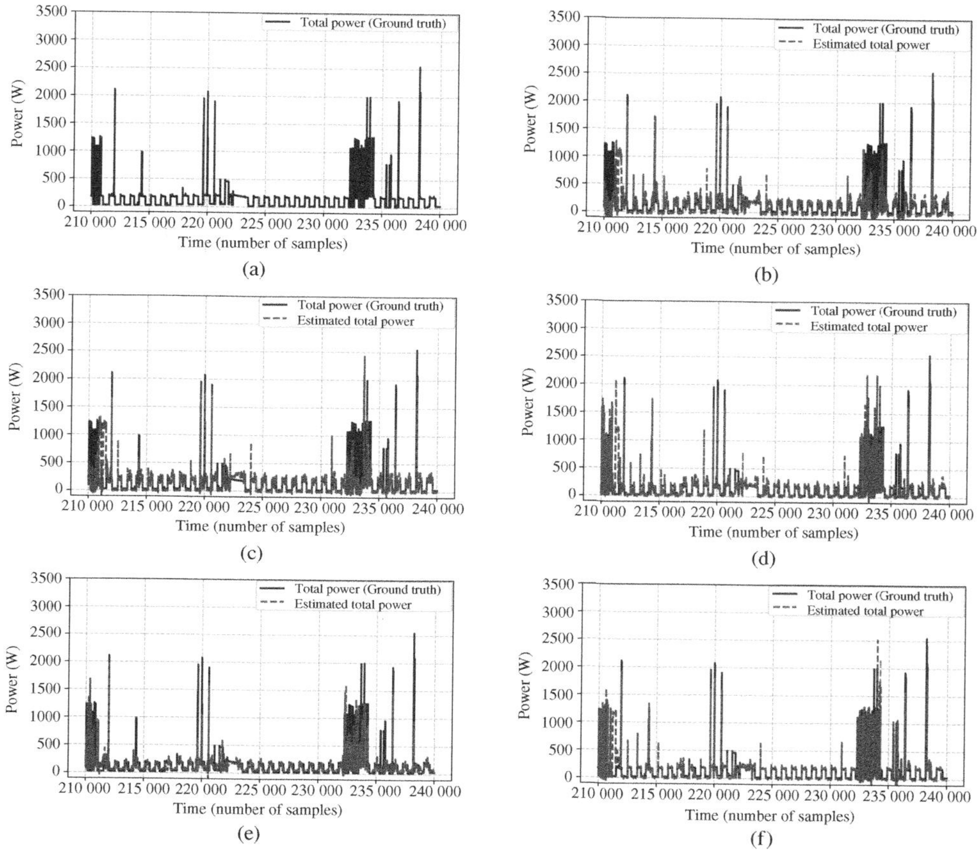

Figure 17.13 Estimated aggregate power (a) Ground truth, (b) DCNN, (c) GLU-Res, (d) BLSTM, (e) AE, (f) PBLSTM.

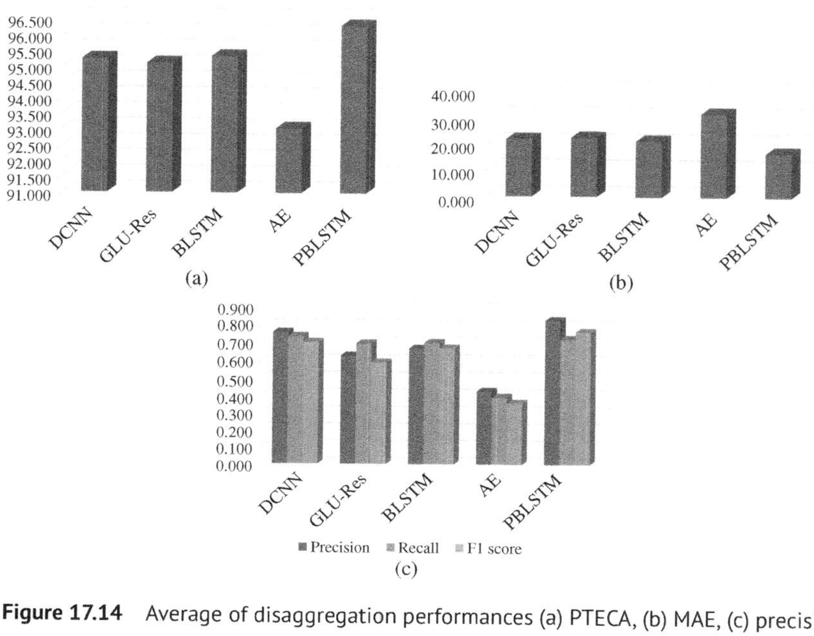

Figure 17.14 Average of disaggregation performances (a) PTECA, (b) MAE, (c) precision, recall, and F1.

PTECA and F1 score and the lowest MAE. AE performs the worst in this research and DCNN performs the second best.

17.5 Conclusion

This chapter has proposed a PBLSTM for energy disaggregation problem. Incorporating SFs in the PBLSTM can significantly improve the disaggregation performance. Parallel connection of BLSTM helps the model to ensemble two individual models and combines the information all together. To validate the performance of the proposed model, it is validated with the REDD dataset. Furthermore, the performance of our proposed method was compared to four different models, and it has been successfully proven to outperform the other models.

References

1 Hart, G.W. (1992). Nonintrusive appliance load monitoring. *Proceedings of the IEEE* 80 (12): 1870–1891.

2 Bhotto, M.Z.A., Makonin, S., and Bajić, I.V. (2017). Load disaggregation based on aided linear integer programming. *IEEE Transactions on Circuits and Systems II: Express Briefs* 64 (7): 792–796.

3 Wittmann, F.M., López, J.C., and Rider, M.J. (2018). Nonintrusive load monitoring algorithm using mixed-integer linear programming. *IEEE Transactions on Consumer Electronics* 64 (2): 180–187.

4 Egarter, D. and Elmenreich, W. (2015). Load disaggregation with metaheuristic optimization. *Energieinformatik 2015* 12 S.

5 Suzuki, K., Inagaki, S., Suzuki, T. et al. (2008). Nonintrusive appliance load monitoring based on integer programming. *2008 SICE Annual Conference*, August 2008, pp. 2742–2747.

6 Biansoongnern, S. and Plangklang, B. (2016). Nonintrusive load monitoring (NILM) using an artificial neural network in embedded system with low sampling rate. *2016 13th International Conference on Electrical Engineering/Electronics, Computer, Telecommunications and Information Technology (ECTI-CON)*, June 2016, pp. 1–4.

7 Zhang, C., Zhong, M., Wang, Z. et al. (2018). Sequence-to-point learning with neural networks for nonintrusive load monitoring. *CoRR*, vol. abs/1612.09106.

8 Kelly, J. and Knottenbelt, W.J. (2015). Neural NILM: deep neural networks applied to energy disaggregation. *CoRR*, vol. abs/1507.06594. http://arxiv.org/abs/1507.06594.

9 Kolter, J.Z. and Jaakkola, T. (2012). Approximate inference in additive factorial HMMS with application to energy disaggregation. In: *Proceedings of the 15th International Conference on Artificial Intelligence and Statistics*, ser. Proceedings of Machine Learning Research, vol. 22 (ed. N.D. Lawrence and M. Girolami), 1472–1482. La Palma, Canary Islands: PMLR. http://proceedings.mlr.press/v22/zico12.html.

10 Kolter, J.Z. and Johnson, M.J. (2011). REDD: A public data set for energy disaggregation research. *SustKDD*.

11 Kim, H., Marwah, M., Arlitt, M. et al. (2011). *Unsupervised Disaggregation of Low Frequency Power Measurements*, pp. 747–758. https://epubs.siam.org/doi/abs/10.1137/1.9781611972818.64 (accessed 19 August 2022).

12 Zoha, A., Gluhak, A., Nati, M., and Imran, M.A. (2013). Low-power appliance monitoring using factorial hidden Markov models. *2013 IEEE 8th International Conference on Intelligent Sensors, Sensor Networks and Information Processing*, April 2013, pp. 527–532.

13 Holmegaard, E. and Kjaergaard, M.B. (2016). NILM in an industrial setting: a load characterization and algorithm evaluation. *2016 IEEE International Conference on Smart Computing (SMARTCOMP)*, May 2016, pp. 1–8.

14 Kim, J., Le, T.-T.-H., and Kim, H. (2017). Nonintrusive load monitoring based on advanced deep learning and novel signature. *Computational Intelligence and Neuroscience* 2017: 4216281.

15 Mauch, L. and Yang, B. (2015). A new approach for supervised power disaggregation by using a deep recurrent LSTM network. *2015 IEEE Global Conference on Signal and Information Processing (GlobalSIP)*, December 2015, pp. 63–67.

16 Ji, T.Y., Liu, L., Wang, T.S. et al. (2019). Non-intrusive load monitoring using additive factorial approximate maximum a posteriori based on iterative fuzzy c-means. *IEEE Transactions on Smart Grid* 10 (6): 6667–6677.

17 Zoha, A., Gluhak, A., Imran, M.A., and Rajasegarar, S. (2012). Non-intrusive load monitoring approaches for disaggregated energy sensing: a survey. *Sensors* 12 (12): 16-838–16-866.

18 Lin, Y. and Tsai, M. (2010). A novel feature extraction method for the development of nonintrusive load monitoring system based on BP-ANN. *2010 International Symposium on Computer, Communication, Control and Automation (3CA)*, Volume 2, May 2010, pp. 215–218.

19 Zhou, C., Liu, S., and Liu, P. (2016). Neural network pattern recognition based non-intrusive load monitoring for a residential energy management system. *2016 3rd International Conference on Information Science and Control Engineering (ICISCE)*, July 2016, pp. 483–487.

20 Krystalakos, O., Nalmpantis, C., and Vrakas, D. (2018). Sliding window approach for online energy disaggregation using artificial neural networks. *Proceedings of the 10th Hellenic Conference on Artificial Intelligence*, ser. SETN '18. New York, NY, USA: ACM, pp.7-1–7-6 http://doi.acm.org/10.1145/3200947.3201011 (accessed 19 August 2022).

21 Chen, K., Wang, Q., He, Z., and He, J. (2018). Convolutional sequence to sequence non-intrusive load monitoring. *The Journal of Engineering* 2018 (17): 1860–1854.

22 Dauphin, Y.N., Fan, A., Auli, M., and Grangier, D. (2016). Language modeling with gated convolutional networks. *CoRR*, vol. abs/1612.08083. http://arxiv.org/abs/1612.08083.

23 Long, W., Lu, Z., and Cui, L. (2019). Deep learning-based feature engineering for stock price movement prediction. *Knowledge-Based Systems* 164: 163–173.

24 Hossain, Z., Rahman, A., Hossain, M., and Karami, J. (2019). Over-differencing and forecasting with non-stationary time series data. *Dhaka University Journal of Science* 67: 21–26.

25 Nabipour, M., Nayyeri, P., Jabani, H. et al. (2020). Deep learning for stock market prediction. *Entropy* 22 (8). https://www.mdpi.com/1099-4300/22/8/840.

26 Gao, P., Zhang, R., and Yang, X. (2020). The application of stock index price prediction with neural network*Mathematical and Computational Applications* 25 (3). https://www.mdpi.com/2297-8747/25/3/53.

27 Lanbouri, Z. and Achchab, S. (2020). Stock market prediction on high frequency data using long-short term memory. *Procedia Computer Science* 175: 603–608. *The 17th International Conference on Mobile Systems and Pervasive Computing (MobiSPC)*, The 15th International Conference on Future Networks and Communications (FNC), The 10th International Conference on Sustainable Energy Information Technology. https://www.sciencedirect.com/science/article/pii/S1877050920317877 (accessed 19 August 2022).

28 Owen, L. and Oktariani, F. (2020). SENN: Stock ensemble-based neural network for stock market prediction using historical stock data and sentiment analysis. *2020 International Conference on Data Science and Its Applications (ICoDSA)*, pp. 1–7.

29 Hao, Y. and Gao, Q. (2020). Predicting the trend of stock market index using the hybrid neural network based on multiple time scale feature learning. *Applied Sciences* 10 (11). https://www.mdpi.com/2076-3417/10/11/3961.

30 Glorot, X. and Bengio, Y. (2010). Understanding the difficulty of training deep feedforward neural networks. In: *Proceedings of the 13th International Conference on Artificial Intelligence and Statistics*, ser. Proceedings of Machine Learning Research, vol. 9. (ed. Y.W. Teh and M. Titterington), 249–256. Chia Laguna Resort, Sardinia, Italy: PMLR. http://proceedings.mlr.press/v9/glorot10a.html.

31 Keskar, N.S., Mudigere, D., Nocedal, J. et al. (2016). On large-batch training for deep learning: generalization gap and sharp minima. *CoRR*, vol. abs/1609.04836. http://arxiv.org/abs/1609.04836.

32 Hoffer, E., Hubara, I., and Soudry, D. (2017). Train longer, generalize better: closing the generalization gap in large batch training of neural networks. *NIPS*.

33 Kingma, D.P. and Ba, J. (2014). Adam: A method for stochastic optimization. *CoRR*, vol. abs/1412.6980. http://arxiv.org/abs/1412.6980.

34 Nelwamondo, F., Golding, D., and Marwala, T. (2013). A dynamic programming approach to missing data estimation using neural networks. *Information Sciences*, 237: 49–58.

18

Reinforcement Learning for Intelligent Building Energy Management System Control*

Olivera Kotevska[1] *and Philipp Andelfinger*[2]

[1]*Computer Science and Mathematics Division, Oak Ridge National Laboratory, Oak Ridge, TN, USA*
[2]*Institute for Visual and Analytic Computing, University of Rostock, Rostock, Germany*

Chapter Objectives

After reading this chapter, you will understand:

- The challenges in the design of modern building energy management systems (BEMS).
- The role of machine learning in modern BEMS.
- The fundamentals of reinforcement learning (RL) and key RL algorithms.
- The applicability and benefits of RL approaches in the BEMS context.
- Open research directions in RL-based approaches for BEMS control.

18.1 Introduction

Since the notion of the Fourth Industrial Revolution was introduced at the 2015 World Economic Forum [1, 2], it has become widely used to describe the development toward increasing automation and connectivity in industry. The four design principles that represent the fourth industrial revolution (Industry 4.0) are interconnection, information transparency, technical assistance, and decentralized decisions. Among the enabling technologies are inter-device connectivity, big data, automatization, intelligent agents, robotics, machine learning (ML), artificial intelligence, and sensors [3]. Closely interlinked with Industry 4.0 is the concept of the *Internet of Things* (IoT) [4], which is the vision of "enabling advanced services by interconnecting (physical and virtual) things based on existing and evolving interoperable information and communication technologies" [5]. In recent years, IoT devices have come into wide use not only in industrial fields such as manufacturing, but increasingly also in our private lives, e.g. in the shape of smart homes.

Fifty billion IoT devices are expected to be connected to the Internet by 2025 [6], collecting and sharing data estimated to reach a total of almost 80 zettabytes [7]. Applications leveraging the expansion of IoT-based systems have the potential to bring about improvements in terms of increased automation, comfort, security, and safety. Beyond the high degree of connectivity

*This manuscript has been authored by UT-Battelle, LLC, under contract DE-AC05-00OR22725 with the US Department of Energy (DOE).

Intelligent Data Mining and Analysis in Power and Energy Systems: Models and Applications for Smarter Efficient Power Systems, First Edition. Edited by Zita Vale, Tiago Pinto, Michael Negnevitsky, and Ganesh Kumar Venayagamoorthy.

of IoT devices, a key enabler is the devices' capability to individually or jointly form "smart" or "intelligent" systems, i.e. systems with the ability to learn from the past and adapt to environmental changes [8]. While many "smart" devices still rely on traditional fixed rule-based logic or on traditional control algorithms requiring simplified models of the target system, increasingly, ML is applied to achieve a larger degree of adaptability. Purely predictive subsystems, e.g. to estimate the time required to heat a room to a certain temperature, can often be trained using supervised learning techniques if sufficient historical data is available.

However, if devices are required to adapt a non-trivial behavior to a changing environment, there is often a need to learn a *policy* that decides what action to take given some observed environment and device state. This is complicated by the fact that data describing optimal behavior may be sparse or unavailable, for instance, because the device is deployed in a new environment. An approach that tackles these challenges is reinforcement learning (RL) [9]. In RL, a so-called agent takes actions and interacts with the environment, relying on trial and error to explore the possible system states and actions. In contrast to supervised learning, RL does not require pairs of input and reference output. Instead, the agent learns by being rewarded or punished for taking an action or reaching a state. While in supervised learning, the goal is to minimize the deviation of a model's output from the reference output, RL aims to maximize the cumulative reward of the actions taken by the agent. With this approach, RL is able to generate complex policies that can form the basis of smart IoT-based systems for instance smart building where appliances and heating, ventilation, and air conditioning (HVAC) system are connected to the Internet [10].

When focusing on the optimization of the energy consumption, IoT-based systems in smart homes and buildings are subsumed under the broader term of building energy management systems (BEMS). The energy consumption of buildings comprises more than 40% of the total energy consumption in the United States (US), with residential buildings alone accounting for around 21% Shaikh et al. [11]. According to the U.S. Energy Information Administration (EIA) report, over 51% of the total annual energy consumption of U.S. households is consumed by HVAC devices [12]. By monitoring the energy consumption and adjusting the energy usage based on user preferences or other conditions and constraints, an intelligent BEMS has the potential to balance aspects such as energy consumption, user comfort, and cost.

This chapter is aimed toward non-experts who would like to learn about RL, as well as RL experts interested in current research challenges in the application domain of BEMS.

The remainder of this chapter is structured as follows: Section 18.2 introduces basic RL concepts, the different RL methods, as well as the use of deep learning in RL. Section 18.3 discusses the use of RL in BEMS. Section 18.4 presents the challenges and research directions in RL for real-world BEMS environments. Section 18.5 summarizes the chapter and points out directions for future research.

18.2 Reinforcement Learning

RL is a computational approach that allows an agent to learn a behavior in a process of trial and error [9]. Unlike supervised ML methods, an RL agent learns directly by interacting with its environment, avoiding the need for labeled reference data [13]. The interaction between an RL agent and its environment can be formalized using the framework of Markov decision processes (MDP) [14]. An MDP is defined by four elements (S, A, R, P): a set of states S, a set of actions A, a reward function $R : S \times A$, and transition probabilities P between the states $P : S \times A \times S' \in [0, 1]$. The goal is to find probabilities P that maximize the cumulative reward.

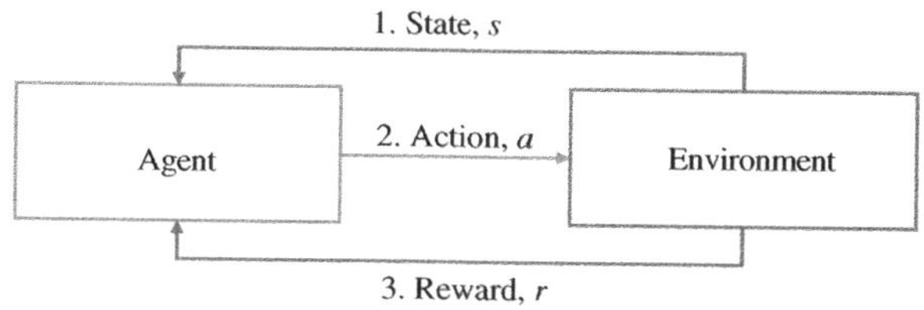

Figure 18.1 Typical reinforcement learning feedback loop.

The feedback loop describing a typical RL process is presented in Figure 18.1: the agent observes the current system state s of the environment and takes an action a based on the policy $\pi(a|s)$. As a result, the agent receives a reward r and the system transitions to the next state s'. The agent's experiences (s, a, r, s') are used to train the policy.

More formally, the goal in RL is to find an optimal policy π^* that maximizes the expected future reward from time t. Almost all RL algorithms either estimate a state-value function $Q^\pi(s)$ to evaluate the "value" of being in state s, or an action-value function $Q^\pi(s, a)$ to evaluate the value of taking the action a at state s. For simplicity, our description of different RL approaches will assume the use of an action-value function. Typically, the estimation of the action-value function obeys the Bellman equation:

$$Q(s, a) = E[r_{t+1} + \gamma r_{t+2} + \gamma^2 r_{t+3} + \cdots |s_t, a_t]$$

where $0 \leq \gamma \leq 1$ is a discount rate parameter that determines how much to prioritize future rewards.

During the learning phase, an estimate of the optimal policy is determined by balancing *exploration*, i.e. taking previously unexplored actions, and *exploitation*, i.e. choosing known high-value actions. Exploration allows the agent to uncover new areas of the state and action space and thus possibly higher cumulative rewards. In contrast, exploitation provides more information along partially known state trajectories, enabling a gradual refinement of the policy. RL algorithms vary in their strategies by which exploration and exploitation are balanced.

In the following, we give a rough taxonomy of RL approaches based on [15] (cf. Figure 18.2). A first categorization is that into model-based and model-free RL approaches, which pertains to the availability of a model of the environment. *Model-based* RL approaches rely on an explicit model of the environment, which may either be available *a priori* or learned. The model reflects partial knowledge about the Markov decision process' transition probabilities and reward function. The reliance on an environment model permits planning by determining the expected reward of possible future actions. This approach combines planning or control techniques such as model predictive control (MPC) with a reward function to calculate optimal candidate actions (e.g. Ref. [16]). In contrast, *model-free* approaches learn a policy without an explicit model, relying on a process closer

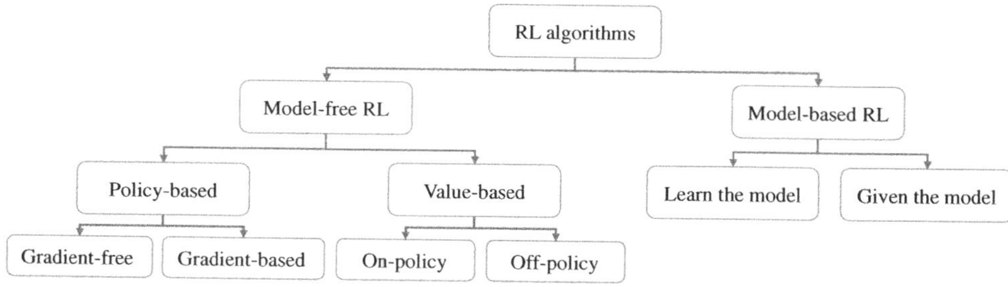

Figure 18.2 Taxonomy of reinforcement learning algorithms. Source: Regattieri et al. [56]/ With permission of Elsevier.

to pure trial and error than the directed actions possible in model-based approaches. The use of model-free approaches is appropriate if a model of the environment is not available and a learned model would be inaccurate or biased, e.g. because of the complexity of the system of interest [17].

Model-free RL approaches differ in their representation of the policy: **policy-based methods** explicitly build a representation of the policy, which is kept in memory during learning. Conversely, **value-based methods** store only a representation of the value function. An appropriate action for a given state is then selected according to the value function.

A key difference between value-based and policy-based approaches lies in the state and action spaces: value-based approaches typically assume discrete state and actions spaces, whereas policy-based approaches typically address continuous spaces as encountered in many control problems. For instance, an intelligent thermostat controlling a building's heating system may sense the current temperature and define a suitable temperature setpoint. Policy-based methods avoid the need for discretization of such continuous variables and exploit the smoothness of the state-action relation for efficient training.

Value-based model-free RL is further categorized into *on-policy* and *off-policy* algorithms. The difference between the two categories is that off-policy algorithms differentiate between a *target policy*, which is the policy being learned, and a *behavior policy*, which determines the actions taken by the agent. In contrast, in on-policy algorithms, a single policy serves both purposes.

More formally, in an **on-policy algorithm**, the value $Q(s, a)$ of taking action a at state s is updated using an action a' taken based on the current policy π. An example is the State-Action-Reward-State-Action (SARSA) algorithm [18], in which given a learning rate α, the policy update takes the following form:

$$Q(s, a) \leftarrow Q(s, a) + \alpha[r + \gamma Q(s', a') - Q(s, a)]$$

To permit exploration, on-policy algorithms typically set $\pi(a|s) > 0$ for all states s and actions a. For instance, in an *ϵ-greedy policy*, with a probability of ϵ, a random action is chosen instead of the "best" action suggested by the value function.

In **off-policy algorithms**, actions are taken based on a behavior policy separate from the target policy being learned. A well-known example is Q-learning [13], where the target policy is updated as follows:

$$Q(s, a) \leftarrow Q(s, a) + \alpha[r + \gamma \max_{a'} Q(s', a') - Q(s, a)]$$

Here, the actions a' are those previously probed according to the behavior policy, e.g. the ϵ-greedy policy. However, the update depends on the action a' with the highest current value $Q(s', a')$, i.e. the target policy is strictly greedy.

The state and action spaces of non-trivial systems are usually too large to be explored in their entirety. Instead, in deep reinforcement learning (DRL), an approximate policy is represented by an artificial neural network.

18.2.1 Deep Reinforcement Learning

Common value-based RL approaches such as Q-learning represent an action-value function $Q^{\pi}(s, a)$ as a so-called Q-table stored explicitly in memory. As the agent explores the environment, the values are updated with better approximations of cumulative rewards.

However, in real-world problems, the state and action spaces may be very large or even infinite. As a consequence, it may be impossible to store the values of state/action pairs in the form of tables in memory. Similarly, this "curse of dimensionality" creates enormous computational demands

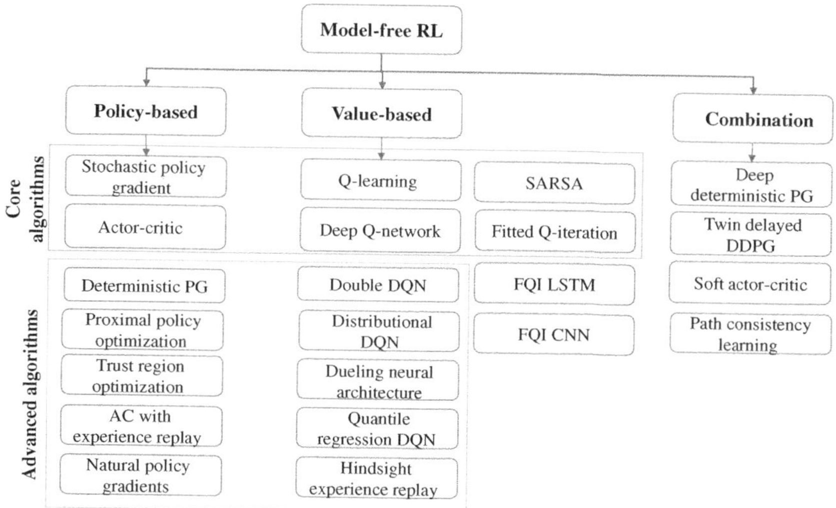

Figure 18.3 Taxonomy of model-free reinforcement learning algorithms. Source: Adapted from Zhang and Yu [15]. *Acronyms: PG: Policy Gradient, AC: Actor Critic, DQN: Deep Q-Network, SARSA: State-Action-Reward-State-Action, FQI: Fitted Q-Iteration, LSTM: Long Short-Term Memory, CNN: Convolutional Neural Network, DDPG: Deep Deterministic Policy Gradient.*

for the trial-and-error interaction of the agent with the environment. To tackle these challenges, instead of representing the value function directly, deep neural networks (DNNs) can be used to capture the relationship between state/actions pairs and values. This combination of RL and DNNs, referred to as **DRL**, avoids the need for manual feature engineering in order to reduce the state space, making it an attractive approach for high-dimensional problems, e.g. in control [10].

In Figure 18.3, we give an overview of common model-free DRL algorithms. We categorize the algorithms into *core algorithms*, which have been refined, extended, or combined, to form *advanced algorithms*.

18.2.2 Advanced Reinforcement Learning

Advanced RL algorithms build on the core algorithms to address particular challenges posed by some real-world environments and tasks. Firstly, real-world systems may violate the common assumption that the environment is stationary. For instance, the response of the environment to an agent's action may depend so strongly on time that a fixed single policy may not be able to capture a near-optimal behavior during different time periods. In the BEMS context, under seasonal changes in outside temperature, an agent may be presented with vastly different regions of an overall state space, and may require entirely different actions depending on the season, e.g. when switching the mode of a HVAC system from heating to cooling. One approach to tackle such non-stationary problems is online, incremental, or continual learning, in which the agent learns constantly or periodically to adapt to changing environments or tasks [19]. Alternatively, if changes are sudden or last for a long time, the agent may switch dynamically among multiple separate models [20]. For this approach to be efficient, a mechanism is required to detect when a switch is appropriate.

A second class of challenging RL problems is given by those involving large numbers of controllable entities. In such scenarios, the high dimensionality of the state and action spaces

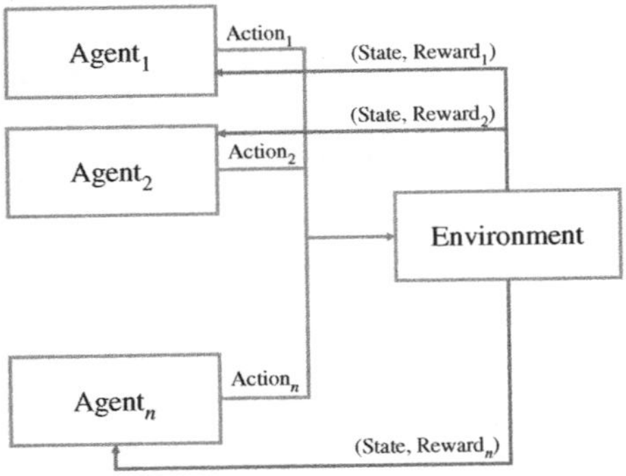

Figure 18.4 Multi-agent reinforcement learning feedback loop.

may prevent a single agent from learning a high-quality overall policy. For example, in a building, there may be different temperature preferences and heating and cooling devices with different characteristics and placement. Such cases may be addressed by *multi-agent DRL* algorithms, in which multiple agents work toward achieving a shared goal, e.g. the minimization of energy consumption. Multi-agent RL problems are formalized in terms of a *Markov game* [21] $(S, \{A_1, \dots, A_n\}, \{R_1, \dots, R_n\}, P)$, where S is the state space observed by all agents, A_i and R_i are the action and reward functions for agent i, and P is the transition probability function (cf. Figure 18.4). Given n agents, a joint action space shared by the agents is given by $A_1 \times A_2 \times \cdots \times A_n$. In this setting, a state transition can be viewed as the result of an overall action taken by all agents jointly. While multi-agent RL can rely on the same fundamental value-based or policy-based approaches as single-agent RL, additional considerations are required for communication among agents, particularly when individual agents only observe parts of the environment, and with respect to scalability toward larger numbers of agents [22].

Multi-agent DRL algorithms are categorized on a spectrum with the two extremes of *cooperative* DRL, in which all agents share the same reward function, and *competitive* DRL, which typically involves zero-sum games, i.e. the agents' individual rewards sum up to zero [23]. Both cooperative and competitive multi-agent RL approaches have been applied in the context of BEMS.

Some real-world problems lend themselves to *hierarchical DRL*, in which a problem is separated into a hierarchy of sub-tasks that can be learned independently and combined to tackle the overall problem [24]. In *multi-task learning*, the information gained throughout the learning process by individual agents working on related tasks is repeatedly combined to improve the agents' further learning progress [25].

In the context of BEMS, the ability of DRL approaches to learn combinations of complex tasks makes it suitable for scenarios such as buildings occupied by multiple types of occupants with different preferences, multi-story buildings with various types of energy-consuming devices, and on a bigger scale when controlling entire communities of homes.

18.3 Applications of Deep Reinforcement Learning in Building Energy Management Systems Control

BEMS collect and analyze data in networked households and commercial facilities to control the buildings' energy needs, while maintaining user comfort preferences, reducing energy consumption, and providing consumers with information about their energy usage. The overall networked system may include thermostats, air conditioning, humidity meters, washing machines, charging stations for electric vehicles, solar panels, lighting sensors, motion sensors, and security cameras. Timely monitoring and efficient control of these devices can improve safety, security, and reduce energy consumption.

The energy consumption of buildings attracts wide interest since it comprises more than 40% of the energy consumption in the United States [11]. BEMS can monitor the energy consumption and adjust the energy usage behavior based on user preferences or other constraints. There are three critical factors to be considered by BEMS: the consumer behavior, the variation in energy demand throughout the day, and the price variation during the day or week. All three of these factors must be accounted for in order to achieve high levels of comfort, low energy consumption, and low cost.

DRL for building energy management has been studied from multiple perspectives: some of the recent studies focused on energy savings [26, 27] and demand-response [28] in system-specific scenarios, while others considered the technical aspect of DRL approaches [29]. The following is a brief survey of recent works in DRL for BEMS, focusing on how DRL has been applied. Thus, we report the DRL method used as well as the chosen objectives, rewards, and the environments. We also list the achieved results and the baseline method used for evaluation. Further, we identify works that report on deployments in a real-world environment as well as articles that considered mixed energy sources, e.g. through the addition of solar panels. Selected are articles that rely on RL, that were published in 2010 or after, and that were cited a minimum of ten times according to Google Scholar.

In the BEMS context, devices can be categorized as (a) devices whose energy consumption can be directly controlled, for example, by adjusting the set-point of a HVAC system or a water heater (WH), and (b) devices whose usage can be shifted in time, which is the case for dishwashers and washing machines.

18.3.1 Heating, Ventilation, and Air Conditioning

HVAC systems are among the largest consumers of energy in buildings. It is estimated that HVAC is responsible for 51% of building energy consumption in the United States [12]. RL has been successfully applied to solve the problem of HVAC control. In this section, we report on recent work focusing on DRL approaches (cf. Table 18.1) and the considered environments (cf. Table 18.2).

It was found that DRL can typically provide energy consumption savings of over 15%, whereas the cost savings varied widely in the range of 10–75%. In the table, we exclude savings reported only visually instead of numerically. Most works relied on single-agent algorithms such as Deep Q-Network (DQN) and DDPG algorithms. Only one study used a multi-agent approach, controlling 30 different zones in a building and achieving up to 75% in cost reduction [31]. HVAC control was approached by some of the groups as a discrete problem using DQN, while others approached it as continuous action problem using the A3C and DDPG algorithms with similar results. Generally, Q-learning-based algorithms have been successful when applied to problems with small, discrete state spaces, but in more realistic systems with larger and possibly continuous state spaces algorithms such as A3C, DDPG, and MAAC are used. As the reward function, most works used energy consumption and user temperature discomfort, with a few works using air quality discomfort, occupancy, heat demand, and price.

Considering the development over time, earlier works typically used Q-learning methods, whereas since 2017, DQN, and actor/critic-based solutions have been used more prominently. Most of the research relies on simulated environments such as Energy Plus or simulation models developed by the research group conducting the study. Only few works showed results from real-world environments [34, 35]. Extensive scalability and robustness testing in a simulated

Table 18.1 Summary of existing work for optimal HVAC control with respect to DRL approaches.

Article	Method[a)]	Objective[b)]	Reward[c)]	Results (%)
Ahn and Park [30]	DQN	EC	EC, ICL	15.7
Yu et al. [31]	MAAC	CM	EC, ITV, CCV	56–75.25
Zhang et al. [32]	MPC	EC	EC, ITV	17–21.8
Gao et al. [33]	DDPG	CM	EC, MTC	—
Zhang et al. [34]	A3C	HE-D	HD, ITV, O	16.7
Zhang et al. [35]	A3C	EC	ITV, PPD, IT	15
Zhang and Lam [36]	A3C	EC	TC, HD, ITV, PPD, IT	16–18.2
Wang et al. [37]	AC-LSTM	EC	TC, ED	2–5
Wei et al. [38]	DQL	CM	ITV, P	20–70
Barrett and Linder [39]	QL	CM	O, A, ITV	10
Li and Xia [40]	MQL	EC	EC, PPD	—

a) DQN: Deep Q-Network, MAAC: Multi Actor Attention Critic, MPC: Model Predictive Control, DDPG: Deep Deterministic Policy Gradient, A3C: Asynchronous Advantage Actor Critic, AC-LSTM: Actor Critic Long Short Term Memory, DQL: Deep Q-Learning, QL: Q-Learning, MQL: Modified Q-Learning.
b) EC: Energy Consumption, CM: Cost minimization, HE-D: Heating Energy/Demand
c) EC: Energy Consumption, ICL: Indoor CO_2 level, ITV: Indoor Temperature Violation, CCV: CO_2 Concentration Violation, MTC: Maintain Thermal Comfort, HD: Heating Demand, O:Occupancy, PPD: Predicted Percentage of Dissatisfied, IT: Indoor Temperature, TC: Thermal Comfort, ED: Energy Demand, P: Price, A: Action.

Table 18.2 Summary of work on optimal HVAC control with respect to environment settings.

Article	Environment[a)]	Baseline[b)]	Methodology[c)]
Ahn and Park [30]	EnergyPlus	RB	C:B, MPC
Yu et al. [31]	SRD	RB, Heuristic	C:B
Zhang et al. [32]	EnergyPlus	PPO	C:B
Gao et al. [33]	S:TRNSYS	QL, DQN, SARSA	C:B
Zhang et al. [34]	EnergyPlus, RD	RB	C:B, D
Zhang et al. [35]	EnergyPlus, RD	RB	C:B, D
Zhang and Lam [36]	EnergyPlus, RD	RB	C:B
Wang et al. [37]	EnergyPlus	RB	C:B
Wei et al. [38]	EnergyPlus	RB	C:B, QL, DAS
Barrett and Linder [39]	SRD	PT	C:B
Li and Xia [40]	EnergyPlus	QL	C:B

a) SRD: Simulation using Real Data, S: Simulation, RD: Real-World Deployment.
b) RB: Rule based, PPO:Proximal Policy Optimization, QL: Q-Learning, DQN: Deep Q-Network, SARSA: State Action Reward State Action, PT: Programmable Thermostat.
c) C:Comparative, B:Baseline, MPC: Model Predictive Control, D:Deployment, QL: Q-Learning, DAS: Different Algorithm Settings.

environment was performed by Yu et al. [31], demonstrating the efficiency of their DRL approach under different environmental conditions such as temperature ranges, variety of agents, tolerance to random temperature distributions, and reward weights. The energy and cost optimization for multi-zone buildings were the focus of [30, 31, 38], while most works focused on single-zone optimization [[37, 39, 40]?].

We point out that out of the considered works, only one [32] used a model-based approach (MPC) and compared their results with a model-free algorithm.

18.3.2 Water Heater

Electric WHs cause 14% of the electricity consumption of residential buildings in the United States [41]. Research efforts have focused on investigating the ways to exploit the demand flexibility in WHs. Tables 18.3 and 18.4 present a summary of the highly-cited papers.

The studies show that DRL reduces the energy consumption by more than 10% and economic cost savings of 15% for the WH. A demonstration on 50 real-world houses was presented by Kazmi et al. [42], where the effectiveness of multi-agent RL compared to single-agent was 20% reduction of energy consumption. Usually, a simulated environment was used for developing and testing the model. The considered factors were price signal, temperature, future demand, outside temperature, as well as load and backup controllers. Similar to the HVAC control problem, the objective function was designed with the goal of reducing energy consumption, economic cost, or reducing the demand and peak load. However, only one of the considered studies included user preferences as part of the reward function [42].

None of the highly-cited papers reported on deployment of the model in a real-world environment, although some studies used real world data for the experiments [42, 45].

Most of the presented studies are based on batch RL, which does not allow for exploration. Still, some works used a MBC [45] or compared model-based and model-free approaches [44]. None of the algorithms commonly used for HVAC were tested for WH control.

Table 18.3 Summary of existing work on RL for optimal water heater control.

Year	Method[a)]	Objective[b)]	Reward[c)]	Results (%)
Kazmi et al. [42]	MARL	EC	EC, OC	20
Ruelens et al. [43]	BRL-FQI-LSTM	CM	EC, OC	10.2
Ruelens et al. [44]	BRL-FQI	CM	FWD, P, BC	15
De Somer et al. [45]	BRL-FQI	MLPU	PVP, PVC	20
Ruelens et al. [46]	BRL-FQI	CM	EC, P	—
Al-Jabery et al. [47]	FQL	PR	T, IC, ECGL	—

a) MARL: Multi Agent Reinforcement Learner, BRL-FQI-LSTM: Batch Reinforcement Learning Fitted Q-Iteration based on Long Short-Term Memory, BRL-FQI: Batch Reinforcement Learning Fitted Q-iteration, FQL: Fuzzified Q-Learning.

b) EC: Energy Consumption, CM: Cost minimization, MLPU: Maximize Local Photovoltaic Usage, PR: Peak reduction.

c) EC: Energy Consumption, OC: Occupant Comfort, FWD: Future Water Demand, P: Price, BC: Backup Controller, PVP: Photovoltaic Production, PVC: Photovoltaic Consumption, T: Temperature, IC: Instantaneous Consumption, ECGL: Estimated Current Grid Load.

Table 18.4 Summary of existing work for optimal water heater control.

Article	Environment[a)]	Baseline[b)]	Methodology[c)]
Kazmi et al. [42]	S, RW	RB	C:B, MPC, SARL, DAS
Ruelens et al. [43]	S	NC	C:B, FQI-CNN, FQI-RNN
Ruelens et al. [44]	S	DTC	C:B, MPC
De Somer et al. [45]	RW	NCL	C:B
Ruelens et al. [46]	S	NC	C:B
Al-Jabery et al. [47]	S	QL	C:B, PSO, DAS

a) S: Simulated, RW: Real-World.
b) RB: Rule based, NC: No-control, DTC: Default Thermostat Controller, NCL: Non-Controllable Loads, QL: Q-Learning.
c) C: Comparative, B:Baseline, MPC: Model Predictive Controller, SARL: Single Agent Reinforcement Learning, DAS: Different Algorithm Settings, FQI-CNN: Fitted Q-iteration Convolutional Neural Network, FQI-RNN: Fitted Q-iteration Recurrent Neural Network, PSO: Particle Swarm Optimization.

18.3.3 Other Devices

Beyond HVAC and WHs, the BEMS literature also considered devices such as electric vehicles, dishwashers, lighting systems, televisions, washing machines, and dryers. In order to reduce the energy consumption of these devices, the design of such devices has been significantly improved in recent years, from the introduction of motion sensors to on-demand control of lights, to intelligent scheduling and control of the different programs of a washing machine. We summarize the highly-cited papers on DRL control of building devices outside of HVAC and WH in Tables 18.5 and 18.6.

The studies report energy cost reductions between 15.3% and 27%, reducing peak demand to under 73% and reducing energy consumption by 14%. Studies that considered controllable devices such as HVAC and devices with schedulable activation times such as washing machines and dish washers in combination, relied on multi-agent DRL. In [52], devices were modeled as time-scaling (i.e. interruptible) or time-shifting (i.e. deferrable) loads or a combination thereof. A multi-objective approach was used only by Mocanu et al. [52]. The most recent works relied on advanced DRL such as multi-agent and hierarchical RL. The charging of electric vehicles was included in the studies by Lee and Choi [24], Xu et al. [48], and Mocanu et al. [52]. Some of the authors considered solar energy as an additional energy source in their multi-agent DRL models [24, 48, 50, 53]. There was only one demonstration in real-world settings [51]. However, most of the studies used simulations based on real-world data.

18.4 Challenges and Research Directions

While many works have demonstrated substantial building energy or cost savings through DRL, often over 15% over rule-based approaches or traditional control methods, there are several avenues to be explored before DRL approaches suited for suited for the real world can reliably be selected and deployed [56].

Table 18.5 Summary of existing method settings on RL to control other devices.

Article	Method[a)]	Objective[b)]	Reward[c)]	Results (%)
Lee and Choi [24]	HDRL-AC	CM	CM, CD, P, W, IT, OT	—
Xu et al. [48]	MARL-QL	CM	EC, DC	—
Lu et al. [49]	MARL-QL	CM	EC, CD	7–73
Lee and Choi [50]	MARL-QL	CM	EC, CD, ED	14
Yu et al. [51]	DDPG	CM	EC, CD, ED	≤ 15.3
Mocanu et al. [52]	DPG	PR, CM	EC, A	26, 27
Wan et al. [53]	AC	CM	P, MBE	11.38
Mahapatra et al. [54]	NFQ	PR	CD, EC	≥ 80
Ruelens et al. [55]	BRL-FQI	CM	CD, P, OT, SR	11–27

a) HDRL-AC: Hierarchical Deep Reinforcement Learning based on Actor Critic, MARL-QL: Multi-Agent Reinforcement Learning Q-Learning based, DDPG: Deep Deterministic Policy Gradient, DPG: Deep Policy Gradient, AC: Actor Critic, NFQ: Neural Fitted Q-learning, BRL-FQI: Batch Reinforcement Learning Fitted Q-iteration.
b) CM: Cost minimization, PR: Peak reduction.
c) CM: Cost minimization, CD: Consumer Discomfort, P: Price, W: Weather, IT: Indoor Temperature, OT: Outside Temperature, EC: Energy consumption, DC: Dissatisfaction cost, ED: Energy Depreciation, A: Actions, MBE: Minimum Battery Energy, SR: Solar Radiance.

Table 18.6 Summary of existing execution settings on RL to control other devices.

Article	Environment[a)]	Baseline[b)]	Methodology[c)]
Lee and Choi [24]	S:AC, WM, ESS, EV	—	C:DES
Xu et al. [48]	SRD:R, AC, H, L, WM, DW, EV	—	C:DES, DAS
Lu et al. [49]	SRD:AC, H, WM, DW, L	MILP	C:w/o DR
Lee and Choi [50]	S:SPS, AC, WM, ESS	MILP	C:B
Yu et al. [51]	SRD:ESS, AC, H	RB	C:B, DQN
Mocanu et al. [52]	SRD	QL	C:DQN
Wan et al. [53]	SRD:ESS, AC, WH, O, EV	RB	C:B, DAS
Mahapatra et al. [54]	S:AC, WM, R, O, DW, St, L, H	RB	C:B
Ruelens et al. [55]	S:WH, H	RB	C:B, DAS

a) S: Simulation, AC: Air Conditioner, WM: Washing Machine, ESS: Energy Storage System, EV: Electric Vehicle, SRD: Simulated using Real Data, R: Refrigerator, H: heating, L:Lightening, DW: dishwasher, SPS: solar photovoltaic system, O: Oven, St: Stove.
b) MILP: Mixed-Integer Linear Programming, RB: Rule-Based, QL: Q-Learing.
c) C: Comparative, DES: Different Environmental Settings, DAS: Different Algorithm Settings, DR: Demand Response, B: Baseline, DQN: Deep Q-Network.

1. **Benchmark scenarios and baseline approaches:** The existing literature on DRL for BEMS relies on a variety of custom scenarios to evaluate the proposed approaches. Typically, the experiments are carried out using a simulation model to represent the environment. The different assumptions made when constructing and parameterizing such models affect the observed results, making it difficult to directly compare results among studies. Establishing standard evaluation scenarios for BEMS would enable clearer insights into the relative performance of different DRL approaches. Similarly, there is currently no well-accepted baseline control method to which the results of DRL-based approaches can be compared. Hence, relative results must always be considered in light of the respective baseline.
2. **Deployment in real-world settings:** As simulated environments can only account for aspects that have been incorporated in the underlying model, experiments in real-world deployments are still needed to increase the confidence in a DRL-based BEMS approach. Only few of the existing studies compared results gathered in simulated and real-world environments (e.g. Refs. [34, 35]).
   If the real-world conditions differ from those in the simulated environment, yet follow some of the same general regularities, transfer learning can be applied to adapt a generic model trained in an idealized simulated setting to the specifics of a real-world deployment, or to adapt models among deployments in different real-world environments [57, 58].
   The deployment of BEMS systems in real-world settings emphasizes the consideration of the building occupants' preferences. While DRL approaches typically operate under some constraints with respect to the users' comfort, existing work focuses on temperature preferences. Additional considerations such as that of visual comfort, which may be affected by the opening and closing of blinds, are likely to be required to account for the users' expectations in practice [59].
   Further, in real-world settings, building occupants may express their preferences by directly interacting with the devices, e.g. by adjusting a thermostat, which can create feedback loops that may not be easy to capture in a simulation model. While human-in-the-loop control problems have been widely considered in the DRL literature (e.g. Ref. [60]), to the best of our knowledge, this is an unexplored avenue in the BEMS domain.
   Finally, since the user preferences, environmental and device conditions as well electricity prices may all dynamically change over time, there is a need for effective online learning capabilities. Open challenges in online RL include the high sample efficiency required to promptly adapt to the dynamic behavior of non-stationary environments and the difficulty of theoretical analysis to determine performance guarantees [19].
3. **Trust**: As the devices operated in a BEMS deployment may not only affect the users' comfort, but may also be safety-critical, high levels of trust are required before a system can be rolled out. As currently discussed in other domains such as autonomous driving, trust in intelligent control can be ensured by explainable and interpretable ML approaches aiming to provide an understanding how the DRL-based system makes its decisions. In explainable ML [61], the model may still be opaque, but additional analyses aim to reveal the basis for the decisions made. In contrast, interpretable ML [62] relies on models that permit a more direct interpretation than more opaque models such as those learned using neural networks. In a DRL setting, an opaque reference model may be substituted by an interpretable surrogate model to achieve interpretability. This approach has recently been explored in the context of BEMS [63] and more specifically for HVAC control [64].

18.5 Conclusions

We presented the current state of research in DRL for BEMS, which are among the largest energy consumers in the United States. After an introduction to fundamental RL techniques and recent advancements to tackle multi-faceted and dynamic problems, we showed how DRL has become a highly promising method to optimize building energy consumption and electricity cost while considering the occupants' preferences.

The energy saving and cost reduction results reported in the literature, often over 15%, are encouraging. We observe that while the existing work covers the key BEMS-controlled devices and most relevant scenarios, few authors considered multi-building or even city-scale settings, which may require multi-agent or multi-objective approaches. Further, there is a need for systematic and targeted studies to generate the broader insights and trust needed before DRL-based BEMS can be rolled out in the real world: Firstly, well-accepted benchmark scenarios and baseline methods are required to enable direct comparisons among the various approaches. Secondly, only few existing works experimented with real-world deployments. Since simulation models may fail to capture the complexity and dynamic behavior of real-world environments, real-world comparisons of different approaches could increase the confidence in DRL-based BEMS and their adaptability to dynamic changes in the environment. Thirdly, as in other application domains of RL, users may expect a certain level of interpretability of the behavior of the devices controlled by BEMS. Techniques from explainable and interpretable ML could be applied to achieve the level of trust demanded from a real-world deployment.

References

1 Schwab, K. (2016). The fourth industrial revolution: what it means, how to respond. https://www.weforum.org/agenda/2016/01/the-fourth-industrial-revolution-what-it-means-and-how-to-respond/ (accessed 19 August 2022).

2 Pentek, T., Hermann, M., and Otto, B. (2016). Design principles for Industrie 4.0 scenarios. *2016 49th Hawaii International Conference on System Sciences (HICSS)*, pp. 3928–3937. https://doi.org/10.1109/HICSS.2016.488.

3 Lee, M.H., Yun, J.H.J., Pyka, A. et al. (2018). How to respond to the fourth industrial revolution, or the second information technology revolution? Dynamic new combinations between technology, market, and society through open innovation.*Journal of Open Innovation: Technology, Market, and Complexity* 4 (3): 21.

4 Ashton, K. (2009). That 'Internet of Things' thing. *RFID Journal* 22 (7): 97–114.

5 ITU (2016). Y.4000: Overview of the Internet of Things. *ITU-T Y.4000/Y.2060*.

6 Ericsson (2020). Cellular networks for Massive IoT. https://www.ericsson.com/en/reports-and-papers/white-papers/cellular-networks-for-massive-iot-enabling-low-power-wide-area-applications (accessed 19 August 2022).

7 IDC (2020). The Growth in Connected IoT Devices is Expected to Generate 79.4ZB of Data in 2025, According to a New IDC Forecast. https://www.idc.com/getdoc.jsp?containerId=prUS45213219 (accessed 19 August 2022).

8 Augusto, J.C., Callaghan, V., Cook, D. et al. (2013). Intelligent environments: a manifesto. *Human-Centric Computing and Information Sciences* 3 (1): 1–18.

9 Littman, M.L. and Moore, A.W. (1996). Reinforcement learning: a survey. *Journal of Artificial Intelligence Research* 4 (1): 237–285. https://doi.org/10.1613/jair.301.

10 Lei, L., Tan, Y., Zheng, K. et al. (2020). Deep reinforcement learning for autonomous Internet of Things: model, applications and challenges. *IEEE Communication Surveys and Tutorials*. 22 (3): 1722–1760.

11 Shaikh, P.H., Nor, N.B.M., Nallagownden, P. et al. (2014). A review on optimized control systems for building energy and comfort management of smart sustainable buildings. *Renewable and Sustainable Energy Reviews* 34: 409–429.

12 EIA (2021). Energy use in homes. https://www.eia.gov/energyexplained/use-of-energy/homes.php (accessed 19 August 2022).

13 Sutton, R.S. and Barto, A.G. (2018). *Reinforcement Learning: An Introduction*. MIT press.

14 Bellman, R. (1957). A Markovian decision process. *Journal of Mathematics and Mechanics* 6 (5): 679–684.

15 Zhang, H. and Yu, T. (2020). Taxonomy of reinforcement learning algorithms. In: *Deep Reinforcement Learning*, 125–133. Springer, Singapore.

16 Nagabandi, A., Kahn, G., Fearing, R.S., and Levine, S. (2018). Neural network dynamics for model-based deep reinforcement learning with model-free fine-tuning. *2018 IEEE International Conference on Robotics and Automation (ICRA)*, pp. 7559–7566. IEEE.

17 Çal?ş?r, S. and Pehlivanoğlu, M.K. (2019). Model-free reinforcement learning algorithms: a survey. *2019 27th Signal Processing and Communications Applications Conference (SIU)*, pp. 1–4. https://doi.org/10.1109/SIU.2019.8806389.

18 Sutton, R.S. (1996). Generalization in reinforcement learning: successful examples using sparse coarse coding. In: *Advances in Neural Information Processing Systems*, 8: 1038–1044.

19 Padakandla, S. (2020). A Survey of Reinforcement Learning Algorithms for Dynamically Varying Environments. *arXiv preprint arXiv:2005.10619*.

20 Padakandla, S., Prabuchandran, K.J., and Bhatnagar, S. (2020). Reinforcement learning algorithm for non-stationary environments. *Applied Intelligence* 50 (11): 3590–3606. https://doi.org/10.1007/s10489-020-01758-5.

21 Littman, M.L. (1994). Markov games as a framework for multi -agent reinforcement learning. In: *Machine Learning Proceedings 1994*, Kaufmann, Morgan (ed.), 157–163. Elsevier.

22 Du, W. and Ding, S. (2020). A survey on multi-agent deep reinforcement learning: from the perspective of challenges and applications. *Artificial Intelligence Review* 54 (5): 3215–3238.

23 OroojlooyJadid, A. and Hajinezhad, D. (2020). A Review of Cooperative Multi-Agent Deep Reinforcement Learning.

24 Lee, S. and Choi, D.-H. (2020). Energy management of smart home with home appliances, energy storage system and electric vehicle: a hierarchical deep reinforcement learning approach. *Sensors* 20 (7): 2157.

25 Vithayathil Varghese, N. and Mahmoud, Q.H. (2020). A survey of multi-task deep reinforcement learning. *Electronics* 9 (9): 1363.

26 Mason, K. and Grijalva, S. (2019). A review of reinforcement learning for autonomous building energy management. *Computers and Electrical Engineering* 78: 300–312. https://doi.org/https://doi.org/10.1016/j.compeleceng.2019.07.019.

27 Yu, L., Qin, S., Zhang, M. et al. (2020). Deep Reinforcement Learning for Smart Building Energy Management: A Survey.

28 Vazquez-Canteli, J.R., Dey, S., Henze, G., and Nagy, Z. (2020). CityLearn: Standardizing Research in Multi-Agent Reinforcement Learning for Demand Response and Urban Energy Management. *arXiv preprint arXiv:2012.10504*.

29 Wang, Z. and Hong, T. (2020). Reinforcement learning for building controls: the opportunities and challenges. *Applied Energy* 269: 115036. https://doi.org/https://doi.org/10.1016/j.apenergy.2020.115036.

30 Ahn, K.U. and Park, C.S. (2020). Application of deep Q-networks for model-free optimal control balancing between different HVAC systems. *Science and Technology for the Built Environment* 26 (1): 61–74.

31 Yu, L., Sun, Y., Xu, Z. et al. (2020). Multi-agent deep reinforcement learning for HVAC control in commercial buildings. *IEEE Transactions on Smart Grid* 12 (1): 407–419.

32 Zhang, C., Kuppannagari, S.R., Kannan, R., and Prasanna, V.K. (2019). Building HVAC scheduling using reinforcement learning via neural network based model approximation. *Proceedings of the 6th ACM International Conference on Systems for Energy-Efficient Buildings, Cities, and Transportation*, pp. 287–296.

33 Gao, G., Li, J., and Wen, Y. (2019). Energy-Efficient Thermal Comfort Control in Smart Buildings Via Deep Reinforcement Learning. *arXiv preprint arXiv:1901.04693.*

34 Zhang, Z., Chong, A., Pan, Y. et al. (2019). Whole building energy model for HVAC optimal control: a practical framework based on deep reinforcement learning. *Energy and Buildings* 199: 472–490.

35 Zhang, Z., Chong, A., Pan, Y. et al. (2018). A deep reinforcement learning approach to using whole building energy model for HVAC optimal control. *2018 Building Performance Analysis Conference and SimBuild*, Volume 3, pp. 22–23.

36 Zhang, Z. and Lam, K.P. (2018). Practical implementation and evaluation of deep reinforcement learning control for a radiant heating system. *Proceedings of the 5th Conference on Systems for Built Environments*, pp. 148–157.

37 Wang, Y., Velswamy, K., and Huang, B. (2017). A long-short term memory recurrent neural network based reinforcement learning controller for office heating ventilation and air conditioning systems. *Processes* 5 (3): 46.

38 Wei, T., Wang, Y., and Zhu, Q. (2017). Deep reinforcement learning for building HVAC control. *Proceedings of the 54th Annual Design Automation Conference 2017*, pp. 1–6.

39 Barrett, E. and Linder, S. (2015). Autonomous HVAC control, a reinforcement learning approach. *Joint European Conference on Machine Learning and Knowledge Discovery in Databases*, pp. 3–19. Springer.

40 Li, B. and Xia, L. (2015). A multi-grid reinforcement learning method for energy conservation and comfort of HVAC in buildings. *2015 IEEE International Conference on Automation Science and Engineering (CASE)*, pp. 444–449. IEEE.

41 Energy Information Administration (2021). Annual Energy Outlook 2020 with Projections to 2050. U.S. Energy Information Administration. https://www.eia.gov/outlooks/aeo/ (accessed 19 August 2022).

42 Kazmi, H., Suykens, J., Balint, A., and Driesen, J. (2019). Multi-agent reinforcement learning for modeling and control of thermostatically controlled loads. *Applied Energy* 238: 1022–1035.

43 Ruelens, F., Claessens, B.J., Vrancx, P. et al. (2019). Direct load control of thermostatically controlled loads based on sparse observations using deep reinforcement learning. *CSEE Journal of Power and Energy Systems* 5 (4): 423–432.

44 Ruelens, F., Claessens, B.J., Quaiyum, S. et al. (2018). Reinforcement learning applied to an electric water heater: from theory to practice. *IEEE Transactions on Smart Grid* 9 (4): 3792–3800.

45 De Somer, O., Soares, A., Vanthournout, K. et al. (2017). Using reinforcement learning for demand response of domestic hot water buffers: a real-life demonstration. *2017 IEEE PES Innovative Smart Grid Technologies Conference Europe (ISGT-Europe)*, pp. 1–7. IEEE.

46 Ruelens, F., Claessens, B.J., Vandael, S. et al. (2014). Demand response of a heterogeneous cluster of electric water heaters using batch reinforcement learning. *2014 Power Systems Computation Conference*, pp. 1–7. IEEE.

47 Al-Jabery, K., Wunsch, D.C., Xiong, J., and Shi, Y. (2014). A novel grid load management technique using electric water heaters and Q-learning. *2014 IEEE International Conference on Smart Grid Communications (SmartGridComm)*, pp. 776–781. https://doi.org/10.1109/SmartGridComm.2014.7007742.

48 Xu, X., Jia, Y., Xu, Y. et al. (2020). A multi-agent reinforcement learning-based data-driven method for home energy management. *IEEE Transactions on Smart Grid* 11 (4): 3201–3211.

49 Lu, R., Hong, S.H., and Yu, M. (2019). Demand response for home energy management using reinforcement learning and artificial neural network. *IEEE Transactions on Smart Grid* 10 (6): 6629–6639.

50 Lee, S. and Choi, D.-H. (2019). Reinforcement learning-based energy management of smart home with rooftop solar photovoltaic system, energy storage system, and home appliances. *Sensors* 19 (18): 3937.

51 Yu, L., Xie, W., Xie, D. et al. (2019). Deep reinforcement learning for smart home energy management. *IEEE Internet of Things Journal* 7 (4): 2751–2762.

52 Mocanu, E., Mocanu, D.C., Nguyen, P.H. et al. (2018). On-line building energy optimization using deep reinforcement learning. *IEEE Transactions on Smart Grid* 10 (4): 3698–3708.

53 Wan, Z., Li, H., and He, H. (2018). Residential energy management with deep reinforcement learning. *2018 International Joint Conference on Neural Networks (IJCNN)*, pp. 1–7. IEEE.

54 Mahapatra, C., Moharana, A.K., and Leung, V. (2017). Energy management in smart cities based on Internet of Things: peak demand reduction and energy savings. *Sensors* 17 (12): 2812.

55 Ruelens, F., Claessens, B., Vandael, S. et al. (2015). Residential Demand Response Applications Using Batch Reinforcement Learning.

56 Dulac-Arnold, G., Mankowitz, D., and Hester, T. (2019). Challenges of Real-World Reinforcement Learning. *arXiv preprint arXiv:1904.12901*.

57 Zhu, Z., Lin, K., and Zhou, J. (2020). Transfer Learning in Deep Reinforcement Learning: A Survey. *arXiv preprint arXiv:2009.07888*.

58 Da Silva, F.L. and Costa, A.H.R. (2019). A survey on transfer learning for multiagent reinforcement learning systems. *Journal of Artificial Intelligence Research* 64: 645–703.

59 Kurte, K., Munk, J., Kotevska, O. et al. (2020). Evaluating the adaptability of reinforcement learning based HVAC control for residential houses. *Sustainability* 12 (18): 7727.

60 Abel, D., Salvatier, J., Stuhlmüller, A., and Evans, O. (2017). Agent-Agnostic Human-in-the-Loop Reinforcement Learning. *arXiv preprint arXiv:1701.04079*.

61 Puiutta, E. and Veith, E.M.S.P. (2020). Explainable reinforcement learning: a survey. In: *International Cross-Domain Conference for Machine Learning and Knowledge Extraction*, 77–95. Springer.

62 Annasamy, R.M. and Sycara, K. (2019). Towards better interpretability in deep Q-networks. *Proceedings of the AAAI Conference on Artificial Intelligence* 33 (1): 4561–4569.

63 Fan, C., Xiao, F., Yan, C. et al. (2019). A novel methodology to explain and evaluate data-driven building energy performance models based on interpretable machine learning. *Applied Energy* 235: 1551–1560. https://doi.org/https://doi.org/10.1016/j.apenergy.2018.11.081.

64 Kotevska, O., Munk, J., Kurte, K. et al. (2020). Methodology for interpretable reinforcement learning model for HVAC energy control. *2020 IEEE International Conference on Big Data (Big Data)*, pp. 1555–1564. https://doi.org/10.1109/BigData50022.2020.9377735.

19

Federated Deep Learning Technique for Power and Energy Systems Data Analysis

Hamed Moayyed[1], Arash Moradzadeh[2], Behnam Mohammadi-Ivatloo[2], and Reza Ghorbani[3]

[1] *ISEP – Instituto Superior de Engenharia do Porto, GECAD – Research Group on Intelligent Engineering and Computing for Advanced Innovation and Development, Polytechnic of Porto (P.PORTO), Porto, Portugal*
[2] *Faculty of Electrical and Computer Engineering, University of Tabriz, Tabriz, Iran*
[3] *Renewable Energy Design Laboratory (REDLab), Department of Mechanical Engineering, University of Hawaii at Manoa, Honolulu, HI, USA*

Nomenclature

Acronyms

AI	artificial intelligence
ANN	artificial neural network
BiLSTM	bidirectional LSTM
CNN	convolutional neural network
ELM	extreme learning machine
EV	electric vehicle
FDIA	false data injection attack
FL	federated learning
GMDH	group method data handling
GRNN	generalized regression neural network
GRU	gated recurrent unit
LSTM	long short-term memory
MAE	mean absolute error
MAPE	mean absolute percentage error
MLP	multilayer perceptron
MSE	mean squared error
non-IID	non-independent and identically distributed
PMU	phasor measurement unit
RESs	renewable energy sources
RMSE	root mean squared error
SCADA	supervisory control and data acquisition
SVM	support vector machine
SVR	support vector regression

Symbols

R^2	correlation coefficient

Intelligent Data Mining and Analysis in Power and Energy Systems: Models and Applications for Smarter Efficient Power Systems, First Edition. Edited by Zita Vale, Tiago Pinto, Michael Negnevitsky, and Ganesh Kumar Venayagamoorthy.

19.1 Introduction

Nowadays, increasing energy consumption, the development of smart grids, and the penetration of different types of energy systems have led to the complexity of power systems and their performance in unstable conditions [1]. On the other hand, the dramatic growth in energy demand has led to the high penetration of renewable energy sources (RESs) such as wind and solar energies in traditional power systems and has increased their complexity. However, performance analysis of power systems requires extensive physical modeling and numerical calculations and requires advanced measurement and monitoring systems, which leads to the production of large volumes of data [2, 3]. In previous decades, the processing of large volumes of data related to power systems was done based on traditional statistical and mathematical techniques, which had a high computational complexity and required experienced people. However, these techniques did not have high performance accuracy and in some cases, the results were not accurate enough. Later, with the efficiency of intelligent techniques based on data mining science, many problems related to data processing in various sciences and industries were solved. Thus, these techniques were able to solve many problems related to the analysis of the behavior of power systems in various areas such as generation, transmission, operation, and power consumption [4, 5].

Intelligent data mining techniques fall into several categories, such as artificial neural networks (ANNs), machine learning, and deep learning algorithms. A closer look at the problem shows that these techniques are used in various issues related to power systems such as condition monitoring of the power systems and its' equipment, cyber-attacks detection, and performing various forecastings such as RESs forecasting, consumption load forecasting, and price forecasting. Reducing computational complexity, high performance speed, and very high accuracy in processing high-dimension data were the obvious advantages of intelligent methods compared to traditional methods [6, 7]. However, these techniques have also in some cases suffered from fundamental problems such as high data dependence, data integrity, and time-series state modeling of data, which have posed many challenges today. In this chapter, after a thorough review of the performance of intelligent techniques in power systems applications, a new technique called federated learning (FL) is proposed for the first time as a trusted solution to solve problems related to intelligent methods.

The continuation of this chapter is organized as follows: Section 19.2 introduces the FL technique. Section 19.3 reviews the performance of artificial intelligence (AI)-based techniques in various applications of power systems. Section 19.4 describes the effectiveness of FL in the various challenges of power systems. Finally, Section 19.5 concludes the chapter.

19.2 Federated Learning (FL)

Applying machine learning and deep learning methods to real-world problems is still a challenge. A proposed model needs to be fed with a complete dataset to support the training process and achieve acceptable performance in the testing process. However, for real-world problems, it would be very difficult to collect a complete dataset. Moreover, the transmission of data sets to a central statistical unit for data analysis is another concern that should be considered in real experiments in terms of intentional or unintentional data modification.

In a real-world application of the distributed machine learning or deep learning model to create an optimal model for electric vehicle (EV) behavior and driving environment, current distributed approaches send a compressed data set to a central server that performs the data analysis. However,

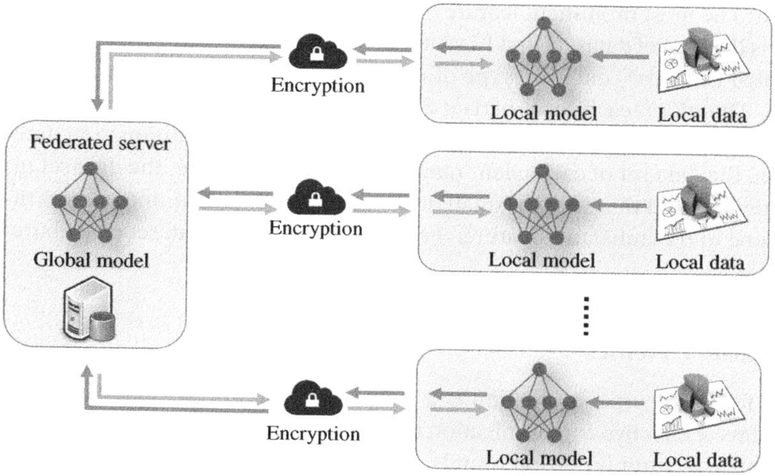

Figure 19.1 Conceptual architecture of FL.

there is a probability that some of the electric vehicles do not generate any data on a certain day, causing difficulties for the central server to train its own model. On the other hand, some of them generated hundreds of gigabytes on the same day, making it difficult to transfer or store these amounts of data on a central server. This issue even raises privacy concerns about cyber threats.

In 2016, Google proposed a solution to these concerns by introducing the FL method [8–10]. In this approach, no data is transferred to the server and only encoded parameters of the ANN are communicated between the central server and the local models. Therefore, the privacy of the data is maintained and there is also no need for data storage on a central server. Moreover, the central server has a sufficient overview of all local models to compensate for the lack of data in certain cases.

Figure 19.1 shows the conceptual architecture of the FL model. On the server side, a suggested model is sent to local models, which subsequently trains their own models on the local data. Then they communicate with the server by sending some coded parameters, and then a new model is pushed down from the central server to the local models. This process continues until all the local models are well trained.

19.2.1 Federated Learning Motivation

Privacy preservation: Data analytics and privacy are now closely intertwined, and cyber-security is now receiving significant attention in technology and operations. The main feature of the FL approach is privacy preservation. In a FL method, the server does not access the data of the individual local devices/clients, only receives temporary parameters to incorporate into its model. Upon completion of the aggregation process on the server, the client would receive the final trained model to apply to its own private data. Therefore, the attack surface in this process is limited only to the local part, unlike the local part and the cloud.

High-performance computing: In the FL approach, the computational overhead on the server is significantly reduced and the responsibility for most of the computation falls on the local devices. The server has the less computationally intensive task of averaging updates from the local devices, rather than constantly retraining a global model on new data sets.

Practical data adaptation: The most prominent feature of FL, which makes it recommended for real-world applications, is the type of datasets that FL can work with [11]. The Non-Independent and Identically Distributed (non-IID) data is a type of data that could be used in each client in a federated approach. Therefore, each local party's data is highly personalized, as the data corresponds to a specific customer, geographic location, and/or time range. In addition to feeding FL through non-IID data, the data set of each client may vary in size. Therefore, the data set of each client may be unbalanced and those with a small data set complete the training process in a short time. Moreover, due to the imbalance feature, the clients with a large dataset contribute more to the overall model.

19.2.2 Performance Evaluation Metrics

The most important point for an accurate evaluation is a good definition of the model's performance. The FL algorithm aims at effective communication between server and clients, so that high performance is achieved with low communication rounds. There are several suggestions in the literature for performance measures that could be used for FL. In the following text, some common measures are described in detail.

- **Classification accuracy:** Classification accuracy refers to the percentage of accurately classified samples in the test portion of the dataset which is commonly used in performance measures within machine learning and deep learning models [12]. Although classification accuracy is considered a well-defined method for evaluating algorithms, it is a poor measure in some specific real-world tasks [13]. Classification accuracy is inadequate in areas where class skewness is prevalent and precisely identifying certain classes is more crucial.
- **Precision and recall:** These measures are also considered in classification problems to evaluate the performance of the model. Precision and recall refer to a binary classification where one class is considered positive and another negative [14]. To reach the final calculation, the confusion matrix should be calculated first, as shown in Figure 19.2. A confusion matrix is an overall summary of the classification results. The number of correctly and incorrectly classified samples is summarized with calculations and broken down by each class. Figure 19.2 shows common metrics derived from the confusion matrix, including accuracy, precision, recall, and FP-rate.
- **Forecasting performance measures:** The evaluation of the performance of a model in a forecasting problem is based on the comparison of the predicted output (y) of the proposed model with the observed value (x), which is also considered as the measured value. The common

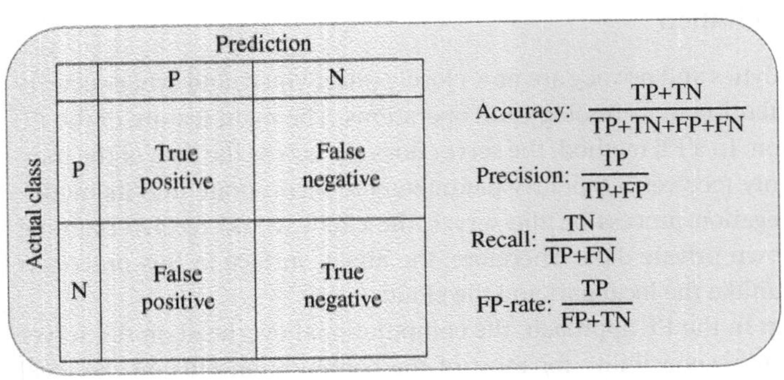

Figure 19.2 Confusion matrix for classification approaches and common performance measures derived from the confusion matrix.

performance indicators in forecasting time series are correlation coefficient (R^2), mean square error (MSE), root mean square error (RMSE), mean absolute error (MAE), and mean absolute percentage error (MAPE) [5, 15]. Each of the above indicators is calculated as follows:

$$R^2 = \frac{\sum_{i=1}^{N}(x_i - \bar{x})(y_i - \bar{y})}{\sqrt{\sum_{i=1}^{N}(x_i - \bar{x})^2 \sum_{i=1}^{n}(y_i - \bar{y})^2}} \tag{19.1}$$

$$MSE = \frac{1}{n}\sum_{i=1}^{N}(y_i - x_i)^2 \tag{19.2}$$

$$RMSE = \sqrt{\frac{\sum_{i=1}^{n}(y_i - x_i)^2}{n}} \tag{19.3}$$

$$MAPE = \frac{1}{N}\sum_{i=1}^{N}\left|\frac{X_i - Y_i}{X_i}\right| \times 100 \tag{19.4}$$

where N is the number of observations, $\bar{y}$ and $\bar{x}$ represents the average of forecasted values and real values, and n shows the sample numbers.

19.2.3 Federated Learning vs. Distributed Machine Learning Approaches

FL differs from similar methods that deal with distributed problems such as data center distribution. Although in both the main goal is to optimize their objective function in the learning process, FL is based on communicating with local devices with simple networks and limited upload speeds. Therefore, FL seeks to reduce communication costs while presenting a desirable model.

Another difference between these two methods is the critical assumptions that must be met for the training dataset to feed the model. Distributed machine learning algorithms require a dataset with the following structure:

- The data distribution over all clients is equal.
- Client data has property of Independently and Identically Distributed (IID).

In contrast, the FL algorithm does not satisfy these assumptions. This means that the client's data could be inequivalent and even non-IID data could be used in FL models.

19.2.4 The Federated Averaging Algorithm

In FL, it is first assumed that the data that is not stored centrally is now distributed among a number K of clients, so that

- Part of the data with a set of indices $\mathcal{P}_k$ data points is dedicated to client k.
- n represents the number of all data points allocated to all clients. Thus, n_k is the number of data points that client k has.
- $n_k = |\mathcal{P}_k|$

Consider the following loss function in a minimization problem:

$$\min_{\theta \in \mathbb{R}^d} f(\theta) \tag{19.5}$$

where

$$f(\theta) \stackrel{\text{def}}{=} \frac{1}{n}\sum_{i=1}^{n} f_i(\theta) \tag{19.6}$$

If (x_i, y_i) is considered as an observation for which the model parameter θ holds, then $f_i(\theta)$ is the loss for predicting this observation. The above loss function can be rewritten in a federated form for K clients as follows:

$$f(\theta) = \sum_{k=1}^{K} \frac{n_k}{n} F_k(\theta) \tag{19.7}$$

where

$$F_k(\theta) = \frac{1}{n_k}\sum_{i \in \mathcal{P}_k} f_i(\theta) \tag{19.8}$$

To simplify the federated loss function, the average loss $F_k(\theta)$ for a given client k could be computed with $\frac{1}{n_k}\sum_{i\in\mathcal{P}_k} f_i(\theta)$, instead of calculating the average loss as an average over n samples within the $\frac{1}{n}\sum_{i=1}^{n} f_i(\theta)$. Then group the loss of all local K parities by calculating a weighted average loss in terms of the number of data points n_k held by each party.

To calculate the gradients of the federated model for each client, first calculate the average gradient g_k using its local data as follows:

$$g_k = \nabla F_k(\theta) \tag{19.9}$$

Then each local parity takes a step of gradient descent and renews each of its parameters, formalized as:

$$\forall k, \theta_k \leftarrow \theta - \eta g_k \tag{19.10}$$

To achieve the final update of the model parameters of the super model stored in the central server, this step can be repeated until the central server then computes the weighted average of these gradients as

$$\theta \leftarrow \sum_{k=1}^{K} \frac{n_k}{n} \theta_k \tag{19.11}$$

This ends a complete round of updates to the central server's super model.

The three main parameters that control the computational effort for one full update in a federated setting are:

1. C; this is the number of clients participating in an assigned update round.
2. E; this is the number of gradient descent steps for each client.
3. B; this is the batch size for the updates for all clients.

Computational burden of a federated averaging algorithm is affected by the above parameters. On the server side, C affects the required computational power at the server level. On the client side, E and B affect the computational effort required to update the client model. The comprehensive pseudocode for this approach was first suggested by [10]. The procedure of the federated averaging algorithm presents in following:

Algorithm 19.1 Federated Averaging

The algorithm is divided into two parts, the server side and the client side. K is the total number of clients, each indexed by k; B is the local minibatch size, E is the number of local epochs, and η is the learning rate. ω denotes the model parameters [10].

Procedure

Server-Side

Initialize ω_0
for each round $t = 1, 2, \ldots$ do
$m \leftarrow \max(C \times K, 1)$
$S_t \leftarrow$(random set of m clients)
for each client $k \in S_t$ in parallel do
$\omega_{t+1}^{k} \leftarrow$ Client Update (k, ω_t)
$\omega_{t+1}^{k} \leftarrow \sum_{k=1}^{K} \frac{n_k}{n} \omega_{t+1}^{k}$

Client-Side (Run on client k)

$B \leftarrow$ (split $\mathcal{P}_k$ into batches of size B)
for each local epoch i from 1 to E do
$\omega \leftarrow \omega - \eta \nabla \ell(\omega; b)$
return ω to server.

The FL algorithm can be used with machine learning or deep learning models. This means that the local model of each client can consist of different AI techniques. In the case where the FL algorithm is equipped with deep learning, we call it federated deep learning.

For a fast convergence of the federated deep learning model, the client models must first be equipped with a powerful deep learning model to perform the proposed task. Then, the server has less overhead in updating the deep learning network parameters of each client.

A standard federated deep learning process consists of a server and K clients, as shown in Figure 19.3. The process can be summarized in four main steps. First, initial training is performed in each local party by receiving the parameters θ_t from the central server. The second step is model aggregation, where the server receives the parameters θ_t^k from each client and performs secure aggregation over the clients. The most commonly used method for aggregation is federated averaging, which is explained in Algorithm 19.1. The third step is parameter broadcasting from the server, where new parameters θ_{t+1} are sent back to each local client so that it can perform new training on its own dataset. Finally, all clients update their individual deep learning models with respect to the aggregated parameters and evaluate the effectiveness of the new models.

19.2.5 Applications of Federated Learning

Regarding salient and interesting features of FL, it is used in a variety of applications. These applications vary considerably in terms of vertical orientation and particular use case, but generally share three main features:

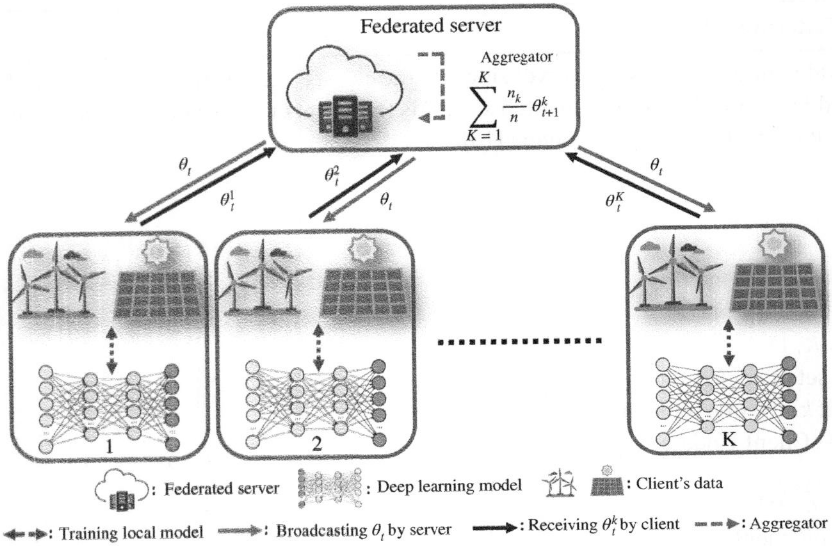

Figure 19.3 Architecture of federated deep learning model consists of a server and K clients.

1. Creating a global super model across multiple local devices.
2. Security of data and its protection against cyber-attacks.
3. Dealing with big data and the difficulty of storing it on a central server.
4. Task labels do not necessarily have to be assigned by humans but can also result from user collaboration.

It is not necessary for an application to satisfy all the above conditions, but these circumstances make FL a prominent technique among other machine learning or deep learning techniques. To demonstrate the potential of this novel technique, we give below a brief overview of some of the recent applications in practice.

Intel Corporation: Patient information is one of the primary roles of health care institutions and collaboration between these institutions is always fraught with challenges in sharing medical data. There has always been a need for a central server to manage this collaboration considering various rights, privacy, technology, and data ownership. As reported in [16], Intel Corporation has applied federated deep learning to multi-institutional collaboration to build deep learning models without sharing patient data. They present the first use of the federated deep learning classifier for brain tumor segmentation in institutional collaboration and find that this novel method can achieve better results compared to traditional distributed methods. This confirms the belief that FL can lead to breakthroughs in the healthcare space without sacrificing patient privacy.

Firefox: Firefox has millions of users every day and there is a need to accurately improve the auto-complete feature to encourage users and increase user retention. As reported in Ref. [17], Firefox used federated deep learning to improve the performance of the Firefox search bar. In this work, they analyzed 360 000 users to build the federated global model without collecting actual user data.

Google: The first practical implementation of federated deep learning on a large scale was done by Google [18]. They applied this technique to build a super global model to improve the quality of virtual keyboard search suggestions. In their work, they faced many of the technical difficulties and challenges of creating a model based on millions of devices. These challenges include connectivity of local devices to the central server, dealing with model training biases across different time zones of local parties, and reducing the consequences of training the model on local devices such as battery life and device speed.

19.2.6 Challenges of Federated Learning

FL is a new technique, and as with any new method, there is room for improvement. This technique is not without its challenges. Below is a summary of the main typical challenges that should be considered when using this method.

Data sampling: FL method that works with non-IID data. Although this feature makes its application close to real-world problems, it might pose some challenges in data sampling tasks. For IID data, the application of statistical and mathematical techniques could be simplified. That is, select a sample that has an acceptable confidence interval and perform these analyses on the sample that is representative of the overall population. This simplification is not satisfactory when working with FL, because the client's datasets are not IID and differ significantly from other clients. Therefore, a random sample of a client is not representative of the global population distribution. To cover this challenge, other sampling methods, such as stratified sampling, can be applied.

Unbalanced: As mentioned earlier, the data volume of each client is not the same. Some clients have very small amounts of data, while other clients generate large amounts of data at the same time. This characteristic sometimes makes it difficult for the server to converge. The solution would be to work with different federated algorithms to achieve fast convergence.

Limited communication: In a practical application, the connection between the clients and the central server is established through the Internet connection. If the server wants to cover a large area, there is a possibility of a poor connection and thus limited communication between the server and all clients. In this situation, converging federated averaging is the server model faces some difficulties and the model of those clients that have a bad connection does not have a good performance.

Maintaining performance: As mentioned earlier, the federated global model on the server is trained using only the parameters obtained from the clients and no raw data helps the server model in this training process. Therefore, special attention should be paid to the global model to achieve good performance and then maintain it while updating the client models. This goal could be achieved by controlling the server performance by adding some additional performance metrics that closely check the server performance and change the federated algorithm in case of performance degradation.

Hardware requirements: The capability of the hardware to apply the model proposed by the server may be an issue in some specific applications on the client side. For example, in applying the distributed deep learning model to build an optimal model for the behavior and driving environment of electric vehicles, the hardware of the vehicles is weaker than typical known hardware such as desktops or laptops, and training their model with complex models could affect the training process and even the performance of the global model. To overcome this problem, clients must be equipped with appropriate hardware to correctly apply the broadcasted model from the server and even to maintain the performance of the global model.

19.3 Power Systems Challenges and the Performance of Artificial Intelligence Techniques in It

Over the past decades, increasing awareness of environmental protection and related government policies to meet the growing energy demand have led to the modernization of power systems and the integration of RESs. While the new and modern structure has greatly helped to distribute sustainable, reliable, and clean energy to users, it has also created new challenges for network security and stability. Thus, the over-reliance of modern power systems applications such as Supervisory Control and Data Acquisition (SCADA), condition monitoring, cyber-security, and phasor measurement unit (PMU) on modern and open communication technologies (like the Internet), has made networks vulnerable and exposed them to various threats. On the other hand, power grid operators are faced with a large amount of data, which has had many adverse effects on scheduling and planning of power systems.

Nowadays, AI-based techniques such as ANN, machine learning, and deep learning algorithms are widely used to solve power system problems. These methods, despite solving many of the problems mentioned and improving the process of optimal operation of power systems, suffer from some applicable problems that could be covered by applying federated deep learning technique. The continuation of this section examines the performance of AI techniques in various applications of power systems.

19.3.1 AI-Based Forecasting in Power Systems

Optimal operation of power systems requires a balance between energy supply and demand, which in the first stage requires knowledge of load consumed. In addition, forecasting the production energy in the coming days is another key factor in establishing supply and demand. Production forecasting should also be accompanied by forecasting the power generation of RESs such as wind and solar energies [19]. In addition to these issues, electricity price forecasting is also one of the most important issues for energy producers and consumers in power systems, which helps both to participate in the electricity markets with careful planning.

Given that 60% of the energy produced is used in residential and commercial buildings, energy management in these buildings is considered as a very ideal way to manage energy in power systems [20]. Cooling and heating loads forecasting in a variety of buildings is suggested and used as an ideal solution to do this. To date, various studies have developed AI-based techniques to perform each of the aforementioned forecasting applications in the power systems.

In Ref. [21], the generalized regression neural networks (GRNNs) as one of the ANN applications has been used to forecast the short-term electrical load and electricity prices in electricity markets. Another application of ANN called neuro-fuzzy model has been introduced in Ref. [22] to forecast the load profile of the day-ahead. In Ref. [23], load forecasting for a power network has been done based on a one of the well-known machine learning techniques called support vector machine (SVM). Short-term load forecasting has been performed based on the regression-based machine learning applications called random forest, multiple linear regression, and gradient boosting in [24]. The electricity demand forecasting for certain days of summer and winter has been done in Ref. [25] via the development of one of the deep learning applications called long short-term memory (LSTM). In another similar study [26], the LSTM technique and gated recurrent units (GRUs) have been proposed for short-term load forecasting in two case studies. Another deep learning technique called bidirectional long short-term memory (Bi-LSTM) has been proposed in Ref. [27] for short-term load forecasting in a microgrid. In Ref. [28], convolutional neural network (CNN) has been proposed as another deep learning application to forecast the probabilistic load of the power grid. In Ref. [29], determining uncertainty in the future load demand for smart grid stability has been performed based on probabilistic load forecasting via a hybrid approach of deep learning algorithms. In Ref. [30], the LSTM and support vector regression (SVR) techniques have been developed as a hybrid model for short-term load forecasting in a microgrid.

Each of the reviewed studies has introduced a load forecasting technique and in each of them, an attempt has been made to increase the forecasting accuracy. Thus, the load forecasting in each of these studies is based on meteorological and historical data as input to AI techniques. A noteworthy point is the time-series of meteorological data and in most cases the privacy of data related to the power grid, which has not been considered in the above studies. In addition, the dependence of all the perfection of the mentioned techniques is another problem related to the use of these methods.

At the same time, so far many challenges have been performed in order to RESs forecasting using AI-based techniques. For example, Multi-Layer Perceptron (MLP) and Group Method of Data Handling (GMDH) are proposed as ANN applications in Ref. [31] for short-term wind speed forecasting. The SVR as a regression method based on machine learning forecasts wind speed in short term time horizons. In Ref. [32], another machine learning application called extreme learning machine (ELM) has been utilized to forecast wind power generation. To obtain the correlation between numerical forecasting of weather and wind power generation, three different machine learning methods, namely SVR, ANN, and Gaussian process have been introduced in Ref. [33]. The LSTM network in [34] was selected to extract the temporal properties of different nodes in the deep learning process from the wind speed and wind direction of adjacent wind shapes. Wind speed forecasting has been performed in [35] to provide a detailed planning of the power generation process based on a hybrid multi-step approach of deep learning techniques called autoencoders and Bi-LSTM, and optimization algorithms. Has been. In Ref. [36], a hybrid network based on a CNN model and a radial basis function neural network with a dual Gaussian function have been developed to forecast 24-hour wind power generation. In Ref. [37], solar radiation forecasting has been done based on other meteorological parameters and the use of ANN algorithms and SVM framework. In Ref. [38], long-term solar radiation forecasting in Australia has been done by introducing the ELM technique. Solar radiation prediction has been performed using six actual datasets collected from various sites in Finland, with the development of LSTM techniques and an aggregation function in Ref. [39]. The CNN technique has been used in Ref. [40] to forecast solar radiation to control the rate of PV power ramps. In Ref. [41], the CNN and autoencoders develop a hybrid model called the convolutional graph autoencoder to forecast solar radiation at multiple locations over a large area of the northern United States.

Renewable energy power generation is directly based on other meteorological parameters such as temperature, humidity, and cloud cover, which contain a very large amount of data. In addition, access to all this data is not possible in all areas, especially in operational areas. Therefore, it is possible to use AI-based techniques for areas that have data available to run the training process. However, modeling the time-series state of weather data and identifying their behavioral patterns at different time intervals is a difficult task for these methods.

19.3.2 AI-Based Condition Monitoring in Power Systems

The emergence of advanced innovations and the integration of advanced monitoring and control approaches have transformed conventional power systems into intelligent power systems around the world. On the other hand, the growing trend of consumer demand for uninterrupted, stable, and high-quality electricity is constantly increasing. This process led to the installation of new transmission lines and distribution networks with intelligent monitoring and control systems, which eventually led to the introduction of a complex power system. Transmission lines and distribution networks are considered to be the two main parts of a system, and the stability and reliability of the electricity distributed to consumers depends on their health and stability [42]. However, sometimes due to various factors, different parts of the power system suffer physical damages and cyber-attacks, which endanger the stability and security of the system. Timely detection of damage and detection of anomalies in different parts of the power system requires proper and effective condition monitoring systems. So far, various methods based on AI have been studied and utilized as fault detection tools and condition monitoring of different parts of a power system.

For example, in Ref. [43], a hybrid method based on the ANN applications and the wavelet transform method for detecting short circuit faults in transmission line has been presented. A hybrid approach based on wavelet-Prony methods and various regression-based machine learning techniques such as GRNN, k-NN, and random forests have been proposed in Ref. [44] to identify transmission line faults. The identification of transmission line short-circuit faults has been performed in Ref. [45] by using the characteristics obtained from wavelet transform and the SVM method. In Ref. [42], the identification of single-phase short-circuit faults in transmission lines has been performed by presenting two well-known applications of deep learning techniques called CNN and deep reinforcement learning. In Ref. [46], a novel hybrid deep learning approach called convolutional-LSTM, which combines CNN and LSTM methods, has been suggested to identify the type and location of all transmission line faults.

Similarly, for stability and fault detection in distribution networks, in Ref. [47], ANN applications and simultaneous measurement systems have been proposed to identify faulty feeders in different operating conditions. Identification and classification of distribution network faults have been done in Ref. [48] using SVM and an innovative AI-based method. In Ref. [49], the detection of short circuit faults in distribution networks by considering the operating modes of scattered generators in the analysis of fault characteristics, based on machine learning-based regression applications. A hybrid approach based on image processing and CNN techniques has been developed to identify and classify distribution network faults in Ref. [50]. In Ref. [51], a fault detection system in a low voltage network has been suggested based on various data mining techniques called binary and multi-class SVM, Fisher diagnostic analysis, and principal component analysis.

Early detection of short-circuit faults in power transformer windings has been performed using the MLP algorithm in Ref. [52]. An unsupervised data mining application called Isometric feature map has been proposed in Ref. [53] to detect short circuit faults in transformer windings. In Ref. [54], a CNN-based image processing approach has been developed to interpret the frequency response curves associated with transformer winding faults to pinpoint the exact location of the fault in the transformer windings.

A review of the literature emphasizes the wide range of applications of AI-based techniques in various parts of power systems, especially in condition monitoring applications. As seen, various AI-based techniques have been used to identify types and locations of faults related to transmission lines and distribution networks. However, these techniques also suffer from high dependence on data quality and data volume in implementing this process. High-volume processing and robust processing systems are requirements for the use of AI-based techniques to analyze real-world data related to power systems, which are obtained by measuring devices installed in transmission lines and distribution networks. In addition, the lack of generalizability and poor performance against cyber-attacks of any of the methods used to date are other major drawbacks.

To address these problems and solve the limitations of AI-based techniques, in this chapter the federated deep learning technique is introduced as a generalizable and cyber-resilient approach to analyze data types in the power and energy systems applications. To examine the performance of the suggested method, the section 19.4 reviews the applications of federated deep learning in power and energy systems.

19.4 Application of Federated Deep Learning in Power and Energy Systems

The FL technique has recently been used in some applications in power and energy systems. Although it has only recently been applied, it has attracted a lot of attention and generated a lot of interest among professionals in the field. These applications have involved Electric Vehicle (EV) networks [55], False Data Injection Attack (FDIA) [56], Solar Radiation Prediction [57], and Heating Load Demand Forecasting in Building Environment [58]. The following is a summary of these works.

19.4.1 Electric Vehicle Networks

Federated deep learning has been applied to EV networks to predict energy demand and reduce communication overhead in EV networks. In this experiment, the charging station provider is considered as a global server and the local EV charging stations are considered as clients of the federated algorithm. The federated approach for energy demand prediction was used to reduce the communication overhead and effectively protect the privacy of EV users, with no data communication between the charging stations and the provider to maintain privacy.

As mentioned earlier, accurate prediction tasks and high-performance computing are among the main features of federated deep learning technique that were investigated in this experiment. In this work, a real-world dataset on the distribution of EV charging stations in the city of Dundee, UK, was analyzed [59]. The performance of FL was measured in predicting the energy demand by analyzing the RMSE.

The results presented in this paper demonstrate the accuracy of the federated model in the prediction task and the RMSE results compared to different learning techniques such as Decision Tree, Random Forest, SVR, k-NN, Stochastic Gradient Descent Regression, and MLP. The analysis of the RMSE results showed that the federated deep learning method provided better prediction of energy demand compared to all the mentioned techniques. This performance was achieved when the communication overhead of the federated deep learning technique was much less than the other techniques.

19.4.2 False Data Injection Attacks in Solar Farms

One of the most common cyber-attacks on smart grids is the FDIA. This type of attack benefits from the failure of traditional bad data detection, and then attackers manipulate smart meter measurements. In this work, federated deep learning was applied to detect FDIA on photovoltaic systems.

In this paper, a federated model for detecting FDIA in solar power DC/DC and DC/AC converters was proposed. The properties of this technique were tested in dealing with IID and non-IID data. Then, the results were compared with classical centralized ML techniques. This analysis has confirmed that federated deep learning significantly outperforms traditional machine learning implementations in terms of communication efficiency, accurate FDIA detection, and privacy protection.

19.4.3 Solar Irradiation Forecasting

Recently, the increasing prevalence of solar energy in power and energy systems around the world has dramatically increased interest in accurately predicting solar irradiation. In this work, the federated deep learning was equipped with a powerful deep learning technique of Bayesian LSTM neural network (BayesLSTM-NN). In this procedure, the federated server updates the parameters of BayesLSTM-NN which is dedicated to each client.

In this work, two real datasets from Solar GIS and National Solar Radiation Database were used to test the suitability of the proposed method. The federated BayesLSTM-NN scheme was applied for solar radiation prediction. The results revealed that the proposed method can achieve competitive performance while protecting privacy. Moreover, the performance of federated BayesLSTM-NN was significantly better than that of the centralized model.

19.4.4 Heating Load Demand Forecasting

The federated deep learning technique has also been used to forecast heating load demand in buildings. The presented method called Cyber-Secure Federated Deep Learning (CSFDL) showed promising performance in predicting heating load demand when the global server is trained and tested considering the heating load demand of 10 different clients in their building environment. To make this technique more reliable, a special security control was applied in CSFDL to protect the parameters that communicate between the server and the clients. The results presented in this work, including a comparative study, have proven the feasibility and accuracy of the CSFDL method when compared its performance with different methods such as SVR, GRNN, LSTM and Bi-LSTM.

19.5 Conclusion

Smart grids, including power and energy systems, are among the critical infrastructures of today's world; they have complex architecture and contain critical devices. If the data generated by these systems is not analyzed accurately, and if security principles are not observed in this analysis, this can lead to national security failures, public order disruptions, loss of life, or major economic damage.

Accordingly, high forecasting accuracy, good classification precision, secure data analysis while maintaining privacy, and across multiple distributed edge devices are among the key requirements for the future of power and energy systems.

The federated deep learning technique was recently introduced by Google with outstanding properties such as decentralized data analysis, accurate performance, privacy preservation, high-performance computation, and the ability to process non-IID data. Although the application of this technique could bring great advances in data analysis of power and energy systems, there are few papers reporting the application of federated deep learning, including electric vehicle networks, FDIA detection in solar farms, and solar radiation prediction. The reported result proves the perfect performance of this technique compared to classical machine learning and deep learning techniques.

There are still a variety of open problems in power and energy systems that needs a powerful data analytics technology, including a global forecasting model for wind energy, solar radiation, and efficient energy consumption. It is foreseeable that federated deep learning will be used in a variety of these applications in the near future and will attract much attention and interest.

In applying this technique, we should consider the few limitations of this technique with monitoring the convergence of the server, periodically checking its performance, and applying this technique with the ability of all clients to connect to the Internet.

References

1 Sadeghian, O., Moradzadeh, A., Mohammadi-Ivatloo, B. et al. (2021). A comprehensive review on energy saving options and saving potential in low voltage electricity distribution networks: building and public lighting. *Sustainable Cities and Society* 72: 103064.

2 Umunnakwe, A., Huang, H., Oikonomou, K., and Davis, K.R. (2021). Quantitative analysis of power systems resilience: standardization, categorizations, and challenges. *Renewable and Sustainable Energy Reviews* 149: 111252.

3 Ardeshiri, A., Lotfi, A., Behkam, R. et al. (2021). Introduction and literature review of power system challenges and issues. In: *Application of Machine Learning and Deep Learning Methods to Power System Problems*, 19–43. https://link.springer.com/10.1007/978-3-030-77696-1_2.

4 Talaat, M., Alsayyari, A.S., Alblawi, A., and Hatata, A.Y. (2020). Hybrid-cloud-based data processing for power system monitoring in smart grids. *Sustainable Cities and Society* 55: 102049.

5 Moradzadeh, A., Mohammadi-Ivatloo, B., Pourhossein, K. et al. (2021). Introduction to machine learning methods in energy engineering. In: *Application of Machine Learning and Deep Learning Methods to Power System Problems*, 61–82. https://www.researchgate.net/publication/260424870_miRNomics_MicroRNA_Biology_and_Computational_Analysis.

6 Barja-Martinez, S., Aragüés-Peñalba, M., Munné-Collado, Í. et al. (2021). Artificial intelligence techniques for enabling Big Data services in distribution networks: a review. *Renewable and Sustainable Energy Reviews* 150: 111459.

7 Zhao, Y., Zhang, C., Zhang, Y. et al. (2020). A review of data mining technologies in building energy systems: load prediction, pattern identification, fault detection and diagnosis. *Energy and Built Environment* 1 (2): 149–164.

8 Konečný, J., McMahan, H.B., Yu, F.X. et al. (2016). Federated Learning: Strategies for Improving Communication Efficiency. arXiv preprint arXiv:161005492. http://arxiv.org/abs/1610.05492.

9 Konečný, J., McMahan, H.B., Ramage, D., and Richtárik, P. (2016). Federated Optimization: Distributed Machine Learning for On-Device Intelligence. http://arxiv.org/abs/1610.02527.

10 McMahan, H.B., Moore, E., Ramage, D. et al. (2016). Federated learning of deep networks using model averaging. *Arxiv* 92 (9): 091118.

11 Zhao, Y., Li, M., Lai, L. et al. (2018). Federated Learning with Non-IID Data. http://arxiv.org/abs/1806.00582.

12 Science, I. (2006). Statistical comparisons of classifiers over multiple data sets. *Journal of Machine Learning Research* 7: 1–30.

13 Provost, F. and Fawcett, T. (2001). Robust classification for imprecise environments. *Machine Learning* 42 (3): 203–231.

14 Buckland, M. and Gey, F. (1994). The relationship between Recall and Precision. *Journal of the American Society for Information Science* 45 (1): 12–19.

15 Moradzadeh, A., Mansour-Saatloo, A., Mohammadi-Ivatloo, B., and Anvari-Moghaddam, A. (2020). Performance evaluation of two machine learning techniques in heating and cooling loads forecasting of residential buildings. *Applied Sciences (Switzerland)* 10 (11): 3829.

16 Sheller, M.J., Reina, G.A., Edwards, B. et al. (2019). Multi-institutional deep learning modeling without sharing patient data: a feasibility study on brain tumor segmentation. In: *Lecture Notes*

in Computer Science (including subseries Lecture Notes in Artificial Intelligence and Lecture Notes in Bioinformatics), 92–104. http://link.springer.com/10.1007/978-3-030-11723-8_9.

17 Hartmann, F. (2018). Federated Learning. Línea.

18 Yang, T., Andrew, G., Eichner, H. et al. (2018). Applied Federated Learning: Improving Google Keyboard Query Suggestions. http://arxiv.org/abs/1812.02903.

19 Moradzadeh, A., Mansour-Saatloo, A., Nazari-Heris, M. et al. (2021). Introduction and literature review of the application of machine learning/deep learning to load forecasting in power system. In: *Application of Machine Learning and Deep Learning Methods to Power System Problems*, 119–135. https://link.springer.com/10.1007/978-3-030-77696-1_6.

20 Moradzadeh, A., Mohammadi-Ivatloo, B., Abapour, M. et al. (2021). A practical solution based on convolutional neural network for non-intrusive load monitoring. *Journal of Ambient Intelligence and Humanized Computing* 12 (10): 9775–9789.

21 Heydari, A., Majidi Nezhad, M., Pirshayan, E. et al. (2020). Short-term electricity price and load forecasting in isolated power grids based on composite neural network and gravitational search optimization algorithm. *Applied Energy* 277: 115503.

22 Malekizadeh, M., Karami, H., Karimi, M. et al. (2020). Short-term load forecast using ensemble neuro-fuzzy model. *Energy* 196: 117127.

23 Zhao, E., Zhang, Z., and Bohlooli, N. (2020). Cost and load forecasting by an integrated algorithm in intelligent electricity supply network. *Sustainable Cities and Society* 60: 102243.

24 Zhang, N., Li, Z., Zou, X., and Quiring, S.M. (2019). Comparison of three short-term load forecast models in Southern California. *Energy* 189: 116358.

25 Bedi, J. and Toshniwal, D. (2018). Empirical mode decomposition based deep learning for electricity demand forecasting. *IEEE Access* 6: 49144–49156.

26 Tang, X., Dai, Y., Wang, T., and Chen, Y. (2019). Short-term power load forecasting based on multi-layer bidirectional recurrent neural network. *IET Generation, Transmission & Distribution.* 13 (17): 3847–3854.

27 Moradzadeh, A., Moayyed, H., Zakeri, S. et al. (2021). Deep learning-assisted short-term load forecasting for sustainable management of energy in microgrid. *Inventions* 6 (1): 15.

28 Huang, Q., Li, J., and Zhu, M. (2020). An improved convolutional neural network with load range discretization for probabilistic load forecasting. *Energy* 203: 117902.

29 Yang, Y., Hong, W., and Li, S. (2019). Deep ensemble learning based probabilistic load forecasting in smart grids. *Energy* 189: 116324.

30 Moradzadeh, A., Zakeri, S., Shoaran, M. et al. (2020). Short-term load forecasting of microgrid via hybrid support vector regression and long short-term memory algorithms. *Sustainability (Switzerland)* 12 (17): 7076.

31 Moradzadeh, A., Mansour-Saatloo, A., Nazari-Heris, M. et al. (2021). Wind speed forecasting using innovative regression applications of machine learning techniques. In: *Application of Machine Learning and Deep Learning Methods to Power System Problems*, 249–263. https://link.springer.com/10.1007/978-3-030-77696-1_12.

32 Wan, C., Xu, Z., Pinson, P. et al. (2013). Probabilistic forecasting of wind power generation using extreme learning machine. *IEEE Transactions on Power Systems* 29 (3): 1033–1044.

33 Du, P. (2018). Ensemble machine learning-based wind forecasting to combine NWP output with data from weather station. *IEEE Transactions on Sustainable Energy* 10 (4): 2133–2141.

34 Khodayar, M. and Wang, J. (2018). Spatio-temporal graph deep neural network for short-term wind speed forecasting. *IEEE Transactions on Sustainable Energy* 10 (2): 670–681.

35 Liu, H. and Chen, C. (2019). Multi-objective data-ensemble wind speed forecasting model with stacked sparse autoencoder and adaptive decomposition-based error correction. *Applied Energy* 254: 113686.

36 Hong, Y.-Y. and Rioflorido, C.L.P.P. (2019). A hybrid deep learning-based neural network for 24-h ahead wind power forecasting. *Applied Energy* 250: 530–539.

37 Meenal, R. and Selvakumar, A.I. (2018). Assessment of SVM, empirical and ANN based solar radiation prediction models with most influencing input parameters. *Renewable Energy* 121: 324–343.

38 Deo, R.C., Şahin, M., Adamowski, J.F., and Mi, J. (2019). Universally deployable extreme learning machines integrated with remotely sensed MODIS satellite predictors over Australia to forecast global solar radiation: a new approach. *Renewable and Sustainable Energy Reviews* 104: 235–261.

39 Abdel-Nasser, M., Mahmoud, K., and Lehtonen, M. (2021). Reliable solar irradiance forecasting approach based on choquet integral and deep LSTMs. *IEEE Transactions on Industrial Informatics* 17 (3): 1873–1881.

40 Wen, H., Du, Y., Chen, X. et al. (2021). Deep learning based multistep solar forecasting for PV ramp-rate control using sky images. *IEEE Transactions on Industrial Informatics* 17 (2): 1397–1406.

41 Khodayar, M., Mohammadi, S., Khodayar, M.E. et al. (2020). Convolutional graph autoencoder: a generative deep neural network for probabilistic spatio-temporal solar irradiance forecasting. *IEEE Transactions on Sustainable Energy* 11 (2): 571–583.

42 Teimourzadeh, H., Moradzadeh, A., Shoaran, M. et al. (2021). High impedance single-phase faults diagnosis in transmission lines via deep reinforcement learning of transfer functions. *IEEE Access* 9: 15796–15809.

43 Idarraga, G., Guillen, D., Ramirez, J.M. et al. (2015). Detection and classification of faults in transmission lines using the maximum wavelet singular value and Euclidean norm. *IET Generation, Transmission & Distribution* 9 (15): 2294–2302.

44 Farshad, M. and Sadeh, J. (2014). Transmission line fault location using hybrid wavelet-Prony method and relief algorithm. *International Journal of Electrical Power & Energy Systems* 61: 127–136.

45 Magagula, X.G., Hamam, Y., Jordaan, J.A., and Yusuff, A.A. (2017). Fault detection and classification method using DWT and SVM in a power distribution network. In: *Proceedings – 2017 IEEE PES-IAS PowerAfrica Conference: Harnessing Energy, Information and Communications Technology (ICT) for Affordable Electrification of Africa, PowerAfrica 2017*, 1–6. IEEE. http://ieeexplore.ieee.org/document/7991190/.

46 Moradzadeh, A., Teimourzadeh, H., Mohammadi-Ivatloo, B., and Pourhossein, K. (2022). Hybrid CNN-LSTM approaches for identification of type and locations of transmission line faults. *International Journal of Electrical Power and Energy Systems* 135: 107563.

47 Ledesma, J.J.G., do Nascimento, K.B., de Araujo, L.R., and Penido, D.R.R. (2020). A two-level ANN-based method using synchronized measurements to locate high-impedance fault in distribution systems. *Electric Power Systems Research* 188: 106576.

48 Baghaee, H.R., Mlakic, D., Nikolovski, S., and Dragicevic, T. (2020). Support vector machine-based islanding and grid fault detection in active distribution networks. *IEEE Journal of Emerging and Selected Topics in Power Electronics* 8 (3): 2385–2403.

49 Zhang, C., Wang, J., Huang, J., and Cao, P. (2020). Detection and classification of short-circuit faults in distribution networks based on Fortescue approach and Softmax regression. *International Journal of Electrical Power & Energy Systems* 118: 105812.

50 Han, J., Miao, S., Li, Y. et al. (2021). Fault diagnosis of power systems using visualized similarity images and improved convolution neural networks. *IEEE Systems Journal* 16 (1): 185–196.

51 Sarwar, M., Mehmood, F., Abid, M. et al. (2019). High impedance fault detection and isolation in power distribution networks using support vector machines. *Journal of King Saud University – Engineering Sciences*. https://linkinghub.elsevier.com/retrieve/pii/S1018363919302004.

52 Moradzadeh, A. and Pourhossein, K. (2019). Early detection of turn-to-turn faults in power transformer winding: an experimental study. In: *Proceedings 2019 International Aegean Conference on Electrical Machines and Power Electronics, ACEMP 2019 and 2019 International Conference on Optimization of Electrical and Electronic Equipment, OPTIM 2019*, 199–204. IEEE. http://dx.doi.org/10.1109/acemp-optim44294.2019.9007169.

53 Moradzadeh, A., Pourhossein, K., Mohammadi-Ivatloo, B., and Mohammadi, F. (2020). Locating inter-turn faults in transformer windings using isometric feature mapping of frequency response traces. *IEEE Transactions on Industrial Informatics* 1. https://ieeexplore.ieee.org/document/9169788.

54 Moradzadeh, A., Moayyed, H., Mohammadi-Ivatloo, B. et al. (2021). Turn-to-turn short circuit fault localization in transformer winding via image processing and deep learning method. *IEEE Transactions on Industrial Informatics* 1. https://ieeexplore.ieee.org/document/9516903.

55 Saputra, Y.M., Hoang, D.T., Nguyen, D.N. et al. (2019). Energy demand prediction with federated learning for electric vehicle networks. In: *2019 IEEE Global Communications Conference (GLOBECOM)*, 1–6. IEEE. https://ieeexplore.ieee.org/document/9013587.

56 Zhao, L., Li, J., Li, Q., and Li, F. (2021). A federated learning framework for detecting false data injection attacks in solar farms. *IEEE Transactions on Power Electronics* 1. https://ieeexplore.ieee.org/document/9547719.

57 Electric vehicle charging sessions dundee. 2021. https://data.dundeecity.gov.uk/dataset/ev-charging-data.

58 Moradzadeh, A., Moayyed, H., Mohammadi-Ivatloo, B. et al. (2021). A secure federated deep learning-based approach for heating load demand forecasting in building environment. *IEEE Access* https://doi.org/10.1109/ACCESS.2021.3139529.

59 Zhang, X., Fang, F., and Wang, J. (2020). Probabilistic solar irradiation forecasting based on variational bayesian inference with secure federated learning. *IEEE Transactions on Industrial Informatics* 1. https://ieeexplore.ieee.org/document/9248598/.

20

Data Mining and Machine Learning for Power System Monitoring, Understanding, and Impact Evaluation

Xinda Ke, Huiying Ren, Qiuhua Huang, Pavel Etingov, and Zhangshuan Hou

Pacific Northwest National Laboratory, Richland, WA, USA

Acronyms

ACF	Auto-correlation function
AI	Artificial intelligence
ANN	Artificial neural network
AR	Autoregressive order
ASOS	Automated surface observing systems
BPA	Bonneville Power Administration
CNN	Convolutional neural network
DFR	Digital fault recorder
FN	False negative
FP	False positive
GIS	Geographical information system
GAF	Gramian angular field
LR	Logistic regression
ML	Machine learning
MRA	Multi-resolution analysis
PACF	Partial autocorrelation function
PCA	Principal component analysis
PMU	Phasor measurement unit
PSS/E	Power system simulator for engineering
ROCOF	Rate of change of frequency
SNR	Signal to noise ratio
TADS	Transmission availability data system
TN	True negative
TP	True positive
VGG	Visual geometry group
WD	Wavelet decomposition
WECC	Western electricity coordinating council

Intelligent Data Mining and Analysis in Power and Energy Systems: Models and Applications for Smarter Efficient Power Systems, First Edition. Edited by Zita Vale, Tiago Pinto, Michael Negnevitsky, and Ganesh Kumar Venayagamoorthy.

20.1 Introduction

In energy society, the traditional electricity grid has been upgraded to the smart grid, which is known as "the next generation power grid." Different from the traditional unidirectional electricity grid, the smart grid has integrated communication and information exchange technologies, therefore, have improved reliability, security, and energy efficiency [1, 2]. In smart grid, the expanding deployments of phasor measurement units (PMUs), smart meters, digital fault recorders, and other contemporary measurement devices and sensors provide the detailed data required for system status and intelligent decision-making support. Therefore, how to store, mine, and analyze the data collected has become one major challenge to ensure the security, efficiency, and reliability of smart grid [3].

There are many challenges at every step of the data management. First of all, the data generated in smart devices impose challenges to the data storage infrastructures. In smart grid, this digital data are frequently unstructured, are recorded over different time scales, and are stored on different servers and databases. The size of the collected data sets is growing rapidly, which complicates data processing and analysis. A new data storing and routing algorithm would be needed to store and route different types of data from different storage nodes to the query points [4]. The direct purpose of collecting and storing big data is effective ways to analyze them to gain the knowledge on system status and support control decisions. With the increase in the amount of data, machine learning (ML) and artificial intelligence (AI) techniques have become more and more essential to data analysis [5]. ML and AI techniques are data-driven analysis algorithms in which the algorithm is trained to produce data-driven decisions or predictions. For different purposes, many different ML and AI techniques have been proposed for the smart grid.

To increase the accuracy of state estimation results, the authors in [6] proposed a robust state estimation method that adapts weight assignment function to eliminate the unwanted disturbed data from PMU measurements. In [7], the authors explored the representative unsupervised clustering methods (partitioning, hierarchical, and density-based methods) to classify all types of disturbances stored in a North American utility from 2007 through 2010. In addition to the purpose of situation awareness, ML- and AI-based forecasting algorithm have also been widely used recently to improve the performance of forecasting algorithms such as load forecast and energy price forecast [8, 9]. In [10], the authors compared the performance of Support-vector-machine-based algorithm and artificial neural network (ANN) for the purpose of the for the electrical load forecast in one year. Based on the results comparison, the authors also analyzed the impacting factors like weather in different forecasting algorithm.

This chapter presents the stepwise analytical procedure and methods to improve power system situation awareness, predict system events, and characterize the impact of extreme weather events on the grid by using ML methodologies. In Section 20.2, we introduce ML methods used for monitoring power systems and detecting abnormal conditions. First, we develop a Wavelet-based PMU anomaly detection and classification framework for detecting abnormal conditions. Then, we evaluate the performance of the proposed PMU anomaly detection and classification frame with actual western interconnection synchrophasor data.

In Section 20.3, we apply a set of ML approaches aiming at deciphering the characteristic behaviors of multiple PMU attributes (e.g. voltage, frequency, rate of change of frequency, phase angle), including their auto-correlation, cross-dependence, similarities, and discrepancies. In Section 20.3, we develop and evaluate a deep learning convolutional neural network (CNN) model [11] to localize and classify faults in the grid. We simulated the Polish 3120-bus system [12] in different spatial zones and locations with four distinct types of faults, and the outputs provided adequate labelled data. The multi-channel synchronous generators' frequency time series obtained from simulation have been encoded to images to enable CNN training and testing. The results show that the designed CNN model can accurately predict the fault locations at all different geophysical zones in the polish system.

After deciphering the characteristic behaviors of multiple PMU attributes in Section 20.3, we develop data mining methods to characterize and model the impact of extreme weather on power transmission system outages in Section 20.4. The network data and historical outage records are spatially and temporally correlated with historical weather data and geographic climate conditions to construct the weather-related outage database for the Bonneville Power Administration (BPA) power system. The weather-related outage database is constructed to (i) analyze the frequency and duration of outages caused by extreme weather events, and (ii) build the probabilistic outage prediction model. Finally, we introduce the fragility curve-based outage rate prediction model that can be used to predict the outage probability by using historical outage data and weather data. The results show that the fragility curve deduced from the weather-related outage database could provide an accurate forecast on the outage rate of transmission lines at the BPA area.

In summary, we developed and applied several statistical and ML methods to monitor system abnormalities, predict system events, and characterize the impact of extreme events on the grid. These applications are crucial for assuring power system security and reliability. Furthermore, it should be noted that the analysis results presented in this section are based on the real synchrophasor data, real line outage data provided by BPA, and publicly available weather data, which demonstrate good applicability and transferability of our methods to real settings.

20.2 Power System Monitoring with Phasor Measurement Unit Data

With the increasing uncertainty and complexity in modern power grid, the system operators are expected to have a better observation of the power grid system status. As a result, PMUs are typically considered as the most promising technology for many applications such as wide area protection, oscillation detection, and voltage/frequency stability assessment [6, 7].

The growing deployment of contemporary measurement devices such as PMUs, smart meters, and digital fault recorders dramatically increases the amount of data collected and stored by electrical utilities [13–15]. This digital information is frequently unstructured, has different time scales, and is stored on different servers and databases. The sizes of the collected data sets are increasing rapidly, which complicate data processing and analysis. However, because the collected information contains many insights about a power system's state and its dynamic behavior, extracting this knowledge can significantly improve situational awareness, detect system-wide or local anomalies (e.g. under-frequency or voltage events), validate system models, and discover/predict equipment malfunctions.

This section provides a brief introduction to the PMU anomaly detection framework and then describes the example of wavelet-based multi-resolution analysis (MRA) anomaly detection and classification.

20.2.1 PMU Anomaly Detection Framework

The framework of wavelet-based anomaly detection and classification is illustrated in Figure 20.1. The entire framework is written in R language. The first signal processing step is MRA with "Haar" wavelet to obtain the decomposed PMU signals, followed by the extraction of the detailed wavelet

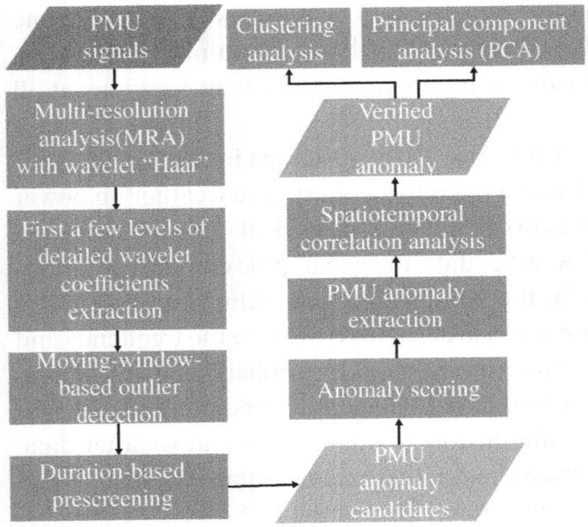

Figure 20.1 Wavelet-based PMU anomaly detection and classification framework.

coefficients at the first 3 resolution levels. The MRA procedure had a time resolution of 1 second due to the down-sampling of raw PMU signals to 1 Hz; thus, Raw PMU signals were down-sampled to 1 Hz, so the time resolution for the MRA procedure was 1 second, and the first three levels of MRAs, D1, D2 and D3, have resolutions of 2, 4, and 8 seconds, respectively. From an event detection perspective, the two-dimensional time-frequency representation of real-world PMUs has significant benefits compared to the regular one-dimensional data in the time domain. Both reliability and sensitivity of the detection performance can be retained by the moving window. If any anomaly was detected at any resolution level, the anomalous score will set to be 1. The anomalous score is used to rank the "likelihood" of events. The score is calculated as the summation of scores at multiresolution levels across different PMU units for each attribute. Since the recorded actual events duration are 5–20 seconds, the detected anomalies were further prescreened by the event durations. Therefore, the spatiotemporal correlation analysis was applied to the anomaly candidates to distinguish between false alarms and real events. The false alarms events have relative weak spatiotemporal correlations due to locality and weak continuity. And strong correlations across units should verify and confirm the real events. For the verified events, we deployed the principal component analysis (PCA) and cluster analysis to identify similarities between the events. Clustering groups multiple objects by putting similar objects into one group. The PCA method is used for similarity identification. By using orthogonal transformation, the PCA converts observations of possibly correlated variables into a set of linearly uncorrelated variables (called principal components). In PCA, the first principal component has the largest possible variance and the most variability. By combining quantitative clustering methods, PCA helps identify useful attributes for discriminating the individual events.

20.2.2 Anomaly Detection and Classification

The developed detection framework was applied to the actual western interconnection synchrophasor data. Figure 20.2 shows the results of MRA and moving-window outlier detection using the frequency attribute for the first unit of PMUs (PMU1). The gray marks highlight the historical event to illustrate the identification of anomalous candidate at each resolution level using our detection framework. By comparison of the signal and detailed wavelet coefficients, the events are more pronounced for the D3, D2, and D1 coefficients than for the original; therefore, it enhances the

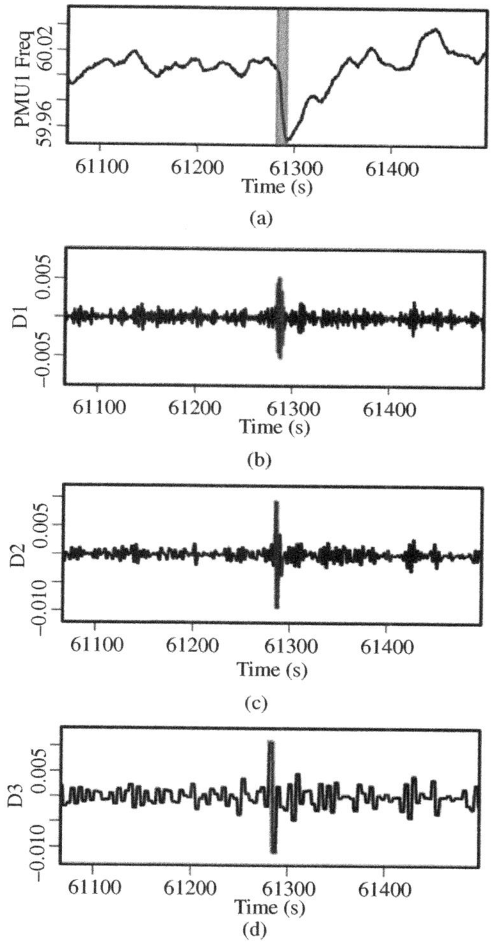

Figure 20.2 An MRA example using PMU1 frequency attribute. The detected events at each scale are marked in gray vertical box.

accuracy of event localization. The key feature of our event detection algorithm was verified by this intensification of coefficients.

As the real-world PMU data sets over 30-days period is tested on our detection framework, 32 historical events were recorded. For each PMU data set, we focused on four attributes: voltage, angle variation, frequency, and rate of change of frequency (ROCOF). Generally, the ROCOF shows noisier signals than the other attributes, whereas the detected candidates have limited spatiotemporal correlations. As observed in [16], ROCOF varies significantly, which reduces the event detection accuracy because of sensor and grid dynamics. The angle variation attribute contains similar and relative redundant information to the frequency attribute because the frequency is derived from the angle variation. Therefore, a focus on the voltage and frequency attributes is warranted for event detection in this study.

In fact, our study detects all recorded events with high scores. For example, Figure 20.3a shows a frequency-based event that occurred at all 12 units at the same time, leaping the frequency amplitude from 59.96 to 60.01 Hz within 10 seconds. The consistent behavior shared by 12 units led to strong spatial correlations implying a systematic behavior and area-wide situation that has to be address on. Figure 20.3b also shows how the same event was detected with strong correlations across all units based on the voltage attribute, in which the voltage amplitude increased by over

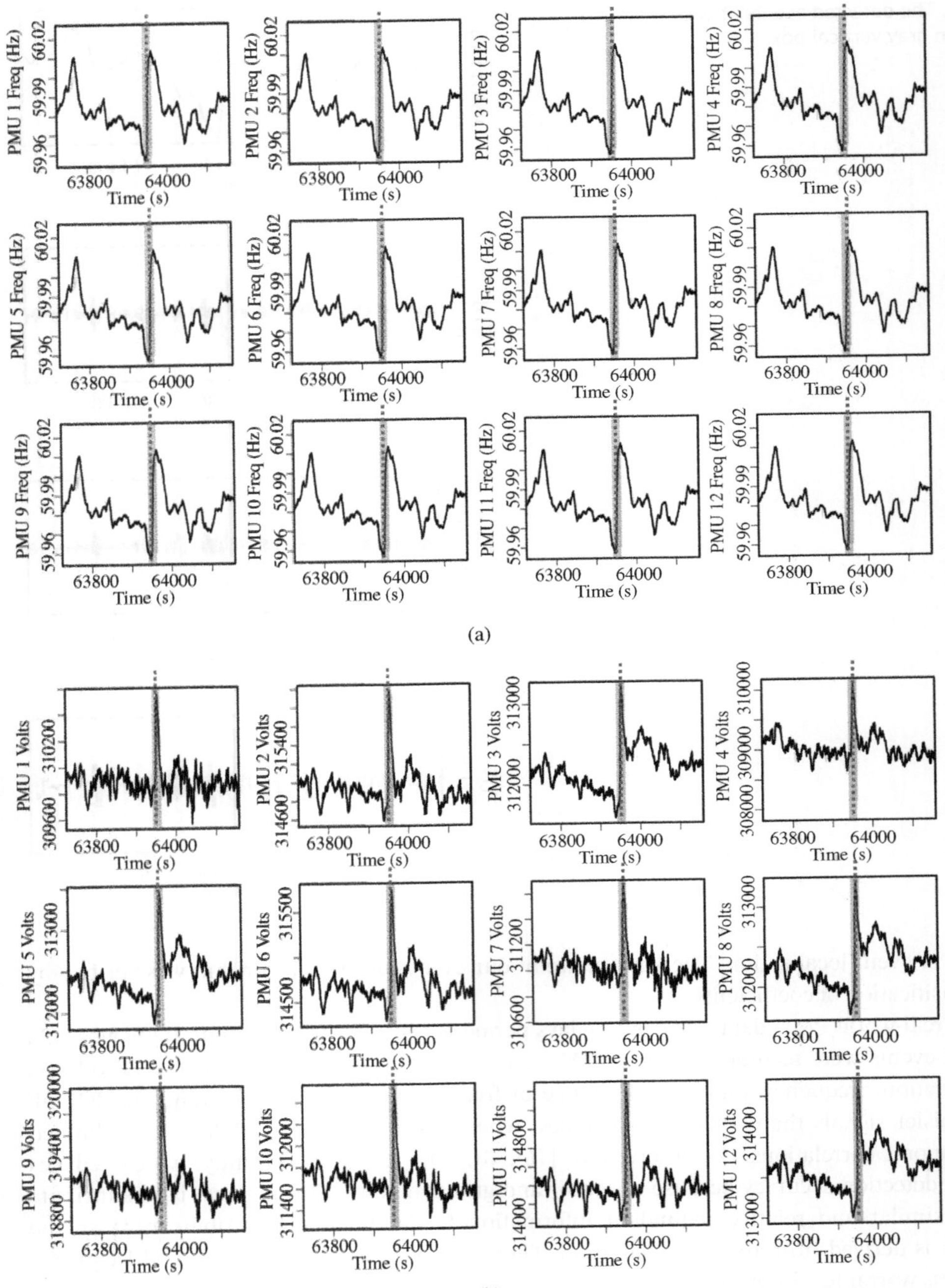

Figure 20.3 An example of abnormal event occurred across all units for (a) frequency attribute; (b) voltage attribute. The detected events for each unit are marked in gray box. The recorded historical events are marked as vertical dash lines.

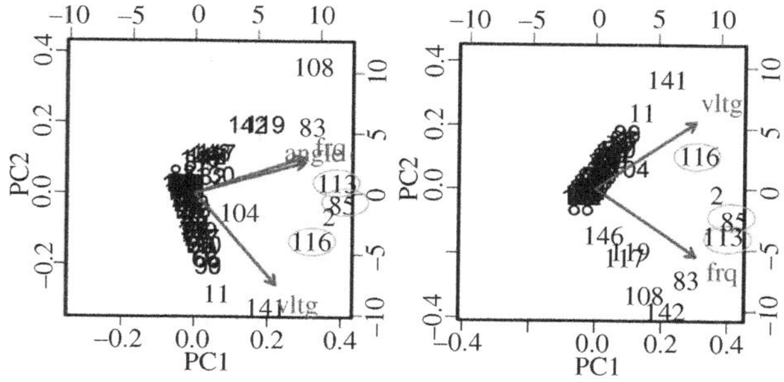

Figure 20.4 PCA biplots of detected events using different PMU attributes. The recorded historical events are circled in gray.

1000 V within 10 seconds. The voltage stabilized following the event, although it did not return to the level prior to the event.

PCA is used to examine and classify the detected events as illustrated in Figure 20.4. The panel in left shows the first two principal components of three attributes (voltage, angle variation, and frequency). The biplot eases group assignment of events; for example, anomalies #83, #108, #119, and #142 exhibit significant variances in both frequency and angle variation, while anomalies #11 and #141 have significant voltage differences compared to the other anomalies. Furthermore, angle variation strongly correlates to frequency, and the two contain redundant information of highly similar behaviors that primarily contribute to the first principal component. The major contributor to the second component was the variability of voltage. When eliminating the redundant angle variation from PCA, voltage and frequency are nearly orthogonal factors, as illustrated in the panel in right (Figure 20.4). Accordingly, PCA aids the classification of the detected events as either frequency-related or voltage-related. There are a few exceptions; for example, three recorded historical events #85, #113, and #116 are distinguished with both frequency and voltage attributes.

20.3 Power System Mechanistic and Predictive Understanding

The flowchart of using PMU data for fault localization and classification is illustrated in Figure 20.5. This section contains two subsections, the first section provides a brief introduction to the spatiotemporal pattern recognition in PMU signals. PMU signals measured from real world were evaluated using a comprehensive set of data analytics approaches to help characterize the system behaviors and their responses to external factors such as weather attributes variations and anomalies. The derived metrics regarding continuity, cross-correlation, spectrum, cluster, signal to noise ratio (SNR), and conditional probability of PMU anomalies, provide quantitative measures to PMU characteristics which can be used for system understanding and PMU data generation purposes, as well as guiding the labeling of PMU anomalies for future supervised learning (e.g. developing surrogate-based predictive models with respect to external influential factors or internal temporal features).

The second section and then describes the usage of deep-learning CNN models for faults classification and localization in Polish 3120-bus system. Four distinct types of faults in different spatial zones and locations were simulated in the Polish system, and the outputs provided adequate and balanced data for CNN training and testing. The PMU time series have been encoded in different ways for generating the image data sets for CNN model training and testing. Different time-series

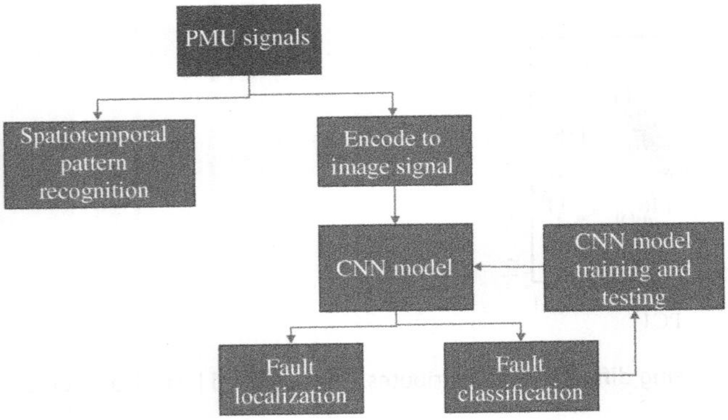

Figure 20.5 Flowchart of using PMU data for fault localization and classification.

encoding approaches have been evaluated in term of the model performance and the benefits have been discussed in term of the fault types and occurring locations.

20.3.1 Spatiotemporal Pattern Recognition in PMU Signals

The characteristic behaviors of multiple PMU attributes (e.g. voltage, frequency, rate of change of frequency, phase angle) are depicted with a set of signal processing and ML approaches. The analysis evaluates auto-correlation, cross-dependence, similarities and discrepancies across the units and temporal scales, and depicts distributions of anomalies and their links to possible external factors such as weather events. PMU measurements from the U.S. Western Electricity Coordinating Council (WECC) system used in data analytics in addition to recorded events and weather extremes are all from real-world data sets. The mechanistic understanding of the PMU dynamics by this study offers guidance on system control and blackout prevention. Additionally, the metrics obtained here is directly applicable to calibrate or filter simulated PMU data for the development of advanced algorithm.

20.3.1.1 Time Series Pattern Recognition

The temporal continuity and periodicity in the PMU data series are examined by the auto-correlogram, the serial correlation in other words. The auto-correlation analysis (ACF) creates a time series sequence with a delayed replica of itself in order to observe the similarity as a function of the time lag between them. Thus, the recurring patterns, the presence of a periodic signal interfered by noise, and the missing fundamental frequencies are identified. In partial autocorrelation function (PACF), a stationary time series is correlated with its own lagged values, regressed at shorter lags, and determines the appropriate lags in the autoregressive patterns. Both the auto-correlation function and PACF shows the relationship strength between an observation in a time series with the observations at prior time steps with different time lags. However, the difference is that the latter intends to remove the indirect correlations.

For the multivariate PMU attributes over different months, the time-series ACF and PACF analyses are applied together in the study to identify the strength of temporal continuity and the existence of periodicity in the PMU signals. Moreover, the generality and transferability of those continuity characteristics in the PMU signals are investigated based on temporal factors such as hourly or seasonal variations. This information directly complements the guidelines on detecting short-term anomalies and/or making long-term forecasts. Figure 20.6 illustrates how two hours on the same

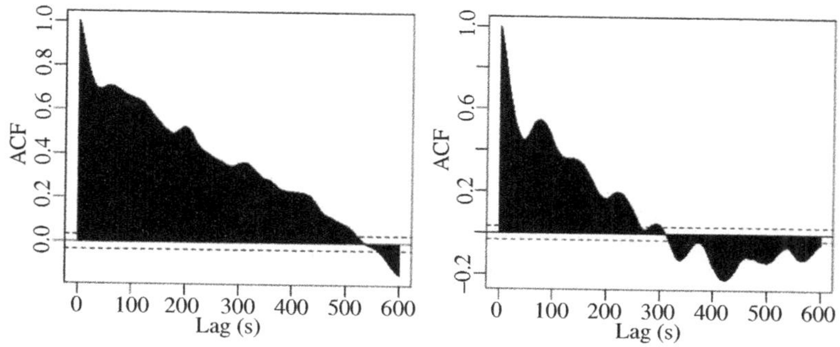

Figure 20.6 ACF for two randomly selected hours on the same day with different temporal continuity. The gray dashed line is the threshold of 0.05.

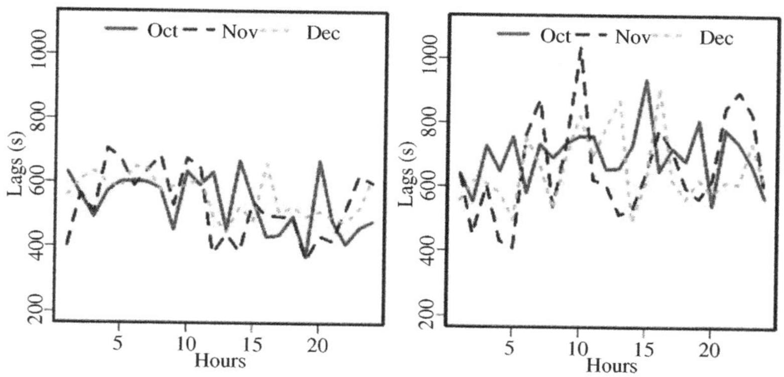

Figure 20.7 Continuity – correlation ranges of PMU frequency in the left panel and voltage in the right panel, based on ACF analysis.

day can have drastically different temporal continuities. As a measure of such continuity, the correlation range is introduced to count the time lag beyond which the auto-correlation values drop below a threshold (i.e. 0.05). In the example, the correlation ranges are 520 and 260 seconds, respectively, meaning that the second one has much weaker temporal continuity.

The correlation ranges of PMU attributes as a function of the hour of a day during different months are summarized in Figure 20.7. The daily pattern of correlation ranges in PMU frequency stretches over 600 seconds; that is, strong continuity in the morning followed by a weaker continuity at 400–500 seconds in the evening. This is due to the relative increase in variability of the power system in the afternoon during the studied months. On the other hand, the voltage which is the other PMU attribute has a different type of consistent pattern with stronger continuity at 600–1000 seconds in daytime but weaker continuity at 400–700 seconds at nighttime. Nonetheless, the correlation range metric of the PMU voltage has hourly variation. Because the PACF guides on the order of autoregressive models for prediction purposes, these PMU attributes are associated with an autoregressive order 1 (AR(1)) model. These correlation ranges, as well as the AR model order, should be used to administer anomaly detection, classification and labeling, verification of extracted temporal patterns, and filtering and evaluation of simulated PMU data.

Wavelet decomposition (WD) is suitable for nonstationary signals and is widely used in speech and image processing as well as time series analysis. The original signals are divided into multiple

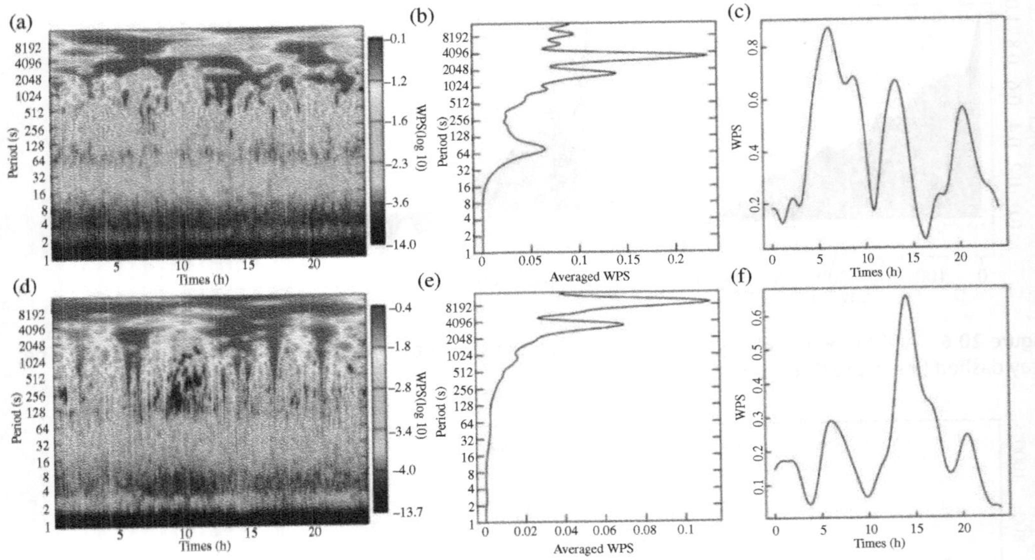

Figure 20.8 The wavelet spectra of PMU attributes at unit #5. (a) the decomposed coefficients at various temporal scales for PMU frequency for the day September 14; (b) the corresponding wavelet power distribution; (c) the extracted dominant hourly component; (d) the decomposed coefficients at various temporal scales for PMU voltage for the day September 14; (e) the corresponding wavelet power distribution; and (f) the extracted dominant hourly component.

components in the forward transformation, and simultaneously decomposed by a high-pass filter h and a low-pass filter g at each step yielding detail coefficients and approximation coefficients (the remaining mixed-signal), respectively. Consequently, WD identifies the dominant component in a mixed signal and locates anomalies at different temporal scales [17]. Additionally, SNR analysis is performed to extract and summarize the SNR attributes to add to the wavelet spectral information in the frequency domain.

Given the result of ACF analysis that shows the correlation ranges are within one hour, it may seem that the detection of short-duration events does not require searching beyond one hour of data. It contributes to removing unnecessary computational burden; however, the wavelet spectra analysis (Figures 20.8b,e) implies that hourly, daily, even weekly signals are necessary for understanding mid-to long-term patterns. As seen in Figures 20.8c,f, the hourly component is sinusoidal featuring high peaks during mid-day time periods associating to the possible external factors such as weather attributes. Both PMU and weather attributes are wavelet-decomposed to verify their associations at matching temporal scales.

In addition, SNR analysis is performed and the extracted SNR attributes are summarized as a complementary metric to the wavelet spectra. The SNR of the PMU frequency and voltage signal is calculated for hours of a day using the autoregressive AR(1) low-frequency noise with the maximum power spectrum 0.08 (Figure 20.9). The SNR ratios of the frequency signals fell within 1 dB range, although there are two significant peaks in the earlier morning and later mid-day as well as lower SNR at mid-day. The SNR ratios are much higher and varies in a wider range for the voltage attribute.

The SNR ratios for both the frequency and voltage attributes vary monthly but are consistent across the units. The SNR is another useful and commonly used measure, particularly for characterizing noise levels in the data. The measure can be used directly for adding noise to PMU data

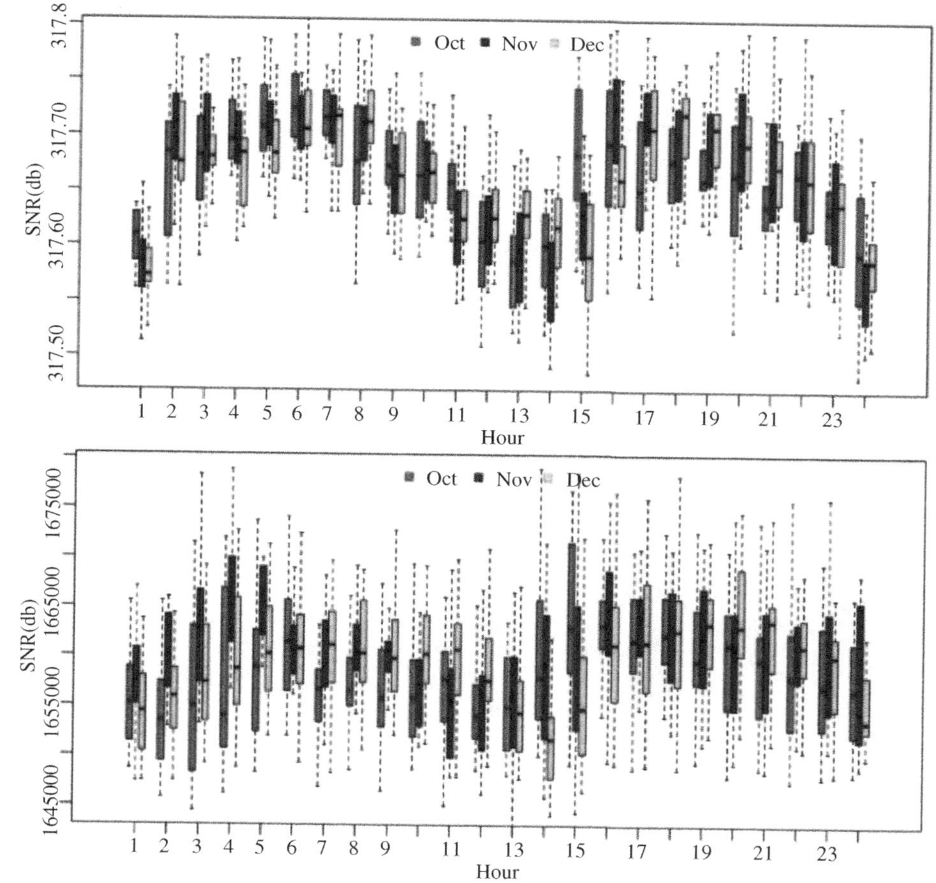

Figure 20.9 The diurnal rhythm of SNR obtained from frequency and voltage PMU attributes for different months.

generated by physics-based data simulators to augment the existing PMU databases, so they are adequate for reliable AI model development and evaluations.

20.3.1.2 Similarities and Variations Across Units

PCA is used to assess the spatial cross-dependence, i.e. the similarities and differences across the PMU units. The number of principal components required to compute the major variability in the data matrix is determined by a scree plot. In a biplot, samples are clustered based on their similarities, along with projections to show how much each characteristic influences a principal component. The findings of block-wise (five-min time window-based) PCA are shown in Figure 20.10 as an example of how the real-time changes of PMU signal behavior are tracked based on the system complexity (number of essential principal components and their relative contributions) and similarity (clustering). In Figure 20.10b, the first two or three components of PMU frequency are typically dominant, and the units are divided into three or four groups by similarity basis as Figure 20.10c. This is because geographically adjacent units tend to be clustered together creating spatially coherence. The findings lead to two implications: (i) PMUs show local/regional behaviors under the impact of local/regional external factors such as weather, despite the inherent

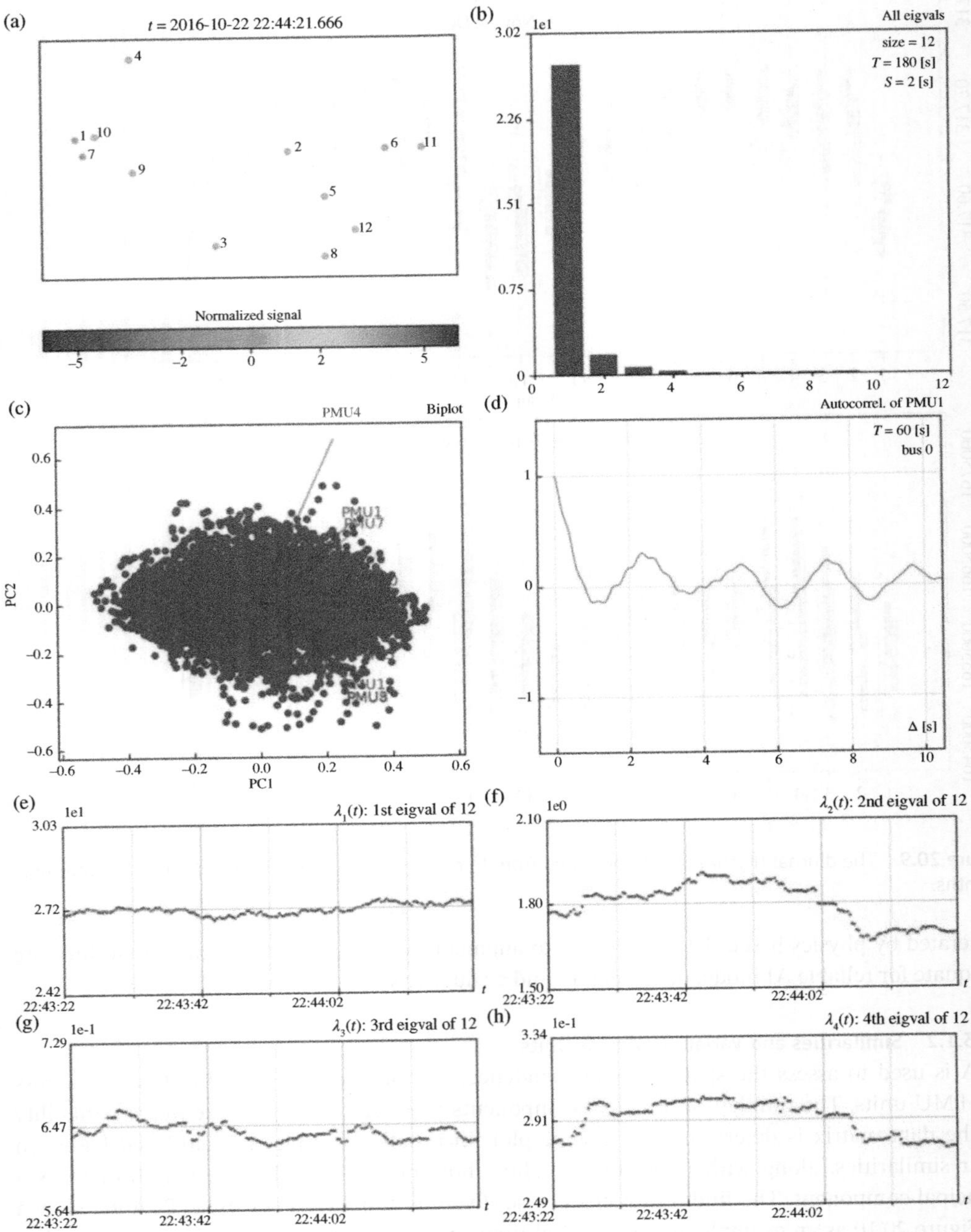

Figure 20.10 Block-wise PCA. The subplots include (a) the rough spatial locations of the units; (b) scree plot; (c) biplot; (c) auto-correlation; (e)–(h) the changes of the first four principal components with respect to time. A five-minute moving window is used to conduct PCA.

strong cross-dependence between units of the power system; (ii) the computational work for simulation-based data augmentation can be simplified by modeling the cluster behaviors and subsequently superimposing within-cluster variability.

Figure 20.10e,h exhibit the relative changes and contribution/dominance of the first four principal components over time. It was plausible that those slight changes correlate to the hourly variations in the original signals, and the occurrences of recorded events that lasts about 20 seconds each have minimal influence. Consequently, it is recommended that PCA be used for anomaly detection on the dominant temporal scales, rather than on real-time PMU pulse event detection as many practices in the area.

20.3.1.3 Similarities/Discrepancies Between Days/Months

In the next part, the time series deviations under varying conditions (e.g. days, seasons) are evaluated with a Taylor diagram to integrate/compare the changes in phase angle and magnitude, as well as Pearson correlation, root-mean-square error, and individual standard deviations. In Figure 20.11, the angle differences are recorded by clustering the time series by distances (combined similarity measure). The daily value was around −10 on November 1, then drastically changed to −25 on November 5, in addition to gradual hourly and daily shifts observed over the week. Therefore, the mid-term continuity and periodicity appear to be lacking in the PMU data, and the long-term behaviors are likely to be attributed to factors associated with weather attributes at the respective scales. Compared to the within-day variation, the between-day variability is patternless and has larger ranges in magnitudes except for November 6 when a change of system status was implemented. The Taylor diagram demonstrates that when a cluster's cross-correlation coefficient reaches 0.6, their deviations in root-mean-square errors are not necessarily small.

20.3.2 Events Classification and Localization Through Convolutional Neural Network

Observation-based event detection, classification, and localization using real-world data are challenging because of the lack of labelled data. Such data inadequacy rises difficulty in deep neural network training; one solution is to augment data by creating ensemble simulation-based training data sets. Herein, we evaluate the viability of ensemble simulation-based data application with various types of faults to give appropriately labelled data for supervised learning. Dynamic simulations are constructed on a Polish 3120-bus system with four types of faults in five geophysical zones. For viability evaluation of the CNN models, the multi-channel time series of machine speed data are collected and encoded to images by classification (fault types) and localization (occurred zones). Time series stacking serves as the benchmark data set in our proposed CNN model because of its straightforward feature requiring the least computational resources. Three approaches – time-domain stacking, wavelet decomposition-based frequency-domain stacking, and polar coordinate system-based Gramian angular field (GAF) stacking – are employed on encoding the time series for investigating the performance of the CNN model. Our model adopts on the visual geometry group (VGG) model architecture to increase accuracy in the model depth [11].

20.3.2.1 Polish System Testbed and Data Preparation

Simulation tests are performed on the Polish 3120-bus system ("case3120sp.m") [12]. Generator, governor, stabilizer, and exciter models for generator dynamics are among the dynamic data generated in the Power System Simulator for Engineering (PSS/E) format. The protection models such as distance relays, generator-based voltage and frequency relays, and under-frequency and under-voltage load-shedding relays are also prepared in the PSS/E format, and incorporated to

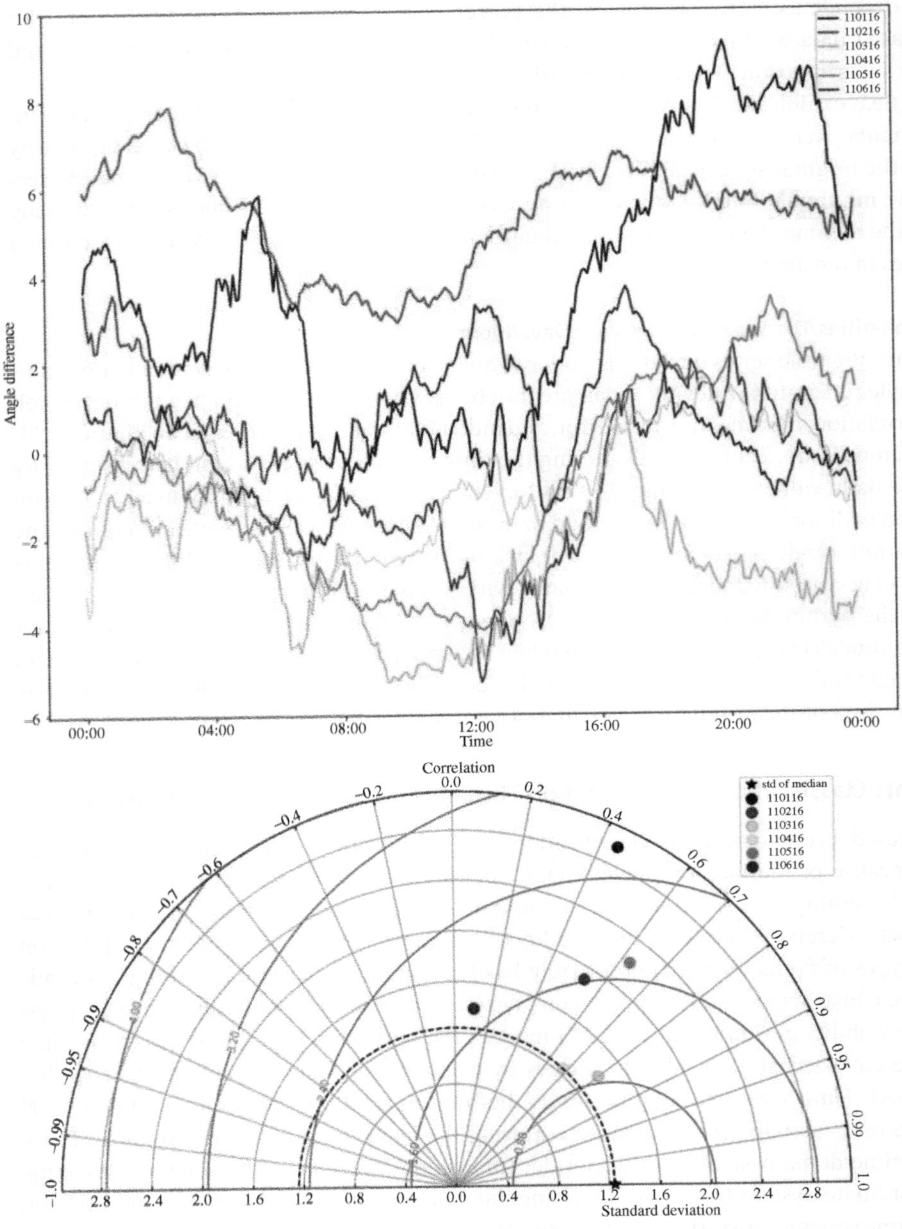

Figure 20.11 PMU angle difference time series during several adjacent days and the corresponding Taylor diagram.

the existing relay protection models of the Polish system. The Polish system has 3120 buses, 3487 branches, 2314 loads, 505 generators, and five zones. The zonal model for the total generation of each zone and the connections between zones are illustrated in Figure 20.12. Zone 3 is the largest zone connected with zone 1 and is followed by zone 2 and 4, while zone 5 is the smallest zone with only 841 MW of generation and connections to zone 1 and 4 only.

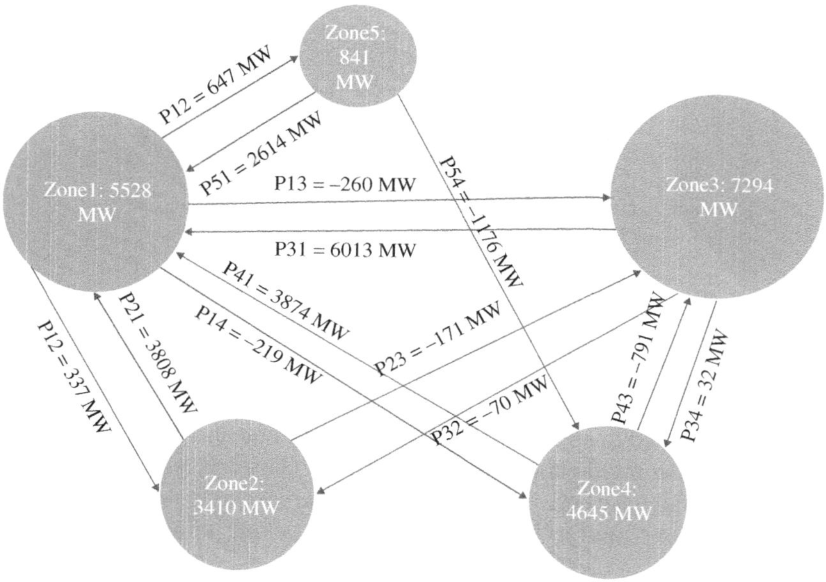

Figure 20.12 Zonal model of Polish system.

Table 20.1 Fault types simulated in the polish system.

Types	Outage of single generator	Outage of two generators	Three-phase bus fault	Multiple line fault	Total contingencies
Fault 1	✓			✓	298
Fault 2			✓		3120
Fault 3	✓				298
Fault 4		✓			298

20.3.2.2 Fault Types and Implementation

The PSS/E software simulated various power system faults based on the four types of distinct outages characteristics as illustrated in Table 20.1. Fault 1 is the scenario of a single generator and multiple line fault, in which each generator and the two transmission lines were affected in the system. Fault 1 has simulated 298 contingencies in total. For the dynamic simulations in Fault 2, a three-phase fault at a bus was considered. Faults 3 and 4 have 298 contingencies, respectively; the former scenario placed each generator out-of-service, and the latter has the two adjacent generators out-of-service.

20.3.2.3 CNN Model Development

Our CNN architecture is designed to train models in the Polish 3120-bus system for faulty type prediction and zonal classification. The CNN model training is first uploaded with the image data sets encoded by the three schemes mentioned in Section 20.3.2. Model training and testing are individually performed on each encoded data scheme to compare the model performances. Each image data set has been divided into three independent subsets (training, validation, and testing), of

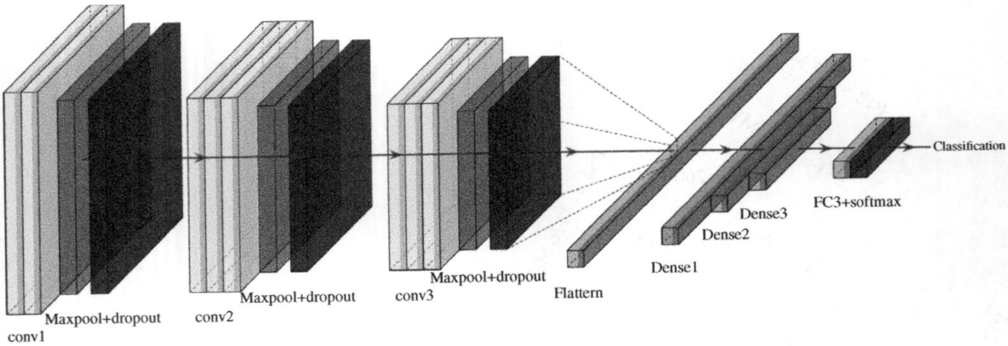

Figure 20.13 The architecture of CNN model.

which training and validation are used to optimize the model configuration and hyperparameters in model development. To note, the multi-class labels data with the classified fault types is also included. The multi-class labels that have been determined during the ensemble simulation setup are useful indicators of fault locations and/or zones for CNN model training with estimated fault locations.

The architecture of CNN model conducted in this study is based on VGG 11 [11], illustrated in Figure 20.13. Inputs to the first two-dimensional convolutional layer are the three image sets with 128×128 RGB fixed-size. Each image is transferred through three blocks of convolutional layers, each containing 2, 3, and 3 layers, respectively. The convolutional blocks have varying numbers of neural nodes, while the receptive field is consistent with each convolutional layer. The convolutional layers are followed by a spatial pooling of max-pooling layer for progressive size reduction of the representation to reduce the network's parameters and computational demand. A stride of 2 is taken across a frame of 2 pixels in max-pooling, and then a dropout layer is added. Dropout is a regularization technique that randomly disables a selected subset of neurons during training to increase robustness and prevent overfitting in model performance [18]. The output from the three convolutional blocks is flattened into a one-dimensional feature array by feeding one into another. Finally, the three fully-connected layers, also called dense layers, are obtained and converted by the soft-max activation to yield the multi-class predictions. The soft-max layer assesses values between 0 and 1 to quantify the probability and confidence to the class that each image belongs to. In the architecture of our CNN model, a rectified linear unit activation function is employed in all layers except for the last dense layer. Because the rectified linear unit activation has a linear property for positive values and zero for negative inputs, it is one of the most used activation functions in neural networks, especially in the CNNs. The stochastic gradient descent is the optimizer by estimating the error gradient for the current state of the model from the training data set to update the model weight through backpropagation. For the multi-label classification problems, the categorical cross-entropy class is used to compute the cross-entropy loss between labels and model predictions. The configuration of last dense layer with the total number of classes is required for the calculation of the loss function so that the soft-max activation can forecast the probability for each class.

20.3.2.4 CNN Model Evaluation

A confusion matrix is introduced for multinomial classification to evaluate the CNN model performance concerning zonal and faulty classifications. The target class values are counted by four types of events in the multi-class confusion matrix: (i) true positive (TP) refers to the positive classes in row and column indicating the correct classification of positive classes. The TP cell is at the top-left corner of the confusion matrix; (ii) False Negative (FN) is a combination of the positive class row and the negative class column, representing the incorrect classification of a positive class as negative. FN cell is at the top-right corner of the confusion matrix; (iii) False Positive (FP) is a combination of the negative class row and the positive class column, which incorrectly assigned the negative class as positive. The FP cell is located at the bottom-left corner of the confusion matrix; (iv) true negatives (TN) is determined by the negative classes in the row and column, that is the contrary of the positive classes, indicating the correct assignment of the negative class as such. It is located at the bottom-right corner of the confusion matrix. The four-count confusion matrix enables calculation and quantification of the class statistics metrics in model performance. As such, sensitivity assesses the model's ability to identify events in the positive class, and specificity measures the degree of accuracy of the positive class assignment. Precision is a metric calculated for the rate of correct classification of the positive class (TP). Accuracy demonstrates the overall ratio of correctly classified assignments in contrast to the total number of assignments in the model.

20.3.2.5 Fault Localization

The well-trained CNN model is applied to the third independent testing data set for zonal classification using time-stacking images. The class statistics are illustrated in Figure 20.14. The overall averaged accuracy reaches 91% with the straightforward time series stacking approach. Zone 1 and 3 have the lowest and highest accuracies at 88% and 96%, respectively. The specificity is high for all zones with an averaged value of 97%, which means the model performs satisfactorily in assigning the positive extract class. For sensitivity and precision measurements, the averaged values are about to 86% for both Zone 1 and 3 outstand among the five zones. All measured metrics show a high-performance level, which means the Polish system has well-defined zonal structures so that the CNN model can accurately predict its fault locations. In Figure 20.14, Zone 1 resulted in the worst values of sensitivity and accuracy due to the model's tendency to predict a small fraction of Zone 1 events compared to the other zones such as Zone 2 and 5. Generally, when there is a physical connection between zones, power flows in the target zone are affected by faults occurred to other

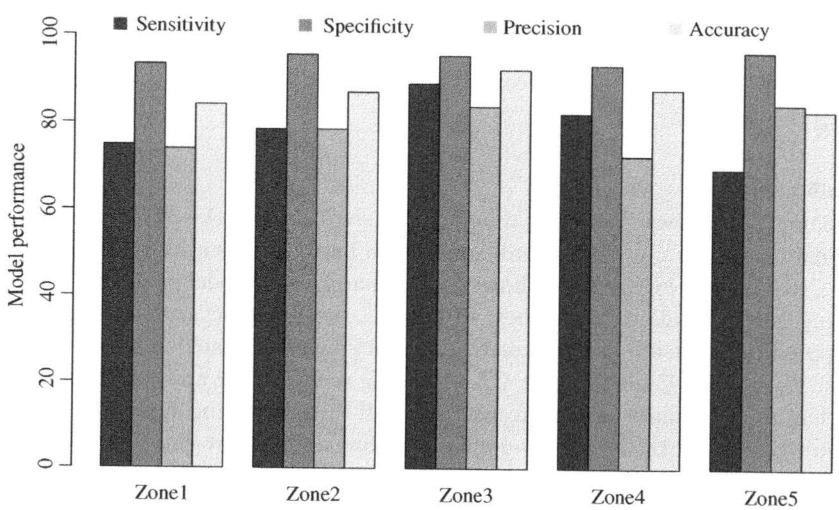

Figure 20.14 Zonal CNN model performance confusion matrix using time-domain stacking encoding approach for the testing set.

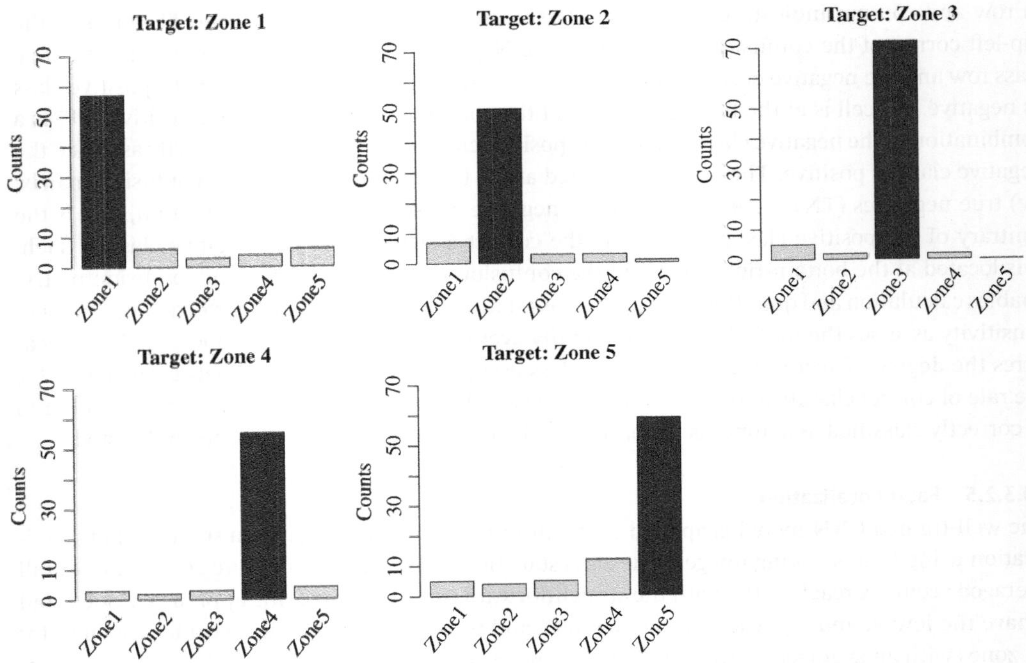

Figure 20.15 Zonal CNN model prediction for each zone using the time-domain-stacked encoding approach. The bars for a given zone shown in dark-gray and light-gray indicate the correct and missed predictions.

zones outside the target zone. The physical connections and impacts between zones cause a slight increase in the number of TNs and FNs in model predictions. However, if there is no connection between the zones, the model accurately predicts TNs. Given that Zone 1 is connected to all the other zones, the flow complexity within Zone 1 can be increased by faults added to other zones. In fact, the zonal prediction of our model (see Figure 20.15) shows that none of the predictions of Zone 3, which is the largest zone of all, falls in Zone 5, where there is no transmission lines connecting the two zones. Similarly, model predictions between Zone 5 and 2 are low as there is no physical connection between those two zones.

20.3.2.6 Fault Classification

Three different encoding approaches, including time-domain-stacking, wavelet decomposition-based frequency-domain stacking and polar coordinate system-based GAF stacking, have been used to evaluate the CNN models. Figure 20.16 shows the comparisons of model predictive accuracy for different time-series-encoding approaches. In general, the frequency-domain-stacking and GAF-stacking approaches have improved predictive accuracy, whereas time-domain-stacking works were better for Fault 3. For Faults 1 and 2, GAF-stacking achieves 97% accuracy, which is 20 and 10% higher than time-domain-stacking, respectively. All three encoding approaches performed poorly for Fault 4; however, the frequency-domain-stacking and GAF-stacking approaches improve the accuracy by 6% and 11% compared with the time-domain-stacking approach, reaching 74% and 79%, respectively. For fault predictions, the various encoding approaches serve different fault types and spatial zonation. Thus, in practice, GAF-stacking is superior in CNN models in terms of distinguishing the patterns from similar fault types.

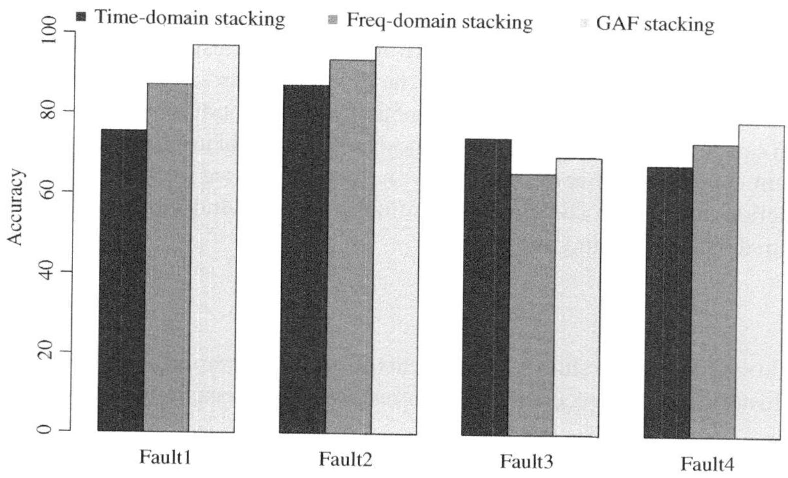

Figure 20.16 Model accuracy comparison from different encoding approaches.

20.4 Characterization and Modelling of Weather and Power Extremes

Wide-area disturbances of electrical power have been monitored for years as the reliability and resilience of electric power infrastructure is a major concern worldwide [19]. A recent report by Climate Central indicates that severe weather and climate events will occur more frequently and intensely under climate change leading to a likely increase in power outage risks. Some large-scale outages may affect millions of people and incur an annual cost of about US$55 [20]. Historical recordings show that the primary causes of the power outages are related to weather and climate. Specifically, approximately 75% of power outages are either caused directly by weather phenomena such as lightning, wind, and heavy rain or indirectly by failures assets when exposed to weather extremes such as overheating and winter storms [20, 21]. In fact, historical monitoring indicates that extreme weather phenomena caused ~80% of large-scale power outages from 2003 to 2012 [22]. Therefore, to identify the weakness in the system and enhance the system reliability against extreme weather, it is imperative to build an adequate weather-dependent outage model and perform risk analysis. However, there are two main challenges: (i) the lack of weather-dependent stochastic modelling for forecasting outage rate and (ii) a lack of sufficient outage data for the weather- and climate-related risk analysis.

Real-world outage data, such as transmission availability data system (TADS); BPA outage data [23]; and data from many public utilities [24, 25] are reliable historical monitoring records with information that describes processes implemented to instill resilience, which potentially could be used as the basis for a comprehensive resilience model. Therefore, many data-driven approaches are used by many researchers to model and calculate the failure probability of transmission assets. Dokic and Pavloski [26] developed a probabilistic model for forecasting weather-related outage probability based on an analysis of BPA outage data and automated surface observing systems (ASOS) weather station data. However, the online outage database usually does not include any geographic conditions or ambient weather conditions that existed before and during outages. Therefore, an additional source of weather data usually is needed for weather- and climate-related risk analysis.

To address the challenges identified above, we first developed a comprehensive weather-related outage database by spatially and temporally correlating network data, historical outage data, historical weather data, and geographic climate conditions in the BPA service area. The constructed weather-related outage database contains comprehensive ambient weather conditions on recorded outages. This information can be used to analyze the frequency and duration of unplanned outages that resulted from different extreme weather events. Based on our analysis, we further introduce the fragility outage prediction model with climate classification to predict wind-induced outages and outages resulting from downed trees and tree cutting.

20.4.1 Data Sources

We used the following data sources for this study: (i) data from the utility geographical information system (GIS), (ii) utility historical outage records, (iii) historical weather measurements, and (iv) geographic climates data.

We used long-term historical outage data collected by BPA [23] for the U.S. Pacific Northwest region. Figure 20.17 shows the location of all the BPA transmission lines and the service area within the Pacific Northwest region. We extracted information for weather-related transmission line outages for the period from January 1, 1999, to May 10, 2018. We identified a total of 16 806

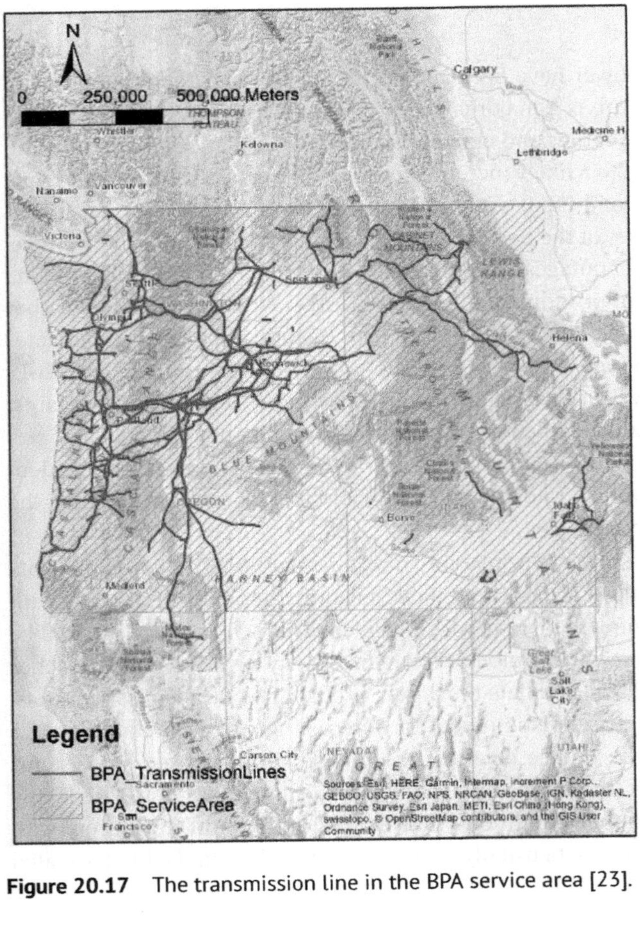

Figure 20.17 The transmission line in the BPA service area [23].

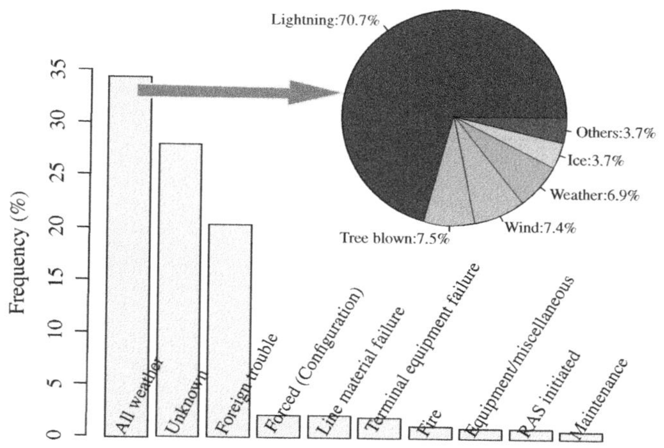

Figure 20.18 Frequency of major power outages (bar chart) and percentage among weather-related outages (pie chart) within the BPA service area, based on the reported outage data during 1999–2018.

weather-related outages in the 20-year historical outage data. Each outage case includes outage information, such as the starting and ending times, durations, outage section descriptions, and causes of the outage. For weather-related outages, weather-related causes are attributed to phenomena such as lightning, ice, tree, tree cut, tree blown, tree growth, wind, and earth slide.

The statistical summary of outage frequency and causes in the BPA territory is summarized in the bar charts in Figure 20.18. As can be seen in the bar chart, the percentage of weather-related cases is about 35%, which is the leading cause of outages for the entire BPA service area, which validated the belief that weather-related outages are the top reasons for unplanned power system outages. The reason for all weather-related outages is shown in the pie chart in Figure 20.18. As can be seen, more than 70% of the weather-related outages are caused by lightning. However, the causes shown in the pie chart are the direct outage reasons rather than the extreme weather conditions that cause those outages. Based on the bar chart of outage reasons, it can also be seen that only partial BPA outage data contain the direct weather cause of outages. The second and the third highest frequency outage causes are Unknown or Foreign Trouble (i.e. errors caused by a neighboring utility or from outside the BPA service area). Because those important ambient weather conditions and the geographical conditions of those historical outages are missing, we established temporal and spatial links between power outage data and environmental data to build an accurate grid outage model for extreme weather conditions. The steps taken to build this model are described below.

20.4.2 Spatiotemporal Analysis

In order to reveal the failure patterns associated with weather extremes, the extreme weather events information is paired with historical power system outages at matching spatial and temporal scales. The locations and the timing of transmission line failure and the recorded weather events are used in the analysis. The two categories of events are mapped to each other by counting the circumstances when the two occurred at a distance shorter than 300 mi and that the difference in timing is within three hours. Totally 3254 cases are matched under these criteria. Table 20.2 shows the statistical summary of extreme weather events that have been matched with outages both spatially and temporally. The weather component "hail" and "thunderstorm wind" stand out to be the top two factors out of the overall 9 event types, accounting for 36.4% and 37.7%, respectively, among all

Table 20.2 The frequency of the extreme weather events matched with outages at matching spatial and temporal scales.

Weather elements	Counts	Percentage (%)
Debris flow	17	0.5
Funnel cloud	52	1.6
Heavy rain	94	2.9
Tornado	94	2.9
Lightning	155	4.8
Flood	197	6.1
Flash flood	236	7.3
Hail	1184	36.4
Thunderstorm wind	1225	37.7

outages. This information enables us to derive conditional probabilities or likelihood of outages, particularly for the hail- or thunderstorm wind-driven events, which can be used in the numerical risk assessment models to quantify the corresponding risks and cost.

The impacts of the hail and thunderstorm wind are then evaluated by zooming into different transmission line types, and such impacts seem to vary significantly between each transmission line type, as illustrated in Figure 20.19. The likelihood of failure is defined as the ratio of the number of outages caused by specific weather extreme type to the total number of outages for each transmission line type. In general, the thunderstorm wind is more likely to trigger outages for all the transmission lines. For the outage caused by hail, the 500 kV transmission line is the most sensitive with a failure likelihood of 50%, followed by the 230 kV transmission line at a failure rate of 40%. Meanwhile, the 161 kV transmission line has a 16% failure rate, which is the lowest risk. When the extreme weather factor is thunderstorm wind, about 60% failure rate is associated with the 230 kV transmission line. Within the failure rate range of 40%~ 50%, there are six types of transmission lines, including 34.5, 115, 138, 287, 345, and 1000 kV. The rest of the transmission lines have a 30% chance to have a thunderstorm wind-driven outage. The 24.9 kV transmission line is at low risk in terms of outage rates. The power system reliability models can be adjusted and improved based on these statistical summaries for different transmission lines under various weather conditions.

In order to have a better understanding of the outage causes provided by BPA, the correlation between the outage causes and the matched extreme weather events are analyzed using cross-tabulation analysis, which is one of the most useful methods to quantitatively analyze the correspondences between multiple variables. In Table 20.3, the outage-causing factor "lighting" is correlated with 85.5% of extreme weather extremes, and the most correlated ones are hail and thunderstorm wind. The factors "tree" and "tree blown" are associated with 2% weather extremes, including "flood" and "thunderstorm." The top three weather elements linked to the unknown outage group are thunderstorm wind, flood and hail, totaling 8% among all extreme weather conditions. The uncertain weather outage category is the most correlated with the flood.

The restoration time associated with extreme weather events is summarized to illustrate the influences of different weather extremes on power system outage durations in Figure 20.20. Here we used all the matched extreme weather events with the power outages with spatiotemporal thresholds of 300 mi in distances and three hours in time differences. As the top two most frequently

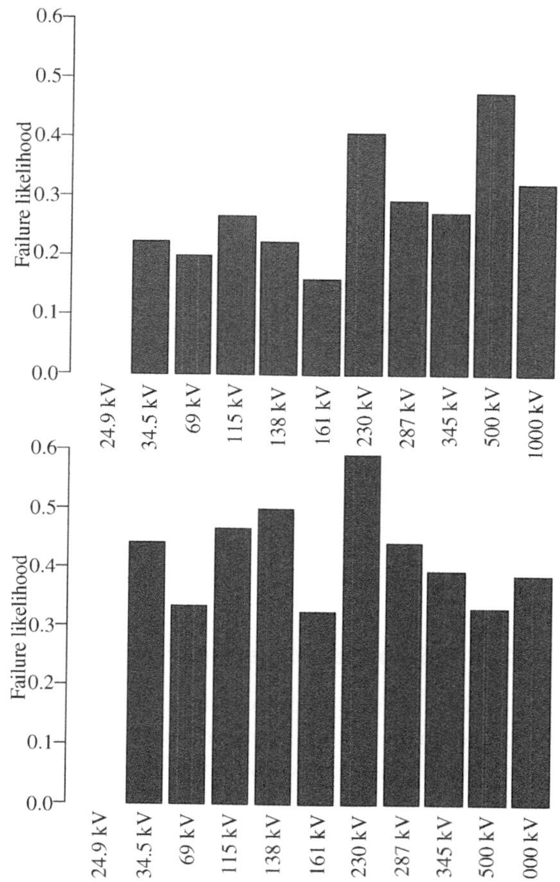

Figure 20.19 Impacts of weather extreme factors for each transmission line type. The top panel is for hail, and the bottom is for Thunderstorm wind.

Table 20.3 The cross-tabulation of extreme weather events types and outage causes.

	Ice	Lightning	Not reported	Tree	Tree blown	Unknown	Weather	Wind	Sum
Debris flow	0	0.4	0	0	0	0.09	0.03	0	0.52
Flash flood	0	6.76	0	0	0	0.49	0	0	7.25
Flood	0.09	1.08	0.03	0	0.8	2.46	1.6	0	6.05
Funnel cloud	0	1.11	0	0	0	0.49	0	0	1.6
Hail	0	33.99	0	0.09	0.18	1.72	0.18	0.22	36.39
Heavy rain	0	1.84	0.03	0	0.06	0.34	0.61	0	2.89
Lightning	0	4.7	0	0	0	0.06	0	0	4.76
Thunderstorm wind	0	33.31	0	0.31	0.77	2.52	0.12	0.61	37.65
Tornado	0	2.34	0	0	0	0.55	0	0	2.89
Sum	0.09	85.53	0.06	0.4	1.81	8.73	2.55	0.83	100

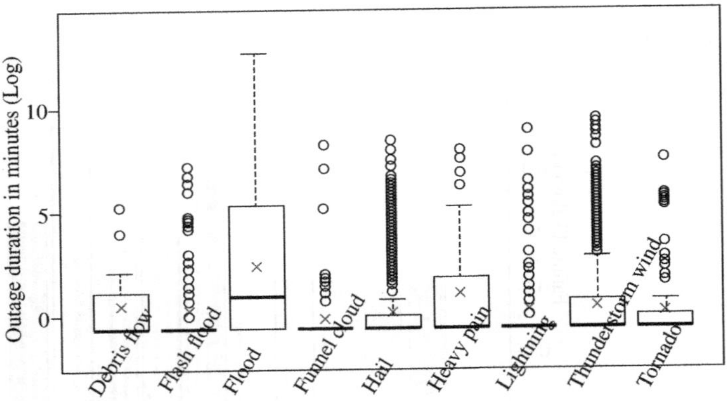

Figure 20.20 Boxplots of outage durations clustered by the extreme weather event type with matched outages both spatially and temporally.

happened extreme weather phenomena, hail and thunderstorm wind, have relatively short outage durations but with large uncertainty ranges due to outliers. The outliers indicate longer restoration times needed under such weather conditions. Those extreme events need to be considered in risk modelling or cascade simulation to help evaluate significant generator violations.

20.4.3 Probabilistic Modelling of Lines Outage

The concept of fragility curves can be used to express the relation of the failure probability of any system component, such as transmission lines, to any individual or combination of weather parameters, such as wind speed and lightning density [27–29]. The shape of the curves depends on the impact of the weather parameter on the failure probability of the component. The weather profile then would be mapped to these curves to obtain the weather-affected failure probabilities of the components, as shown in Figure 20.21. These curves can be developed either through modelling the weather aspects that affect the failure rate of components, such as wind speed, or using existing empirical data to determine the failure rate of the components under different weather conditions.

The approaches of fragility curves development have four broad classifications: judgmental, empirical, analytical, and hybrid [30]. Judgmental approaches are established on engineering judgment while empirical approaches are derived from experts' observations. Analytical approaches are developed from models, and at last, hybrid approaches are the combination of two or more of the other approaches. Each approach varies by the data availability for modelling and the suitable

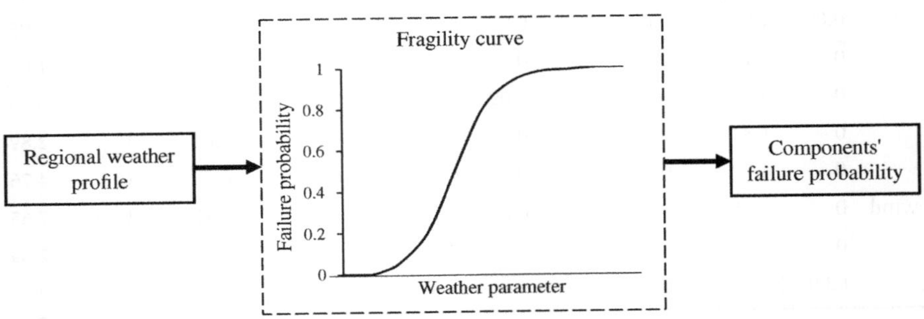

Figure 20.21 Conceptual illustration of estimating components failure probability using fragility curves.

condition of implementation. However, the choice of which approach to use involves making a trade-off between cost, data and precision that is appropriate for the application.

The objective of this study is to forecast the probability of an outage event, given its various weather-related parameters as well as the geographic climate condition. The forecasted outage probability of the transmission line will be cross-validated with the realistic outage probability obtained from the historical outage database.

We used long-term historical outage data collected by BPA [23] for the U.S. Pacific Northwest region for the period from January 1, 1999, to May 10, 2018. Two sets of outage data (wind and tree cut) and the associated weather and climate condition data were extracted from the BPA historical outage data set we presented in Section 20.4.2. The weather parameters include air temperature, humidity, wind speed, air density, pressure and precipitation.

We evaluated and compared the prediction performance of the fragile curve model against the logistic regression (LR) model proposed in [31, 32]. In the LR forecasting model, a binary classification model y is used to classify outage and no-outage events. The LR model is fitted using the conditional distribution of the labels y, given the observations X, where input X is the set of weather attributes (temperature, air pressure, wind speeds, etc.) The model is optimally fitted by determining the optimal coefficients w that maximize the logistic loss function. A more detailed introduction on the LR model can be seen in [31, 32].

The prediction capability of the fragile curve was assessed for the task of classifying whether an outage event occurred at a certain substation or not based on the probability estimates of its occurrence. Its performance was assessed using the previously mentioned metrics. The results obtained are summarized in Table 20.4 for the prediction performance of wind caused outages and in Table 20.5 for the prediction performance of Tree/Tree Cut caused outages.

From the results in Tables 20.4 and 20.5, the following observations can be made: (i) Both logistic regression and the fragile curve model have very good performance for both Tree Cut and Wind caused outages. (ii) The fragile curve-based method has better forecast accuracy than Logistic regression. This supports the hypothesis that climate information is truly relevant to the occurrence of outage probability and can be further utilized to improve the accuracy of prediction. These results also verify the necessity to integrate weather data into the proposed outage forecast model.

Table 20.4 Prediction performance of different forecast models for Wind outages.

	Forecast accuracy	Bias
Logistic regression	0.76	0.56
Fragile curve	0.79	0.62

Table 20.5 Prediction performance of different forecast model for Tree/Tree Cut outages.

	Forecast accuracy	Bias
Logistic regression	0.68	0.62
Fragile curve	0.72	0.71

20.5 Conclusion

In this chapter, the implementations of ML methodologies for improving power system situation awareness and system event prediction, as well as for the characterizing the impact of extreme weather events on the grid have been demonstrated. In Section 20.2, the wavelet-based anomaly detection and classification framework are applied on the real-world PMU data sets to facilitate power system monitoring and abnormal condition detection. In Section 1.3, the ML-based approaches are used for the spatiotemporal pattern recognition of PMU. Then, a deep-learning CNN models is applied for faults classification and localization in the Polish 3120-bus system. The results show that the proposed CNN model can accurately predict the fault locations at all different geophysical zones in the polish system. Finally, we first established a model to characterize the impacts of extreme weather on the power transmission system outages with data mining methods in Section 20.4. Then we proposed a fragility curve-based outage rate prediction model by using historical outage data and weather data. The results show that the fragility curve deduced from the weather-related outage database could provide an accurate forecast on the outage rate of transmission lines.

References

1 Gungor, V.C., Sahin, D., Kocak, T. et al. (2011). Smart grid technologies: communication technologies and standards. *IEEE Transactions on Industrial Informatics* 7 (4): 529–539.

2 Jenkins, N., Liyanage, K., Wu, J., and Yokoyama, A. (2012). *Smart Grid*. Wiley.

3 Liu, J., Li, X., Liu, D. et al. (2011). Study on data management of fundamental model in control center for smart grid operation. *IEEE Transactions on Smart Grid* 2 (4): 573–579.

4 Luo, F., Dong, Z.Y., Zhao, J. et al. (2015). Enabling the big data analysis in the smart grid. *2015 IEEE Power & Energy Society General Meeting*.

5 Hossain, E., Khan, I., Un-Noor, F. et al. (2019). Application of big data and machine learning in smart grid, and associated security concerns: a review. *IEEE Access* 7: 13960–13988.

6 Zhao, J., Zhang, G., Das, K. et al. (2016). Power system real-time monitoring by using PMU-based robust state estimation method. *IEEE Transactions on Smart Grid* 7 (1): 300–309.

7 Dahal, O.P., Brahma, S.M., and Cao, H. (2014). Comprehensive clustering of disturbance events recorded by phasor measurement units. *IEEE Transactions on Power Delivery* 29 (3): 1390–1397.

8 Deihimi, A., Orang, O., and Showkati, H. (2013). Short-term electric load and temperature forecasting using wavelet echo state networks with neural reconstruction. *Energy* 57: 382–401.

9 Diamantoulakis, P.D., Kapinas, V.M., and Karagiannidis, G.K. (2015). Big data analytics for dynamic energy management in smart grids. *Big Data Research* 2 (3): 94–101.

10 Xinxing, P. and Lee, B. (2012). A comparison of support vector machines and artificial neural networks for mid-term load forecasting. *2012 IEEE International Conference on Industrial Technology*.

11 Simonyan, K. and Zisserman, A. (2014). Very deep convolutional networks for large-scale image recognition. arXiv preprint arXiv:1409.1556.

12 MATPOWER (2008). MATPOWER case archive – "Case3120sp.m" – 2008 summer peak Polish Case.

13 Akhavan-Hejazi, H. and Mohsenian-Rad, H. (2018). Power systems big data analytics: an assessment of paradigm shift barriers and prospects. *Energy Reports* 4: 91–100.

14 Kezunovic, M., Pinson, P., Obradovic, Z. et al. (2020). Big data analytics for future electricity grids. *Electric Power Systems Research* 189: 106788.

15 Wang, B., Fang, B., Wang, Y. et al. (2016). Power system transient stability assessment based on big data and the core vector machine. *IEEE Transactions on Smart Grid* 7 (5): 2561–2570.

16 Konakalla, S.A.R. and de Callafon, R. (2016). Optimal filtering for grid event detection from real-time synchrophasor data. *Procedia Computer Science* 80: 931–940.

17 Daubechies, I. (1992). *Ten Lectures on Wavelets*. SIAM.

18 Hinton, G.E., Srivastava, N., Krizhevsky, A. et al. (2012). Improving neural networks by preventing co-adaptation of feature detectors. arXiv preprint arXiv:1207.0580.

19 Kezunovic, M., Dobson, I., and Dong, Y. (2008). Impact of extreme weather on power system blackouts and forced outages: New challenges. *7th Balkan Power Conf, Citeseer*.

20 Campbell, R.J. and Lowry, S. (2012). Weather-related power outages and electric system resiliency, Congressional Research Service, Library of Congress Washington, DC.

21 Pereira, M.G., Fernandes, L.F.S., Macário, E.M.B., and Gaspar, S.M. (2015). Climate change impacts in the design of drainage systems: case study of Portugal. *Journal of Irrigation and Drainage Engineering* 141 (2): 05014009.

22 Kenward, A. and Raja, U. (2014). Blackout: Extreme weather, climate change and power outages. *Climate Central* 10: 1–23.

23 Bonneville Power Administration. BPA Geospatial Open Data. https://data-bpagis.hub.arcgis.com/ (accessed 21 October 2022).

24 Kelly-Gorham, M.R., Hines, P., and Dobson, I. (2019). Using historical utility outage data to compute overall transmission grid resilience. arXiv preprint arXiv:1906.06811.

25 Koval, D. and Chowdhury, A. (2005). An investigation into extreme-weather-caused transmission line unavailability. *IEEE Power Engineering Society General Meeting*, 2005, IEEE.

26 Dokic, T., Pavlovski, M., Gligorijevic, D. et al. (2019). Spatially aware ensemble-based learning to predict weather-related outages in transmission. *Proceedings of the 52nd Hawaii International Conference on System Sciences*.

27 Panteli, M. and Mancarella, P. (2015). Influence of extreme weather and climate change on the resilience of power systems: Impacts and possible mitigation strategies. *Electric Power Systems Research* 127: 259–270.

28 Panteli, M. and Mancarella, P. (2015). Modeling and evaluating the resilience of critical electrical power infrastructure to extreme weather events. *IEEE Systems Journal* 11 (3): 1733–1742.

29 Winkler, J., Duenas-Osorio, L., Stein, R., and Subramanian, D. (2010). Performance assessment of topologically diverse power systems subjected to hurricane events. *Reliability Engineering & System Safety* 95 (4): 323–336.

30 Jeong, S.-H. and Elnashai, A.S. (2007). Probabilistic fragility analysis parameterized by fundamental response quantities. *Engineering Structures* 29 (6): 1238–1251.

31 Dokic, T. and Pavlovski, M. (2019). Spatially aware ensemble-based learning to predict weather-related outages in transmission. *The Hawaii International Conference on System Sciences–HICSS*, Maui, Hawaii, January 2019.

32 Kezunovic, M., Obradovic, Z., Djokic, T., and Roychoudhury, S. (2018). Systematic framework for integration of weather data into prediction models for the electric grid outage and asset management applications.

Conclusions

Zita Vale[1], *Tiago Pinto*[1,2], *Michael Negnevitsky*[3], *and Ganesh Kumar Venayagamoorthy*[4]

[1] GECAD Research Group on Intelligent Engineering and Computing for Advanced Innovation and Development (GECAD), Polytechnic Institute of Porto (ISEP/IPP), Porto, Portugal
[2] University of Trás-os-Montes e Alto Douro, Vila Real, Portugal
[3] School of Engineering, University of Tasmania, Hobart, Tasmania, Australia
[4] Holcombe Department of Electrical and Computer Engineering, Real-Time Power and Intelligent Systems Laboratory, Clemson University Clemson, SC USA and School of Engineering, University of KwaZulu-Natal, Durban, South Africa

The increasing penetration of distributed renewable energy sources and the consequent empowerment of consumers to become active players in mitigating the lack of generation flexibility with demand flexibility are driving the power and energy system toward an historic paradigm shift. The small scale, diversity, and especially the number of new players involved in the power and energy system potentiate a significant growth of generated data. Moreover, advances in telecommunications and digitalization hugely increased the volume of data that result from power and energy components, installations, and systems operation. These data are becoming more and more important for power and energy systems operation and planning, with relevant impact on all involved entities, from producers, consumers, and aggregators to market and system operators. However, although the power and energy community are fully aware of the intrinsic value of those data, the methods to deal with it still require significant improvements and research. Data mining and intelligent data analysis are thereby playing a crucial role in this domain, by enabling players to improve their decision-making process and gain awareness on the power and energy environment. This book brings together the state-of-the-art advances in intelligent data mining and analysis as drivers for the needed evolution of power and energy systems.

This book overviews the fundamentals of data mining and analysis, as means to introduce the reader to the main concepts of the domain, and reviews the most relevant works already developed in the field of power and energy systems. The applications of data mining and analysis models and methods discussed in this book cover the fields of supervised and unsupervised data mining, as well as other data analysis approaches. The two main problems using supervised data mining that are addressed are regression-based methods, from which we highlight works that apply linear regression, artificial neural networks, support vector regression, as well as stochastic and decomposition models; and classification-based methods, including support vector machines, K-nearest neighbors, and once again artificial networks. Regarding unsupervised models, we describe some association rule-based approaches, and clustering-based approaches such as k-means, G-means and hierarchical clustering. The description of these applications is completed by some examples of use of other machine learning and data analysis approaches, from which we can specify reinforcement learning models, deep learning models, probabilistic models and federated learning, besides

Intelligent Data Mining and Analysis in Power and Energy Systems: Models and Applications for Smarter Efficient Power Systems, First Edition. Edited by Zita Vale, Tiago Pinto, Michael Negnevitsky, and Ganesh Kumar Venayagamoorthy.

several specific ad-hoc data analysis methodologies that are proposed to deal with problems with very specific characteristics.

Although the recent advances in data mining and analysis models, methods, and applications described in this book are providing important contributions in dealing with many of the most relevant challenges in the power and energy domain, there are still significant issues that need to be addressed in order to cope with such a dynamic and complex sector such as power and energy systems. On one hand, we can identify the lack of users' trust as one of the main barriers for the widespread of intelligent solutions in the sector. Most of the applied solutions are seen as black boxes, often with very good results but with a very low level of interpretability. Significant advances in the field of models explainability are required in order to improve the general acceptance over automatically generated decisions and recommendations. Dealing with data streams and real time data is still an important issue, although many solutions have already been proposed in this field. The volume, variability, and speed at which data is collected keep on increasing, thus turning the solution to this problem a road without an end at sight. The challenges of dealing with today's and tomorrow's data characteristics extend to the issues of data fusion, considering the diversity of data sources, and the extraction of value from the correlation and synergies between multiple data sources. Data extraction from these data sources, in turn, is another focus of current and future research and development. It requires the continuous improvement of the available internet of things technologies in a way to enable dealing with the extraction of data from multiple sources, in different formats and with different characteristics, but also with the communications between distinct devices, being sensors, meters, actuators, or any other type of device, and finally with the actual control or user interaction actions that reflect the results from the data mining and analysis process. Dealing with such variety of data, especially in an internet of things context, brings out yet another problem, which refers to the data privacy and security issues. Significant effort is being put into these matters, finding suitable ways of dealing with data treatment, communications, and storage in order to guarantee the necessary data privacy and security. However, and given the constantly evolving nature of data characteristics and of new business models and involved players in the power and energy sector, this is a challenge that will continue requiring dedicated attention for future development. Thereby, although the advances already accomplished in intelligent data mining and analysis models are considerable, as discussed by this book, there are still many relevant paths for future research and development, which will require our continuous venture in this domain.

Index

Intelligent Data Mining and Analysis in Power and Energy Systems: Models and Applications for Smarter Efficient Power Systems, First Edition. Edited by Zita Vale, Tiago Pinto, Michael Negnevitsky, and Ganesh Kumar Venayagamoorthy.

d

g

h

i

j

k

l

m

q

r

t

IEEE Press Series on Power and Energy Systems

Series Editor: Ganesh Kumar Venayagamoorthy, Clemson University, Clemson, South Carolina, USA.

The mission of the IEEE Press Series on Power and Energy Systems is to publish leading-edge books that cover a broad spectrum of current and forward-looking technologies in the fast-moving area of power and energy systems including smart grid, renewable energy systems, electric vehicles and related areas. Our target audience includes power and energy systems professionals from academia, industry and government who are interested in enhancing their knowledge and perspectives in their areas of interest.

1. *Electric Power Systems: Design and Analysis, Revised Printing*
   Mohamed E. El-Hawary
2. *Power System Stability*
   Edward W. Kimbark
3. *Analysis of Faulted Power Systems*
   Paul M. Anderson
4. *Inspection of Large Synchronous Machines: Checklists, Failure Identification, and Troubleshooting*
   Isidor Kerszenbaum
5. *Electric Power Applications of Fuzzy Systems*
   Mohamed E. El-Hawary
6. *Power System Protection*
   Paul M. Anderson
7. *Subsynchronous Resonance in Power Systems*
   Paul M. Anderson, B.L. Agrawal, and J.E. Van Ness
8. *Understanding Power Quality Problems: Voltage Sags and Interruptions*
   Math H. Bollen
9. *Analysis of Electric Machinery*
   Paul C. Krause, Oleg Wasynczuk, and S.D. Sudhoff
10. *Power System Control and Stability, Revised Printing*
    Paul M. Anderson and A.A. Fouad

11. *Principles of Electric Machines with Power Electronic Applications*, Second Edition
Mohamed E. El-Hawary

12. *Pulse Width Modulation for Power Converters: Principles and Practice*
D. Grahame Holmes and Thomas Lipo

13. *Analysis of Electric Machinery and Drive Systems*, Second Edition
Paul C. Krause, Oleg Wasynczuk, and S.D. Sudhoff

14. *Risk Assessment for Power Systems: Models, Methods, and Applications*
Wenyuan Li

15. *Optimization Principles: Practical Applications to the Operations of Markets of the Electric Power Industry*
Narayan S. Rau

16. *Electric Economics: Regulation and Deregulation*
Geoffrey Rothwell and Tomas Gomez

17. *Electric Power Systems: Analysis and Control*
Fabio Saccomanno

18. *Electrical Insulation for Rotating Machines: Design, Evaluation, Aging, Testing, and Repair*
Greg C. Stone, Edward A. Boulter, Ian Culbert, and Hussein Dhirani

19. *Signal Processing of Power Quality Disturbances*
Math H. J. Bollen and Irene Y. H. Gu

20. *Instantaneous Power Theory and Applications to Power Conditioning*
Hirofumi Akagi, Edson H. Watanabe, and Mauricio Aredes

21. *Maintaining Mission Critical Systems in a 24/7 Environment*
Peter M. Curtis

22. *Elements of Tidal-Electric Engineering*
Robert H. Clark

23. *Handbook of Large Turbo-Generator Operation and Maintenance*, Second Edition
Geoff Klempner and Isidor Kerszenbaum

24. *Introduction to Electrical Power Systems*
Mohamed E. El-Hawary

25. *Modeling and Control of Fuel Cells: Distributed Generation Applications*
M. Hashem Nehrir and Caisheng Wang

26. *Power Distribution System Reliability: Practical Methods and Applications*
Ali A. Chowdhury and Don O. Koval

27. *Introduction to FACTS Controllers: Theory, Modeling, and Applications*
Kalyan K. Sen and Mey Ling Sen

28. *Economic Market Design and Planning for Electric Power Systems*
James Momoh and Lamine Mili

29. *Operation and Control of Electric Energy Processing Systems*
James Momoh and Lamine Mili

30. *Restructured Electric Power Systems: Analysis of Electricity Markets with Equilibrium Models*
Xiao-Ping Zhang

31. *An Introduction to Wavelet Modulated Inverters*
S.A. Saleh and M.A. Rahman

32. *Control of Electric Machine Drive Systems*
Seung-Ki Sul

33. *Probabilistic Transmission System Planning*
Wenyuan Li

34. *Electricity Power Generation: The Changing Dimensions*
Digambar M. Tagare

35. *Electric Distribution Systems*
Abdelhay A. Sallam and Om P. Malik

36. *Practical Lighting Design with LEDs*
Ron Lenk and Carol Lenk

37. *High Voltage and Electrical Insulation Engineering*
Ravindra Arora and Wolfgang Mosch

38. *Maintaining Mission Critical Systems in a 24/7 Environment*, Second Edition
Peter Curtis

39. *Power Conversion and Control of Wind Energy Systems*
Bin Wu, Yongqiang Lang, Navid Zargari, and Samir Kouro

40. *Integration of Distributed Generation in the Power System*
Math H. Bollen and Fainan Hassan

41. *Doubly Fed Induction Machine: Modeling and Control for Wind Energy Generation Applications*
Gonzalo Abad, Jesús López, Miguel Rodrigues, Luis Marroyo, and Grzegorz Iwanski

42. *High Voltage Protection for Telecommunications*
Steven W. Blume

43. *Smart Grid: Fundamentals of Design and Analysis*
James Momoh

44. *Electromechanical Motion Devices*, Second Edition
Paul Krause, Oleg Wasynczuk, Steven Pekarek

45. *Electrical Energy Conversion and Transport: An Interactive Computer-Based Approach*, Second Edition
George G. Karady and Keith E. Holbert

46. *ARC Flash Hazard and Analysis and Mitigation*
J.C. Das

47. *Handbook of Electrical Power System Dynamics: Modeling, Stability, and Control*
Mircea Eremia and Mohammad Shahidehpour

48. *Analysis of Electric Machinery and Drive Systems*, Third Edition
Paul C. Krause, Oleg Wasynczuk, S.D. Sudhoff, and Steven D. Pekarek

49. *Extruded Cables for High-Voltage Direct-Current Transmission: Advances in Research and Development*
Giovanni Mazzanti and Massimo Marzinotto

50. *Power Magnetic Devices: A Multi-Objective Design Approach*
S.D. Sudhoff

51. *Risk Assessment of Power Systems: Models, Methods, and Applications*, Second Edition
Wenyuan Li

52. *Practical Power System Operation*
Ebrahim Vaahedi

53. *The Selection Process of Biomass Materials for the Production of Bio-Fuels and Co-Firing*
Najib Altawell

54. *Electrical Insulation for Rotating Machines: Design, Evaluation, Aging, Testing, and Repair*, Second Edition
Greg C. Stone, Ian Culbert, Edward A. Boulter, and Hussein Dhirani

55. *Principles of Electrical Safety*
Peter E. Sutherland

56. *Advanced Power Electronics Converters: PWM Converters Processing AC Voltages*
Euzeli Cipriano dos Santos Jr. and Edison Roberto Cabral da Silva

57. *Optimization of Power System Operation*, Second Edition
Jizhong Zhu

58. *Power System Harmonics and Passive Filter Designs*
J.C. Das

59. *Digital Control of High-Frequency Switched-Mode Power Converters*
Luca Corradini, Dragan Maksimoviæ, Paolo Mattavelli, and Regan Zane

60. *Industrial Power Distribution*, Second Edition
Ralph E. Fehr, III

61. *HVDC Grids: For Offshore and Supergrid of the Future*
Dirk Van Hertem, Oriol Gomis-Bellmunt, and Jun Liang

62. *Advanced Solutions in Power Systems: HVDC, FACTS, and Artificial Intelligence*
Mircea Eremia, Chen-Ching Liu, and Abdel-Aty Edris

63. *Operation and Maintenance of Large Turbo-Generators*
Geoff Klempner and Isidor Kerszenbaum

64. *Electrical Energy Conversion and Transport: An Interactive Computer-Based Approach*
George G. Karady and Keith E. Holbert

65. *Modeling and High-Performance Control of Electric Machines*
John Chiasson

66. *Rating of Electric Power Cables in Unfavorable Thermal Environment*
George J. Anders

67. *Electric Power System Basics for the Nonelectrical Professional*
Steven W. Blume

68. *Modern Heuristic Optimization Techniques: Theory and Applications to Power Systems*
Kwang Y. Lee and Mohamed A. El-Sharkawi

69. *Real-Time Stability Assessment in Modern Power System Control Centers*
Savu C. Savulescu

70. *Optimization of Power System Operation*
Jizhong Zhu

71. *Insulators for Icing and Polluted Environments*
Masoud Farzaneh and William A. Chisholm

72. *PID and Predictive Control of Electric Devices and Power Converters Using MATLAB®/Simulink®*
Liuping Wang, Shan Chai, Dae Yoo, and Lu Gan, Ki Ng

73. *Power Grid Operation in a Market Environment: Economic Efficiency and Risk Mitigation*
Hong Chen

74. *Electric Power System Basics for Nonelectrical Professional*, Second Edition
Steven W. Blume

75. *Energy Production Systems Engineering*
Thomas Howard Blair

76. *Model Predictive Control of Wind Energy Conversion Systems*
Venkata Yaramasu and Bin Wu

77. *Understanding Symmetrical Components for Power System Modeling*
J.C. Das

78. *High-Power Converters and AC Drives*, Second Edition
Bin Wu and Mehdi Narimani

79. *Current Signature Analysis for Condition Monitoring of Cage Induction Motors: Industrial Application and Case Histories*
William T. Thomson and Ian Culbert

80. *Introduction to Electric Power and Drive Systems*
Paul Krause, Oleg Wasynczuk, Timothy O'Connell, and Maher Hasan

81. *Instantaneous Power Theory and Applications to Power Conditioning*, Second Edition
Hirofumi, Edson Hirokazu Watanabe, and Mauricio Aredes

82. *Practical Lighting Design with LEDs*, Second Edition
Ron Lenk and Carol Lenk

83. *Introduction to AC Machine Design*
Thomas A. Lipo

84. *Advances in Electric Power and Energy Systems: Load and Price Forecasting*
Mohamed E. El-Hawary

85. *Electricity Markets: Theories and Applications*
Jeremy Lin and Jernando H. Magnago

86. *Multiphysics Simulation by Design for Electrical Machines, Power Electronics and Drives*
Marius Rosu, Ping Zhou, Dingsheng Lin, Dan M. Ionel, Mircea Popescu, Frede Blaabjerg, Vandana Rallabandi, and David Staton

87. *Modular Multilevel Converters: Analysis, Control, and Applications*
Sixing Du, Apparao Dekka, Bin Wu, and Navid Zargari

88. *Electrical Railway Transportation Systems*
Morris Brenna, Federica Foiadelli, and Dario Zaninelli

89. *Energy Processing and Smart Grid*
James A. Momoh

90. *Handbook of Large Turbo-Generator Operation and Maintenance*, Third Edition
Geoff Klempner and Isidor Kerszenbaum

91. *Advanced Control of Doubly Fed Induction Generator for Wind Power Systems*
Dehong Xu, Dr. Frede Blaabjerg, Wenjie Chen, and Nan Zhu

92. *Electric Distribution Systems*, Second Edition
Abdelhay A. Sallam and Om P. Malik

93. *Power Electronics in Renewable Energy Systems and Smart Grid: Technology and Applications*
Bimal K. Bose

94. *Distributed Fiber Optic Sensing and Dynamic Rating of Power Cables*
Sudhakar Cherukupalli and George J Anders

95. *Power System and Control and Stability,* Third Edition
Vijay Vittal, James D. McCalley, Paul M. Anderson, and A. A. Fouad

96. *Electromechanical Motion Devices: Rotating Magnetic Field-Based Analysis and Online Animations,* Third Edition
Paul Krause, Oleg Wasynczuk, Steven D. Pekarek, and Timothy O'Connell

97. *Applications of Modern Heuristic Optimization Methods in Power and Energy Systems*
Kwang Y. Lee and Zita A. Vale

98. *Handbook of Large Hydro Generators: Operation and Maintenance*
Glenn Mottershead, Stefano Bomben, Isidor Kerszenbaum, and Geoff Klempner

99. *Advances in Electric Power and Energy: Static State Estimation*
Mohamed E. El-hawary

100. *Arc Flash Hazard Analysis and Mitigation*, Second Edition
J.C. Das

101. *Maintaining Mission Critical Systems in a 24/7 Environment,* Third Edition
Peter M. Curtis

102. *Real-Time Electromagnetic Transient Simulation of AC-DC Networks*
Venkata Dinavahi and Ning Lin

103. *Probabilistic Power System Expansion Planning with Renewable Energy Resources and Energy Storage Systems*
Jaeseok Choi and Kwang Y. Lee

104. *Power Magnetic Devices: A Multi-Objective Design Approach*, Second Edition
Scott D. Sudhoff

105. *Optimal Coordination of Power Protective Devices with Illustrative Examples*
Ali R. Al-Roomi

106. *Resilient Control Architectures and Power Systems*
Craig Rieger, Ronald Boring, Brian Johnson, and Timothy McJunkin

107. *Alternative Liquid Dielectrics for High Voltage Transformer Insulation Systems: Performance Analysis and Applications*
Edited by U. Mohan Rao, I. Fofana, and R. Sarathi

108. *Introduction to the Analysis of Electromechanical Systems*
Paul C. Krause, OlegWasynczuk, and Timothy O'Connell

109. *Power Flow Control Solutions for a Modern Grid using SMART Power Flow Controllers*
Kalyan K. Sen and Mey Ling Sen

110. *Power System Protection: Fundamentals and Applications*
John Ciufo and Aaron Cooperberg

111. *Soft-Switching Technology for Three-phase Power Electronics Converters*
Dehong Xu, Rui Li, Ning He, Jinyi Deng, and YuyingWu

112. *Power System Protection*, Second Edition
Paul M. Anderson, Charles Henville, Rasheek Rifaat, Brian Johnson, and Sakis Meliopoulos

113. *High Voltage and Electrical Insulation Engineering*, Second Edition
Ravindra Arora and Wolfgang Mosch

114. *Modeling and Control of Modern Electrical Energy Systems*
Masoud Karimi-Ghartemani

115. *Control of Power Electronic Converters with Microgrid Applications*
Arindam Ghosh and Firuz Zare

116. *Coordinated Operation and Planning of Modern Heat and Electricity Incorporated Networks*
Mohammadreza Daneshvar, Behnam Mohammadi-Ivatloo, and Kazem Zare

117. *Smart Energy for Transportation and Health in a Smart City*
Chun Sing Lai, Loi Lei Lai, and Qi Hong Lai

118. *Wireless Power Transfer: Principles and Applications*
Zhen Zhang and Hongliang Pang

119. *Intelligent Data Mining and Analysis in Power and Energy Systems: Models and Applications for Smarter Efficient Power Systems*
Zita Vale, Tiago Pinto, Michael Negnevitsky, and Ganesh Kumar Venayagamoorthy

Printed and bound by CPI Group (UK) Ltd, Croydon, CR0 4YY
17/09/2026
14964257-0005